EEG: The First 100 Years

Francesco Brigo • Oriano Mecarelli
Editors

EEG: The First 100 Years

Past, Present and Future of Electroencephalography

Springer

Editors
Francesco Brigo
Innovation, Research and Teaching Service; Teaching Hospital of the Paracelsus Medical Private University (PMU)
Azienda Sanitaria dell'Alto Adige
Bolzano-Bozen, Italy

Oriano Mecarelli
Department of Human Neurosciences
Sapienza University
Rome, Italy

ISBN 978-3-031-86877-1 ISBN 978-3-031-86878-8 (eBook)
https://doi.org/10.1007/978-3-031-86878-8

This Springer imprint is published by the registered company Springer Nature Switzerland AG
The registered company address is: Gewerbestrasse 11, 6330 Cham, Switzerland

Preface

On July 6, 1924, Hans Berger recorded the first electroencephalogram (EEG) on a human being, marking the beginning of a revolutionary diagnostic method that in 2024 celebrated its centenary. Although researchers in the late nineteenth century had succeeded in recording bioelectrical brain activity in animals, it was Berger who first demonstrated that the human brain also generates characteristic electrical rhythms. Using rudimentary equipment—capable of recording on a moving roll of paper with just two electrodes placed on the scalp—he identified patterns that remain central to human EEG studies today.

To confirm the authenticity of these phenomena, Berger performed numerous additional recordings, ensuring that what he observed was truly generated by the brain and not artifacts. His meticulous approach delayed publication until 1929, when he released his findings in a series of 14 articles written in German and published in psychiatric journals. This limited their initial impact, as skepticism within the international scientific community hampered wider recognition. However, in 1934, Nobel laureate Edgar Douglas Adrian and bioengineer Brian Harold Matthews from the University of Cambridge validated Berger's results in the journal Brain. Their endorsement sparked global interest, establishing EEG as a foundational technique in brain research.

In the 1940s and 1950s, despite the challenges of the post-war period, EEG experienced remarkable growth, becoming the primary tool for studying brain function. The introduction of digital technology in the late 1980s revitalized the field, bridging gaps left by the rise of neuroimaging and reaffirming EEG's role as the gold standard in epileptology. Today, it remains indispensable for diagnosing, classifying, and managing epilepsy.

For the past three decades, EEG has expanded further into critical areas, including neurological emergency management, intensive care neuromonitoring, and the evaluation of acute and chronic encephalopathies. When integrated with other neurophysiological and neuroimaging techniques, digital EEG provides a multidimensional approach to understanding and treating brain disorders, offering unparalleled diagnostic and prognostic insights.

Understanding the present and planning for the future require a deep appreciation of the past. This book aims to honor the pioneering spirit that brought EEG to life and celebrate its century-long journey of innovation and discovery. As we celebrate 100 years of EEG, we are reminded of its profound impact on neuroscience and medicine. Beyond reflecting on its achievements, let us embrace the challenges and opportunities ahead, ensuring that this remarkable tool continues to illuminate the mysteries of the human brain and pave the way for groundbreaking advancements in science and clinical practice.

Bolzano-Bozen, Italy Francesco Brigo
Rome, Italy Oriano Mecarelli

Contents

1 Precursors of Clinical Electroencephalography: From Animal Electricity to Brain Wave Recording

Francesco Brigo

1.1 Introduction

The journey toward the discovery and application of clinical electroencephalography (EEG) is rooted in centuries of scientific inquiry into electricity, magnetism, and the nervous system. This chapter examines the key milestones that served as precursors to modern EEG technology, tracing the evolution of scientific thought and instrumentation. It highlights pivotal moments, including the discovery of animal electricity, the invention of the Leyden jar, and the development of precision instruments such as the string galvanometer. These advancements enabled researchers to measure and analyze the brain's electrical oscillations, despite the technological limitations of their time, and shaped the modern field of neurophysiology.

1.2 The Origins of Electricity

The history of the EEG is closely intertwined with the history of electricity, a physical phenomenon involving electric charges and their interactions. Electricity is one of nature's fundamental forces, underpinning various phenomena such as lightning, magnetism, chemistry, and electronics.

The term "electricity" was introduced by William Gilbert (1544–1603), an English physician and physicist, in 1600. Gilbert systematically studied magnets and static electricity, which involves the accumulation of electric charges on object surfaces [1]. He observed that magnets could attract or repel other magnets and materials such as iron, nickel, and cobalt. Additionally, he noted that objects such as amber, glass, and sulfur, when rubbed with cloth or fur, could attract or repel items such as feathers, paper, or hair—phenomena first recorded by the ancient philosopher Thales of Miletus. Gilbert, who coined the term "electrick" (derived from the Greek word *elektron*, meaning amber), also developed the first electroscope, the *versorium*, which detected static electricity. In his influential work *De Magnete, Magneticisque Corporibus, et de Magno Magnete Tellure* (*On the Magnet and Magnetic Bodies, and on the Great Magnet the Earth*), he dedicated a chapter to "electrica," clearly distinguishing it from magnetism [2]. This work was the first scientific account of static and frictional electricity and laid the foundation for discovering intrinsic electricity in nerves.

F. Brigo (✉)
Innovation, Research and Teaching Service; Teaching Hospital of the Paracelsus Medical Private University (PMU), Azienda Sanitaria dell'Alto Adige, Bolzano-Bozen, Italy
e-mail: Francesco.Brigo@sabes.it

F. Brigo, O. Mecarelli (eds.), *EEG: The First 100 Years*,
https://doi.org/10.1007/978-3-031-86878-8_1

1.3 The First Electrostatic Generators

In 1666, Otto von Guericke (1602–1686), a German physicist and inventor, created the first electrostatic generator—a frictional machine that produced sparks by rubbing a sulfur globe [3]. By rotating the globe with a crank and using a metal rod to collect the charge, Guericke demonstrated its ability to repel floating feathers and produce sparks capable of igniting gunpowder, melting metal, and shocking individuals. He noted that sparks could jump gaps, creating electric arcs, and theorized the existence of two electric fluids in the globe and rod that attracted or repelled each other [3]. Guericke's device served as a precursor to later machines used in nervous system studies until the Leyden jar emerged in the mid-1700s [1].

In 1729, Stephen Gray (1666–1736), an English dyer and astronomer, demonstrated that electricity could travel through wires [4]. He suspended a brass ball from a silk thread connected to a charged glass tube, observing that the ball attracted light objects such as feathers even at a distance. When Gray replaced the silk thread with a metal wire, the ball continued to attract objects, leading him to conclude that the wire transmitted electric charge. He experimented with various materials, discovering that some, like metals, conducted electricity, while others did not. Gray's findings, presented to the Royal Society, earned him membership in 1732.

Fig. 1.1 Illustration from *Elementary Treatise on Natural Philosophy, Part 3: Electricity and Magnetism* by Augustin Privat Deschanel (D. Appleton and Co., New York, 1876), depicting the discovery of the Leyden jar in Pieter van Musschenbroek's laboratory. In 1746, Andreas Cunaeus, a student of van Musschenbroek in Leyden, conducted an experiment to "condense" electricity using a glass of water as a capacitor. A rotating glass sphere served as an electrostatic generator, transferring charge via a chain to a metal bar and then through a wire into the water. Cunaeus inadvertently discovered the jar's storage potential when he received a powerful shock after touching the wire. This groundbreaking experiment demonstrated the ability to store large electric charges, far exceeding the capacity of an electrostatic machine, leading to widespread interest in Leyden jars

1.4 The Invention of the Leyden Jar

In 1746, inspired by earlier work by Ewald George von Kleist (1700–1748), Dutch physicist Pieter van Musschenbroek (1692–1761) invented the Leyden jar, a device for storing electric charge [5] (Fig. 1.1). The Leyden jar consisted of a partially water-filled glass jar with external metal foil. A metal rod inserted through the cork connected the interior water to an electrostatic generator, allowing the jar to accumulate and release electric charges. This device, the first capacitor, could store significantly more charge than earlier generators, producing stronger sparks and shocks that facilitated important discoveries in electricity.

French physicist and clergyman Jean-Antoine Nollet (1700–1770) popularized the Leyden jar and demonstrated that electricity could pass through water and living organisms, suggesting it was a fluid capable of high-speed flow [6]. In 1752, Benjamin Franklin (1706–1790) conducted his famous kite experiment, proving that lightning was a form of electricity [7]. By flying a kite with a metal key attached during a thunderstorm, Franklin captured the electrical charge in a Leyden jar. Observing sparks and shocks similar to those produced by electrostatic generators, Franklin proposed that electricity was a property of matter transferable between bodies. He introduced the terms "positive" and "negative" to describe electric charge and developed the concept of "grounding" to discharge objects safely [7]. Franklin's invention of the lightning rod further demonstrated practical applications of electricity, providing a method to divert electric currents safely into the ground and protect structures from lightning strikes.

1.5 The Discovery of Animal Electricity

In 1780, Italian physician and anatomist Luigi Galvani (1737–1798) observed that a frog's leg muscles twitched when connected to a metal wire and an electrostatic generator [8, 9]. This discovery led Galvani to propose the existence of "animal electricity," a vital force he believed was electrical in nature. He theorized that nerves conducted this electricity and muscles acted as receivers, publishing his findings in *De Viribus Electricitatis in Motu Musculari Commentarius* (*Commentary on the Effect of Electricity on Muscular Motion*) in 1791 [10] (Fig. 1.2).

Galvani's work inspired his nephew, Giovanni Aldini (1762–1834), who expanded the research by conducting experiments on human corpses [11]. In a 1804 publication, Aldini documented his findings and demonstrations performed across Europe [12]. Using electric currents to stimulate brains and muscles, Aldini famously conducted public experiments where he caused the limbs and faces of executed criminals to move (Fig. 1.3). He also championed electrotherapy, advocating for its potential to treat conditions such as paralysis, epilepsy, and apoplexy.

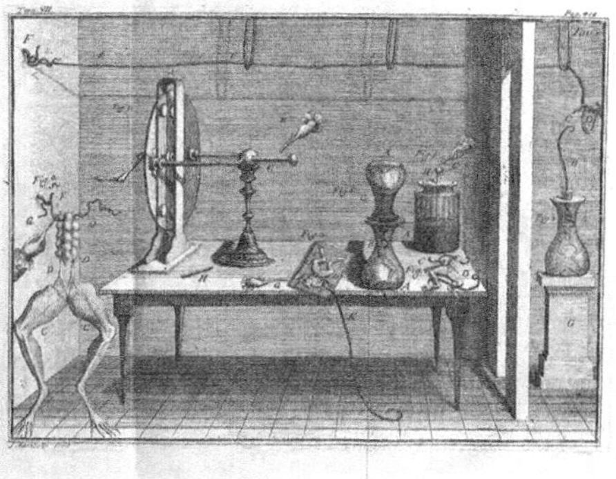

Fig. 1.2 Plate I from Luigi Galvani's *De Viribus Electricitatis in Motu Musculari Commentarius* (1791), illustrating his experiments on the effects of electricity on muscle contraction. Galvani's work revealed the phenomenon of "animal electricity," demonstrating that nerves and muscles responded to electrical stimulation

Fig. 1.3 Steel engraving from the revised edition of *Frankenstein* by Mary Shelley (Colburn and Bentley, London, 1831). Shelley was likely inspired by Giovanni Aldini's public demonstrations of using electrical currents to stimulate muscle activity in animal and human corpses. Aldini's belief in electricity's potential to reverse death influenced Shelley's portrayal of Dr. Victor Frankenstein's attempt to reanimate dead tissue, blending scientific intrigue with literary imagination

1.6 The Invention of the Electric Battery

In 1791, Italian physicist Alessandro Volta (1745–1827) challenged Luigi Galvani's theory of animal electricity. Volta proposed that the contraction of the frog's leg was not due to an inherent "animal electricity" but rather resulted from the contact between two different metals [9]. Through extensive experimentation with various combinations of metals and fluids, Volta invented the voltaic pile in 1800—the first electric battery. This device consisted of alternating zinc and copper discs separated by cloth soaked in saltwater, generating a continuous electric current.

Volta announced his groundbreaking discovery to the Royal Society that same year, heralding the advent of direct current electricity [13]. The voltaic pile revolutionized the study of electricity and enabled transformative experiments and discoveries in physics and neurophysiology.

1.7 The Measurement of the Electrical Activity of Muscles and Nerves

Emil du Bois-Reymond (1818–1896), a German physiologist of Huguenot origin, made significant contributions to neurophysiology by building on the work of Luigi Galvani and Alessandro Volta [14]. He enhanced the sensitivity and accuracy of the galvanometer, enabling the precise recording of changes in electric potential within muscles and nerves during stimulation and contraction. Du Bois-Reymond introduced the term "negative Schwankung" (negative variation) to describe the electrical impulses traveling along nerve fibers, correlating these impulses with changes in action potential during muscle contraction. He demonstrated that the electrical activity of muscles and nerves was intrinsic to living tissue rather than a mere response to external stimuli. His influential research, conducted between 1848 and 1884, profoundly shaped the development of neurophysiology [15].

During the same period, Hermann von Helmholtz (1821–1894), a German physician and physicist, made another groundbreaking discovery by calculating the speed of nerve impulses. Inspired by ballistics, Helmholtz used a frog's sciatic nerve and a galvanometer in his experiments, determining in 1849 that nerve impulses traveled at approximately 30 meters per second—far slower than the speed of light [16]. This finding challenged the prevailing belief that nerve impulses were simple electric currents, instead revealing their complexity as physiological processes essential for information transmission.

1.8 Richard Caton (1842–1926) and the First Recording of Electrical Activity in the Brain

Richard Caton (1842–1926), an English physician and physiologist, is credited as the first to record the electrical activity of the brain in animals [17–20] (Fig. 1.4). After graduating from the University of Edinburgh in 1866, Caton moved to Liverpool, where he practiced medicine and lectured at the Royal Infirmary School of Medicine. Influenced by the neurophysiological work of Emil du Bois-Reymond and Hermann von Helmholtz, Caton sought to determine whether the brain generated electrical currents.

Fig. 1.4 Portrait of Richard Caton (1842–1926), a pioneer in electrophysiology and the first to record electrical activity of cerebral origin in animals. Caton's groundbreaking studies in 1875 demonstrated the existence of electrical waves in the brain, paving the way for modern electroencephalography

In 1875, Caton successfully recorded the brain's electrical activity in rabbits and monkeys. Using a galvanometer connected to electrodes placed on the exposed cerebral hemispheres, he employed a Thomson reflecting galvanometer to visualize the signals. This device cast light beams onto a wall, allowing the waveforms to be observed. Caton noted distinct waveforms corresponding to sensory stimulation and anesthesia, which varied across different brain regions.

Caton presented his groundbreaking findings on August 4, 1875, at the British Medical Association meeting in Edinburgh. His concise account, later published in the British Medical Journal under the title The Electric Currents of the Brain, remains historically significant despite its brevity [21]:

> *In every brain hitherto examined, the galvanometer has indicated the existence of electric currents. The external surface of the grey matter is usually positive in relation to the surface of a section through it. Feeble currents of varying direction pass through the multiplier when the electrodes are placed on two points of the external surface, or when one electrode on the grey matter, and one on the surface of the skull. The electric currents of the grey matter appear to have a relation to its function. When a part of the grey matter is in a state of functional activity, its electric current usually exhibits negative variation. For example, on the areas shown by Dr. Ferrier to be related to rotation of the head and to mastication, negative variation of the current was observed to occur whenever those two acts respectively were performed. Impressions through the senses were found to influence the currents of certain areas;* e.g., *the currents of that part of the rabbit's brain which Dr. Ferrier has shown to be related to movements of the eyelids, were found to be markedly influenced by stimulation of the opposite retina by light.*

In 1877, Richard Caton published a more detailed account of his experiments in the British Medical Journal, documenting studies conducted on over 40 animals, primarily rabbits [22]. He demonstrated both spontaneous and evoked electrical activity in the brain, observing changes in oscillations with light exposure. Caton proposed that these electrical currents reflected mental states and cerebral function, modulated by external stimuli.

Despite the groundbreaking nature of his findings, Caton's work received limited recognition during his time. This was partly due to the brevity of his initial publication, which was presented as an abstract. Methodological challenges, including the low sensitivity of available galvanometers, electrical interference, and difficulties in interpreting signals, further hindered the broader acceptance of his research. Eventually, Caton abandoned brain studies and shifted his focus to politics, later serving as Liverpool's mayor in 1907.

Although his contributions were overlooked initially, they were later acknowledged by Hans Berger (1873–1941), who built upon Caton's foundational work. Today, Caton is recognized as a pioneer in modern neuroscience and a precursor to the development of EEG [23].

1.9 Adolf Beck (1863–1942)

Adolf Beck (1863–1942), a Polish physiologist, advanced the study of brain electrical activity while working under neurologist Napoleon Nikodem Cybulski (1854–1919) [24, 25]. In 1887, independently of Richard Caton, Beck recorded brain electrical activity in dogs and cats as part of his doctoral thesis. Using non-polarizable clay electrodes and a galvanometer, he discovered that brain waves varied according to the animals' levels of consciousness and sensory stimulation, independent of movements, respiration, and pulse.

Beck observed that illuminating the eye with magnesium light evoked strong electrical oscillations in the contralateral occipital lobe—a phenomenon Austrian physiologist Fleischl von Marxow (1846–1891) had previously documented in 1883. Marxow's findings had been reported in a sealed letter to the Imperial Academy of Science of Vienna, following the practice of reserving priority for provisional discoveries requiring validation [26]. Beck's results demonstrated the existence of evoked potentials arising from sensory stimulation.

Unlike Caton's work, Beck's research gained considerable attention and sparked debates regarding the priority of discoveries in brain electrical activity. These disputes were eventually resolved when Caton referenced his earlier publications, affirming his precedence.

In subsequent collaboration with Cybulski, Beck documented slower, larger brain waves during sleep or anesthesia. In contrast, electrical changes were more pronounced in cortical areas corresponding to specific stimuli. For example, forelimb stimulation produced oscillations in the area of the cruciate sulcus, while visual stimuli triggered changes in the occipital lobe [27].

1.10 The Invention of the String Galvanometer

The string galvanometer, invented in 1901 by Willem Einthoven (1860–1927), a Dutch physiologist and Nobel laureate (1924), revolutionized the measurement of small electrical currents. Einthoven used this innovative device to pioneer the recording of electrical activity in the heart, leading to the development of the electrocardiogram [28].

The galvanometer consisted of a thin, silver-plated quartz string suspended between two powerful electromagnets. When an electric current passed through the string, it interacted with the magnetic field, causing the string to move. This movement was recorded by a beam of light reflected from a mirror attached to the string, providing precise visual data.

Compared to earlier instruments used by researchers such as Richard Caton and Adolf Beck for measuring brain electrical activity, the string galvanometer offered significantly greater sensitivity and precision. It could detect extremely small currents and accurately capture the frequency and duration of signals, laying the groundwork for future advancements in neurophysiological recording and scientific instrumentation.

1.11 Vladimir Vladimirovich Pravdich-Neminsky (1879–1952)

Vladimir Vladimirovich Pravdich-Neminsky (or W. W. Prawdicz-Neminski in Ukrainian) was a Ukrainian physiologist who studied medicine at the University of Kiev. He later worked at the university and the Ukrainian Academy of Sciences. Under the mentorship of Vasyl Chagovets (1873–1941), a prominent Ukrainian and Russian physiologist, Pravdich-Neminsky utilized his expertise in physics and chemistry to achieve a historic milestone in 1913: recording spontaneous electrical brain activity in living dogs [29].

Using non-polarizable clay electrodes initially developed by Du Bois-Reymond and an Edelmann string galvanometer, Pravdich-Neminsky successfully captured brain waves. To confirm that the recordings reflected genuine brain activity, he administered curare to suppress voluntary muscle activity, demonstrating that the electrical fluctuations were independent of cardiac pulses. Notably, some recordings were obtained from the intact surfaces of dogs' skulls [29]. The results were documented using photographic techniques, with a light beam from the galvanometer projected onto moving photographic film.

In subsequent experiments, Pravdich-Neminsky identified two distinct patterns of brain electrical rhythms, which he initially described as "waves of the first and second order" [30]. These rhythms were later adopted and reclassified by Hans Berger as alpha and beta waves.

Pravdich-Neminsky coined the term *Elektrocerebrogramm* to describe his method of recording electrical brain activity [30]. However, Hans Berger criticized the term as a linguistic "barbarism" due to its mix of Greek and Latin roots. Berger proposed the term *Elektrenkephalogramm*, which was later anglicized to "electroencephalogram (EEG)" [23]. Interestingly, in Pravdich-Neminsky's original photographic records of the *Elektrocerebrogramm,* the suffix "gramm," derived from the Greek *graphein* (to write), could be considered a misnomer, as it traditionally refers to recordings made on paper or other tangible media [19].

1.12 Conclusions

The development of EEG prior to Hans Berger's discoveries was a cumulative process, shaped by the contributions of numerous scientists and key advancements in electricity and magnetism. Milestones such as the discovery of animal electricity, the measurement of electrical activity in muscles and nerves, and the invention of the string galvanometer provided the foundational groundwork for studying brain electrical activity.

Despite these achievements, early pioneers faced significant challenges. The lack of sensitive instruments, interference from the skull and scalp, difficulties in interpreting signals, and skepticism from the scientific community hindered the widespread adoption and replication of their methods. Consequently, many of their findings were overlooked or underappreciated.

It was not until the early twentieth century, with technological advancements and the dedicated efforts of Hans Berger, that EEG research reached a turning point. Berger's pioneering work represented a major leap forward in understanding brain activity and applying EEG technology. His contributions ultimately secured EEG's recognition as a crucial tool in neuroscience and medicine.

References

1. Brazier MAB. The Forerunners: experimentation challenges the ancients. In: Brazier MAB, editor. A history of neurophysiology in the 17th & 18th centuries. New York: Raven Press; 1984. p. 6–17.
2. Gilbert W. De magnete, magneticisque corporibus, et de magno magnete tellure. London: Peter Short; 1600.
3. Guericke O. Experimenta nova (ut vocantur) magdeburgica de vacuo spatio. Amsterdam: J. Janssonium Waesbergium; 1672.
4. Clark DH, Clark SPH. Newton's Tyranny: the suppressed scientific discoveries of Stephen Gray and John Flamsteed. W.H. Freeman & Co.; 2001.
5. Musschenbroek PV. Institutiones physicae conscriptae in usus academicos. Lugdunum Batavorum: Apud Samuelem Luchtmans et filium; 1748.
6. Nollet JA. Recherches sur les causes particulières des phénomènes électriques, et sur les effets nuisibles ou avantageux qu'on peut en attendre. In: Mémoires de l'Académie royale des sciences. Paris: Imprimerie royale; 1746. p. 209–64.
7. Franklin B. Experiments and observations on electricity, made at Philadelphia in America, by Mr. Benjamin Franklin, and communicated in several letters to Mr. P. Collinson, of London, F.R.S. London: Printed and sold by E. Cave, at St. John's Gate; 1751.
8. Piccolino M. Luigi Galvani and animal electricity: two centuries after the foundation of electrophysiology. Trends Neurosci. 1997;20(10):443–8.
9. Piccolino M, Bresadola M. Shocking Frogs: Galvani, Volta, and the electric origins of neuroscience. Oxford: Oxford University Press; 2013.
10. Galvani L. De viribus electricitatis in motu musculari commentarius. In: De Bononiensi Scientiarum et Artium Instituto atque Academia Commentarii, vol. VII. Bologna: Typographia Instituti Scientiarum; 1791. p. 363–418.
11. Parent A. Giovanni Aldini: from animal electricity to human brain stimulation. Can J Neurol Sci. 2004;31(4):576–84.
12. Aldini J. An account of the late improvements in Galvanism, with a series of curious and interesting experiments performed before the commissioners of the French National Institute, and Repeated Lately in the Anatomical Theatres of London. London: Printed for Cuthell and Martin, and J. Murray; 1804.
13. Volta A. On the electricity excited by the mere contact of conducting substances of different kinds. In a letter from Mr. Alexander Volta, F. R. S. Professor of Natural Philosophy in the University of Pavia, to the Rt. Hon. Sir Joseph Banks, Bart. K. B. P. R. S. Philos Trans R Soc Lond. 1800;90:403–31.
14. Pearce JM. Emil Heinrich Du Bois-Reymond (1818–96). J Neurol Neurosurg Psychiatry. 2001;71(5):620.
15. du Bois-Reymond EH. Untersuchungen über thierische Elektricität. Berlin: G. Reimer; 1848–1884.
16. von Helmholtz H. Messungen über den zeitlichen Verlauf der Zuckung animalischer Muskeln und die Fortpflanzungsgeschwindigkeit der Reizung in den Nerven. In: Archiv für Anatomie, Physiologie und wissenschaftliche Medicin. Berlin: G. Reimer; 1850. p. 276–364.
17. Cohen Of Birkenhead. Richard Caton (1842–1926) Pioneer Electrophysiologist. Proc R Soc Med. 1959;52(8):645–51.

18. Brazier MA. The history of the electrical activity of the brain as a method for localizing sensory function. Med Hist. 1963;7(3):199–211.
19. Niedermeyer E. Historical aspects. In: Niedermeyer E, Lopes da Silva F, editors. Electroencephalography: basic principles, clinical applications, and related fields. Philadelphia: Williams & Wilkins; 2005. p. 1–15.
20. Ormerod W. Richard Caton (1842–1926): pioneer electrophysiologist and cardiologist. J Med Biogr. 2006;14(1):30–5.
21. Caton R. The electric currents of the brain. Br Med J. 1875;278
22. Caton R. Interim report on investigations of the electric currents of the brain. Br Med J. 1877;1:62–5.
23. Berger H. Über das Elektrenkephalogramm des Menschen. Arch Psychiatr Nervenkr. 1929;87: 527–70.
24. Brazier MAB. A history of the electrical activity of the brain. The first half-century. London: Pitman; 1961.
25. Coenen A, Fine E, Zayachkivska O. Adolf Beck: a forgotten pioneer in electroencephalography. J Hist Neurosci. 2014;23(3):276–86.
26. von Marxow EF. Mittheilung betreffend die Physiologie der Hirnrinde. Centralblatt für Physiologie. 1890;4:537–40.
27. Beck A, Cybulski NN. Weitere Untersuchungen über die elektrischen Erscheinungen in der Hirnrinde der Affen und Hunde. Zentralbl Physiol. 1892;6:1–6.
28. Fisch C. Centennial of the string galvanometer and the electrocardiogram. J Am Coll Cardiol. 2000;36(6):1737–45.
29. Pravdich-Neminsky VV. Über die elektrischen Erscheinungen im Gehirn des Menschen. Arch Psychiatr Nervenkr. 1914;54(1):1–32.
30. Práwdicz-Neminski WW. Zur Kenntnis der elektrischen und der Innervationsvorgänge in den funktionellen Elementen und Geweben des tierischen Organismus. Elektrocerebrogramm der Säugetiere. Pflügers Archiv für die Gesamte. Physiologie. 1925;209:362–82.

2 Hans Berger (1873–1941): The Search for "Mental Energy" That Brought the Electroencephalography

Walter van Emde Boas and Peter J. Koehler

2.1 Short Biography [1–8]

Hans Heinrich Wilhelm Ernst Berger (Fig. 2.1) was born, on May 21, 1873, in Neuses, a small village, now part of Coburg, in northern Bavaria, Germany. He was the only son among five children of Paul Friedrich Berger (1834–1900), a local physician, and Anna Rückert (1839–1919), daughter of the German Poet and Orientalist Friedrich Rückert (1788–1866).

After finishing the local grammar school and the Gymnasium Casimirianum in Coburg, Berger entered the University of Berlin in 1892, for half a year of study of natural sciences, mathematics, and astronomy, leaving early for a year in the military. A serious accident during a training maneuver caused him to change his mind, and after temporary discharge from the army, Berger entered the Medical University in 1893 and passed the medical state exam there, in 1897. He then took up a position as junior assistant in the Psychiatric and Neurological Clinic of the University of Jena, the clinic where, apart from the war years 1914–1918, he was to remain professionally active until his retirement.

Fig. 2.1 Hans Berger. (From Archives of the Max Planck Society, Berlin, VI. Abt., Rep. 1, Berger, Hans 2)

Already in 1900, Berger was promoted to "Oberarzt," and in 1901, he was appointed "Privat Dozent," followed by an extraordinary professorship in 1906. In 1911, he married Baroness Ursula Erna Marie von Bülow (1886–1970), who had been working for him as an assistant in his laboratory since 1910, and gave birth to their first and only son, Klaus, in 1912, followed by three daughters, Ruth (1914), Ilse (1917), and Rosemarie (1921). In 1918, Berger took over leadership of the clinic, first acting for and after Otto Ludwig Binswanger's (1852–1929) retirement in 1919, as regular director and full professor, a dual position he continued to hold until his

W. van Emde Boas
Deptartment of EEG and Epilepsy Monitoring Units, Dutch Epilepsy Foundation S.E.I.N. Clinics, Heemstede & Zwolle, the Netherlands

P. J. Koehler (✉)
Faculty of Health, Medicine and Life Sciences, Maastricht University, Maastricht, The Netherlands
e-mail: pkoehler@neurohistory.nl

F. Brigo, O. Mecarelli (eds.), *EEG: The First 100 Years*,
https://doi.org/10.1007/978-3-031-86878-8_2

academic retirement at the age of 65 years in October 1938. In the 1927–1928 academic year, he was rector of the university and pro-rector from 1935 to 1938 [6].

Berger, who was denied the opportunity to continue clinical or scientific work in the university clinic after his retirement, first started a private practice in Jena [5] but was then called up for a short period of military service, as a psychiatric consultant for the *fourth Wehrmachtkreise* in Saxony [4]. He then moved to resume private practice at the former Walda sanatorium in Bad Blankenburg, a small town in the Thuringian Forest, some 40 km southwest of Jena. He was again thwarted in his efforts, however, when the sanatorium was taken over for military use [4]. Eventually, he was called back to the clinic from September 1939 to early January 1940 to temporarily replace his successor Berthold Kihn (1895–1964), but it proved impossible to resume his scientific work there and Berger returned to Bad Blankenburg. There his last years were filled with reading, nature walks, and work on philosophy and parapsychology [2]. Following a viral infection in May 1941, Berger progressively sank into depression, a second episode after an earlier one in 1921 [4] that he himself did not recognize as such and mistook for a cardiac condition. On his own initiative, he was admitted to the Medical Clinic in Jena, where after a few days, on June 1, 1941, he took his own life by hanging. On June 4, 1941, he was buried at the Johannis Cemetery, near the Neuropsychiatric University Clinic, where his grave, restored in July 2024, can still be visited [8].

2.2 Medical Career

In March 1891, Berger, still a high school student, wrote his first entry in his private diary, quoting a short verse by his grandfather Rückert, whom he never knew personally but admired immensely and whose portrait he later kept in his office in Jena [4, 8]: "*Nicht alles, was Du weißt, darfst Du allen vertrauen, noch minder alle, was Du nicht weißt, lassen schauen.*" Roughly translated as "*You should not share everything that you know with all, but even less so all that which you do not know,*" it reflects the scientific and philosophical approach to life and work that characterized Berger throughout his life [3]. In further entries in these early diaries, he pondered his expectations of a scientific career, initially and despite his father's wish for him to opt for medicine or physiology, either zoology, astronomy, or mathematics, prompting him to actually follow a half year of lectures in the latter two in Berlin, in late 1892. Disappointed in both the content of these studies and the atmosphere in the big city, he returned home and then decided for a military career, aiming to attain a rank as officer, while actually hoping in that way to obtain the financial security to pursue a career mostly focused on less exact sciences [8]. In 1893, however, a serious accident caused Berger to change his mind.

In his last publication, *Psyche* [9], he wrote.

> *True transmission of thought is one of the still controversial phenomena in parapsychology, which, however, in my opinion should be scientifically recognized as a true phenomenon. Contrary to other parapsychological phenomena such as telekinesis, materialization and ghost appearances, all of which have been proven as fraudulent and/or based on self-deception, this true transmission of thought has been proven to exist ... as much as such a phenomenon can be proven.*

And

> *I convinced myself already early that something like this must exist. As a 19-year-old student and during a military exercise near Würzburg, I suffered a serious accident and barely escaped certain death. Riding a rearing and stumbling horse on the rim of a hollow road I fell into the ditch of the road in front of a moving battery and ended up under the wheels of a cannon. At the last moment the six-horse spanned Cannon Tow was brought to a halt and I came away with just the scare. All this took place in the morning hours of a beautiful spring day. On the evening, that same day, I received a telegram from my father, inquiring about my wellbeing. This was the first and the only time during my life that I got such a message. It was my oldest sister, with whom I held a special close family bond, who brought on this telegraphic inquiry, as she had suddenly informed my parents that she was certain that I had suffered an accident. The family lived in Coburg at the time. This*

constitutes a case of spontaneous transfer of thoughts, whereby I, in a moment of extreme danger and facing certain death, must have acted as sender, with my sister, with whom I was very close, as receiver. (Transl. WvEB: Berger does not use the term Telepathy here)

This experience decided his future. Shortly thereafter, Berger was temporarily discharged from the military, began medical studies, and embarked on what would become his quest for "Mental-" or "Psychic Energy." Berger took up medical studies first in Würzburg during the summer of 1893, then moved to the University in Jena for the winter semester 1893/1894, and subsequently, as was customary at the time, moved to other universities, including Kiel and again Würzburg. In the summer of 1896, he returned to Jena, where he felt most at home, joined the traditionally oriented student society *Burschenschaft Arminia auf dem Burgkeller* [1897], and passed the medical state exam on June 3, 1897 [6].

Following this "Abitur," he took up the position of junior assistant to Theodore Ziehen (1862–1950) at the Psychiatric and Neurologic Clinic of the University of Jena, at that time headed by Binswanger, where he had already worked as an Intern during his medical studies. In December 1897, Berger finished his dissertation on the spinal cord pathology in neurosyphilis and was offered a position in the clinic by Binswanger, initially as a so-called "volunteer-physician." Following Ziehen's departure to Utrecht (Netherlands) in 1900, Berger succeeded him as "Oberarzt" [senior physician]. Following the defense of his habilitation thesis in July 1901 [10], he was admitted first as "Privat Dozent" [lecturer], followed by the appointment as extraordinary Professor in 1906.

From April to October 1898, Berger had to take a leave of absence from the clinic in order to complete his tour of army duty. During the war years 1914–1918, he served first as a military staff physician in Erfurt and then from June 1915 to November 1918 on the Western Front as a neuropsychiatrist at the front hospital in Rethel and the military hospital in Sedan, both just northeast of Reims, France. For his military duties, he obtained several awards, including the Iron Cross second Class [6].

On his return to Jena in 1918, Berger was first appointed acting director of the clinic, and, following the retirement of Binswanger and at the latter's express wish [8] as his successor as director and full professor of Psychiatry. Both positions remained unchanged until his age-appropriate retirement on September 2, 1938.

Kurt Kolle (1898–1975), who trained under Berger as an assistant, recounted the strict and methodical way Berger ran the clinic and his own workday, following a strict time schedule that was not to be disrupted unless neurosurgical procedures were in progress [1]. The latter very much reflected the main interest, apart from his psychophysiological and later the electrophysiological studies, of Berger, who, despite his official position as a professor of psychiatry, had relatively little interest in the typical psychiatric aspects of his patients [1] and in practice and as a teacher impressed more as a neurologist [3]. In this, he conformed to the general tendency at the time of psychiatrists in the German language world to study the brain, rather than to conduct scientific research and formulate theories about psychiatric diseases [11].

In Berger's bibliography of 123 papers, chapters, and books, some of which are double publications [6] (the bibliography in [4] is incomplete), only about 30 deal with psychiatric disorders. The majority was published before his appointment as a professor of psychiatry and most of them are on neuropathology or typical "organic" psychiatric disorders such as neurosyphilis, postinfectious, toxic or otherwise organic psychosis, alcoholism, and sleep disorders. His inaugural lecture, on taking over the chair of psychiatry in 1919, was titled "Hirn und Seele" and dealt mainly with the Mind-Brain-Psyche problem, rather than with clinical psychiatry [12]. Only five papers deal with psychiatric topics such as Dementia Precox [the old term for schizophrenia], Melancholia, Paranoia, or Psychotherapy. Clinically, he was more interested in brain function and cortical localization, an interest that he shared with two colleagues from his early training years under Ziehen, Oskar Vogt (1870–1959) and Korbinian Brodmann (1868–1918), and which, tellingly, was to be the subject of his 1927 Rector's Lecture: "On Cerebral Localization" [13].

In this lecture, reprinted in [6], Berger presented a short historical overview of ideas about cerebral localization from the ancient Greek and Roman philosophers and physicians to the concept of Phrenology by Franz Joseph Gall (1758–1828). From there, he continued to the beginning of modern insights on localization, pioneered by the clinical findings of Paul Broca (1824–1880) and Carl Wernicke (1848–1905) and the early electrical stimulation studies by Gustav Fritsch (1838–1927) and Eduard Hitzig (1838–1907) and the more recent ones by Victor Horsley (1857–1916), Fedor Krause (1857–1937), Otfried Foerster (1873–1941), and Harvey Cushing (1869–1939). He correlated these findings with the then-current neuroanatomical concepts of cortical fields and neuronal structures by Vogt, Brodmann, Vladimir Betz (1834–1894), and others and discussed the various concepts concerning sensory impressions and higher nervous processes such as thinking or remembering. He distanced himself from the concept of Constantin von Monakow (1853–1930) and others, who claimed that such functions are diffusely represented throughout the whole cortex. Although he agreed that, in diffuse brain disorders, higher psychical functions are often impaired, he argued for more localized "Engrams," based on his own clinical experience and the work of others. He emphasized the importance of the extensive network of nerve fibers connecting all parts of the brain, as studied by Paul Flechsig (1847–1929) and others. He briefly mentioned how "attention" is often impaired as a first sign of cerebral dysfunction, while other higher functions remain intact, and that such loss of attention can also be observed during temporary oxygen deprivation. Eventually, he concluded that simple functions, that is, motor, primary sensory, visual, and so on have a strict cortical localization with related functions in related areas. He believed that this was less clear for more complex functions, but that even these are located in or dependent on widespread cortical connections. All physical or psychological activities in these connections would be dependent on and thus an expression of material energetic processes. Apart from a brief mention of psychiatric disorders as yet another expression of higher mental functional impairment, there is no reference to psychiatry and Berger's few self-references are to neurological articles or psychophysiological studies. His final conclusion, however, clearly reflects his persisting interest in "Psychic Energy" even if the actual term is not mentioned here.

Kolle suggested that the Rector's lecture [13] represented a kind of creed for Berger and described the meticulous attention to localization during the personal patient examinations by Berger in surgical cases, in cooperation with the neurosurgeon Nicolai Guleke (1878–1951), with whom Berger also published on the surgical treatment of brain tumors [14]. Other publications, highlighting this neurological orientation, deal, apart from a number of papers on localization, with including traumatic, vascular, or infectious brain lesions, epilepsy, and neuropathologic findings in various conditions.

His lifelong prime interests, however, already stemming from his high school years and early study period, were the mind-brain problem and the physical and chemical processes underlying higher brain functions. Although appreciated as an excellent clinician and teacher, it was his experimental work in these areas that was to be his main contribution to medicine, even if least appreciated at the time.

2.3 Berger's Scientific Work

2.3.1 The Formative Years

Hans Berger grew up in a rich intellectual environment, his paternal grandfather, a barber and small-time surgeon, his father, a physician in the local hospital, and his mother, the daughter of the poet, philosopher, and orientalist Friedrich Rückert. Moreover, as a student at the local Gymnasium, he was taught the mix of natural science and humanities, philosophy, and theology, common in Germany at the time. In his early diaries, he pondered about his future intentions, to become a scientist, maybe an astronomer, but not a simple-minded one, ignoring the humanities [Berger, diaries I/23; Dec. 2 1891, quoted by 2, 3]. Following his deception with early studies in Berlin and the decision, at his father's wish, to choose medicine, he wrote: "*... thus a physician. But I can always practice natural science as a sideline, notably I would like to focus on zoology. Father would like me to become a physiologist. Maybe that as well*" [Berger, diaries I/50; Oct. 15 1893, quoted by 2].

Already during his high school days, he had read *Seele und Leib in Wechselbeziehung zu einander* (1865) by Jacobus LC Schroeder van der Kolk (1797–1862), a book he found in his mother's library [3], kindling his interest in the relation between the physical body and the workings of the mind. Furthermore, he must surely have studied the work of Ernst Haeckel (1834–1919), who was an important local figure in academic circles in Jena at that time [5] and a proponent of the concept of Spiritual Monism. Berger later rejected this concept in his lectures on Psychophysiology [15, reprinted in 6].

The formative years of the young student and physician Berger were generally marked by great advances in knowledge and theories in the fields of (neuro)anatomy—microscopy and—physiology and, as an avid reader, he was well aware of these. Of special influence were physician and physicist Julius Robert von Mayer (1814–1878), who had formulated the first version of the law of conservation of energy half a century earlier, and Wilhelm Ostwald (1853–1932), a pioneer in physical chemistry and Nobel prize recipient (1909), who had proposed a rather monistic "universal" concept of psychic energy in 1908. Another, more brain-oriented concept of psychic energy had been proposed by Alfred Lehmann (1858–1921), a Danish physiologist, who would inspire much of Berger's early experimental work. Max Verworn (1863–1921), professor of physiology in Jena during Berger's medical training, who had devised the theory of "Biogen" molecules for contact between the dendrites and neurons, then known as discontinuous by the work of Santiago Ramon y Cajal (1852–1934), also made an important mark and is often quoted by Berger in his lectures and other works [5, 6]. Based on these as well as other works, Berger became convinced early on that mental functions and consciousness had their basis in physiological processes in the brain and sensory organs, an idea that drove his thinking and research throughout his professional life, until his last publication [9], where he wrote:

> *In any case we can assume as certain, that activities in the cerebral cortex may be related with phenomena of consciousness. Additionally, we must assume that all mental functions are unequivocally bound with material processes in the cerebral cortex.*

And

> *... based on our clinical experience we must assume that all mental functions, from mere perceptions up to three of the most intricate thinking processes, are unambiguously dependent on material processes in the human brain cortex.*

And where he may even have had earlier thoughts on the subject, it would be these "material processes" in the supposed form of a *"Psychischen Energie"* that Berger would persistently search for, from his first studies on brain temperature [16] to this last publication, *Psyche* [2, 9]. His first professional years, however, focused on different studies.

The Jena Psychiatric Clinic, at that time headed by Otto Binswanger, represented one of the leading European clinics in the field, but psychology and early psychophysiology, as studied by the pioneer scientists of the nineteenth century, such as Ernst Heinrich Weber (1795–1878), Gustav Theodor Fechner (1801–1887), and Hermann von Helmholtz (1821–1894), a.o., had fallen out of fashion. Emphasis in neurology and psychiatry had shifted to the realms of either neuroanatomy and neuropathology, as practiced by Bernhard von Gudden (1824–1886), Theodor Meynert (1833–1892), Constantin Von Monakow, Alois Alzheimer (1864–1915), and others, or, alternatively, to theoretical approaches such as Emile Kraepelin's (1856–1926) characterology, Eugen Bleuler's (1857–1939) emphasis on personality disorder, or, somewhat later, the phenomenological approach of Karl Jaspers (1883–1969) or the psychotherapeutic systems of Sigmund Freud (1856–1939), Alfred Adler (1870–1937), and others [2, 3]. Berger felt the former was not sufficiently functional to suit his psychophysiological interests, while the latter, in his view, lacked sufficient scientific basis [3].

Fortunately for Berger, however, Ziehen, who was still primarily interested in psychophysiology, was the chief physician in the Jena clinic, responsible for the direct supervision of the junior assistants. He was -later- appointed as such, first in Utrecht (Netherlands) and then Berlin, before finally turning to philosophy and ending his career as professor in that field in Halle. Ziehen had investigated -without much success- whether psychiatric conditions correlated with specific changes in pulse waves and stimulated Berger and other upcoming neuropsychiatrists, among whom Korbinian Brodmann, to further investigate the functions of the living brain [5].

As aforementioned, Berger's dissertation of 1897 was a neuropathological study on the degeneration of anterior horn spinal cord neurons in syphilitic dementia, which was also the subject of his first published article in 1898 [17]. However, his 1900 article on experimental anatomic studies on impaired development of the occipital cortex in dogs and cats following sensory deprivation already illustrates his interest in the relation between sensory and mental processes and the material substance of the brain [3, 18]. Berger also briefly self-experimented on the possible transferability of acute psychiatric conditions, injecting himself with blood samples taken from patients with acute episodes of psychosis [5, 19]. However, most of his experimental work from the beginning focused on psychophysiology and measurement of bodily functions in mental conditions.

2.3.2 Studies on Cerebral Circulation

Ziehen's own work included sphygmometric studies of the blood pulse and was very much in line with the work of the Danish physiologist Lehmann, whose work on the pulse in the arm and in the carotid arteries as well as its fluctuations during various mental states [20] also significantly influenced Berger and were often was quoted by him [2]. For Berger's dissertation work, however, Ziehen encouraged him to investigate fluctuations in the cerebral circulation as well, using the method already described by Angelo Mosso (1846–1910), who had performed plethysmographic studies on animals and patients with surgical cranial defects [21]. Berger also had access in his clinic to a patient with such a surgically created defect, following surgery for a brain tumor. Using a special rubber cap, connected to a system of pneumatic hoses and eventually writing out fluctuations, caused by pulse and respirations on a black kymometer drum, Berger was able to study the effects of various pharmaceutical substances on cerebral circulation.

This study line was initiated by a chance observation of a marked improvement, induced by cocaine, in a psychotic patient in a stuporous and mute state and resulted first in his *Habilitations thesis* in 1901 [10] and later in two more extensive publications [22]. In these studies, illustrated by curves in the text and with illustrated atlases added to the 1904 and 1907 publications, Berger described the effects on the recorded pulse forms of substances such as cocaine, morphine, chloroform, camphor, caffeine, or amyl nitrite but, in the later studies, which included additional subjects with accessible cranial defects, also the effect of sensory or emotional stimuli.

This work provided further evidence that reliable recording of cerebral circulation pulsations, in humans was possible, that cardiac pulsations, respiratory fluctuations, and stimulus-related changes could be distinguished, and that the latter were limited to the cerebral cortical regions. However, Berger's hope to find specific psychophysiological and or localizable changes in the recorded pulses that could be correlated to highly specific functions, that is, arithmetic, joy, fear, and so on remained unfulfilled [6].

2.3.3 First Electro-Cortical Recordings, Studies on Cerebral Temperature, and Other Experiments [2, 3, 5, 6]

Possibly prompted by this failure—since the diaries over the period 1897–1910 are missing we have no certain personal information on his motivations—and his continued search for the energetic processes Berger assumed to underlay mental activity, he discontinued cranial plethysmography and in 1902 began experiments to record electrical activity from the cerebral cortex in cats and dogs. As in his later work on the electroencephalogram, these experiments were inspired by the pioneering studies of predecessors such as Richard Caton (1842–1926), Adolf Beck (1863–1942), Napoleon Cybulski (1854–1919), and Ernst von Fleischl-Marxow (1846–1891), all of whom had already succeeded in recording fluctuations in electrical currents from the exposed cortex in experimental animals. The experiments were usually conducted only by Berger himself in a small laboratory, situated in the basement of the clinic. Due to the limited sensitivity of his equipment, initially a capillary electrometer and later a string galvanometer, these early studies focused on possible changes following sensory stimuli and observations of the effect of electrical stimulation and not yet on spontaneous cortical rhythms. Yet they remained limited and largely unsuccessful, partly due to the limitations of the measurement equipment, but also partly due to Berger's erroneous—and thus unfulfilled—expectation that there would be a quantitative relationship between stimulus intensity and cortical response [5]. In diary notes of Nov 30 and Dec 3, 1910 [Berger, diaries II, 148 and 152, quoted and illustrated in 2], Berger summarized nine experiments on dogs, performed in the period 1902–1910, noting

> "*Out of 9 experiments only 1 result and that one actually unreliable since skin-potentials could not unequivocally be excluded during that experiment. Yet it cannot be said that I have given up easily. 8 years! All the time trying over, again.*" And 4 days later: "*I think that I will definitely stop the experiments on the dog cerebral cortex—Observations in Humans.*"

Another line of experimental work during this period focused more on investigating possible energetic processes, related to mental functions by measuring the brain temperature. In humans, such studies were possible in surgical patients, where the neurosurgeon would use thin probes to "feel out" the location, depth, and extent of a lesion and Berger could use a specially made thermometer with a long tip to make measurements along the puncture path already made by the neurosurgeon. In addition to recordings in humans, Berger also made experimental measurements in a chimpanzee, acquired specifically for these studies. Contrary to the electro-cortical studies, the work on cerebral temperature resulted in a publication, however, again with a largely negative result in so far that Berger had to conclude that he could not unequivocally prove that there was any change in cortical temperature during sensory or mental stimuli, other than what could be explained by changes in blood flow [6, 16]. Nevertheless, Berger did continue some observations on temperature and brain function, even during his work as a military physician in France during World War I, now focusing more on external (skin of the arm or other locations) measurements and the effects of mental activities, stimuli, and even hypnosis [6, 23].

These observations were as much part of both Berger's scientific and clinical work as of his academic duties which, apart from the clinical teaching and lectures on psychiatry, included lectures on the anatomy and physiology of the brain, psychology, and notably "physiological psychology," the official name, listed for his lectures that later would be published as his *Lectures on Psychophysiology* [15].

In addition to the methods, already mentioned, Berger was very interested in the study of the Psychogalvanic Skin Responses, as pioneered by Ivan Tarkhanov—or Ivane Tarkhnishvili, his original Georgian name—

(1846–1908). After giving up on the electrocortical recordings in 1910, he started investigating the Galvanic skin response in the Jena Clinic, initially predominantly as an experimental test. And where other investigations had been rather disappointing, work on Galvanic skin responses at least brought him happiness: in 1910, he was joined by Ursula von Bülow as assistant and she turned out to be not only a good test-assistant and—subject—some rather touching and suggestively intimate test protocols are quoted by Borck—but also, in 1911, became his wife for life and mother of their four children [2, 5].

2.3.4 The Electroencephalography Studies [2, 3, 5, 24]

Although the animal studies with cortical recordings thus had remained unsuccessful and Berger's activities for a time focused on other areas he did not totally give up on the idea. In his War diaries (1915–1918), between notes on philosophical or religious readings, he writes on possible further experimental studies, including again measurements and recordings from the cerebral cortex [2]. On returning to the clinic and taking over the directorship from Binswanger, clinical and administrative duties initially prevented Berger from resuming his experimental work. In 1920, he started renewed studies, including an unsuccessful attempt to record electrical brain activity from a bald medical student. Only in 1924, he found time to resume more systematic experimental work, initially mostly involving indirect cortical stimulation through the skin in patients with surgical cranial defects, supplied by Guleke, and aimed at observing sensory effects [3].

Berger's real main interest however was the intrinsic cortical activity as an expression of his hypothetical "Psychic Energy" and possible changes in that activity, induced by sensory or mental input.

On June 2, 1924, a diary note mentioned the idea of recording, rather than stimulating, in patients with trepanations, and soon after Berger invited a teenage boy who repeatedly had been the subject of stimulation experiments, to come to the hospital on a Sunday for recordings during mathematical tasks [5, 8].

Using a small Edelmann String Galvanometer, based on the instrument that Willem Einthoven (1860–1927, Nobel prize 1924) had developed and used for his studies on the electrocardiogram [25], Berger on July 6, 1924, succeeded for the first time to record some fluctuations which he considered as possible, albeit certainly not yet proven, electrocortical fluctuations.

Soon afterward Berger obtained a more advanced Edelman String Galvanometer, Berger also co-recorded the ECG using the Einthoven technique [24]. In 1926 an even more sensitive recording instrument, the Siemens coil galvanometer became available. Using these instruments, Berger performed a long series of meticulous and carefully documented recordings, both in patients with trepanations and in healthy humans, including his son Klaus and himself, usually recording from subcutaneous needle electrodes but also with scalp electrodes [24]. Still, it took another 5 years before Berger, often uncertain about the reliability of his own findings, and at times, even as late as 1928, he was on the verge of giving up [2, 8]. Then in 1929, he published his first paper on electroencephalography (EEG): *"Über das Elektrenkephalogram des Menschen,"* where, now with sufficient self-confidence, he wrote "... *that I finally, indeed, believe to have discovered and for the first time published the Human Electroencephalogram*" [24].

In the period of 1930–1938, this first report would be followed by thirteen more successive reports as well as a great number of related papers, already documenting much of the essentials, including artifacts and the effect of drugs and anesthetics, of the human EEG in wakefulness and sleep, in adults and children and in healthy subjects as well as in patients with epilepsy, brain tumors, trauma, and other brain disorders [2, 4]. In 1938, he summarized these findings in a paper, published at the occasion of his admission to the German Academy of Natural Scientists "Leopoldina" [26] (Fig. 2.2).

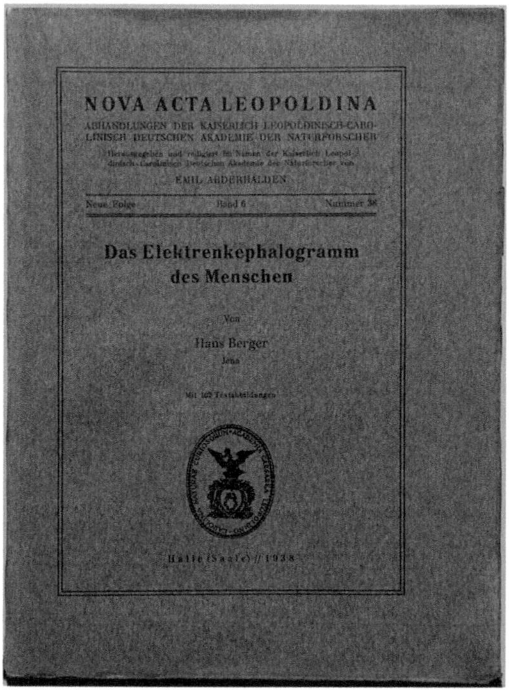

Fig. 2.2 Berger's publication on the occasion of his admission to the German Academy of Natural Scientists "Leopoldina" [26], summarizing the contents of his 14 Reports. (Public domain)

2.4 Hans Berger, the Man

All biographers describe Berger's personality as complex: of athletic build and physically impressive, pedantic and overbearing to his residents, highly intelligent and knowledgeable but at the same time sensitive, shy, uncertain, fearful for criticism, and then also stubborn and precise to the point of obsessiveness [1–3, 8]. He was totally focused on his work and family life, did not drink or smoke [1] and, apart from his reading and outdoor activities, hardly seems to have had other interests.

The diaries and experimental protocols reflect his excellent clinical and scientific knowledge as well as his wide and erudite interest in many fields outside direct neuropsychiatry or neurophysiology but also, continuously, his uncertainties, doubts, and misgivings about his work and life [2].

Kurt Kolle, who trained under Berger as an assistant and Pierre Gloor (1923–2003), quoting Raphael Ginzberg, another assistant, described the strict and methodic as well as somewhat pedantic way how his professor worked, spent the day in the clinic and dealt with the patients and with his assistants. Following breakfast at 06:30, Berger would enter the clinic every day, 6 days a week, at exactly 07:00 in the morning for work private work on his diaries or other matters. On Sundays only, the starting time would be 09:00. All patients who had been admitted the day before should have been thoroughly examined and have their complete records ready for presentation at 09:00 followed by the morning rounds from 10:00 to 12:00. The final medical report as well as the letter to the referring physician of all patients that had been discharged the day before also had to be ready to be checked and discussed, with Berger paying attention to even the smallest details.

Berger would then leave the clinic at 12:00 sharp for a walk in the neighboring forest or, in the summertime, a swim, with his wife and children, then have lunch at 13:30, take some rest, and return to the clinic at 15:00 sharp for his private practice until 17:00. From there on, unless prevented by acute clinical duties, the evening would be free for experimental studies, dinner with the family and, as from 1930 and often until late in the evening, dictation to a former patient, Schlömilch, who for many years thus assisted Berger as a secretary [1, 2]. Morning rounds had to remain focused strictly on the cases at hand, but, on Saturdays from 17:00 until 19:00, the residents had to present and discuss interesting or difficult cases during a session called *"Klughusten,"* strictly to be translated as "clever coughing" [1].

In his work on the EEG, Berger, essentially an autodidact as a neurophysiologist, always remained reclusive, working alone with only Mr. W. Keuscher as a technician and with only limited assistance from his residents, other than as test subjects. While Berger did have contact with and got advice and help from the Department of Physics of the Jena University, he neither requested nor got any support from Wilhelm Biedermann (1852–1929), an outstanding electrophysiologist who at that time directed the Physiological Laboratory [2, 3].

During the years before the publication of the first report, he hardly shared his findings with anyone [3], and before 1937, he did not attend or participate in any of the international congresses and only in few, mostly small local meetings [2, 4]. His reports on the EEG were published in German language journals that were hardly read by neurophysiologists who at the time were mostly working on action potentials, brain stimulation, or cortical ablation studies [4]. Word about and the introduction of EEG in other laboratories thus came only slowly, even in Germany but certainly abroad, where Berger's work initially was greeted with severe doubts [27].

However, after Berger's findings were confirmed by Edgar Douglas Adrian (1889–1977) and Bryan Harold Cabot Matthews (1906–1986) [28] national and international recognition came fast.

Initially and erroneously, Adrian and Matthews were credited with the invention of the EEG, but this misunderstanding was quickly corrected by Adrian who actually proposed the eponym "Berger Rhythm." Berger, however, preferred to stick to his original nomenclature: alpha rhythm [2, 5] Starting in the early 30s, Berger corresponded with and was visited by colleagues from abroad, who had started to experiment with the EEG themselves, including Herbert Henri Jasper (1906–1999), Frederick Andrew Gibbs (1903–1992), and William Grey Walter (1910–1977) [5]. In 1937, Berger was received with great honors at the International Congresses of Psychology in Paris and at the Galvani Memorial Meeting in Bologna. An invitation to come to the USA in 1939 had to be canceled due to the war/political conditions [29].

2.5 Berger a Nobel Prize Candidate?

In his biographical chapter, Kolle states quite explicitly that Berger, should he have survived the years of World War II, would certainly have the Nobel Prize bestowed on him and most biographers either quote this statement or otherwise agree that Berger certainly would have deserved this honor. About his actual candidacy, however, doubts and controversy remain. Even before this international acceptance of the EEG, Berger was apparently sufficiently known and esteemed, to be invited, in 1929, to lecture in Stockholm—an invitation that he first accepted, but then had to reject [5, 6]. In 1930, this was followed by an invitation by the Nobel committee to propose a potential name for the Nobel Prize of that year, an invitation that usually indicates that the invited scientist is himself among his peers of possible Nobel candidates [5].

In his diaries of 1930 and later, even in 1938, Berger indeed expressed thoughts and hope about such an award [5], and in many of the biographies, quoted before, it was suggested that Berger was, indeed, nominated for and awarded the Nobel Prize in 1936 but was denied the option to accept by the Nazi Government, which had forbidden any German scientist to accept that prize after 1936 when the Peace Prize of 1935 had been awarded to the left-oriented and pacifist Journalist Carl von Ossietzky (1889–1938). More recent research in a.o. The Archives of the Swedish Academy, however, has not provided any evidence that Berger, indeed, was nominated for that year. There were three nomination proposals in 1940 when the prize was canceled, due to the war circumstances. Two more nominations, in 1942 and 1947, were disregarded since Berger had died by then [4–6, 30].

2.6 Berger and the National Socialist Regime

Early biographies and reminiscences either ignored the political background during Berger's later years [1, 2, 4, 29] or tended to describe Berger as opposed to or, notably in his rather abrupt and unceremonious "forced" retirement in 1938, even as a victim of the 1933 National Socialist takeover and the subsequent Nazification of the German academic and medical institutions [1, 3, 27]. More recent studies paint a different picture and suggest that Berger, although probably not an outright Nazi adept and not a member of the official NSDAP, otherwise remained quite in line with the political developments after 1933. Berger had been a member of the *"Deutschnationalen Volkspartei"* and later of the *"Deutschen Volkspartei"* [6, 31], indicating conservative and nationalist convictions. In 1934, he also enlisted as a supporting member of the SS, possibly in order to secure his position in the otherwise Nazified University organization [7, 8, 32], and also joined the *"Reichsbundes der Kinderreichen,"* an agency, instituted to promote and preserve large and healthy Aryan families [31, 33].

It can thus be assumed that Berger was a firm believer in the then-current theories about race, degenerative disorders, and eugenics [33], popular not only in Germany but throughout the Western academic world, well summarized by Zeidman [31].

According to Zeidman, quoting the extensive study of S. Zimmermann and T. Zimmerman [34], the Jena University was even more Nazified than the national average; that Berger was not an active Party member did apparently not bother the Rector, Karl Astel (1898–1945), himself an outspoken Nazi who reported Berger as "... still of national socialist inclination and politically completely reliable" [32].

In the period 1936–1938, Berger, together with some of his staff members actively and willingly, took part in the activities of the Jena Branch of the "Erbgesundheitsgericht," the "Court for Genetic Health" [7, 32, 33]. This court decided in cases of protest against forced sterilization of mentally or physically impaired persons, patients with epilepsy, or psychiatric patients who according to the Nazi Racial Laws fell under the "Law for the prevention of hereditary offspring." In Jena, this court was quite active and responsible for a relatively large number of forced sterilizations. When Berger had to decide, he rarely decided against the procedure [32, 33]. Following his retirement in 1938, Berger stepped down from his function as a consultant to the court but when asked by Astel in March 1941 to resume this position Berger wrote that he was pleased to accept the invitation. Physical and mental health conditions, however, prevented him from further participation [32].

Berger was actively involved in the selection of Kihn, also an outspoken Nazi, as his successor [32] and, in the period 1933–1938, did not protest when overruled by the Nazi-led administrators in decisions on clinical positions in his department [33]. Also, in his diary, he already considered the option of moving his material and work to the clinic of Vogt, in Berlin-Buch, where Jan Friedrich Tönnies (1902–1970) together with others had already started EEG studies in 1930 [5].

Thus, the story of the abrupt and shameful notification of forced retirement, given to him as a surprise at the end of a clinical round and then banishment from further work as a revenge for his anti-Nazi stance [2–4], appears as apocryphal as the myth concerning the 1936 Nobel nomination. The fact that Berger was recalled to the clinic from September 1939 until January 1940, to temporarily replace Kihn [2] and that, also in September 1939 "... *notwithstanding advanced age but considering the current war* ...," he actively made himself available and was appointed consultant psychiatrist in the "*Wehrkreis IX*" [6] provide further evidence that Berger was not ostracized by the Nazi managers. That he was unable to resume his experimental work and eventually did not receive an advanced EEG recorder from Siemens, ordered shortly before the war and paid by the Zeiss foundation [2] thus appears primarily a result of the war conditions, not as a personal attack on Berger.

2.7 The Aftermath

2.7.1 Berger and the EEG

Thus, from the early 1930s, Berger and his work gradually gained recognition, even more internationally than in German neurological circles [5]. At the same time, EEG was gradually accepted as an experimental, but also a clinical tool, and although Berger continued his own studies until his retirement in 1938, he was—in fact—quickly overtaken by others in both fields.

Technical restraints played a role: Berger continued to use the Siemens Galvanometer, obtained in 1926 and a slightly improved one, obtained in 1932, coupled with an optical writing system [2, 5] rather than the more sensitive systems that were developed by his followers and competitors. But also, Berger's method of recording: observing the fluctuations between electrodes positioned at maximum distance over the entire neurocranium and his persistent belief that the alpha rhythm thus reflected a diffuse cerebral activity, generated throughout the cortex as a whole, may have limited his understanding of his own findings [5].

Adrian and Matthews had already used differential amplifiers for their replication studies.

In Berlin-Buch at the *Kaiser Wilhelm Institut für Hirnforschung*, headed by Oskar Vogt, the engineer Tönnies built his five-channel differential amplifier-ink writing Neurograph system for animal and human studies, and together with Alois Kornmüller (1905–1968) showed that specific activities or responses could be recorded from specific cortical regions [5]. Grey Walter, himself an engineer rather than a neurologist, described the localizing value of delta slow activity for the diagnosis of brain tumors. In the USA, William Gordon Lennox (1884–1960), Gibbs, and Jasper all quickly accepted the EEG as a practical tool for the study, diagnosis, and/or management of epilepsy and in many places, both in Europe and the USA, early EEG laboratories and "schools" started to develop already in the early pre-World War II years. In the Post-War decades, EEG firmly established itself as an indispensable tool in neurology and neuroscience [35].

Ironically and contrary to Berger's hope and expectations, his findings in patients with acute psychosis and some other psychiatric conditions did not yield any significant results [6], and although he was able to describe the effects of attention and mental activities, Berger did not succeed to establish a relation between specific mental activities and specific EEG phenomena [5].

Nor did his EEG observations solve his quest for Psychic Energy. Following his retirement, he had hoped to resume further studies, together with his former pupil Rudolf Lemke (1906–1957), notably regarding higher EEG frequencies, but, as noted earlier, war circumstances interfered with the delivery of the necessary recording equipment. Berger spent his last active year in Bad Blankenburg, writing *Psyche* [9], a short monograph where he returns to the subject of telepathy, a phenomenon in which he firmly believed but which he did not relate to the EEG [3, 36] (Fig. 2.3). These relative failures notwithstanding, Berger's pioneering role in the discovery and early development of EEG as a clinical and experimental tool was established and remained so until today.

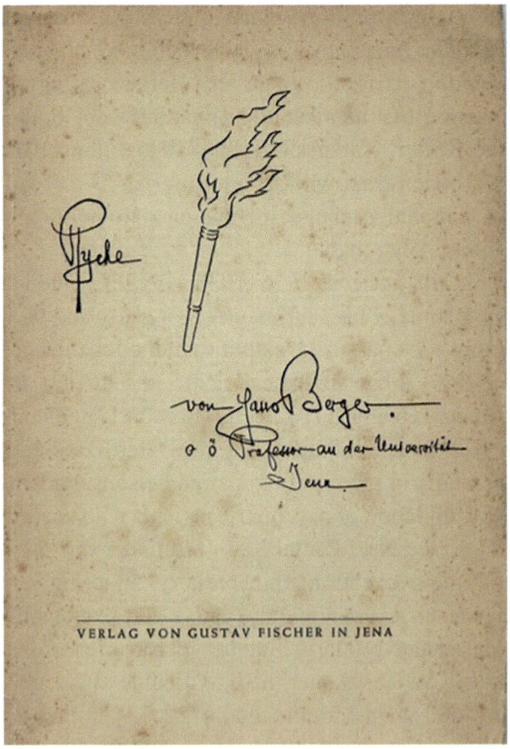

Fig. 2.3 Berger's *Psyche* (1940). (Public domain)

2.7.2 The Person Berger

In the post-World War II years, early biographers, including his former residents, Lemke who in 1945 succeeded Berger in Jena, first as acting director and, as from 1949, also as a full professor, and Kolle, who was appointed to the chair in Munich in 1952, all had undergone and, despite more or less active involvement in national socialist medical practice, survived an intensive de-Nazification process [31] and may have used their writings as a posthumous way to do the same for their esteemed tutor. Moreover, after 1945 when EEG was gradually introduced and accepted worldwide, Berger's importance as the actual inventor of the EEG was internationally acknowledged. His suicide in 1941 suggested an anti-Nazi attitude and his other clinical work appears initially to have been largely ignored. In 1956, at the instigation of Lemke, the Neuropsychiatric University Clinic in Jena was named after Berger, a name that was continued when in 1994 the clinic was split into separate units for neurology, psychiatry, and pediatric neuropsychiatry. The German Society for Clinical Neurophysiology (DGfKN) instituted the "Hans Berger-Prize" that was triennially awarded to recipients under that name from 1960 to 2019 [8].

Fig. 2.4 The old building where both Binswanger and then Berger worked and which, internally renovated, still is fully operational. Although the name of Hans Berger officially has been struck from the University Clinic his name remains inscribed over the clinic's main entrance. (Photo: WvEB, June 18, 2024)

In the wake of renewed historical interest in the period 1933–1945, focusing on German and Austrian Universities and scientists, including those cited in the preceding paragraphs, many illustrious names became tainted, including that of Berger. In 2021, the DGfKN struck the name of Berger from the prize, and in 2022, the directorship of the Jena University also decided to strike his name from the University Neurological Clinic [8, 37]. At the time of writing (June 2024), however, the name of Hans Berger remains inscribed over the main entrance of the old but renovated clinic building that remains in full use and also carries commemorative panels dedicated to Otto Binswanger and Hans Berger (Fig. 2.4; personal observation WvEB).

A proposal in 2020 to rename the "Hans Berger Street" in Jena was rejected, but in May 2022, it was decided to add an information sign to the street sign [37]. Only after the active intervention of a dedicated student group, this was eventually realized in 2023. At the instigation of the same group, an even more elaborate memorial plaque was installed on the pedestal of the portrait bust of Berger by the German sculptor Gabrielle Reinemer (*1948) that stands in front of the clinic [38] (Fig. 2.5).

Thus, in 2024, the year in which the DGfNK organized a centennial memorial symposium in Jena in his honor (https://dgkn.de/fuer-experten/termine/Eventdetail/522/-/vortraege-und-podiumsdiskussion), there is neither a Berger rhythm nor—officially—a Hans Berger Clinic. What remains, however, is an impressive series of publications, archives, and diaries documenting and illustrating the tenacious and persisting work of a dedicated and focused physician and neuroscientist: Hans Berger, the father of the electroencephalogram.

Fig. 2.5 Portrait Bust of Hans Berger in front of the Clinic Building at the Philosophenweg 3 in Jena. (Photo: WvEB, June 18, 2024) The explicatory signboard reads
Physician, Scientist and Fellow Traveler
Hans Berger, neurologist and psychiatrist from Jena, in 1924 recorded, for the first time, the electrical activity from the human brain. Thereby he paved the road for [electro] encephalography as a method to study brain activity and he is considered as [a] founder of Clinical Neurophysiology
During the N[ational] S[ocialist] period, as consultant and juror for the Erbgesundheitsgericht [Court for Genetic Health] he was co-responsible for a large number of forced sterilizations, some of them at his own instigation. In doing so he acted against medical values and human dignity
His memory thus remains ambiguous and a warning call

References

1. Kolle K. Hans Berger 1873–1941. In: Kolle K, editor. Grosse Nervenärzte. Stuttgart: Georg Thieme Verlag; 1956. p. 1–6.
2. Jung R. Hans Berger und die Entdeckung des EEG nach seinen Tagebüchern und Protokollen. In: Werner R, editor. Jenenser EEG-Symposion 30 Jahre Elektroenzephalographie. Berlin: VEB Verlag Volk und Gesundheit; 1963. p. 20–53.
3. Gloor P. Hans Berger and the discovery of the electroencephalogram. In: Gloor P, editor. Hans Berger on the Electroencephalogram of Man. Electroenceph. Clin. Neurophysiol. 1969, Suppl.28 pp 1–36.
4. Karyofilis A. Hans Berger, Eine biografische Studie. Hildesheim: Im Selbstverlag; 1974.
5. Borck C. Hirnströme. Eine Kulturgeschichte der Elektroenzephalographie. Göttingen: Wallstein Verlag; 2005.
6. Bauer J, Kluge H. Das wissenschaftliche Gesamtwerk des Jenaer Nervenarztes Hans Berger. Stuttgart: Franz Steiner Verlag; 2011.
7. Borck C. „Berger, Hans" in: NDB-online, veröffentlicht am 01.07.2022, https://www.deutsche-biographie.de/118850709.html#dbocontent.
8. Lemke JR, Kluger G, Krämer G. Hans Berger und 100 Jahre Elektroenzephalogram. Clin Epileptol. 2024;37:103–11. https://doi.org/10.1007/s10309-024-00664-x.
9. Berger H. Psyche. Jena: Gustav Fischer Verlag; 1940.
10. Berger H. Zur Lehre von der Blutzirkulation in der Schädelhöhle des Menschen, namentlich unter den Einflusz von Medicamenten: Experimentelle Untersuchungen. Jena: Gustav Fischer Verlag; 1901.
11. Koehler PJ. Development of neurology as a specialty in Europe. In: Education program syllabus: history of neurology. American Academy of Neurology; 2007. p. 72–8.
12. Berger H. Hirn und Seele. Antrittsvorlesung bei Übernahme des Lehrstuhls in Jena. Jena: Gustav Fischer Verlag; 1919.
13. Berger H. Ueber die Lokalisation im Groszhirn. Jena: Gustav Fischer Verlag; 1927.
14. Berger H, Guleke N. Über Hirntumoren und ihre operative Behandlung. Deutsche Zeitschrift f Chirurgie. 1927;203:104–67.
15. Berger H. Psychophysiologie in 12 Vorlesungen. Jena: Gustav Fischer Verlag; 1921.
16. Berger H. Untersuchungen über die Temperatur des Gehirns. Jena: Gustav Fischer Verlag; 1910.
17. Berger H. Degenerationen der Vorderhornzellen de Rückenmarks bei Dementia paralytica. Monatsschr f Psychiat Neurol. 1898;3:1–30.
18. Berger H. Experimentell-anatomische Studien über die durch den Mangel optischer Reize varanlassten Entwicklungshemmungen im Occipitallappen des Hundes und der Katze. Arch f Psych Nervenkrankh. 1900;33:521–67.

19. Berger H. Experimentelle Studien zur Pathogenese acuter Psychosen. Berlin Klin Wochenschr. 1903;40:693–5.
20. Lehmann A. Die körperliche Äusserungen psychischer Zustände, 1 Teil: Plethysmographische Untersuchungen. Leipzig: Reisland Verlag; 1899.
21. Koehler P. The study of brain movements. World Neurol. 2023;38:9–11.
22. Berger H. Ueber die körperliche Aeusserungen psychischer Zustände. Weitere experimentelle Beiträge zur Lehre von der Blutzirculation in der Schädelhöhle des Menschen. Jena: Gustav Fischer Verlag; 1 Teil, 1904; 2 Teil 1907.
23. Berger H. Untersuchungen über die psychische Beeinflussung der Hauttemperatur. J f Psychol Neurol. 1922;27:209–21.
24. Berger H. Über das Elektrenkephalogramm des menschen. I Mitteilung Arch f Psych Nervenkrank. 1929;87:527–70.
25. Einthoven W. Die galvanometrische Registrierung des menschlichen Elektrokardiogramms, zugleich ein Beurteilung der Anwendung des Capillar-Elektrometers in der Physiologie. Pflügers Arch ges Physiol. 1903;99:472–80.
26. Berger H. Das Elektrenkephalogramm des menschen. Nova Acta Leolpoldina Halle(Saale). 1938;6–38:173–309.
27. Jasper HH. Preface. In: Gloor P, editor. Hans Berger on the Electroencephalogram of Man. Electroenceph. Clin. Neurophysiol. 1969, Suppl. 28 pp V–VII.
28. Adrian AD, Matthews BHC. The Berger Rhythm. Potentialchanges from the occipital lobes in man. Brain. 1934;57:355–85.
29. Werner R. Hans berger zum Gedächtnis. In: Werner R, editor. Jenenser EEG-Symposion 30 Jahre Elektroenzephalographie. Berlin: VEB Verlag Volk und Gesundheit; 1963. p. 13–9.
30. Gerhard U-J, Schönberg A, Blanz B. "Hätte Berger das Ende des Zweiten Weltkrieges noch erlebt -gewiss wäre er ein Anwärter auf den Nobelpreis Geworden" Hans Berger und die Legende vom Nobelpreis. Fortschr Neurol Psychiatr. 2005;73:156–60.
31. Zeidman LA. Brain science under the Swastica ethical violations, resistance, and victimization of neuroscientists in Nazi Europe. Oxford: Oxford Universtit Press; 2020.
32. Zeidman LA, Stone J, Kondziella D. New revelations about Hans Berger, father of the electroencephalogram (EEG), and his ties to the Third Reich. J Child Neurol. 2014;29:1002–10.
33. Tolk K. Therapeutische Unzulänglichkeiten und nationale Überzeugungen. Wie die Jenaer Psychiater um Hans Berger in der Zwischenkriegszeit ihre Patienten behandelten PhD Thesis, Jena: Friedrich Schiller Universität; 2018.
34. Zimmermann S, Zimmerman T. Die Medizinsche Facultät der Universität Jena im "Dritten Reich" – ein Überblick. In: Hossfeld U, John J, Lemuth O, Stutz R, editors. Kämpferische Wissenschaft: Studienzur Universität Jena im Nationalsozialismus. Köln: Böhlau Verlag; 2003. p. 401–36.
35. Niedermeyer E, Shomer DL. Historical aspects of EEG. In: Shomer DL, Lopes da Silva FH, editors. Niedermeyer's electroencephalography. Basic principles, clininical applications, and related fields. 6th ed. Philadelphia: Wolters Kluwer/Lippincot Williams & Wilkins; 2011. p. 1–16.
36. Berger H. Ueber das Elektrenkephalogram des Menschen. XIV Mitteilung Arch f Psychiat Nervenkrankh. 1938;108:407–31.
37. Stridde T. Klinik-Namen ohne Hans Berger. UKJ tilgt Benennung nach groszem Erfinder wegen seine Verstrickung in NS-System. Ergänzungstext für Strasse in Lobeda-West geplant. Thurische Landeszeitung 31.05.2022, ZAJE2/125: 14.
38. Bachman J. Jenaer Koryphäe und Nazi-Täter: Mit die Geduld am Ende sind Akteure, die seit vielen jahren für eine Erlärtafel an de Hans-Berger Büste von der Jenaer Psychiatrie kämpfen. Thurische Landeszeitung 27.09.2023, ZAJE3/225: 15.

3 Hans Berger (1873–1941) and The Discovery of Clinical Electroencephalography

Francesco Brigo

3.1 Introduction

Hans Berger (1873–1941), a German psychiatrist, was the first to record electrical brain waves in humans, marking the inception of clinical electroencephalography (EEG) [1]. His pioneering work was fueled by a deep personal interest in psychic phenomena and a scientific curiosity about the brain's electrical nature. This chapter explores the historical context of Berger's discovery, the methodologies he employed, and the profound scientific impact of his contributions to neuroscience and clinical medicine.

3.2 From "Transfer of Thoughts" to Medicine

Hans Berger harbored a lifelong fascination with what might today be termed telepathy. He believed in the possibility of the human mind transmitting and receiving psychic energy and dedicated his career to uncovering the physiological basis of this elusive phenomenon. Berger's interest was sparked by a near-fatal accident he experienced at the age of 19 years during his military service [2]. On the same day as the incident, Berger's sister, living in another town, experienced a powerful premonition of his peril and urged their father to send a telegram to check on his well-being. Berger received the telegram shortly after the accident and was deeply struck by his sister's intuition.

Reflecting on this experience, Berger speculated that in a moment of mortal danger, as he contemplated certain death, he might have acted as a "sender," transmitting his thoughts to his sister, who, being emotionally close to him, acted as a "receiver" [2]. Fascinatingly, Berger's own terminology when recounting this event paralleled the functioning of a telegraph, which communicates over distances without requiring direct, tangible connections.

This profound episode convinced Berger of a mysterious connection between individuals with deep emotional bonds. He began to wonder if this connection was mediated by a form of psychic energy and, if so, how it might be measured and understood scientifically. Determined to unravel the mechanisms underlying the transmission of thoughts without physical intermediaries, Berger pursued a career in medicine, hoping to find answers to these enigmatic questions.

F. Brigo (✉)
Innovation, Research and Teaching Service; Teaching Hospital of the Paracelsus Medical Private University (PMU), Azienda Sanitaria dell'Alto Adige, Bolzano-Bozen, Italy
e-mail: Francesco.Brigo@sabes.it

F. Brigo, O. Mecarelli (eds.), *EEG: The First 100 Years*,
https://doi.org/10.1007/978-3-031-86878-8_3

3.3 At the Search of Energy in the Brain

At the turn of the nineteenth century, scientific attempts to understand brain function largely relied on the principle of the conservation of energy [3]. This principle was interpreted and applied by prominent figures such as Gustav Theodor Fechner (1801–1887), Theodor Meynert (1833–1892), August Forel (1848–1931), Sigmund Freud (1856–1939), Wilhelm Ostwald (1853–1932), and Alfred Lehmann (1858–1921). Within this intellectual framework, Hans Berger began his pioneering work in psychophysiology.

Berger strongly advocated for the principles of interaction and energy (*Wechselwirkung und Energieprinzip*), which he regarded as established empirical facts [4]. He was particularly influenced by the theories of Danish psychophysiologist Alfred Lehmann, who proposed that the brain's chemical energy and metabolic supply are transformed into heat, electricity, and psychic activity. Berger speculated on the possibility of precisely measuring the energy transformed into psychic processes. Rejecting Cartesian dualism and vitalism, he sought to investigate mental phenomena through physiological processes, integrating them into the biological sciences. Referring to Lehmann's theory, Berger wrote:

> [Psychological energy] is created alongside several other forms of energy during the work of the brain; it is a form of energy of its own kind. … It must firstly have physical properties and can therefore also be measured as physical energy and secondly it must be subject to the law of conservation of energy. When psychic energy arises, an equivalent amount of other types of energy must disappear. The psychological energy is created through transformation from physical energy and can be converted back into it [4].

Guided by Lehmann's ideas, Berger embraced the concept of "internal psychic energy" (*eigene psychische Energie*). He argued that there are quantitative differences between nervous and psychophysiological processes, the latter being associated with parallel psychological phenomena under the principle of psychophysical parallelism [4]. According to Berger, psychophysiological processes emerged from specialized nerve cells capable of performing "rapid and intensive dissimilation processes," in which complex chemical compounds broke down into simpler substances, releasing energy. These nerve cells were uniquely adapted for such functions.

Berger further posited that these enhanced metabolic (specifically catabolic) properties made nerve cells highly sensitive to disruptive influences such as narcotics or restricted blood flow to the cerebral cortex. He believed that this sensitivity was intricately linked to the brain's psychophysiological functions [4].

Ultimately, Berger was a staunch advocate of psychophysical parallelism, asserting that psychological processes correspond to physiological processes in the cerebral cortex with a one-to-one relationship. He found this theory compatible with the law of conservation of energy, providing a scientific basis for integrating psychological phenomena into the broader framework of biology.

3.4 The Inspiration From Caton's Research

After graduating from the University of Jena in 1897, Hans Berger began his career at the Jena Psychiatric University Clinic, where he worked until his retirement in 1938. His research was fueled by a deep interest in psychophysiology, with a focus on uncovering the cerebral mechanisms underlying mental processes.

Initially, Berger employed cerebral plethysmography, a technique developed by Italian physiologist Angelo Mosso (1846–1910) [5], to investigate blood circulation in patients with skull defects, a condition that became relatively common following World War I. He examined how cerebral blood flow responded to various drugs and psychological states, such as attention and sensory stimulation [6]. Through his systematic studies of how cerebral blood volume fluctuated in response to psychic processes, Berger sought to objectively record mental states using physiological methods.

During this period, Berger's attention turned to the work of Richard Caton (1842–1926), a

British physiologist who, in 1875, recorded electrical currents in animal brains using a galvanometer, an instrument for measuring electric current [7]. Caton's experiments involved attaching electrodes to partially exposed brains of rabbits and monkeys, linking them to a galvanometer. The device's needle moved in response to stimuli such as light, sound, or touch, demonstrating that the brain generated electrical signals corresponding to sensory input. Caton also observed variations in these signals based on the animals' states of consciousness [7]. However, despite proving the existence of electrical brain activity in animals, Caton's findings garnered little attention or were dismissed by the broader scientific community.

The precise reasons behind Berger's transition to cerebral electrophysiology remain unclear, as the volumes of his diary from 1897 to 1910 have been lost [8]. Nevertheless, it is plausible that Berger recognized the limitations of cerebral plethysmography in elucidating the connection between physiological and psychological processes. Determined to overcome these shortcomings, he sought a device capable of objectively and reliably detecting and recording psychological phenomena as measurable and visualizable physical states.

3.5 Berger's Discovery of the Human Electroencephalogram

In 1902, Hans Berger began his experiments on cerebral electrical activity, studying four dogs and one cat [1]. During this period, and until 1910, he employed a Lippmann capillary electrometer, a sensitive instrument designed to measure small changes in electric potential when electrodes were applied to the brain surface of unanesthetized animals. Invented by Gabriel Lippmann (1845–1921) in 1873, the electrometer operated on the principle that the surface tension of mercury in a capillary tube changes with the electric potential difference between two electrodes: one in contact with the mercury and the other with a solution of dilute sulfuric acid above it. The movement of the mercury meniscus allowed the detection of voltage changes as small as 25 μV.

Despite its sensitivity, Berger found the Lippmann capillary electrometer disappointing. He observed electrical oscillations in the dog brain in only one out of five experiments, prompting him to redirect his research focus to investigating temperature changes in the cerebral cortex during sensory and emotional stimulation and mental tasks [9].

In 1924, at the age of 51 years, Berger resumed his research on cerebral electrophysiology, this time using the Edelmann string galvanometer. This instrument significantly improved the detection of brain electrical activity in animals and enhanced the quality of his recordings. The exact reasons for his renewed interest in cerebral electrophysiology remain unclear, but over the years, Berger increasingly recognized the profound correlation between electrical stimulation of the cortex and psychophysical events associated with mental processes [8].

By 1926, Berger had adopted the more advanced Siemens double-coil galvanometer, which offered a sensitivity of 130 μV/cm (Figs. 3.1 and 3.2). Using this instrument and non-polarizable clay electrodes, he successfully recorded human EEG tracings, which he first presented in his landmark 1929 report.

Berger began his human experiments by recording cortical activity in patients with skull defects, such as fractures or trepanations, as he believed these conditions would reduce interference from bone and skin (Fig. 3.3). On July 6, 1924, Berger achieved what he considered his first successful recording [10]. He placed non-polarizable electrodes on the exposed brain tissue of a 17-year-old college student named Carl Zedel, who had undergone a palliative trepanation over the left hemisphere performed by neurosurgeon Nikolai Guleke (1878–1958), director of the Surgical University Clinic in Jena. As Berger recalled:

> There have been no prior investigations into electrical events in the human brain, nor any published records that correspond to what I am about to report. After several unsuccessful attempts, I

Fig. 3.1 Hans Berger's EEG laboratory at the University of Jena, operational from 1926 to 1931 (reproduced from Gloor, 1969 [8]). The technical setup for recording human brain electrical activity featured a double-coil galvanometer, positioned on the left side of the laboratory. To the right of the galvanometer was the recording apparatus, equipped with a camera for capturing EEG tracings. Additional equipment for resistance measurements was located at the far right. The subject under examination reclined on a couch positioned to the left of the galvanometer

Fig. 3.2 Initial EEG recording apparatus employed by Hans Berger to capture string movements using the original Edelmann String Galvanometer (reproduced from Gloor, 1969 [8]). The components of the apparatus were meticulously labeled by Berger's technician, W. Keuscher, as follows: "In the recording apparatus shown in the picture, *no film* was used, but a roll of silver bromide paper, 5 to 6 mc in width. These exposed curves ran into a metal box and were unrolled and developed in the dark room. [...] *Number 1* is the crank driving the motor. *Number 2* indicates the markers made of fibers, I believe they were blades of straw cut to the desired size. *Number 3* was a switch. If I turned it to the right, the recording paper moved and ran into the metal box. *Number 4* was a graduated lens. Through it when it was opened the string of the galvanometer was projected upon the photographic paper. *Number 5* was the diaphragm and the marker for the individual curves from 1 to ?. *Number 6* was the famous box into which the exposed paper glided. There was a sliding bar fixed to the box which after the end of the experiment simultaneously cut off the record and closed the slit. *Number 7* was the time marker, a tuning fork with a period o 1/10 of a second, which, however, was provided with a clock work and which when necessary had to be wound up." [8]

finally succeeded on July 6, 1924, in making the first relevant observations in a 17-year-old young man. This individual had undergone a palliative trepanation on the left cerebral hemisphere, performed by Guleke, due to a suspected brain tumor. When signs of increased intracranial pressure recurred following an initial remission, the original trephine opening was enlarged posteriorly, resulting in the alleviation of symptoms. About a year after the second operation, I attempted to demonstrate currents in the area of the trephine opening, where bone was missing, using non-polarizable boot-shaped clay electrodes [*Tonstiefelektroden*; these were non-polarizable electrodes made of cotton soaked in clay and protruding from a glass tube filled with zinc sulfate. This tube touched a zinc wire and connected to the galvanometer] and the small Edelmann string galvanometer.

Initially, the experiments were unsuccessful. It was only when I placed the two clay electrodes 4 cm apart in the vicinity of a scar running vertically from above downwards through the middle of the enlarged trephine opening that I could, with significant magnification [the original term *Vergrößerungen* denotes an optical way of enlarging the deflection of the galvanometer beam without any actual amplification involved], observe the continuous oscillations of the galvanometer string. This was achieved either by inserting a platinum thread with a resistance of 5200 Ohms or a quartz thread with a resistance of 3200 Ohms. No oscillations could be demonstrated with the clay electrodes in the region of the trephine opening away from the very firm scar. This initial result suggested that, probably in humans as in rabbits, dogs, and monkeys, continuous electrical currents [*stän-*

Fig. 3.3 Photograph of one of the earliest EEG recording attempts in humans (reproduced from Gloor, 1969 [8]). According to W. Keuscher, Hans Berger's technician, this recording was conducted on a patient who had undergone left-sided trepanation. Silver electrodes were affixed to the patient's scalp using adhesive tape. Notably, this appears to be the only instance where EEG was recorded with the patient seated; subsequent recordings were conducted with patients lying down for greater stability and comfort

dige elektrische Ströme] can be recorded from the surface of the intact cerebral cortex [1].

Unfortunately, Hans Berger did not include any illustrations of the EEG recording from his first patient. For nearly five years, Berger grappled with doubts about the cerebral origin of the electrical oscillations he had recorded, even as his eagerness to pursue this line of research persisted. On July 11, 1928, just months before submitting his first article on the EEG in April 1929, Berger confided in his diary: "Plans! I feel the need for creative scientific work. For many years I have worked in vain on the presumed EEG. What now? Abandon EEG!" [quoted in: 8].

Despite his frustrations, Berger had a breakthrough a few months later. On September 28, 1928, he noted: "The curves from … show that I actually have found the human *Elektrenkephalogramm*: *Eureka* [noted down in Greek letters] is right!—The resistance of the skin is very great, so even small displacements of the needle show a great effect on the quality of the exposure. The curves show with definitiveness & certainty that they do not form in the skin. The curves agree exactly with the best curves from P[.]! They also indicate far-reaching agreement with the curves taken from the bone gap by Dr. G[.] 'I've got it, I've found it!', just when I had already given up the matter in despair!" [translation in: 11].

Between 1924 and 1929, Berger conducted a series of meticulous recordings to rule out interference from various sources, including skin conductance, psychogalvanic reflexes, scalp vessel pulsations, heart activity, muscle contractions, and eye movements. To validate the authenticity of brain signals, Berger juxtaposed the electrical waves with cerebral plethysmograms and electrocardiograms in his initial EEG photographs, highlighting their distinct nature [1, 12]. By repeating his experiments and eliminating artifacts, Berger ensured that the observed waves were genuine brain signals. His rigorous approach not only aimed to achieve a technically perfect physiological recording but also sought to distinguish psychic activity from deviations in the physiological curves [13].

Berger experimented with various techniques and eventually determined that a bipolar montage with fronto-occipital leads for one-channel EEG tracings, combined with simultaneous electrocardiogram recordings and a time marker, yielded the best results. He tested this configuration on himself, his son Klaus, his daughter Ilse, and several patients.

In 1929, Berger published his landmark article *Über das Elektrenkephalogramm des Menschen* (*On the Electroencephalogram of Man*) in the journal *Archiv für Psychiatrie und Nervenkrankheiten* (Archives of Psychiatry and Neurology) [1]. This first article was followed by 13 others with the same title, in which Berger elaborated on his discoveries and applications of the EEG. In his initial publication, Berger acknowledged the foundational work of predecessors such as Richard Caton (1842–1926), Adolf Beck (1863–1942), Fleischl von Marxow (1846–1891), and Vladimir Pravdich-Neminsky (1879–1952), demonstrating his awareness of the current state of research on electrical currents in animal brains.

Through his epidural recordings, Berger realized that skull defects did not always enhance recording quality due to dural thickening and post-surgical adhesions. Consequently, he shifted to placing electrodes on the intact human skull. As he wrote: "From the beginning, this was my hope: that it would become possible to record from the human scalp with an intact skull the oscillations of the electrical current which can be obtained in animals from the surface of the brain and in humans with bone defects from the epidural space ..." [1].

In 1931, Berger presented EEG recordings obtained from needle electrodes placed directly on the cerebral cortex. These recordings revealed that the electrical oscillations originated in the cortex, as the signals became flat when the electrodes were inserted approximately 4 cm into the underlying white matter [14]. Reflecting on this finding, Berger concluded: "Based on my detailed experiences reported here, I believe it is reasonable to assume that the cerebral cortex primarily generates these electrical currents [*Ströme*] or potential oscillations [*Spannungsschwankungen*], with the latter being the more accurate physical term, although other influences may also play a role" [14].

3.6 Alpha and Beta Waves, Active and Passive EEG

Early in his research, Berger "had been struck early by the fact that in many experimental subjects, opening their eyes while recording the curve from the skull surface caused an immediate change in the EEG, and that during mental tasks, such as solving an arithmetic problem, simply naming the task sometimes caused the same EEG change" [12]. In a letter to a colleague, Berger expressed concerns about the close relationship between EEG amplitude deflections and eye opening or closing, stating, "The changes in the EEG are merely accompanying attention, which is crucial for our mental processes" [15]. These observations revealed Berger's reluctance to accept that a seemingly mechanical act like eye opening—devoid of significant psychic content—could profoundly affect EEG recordings. His concerns were further compounded by the lack of consistent EEG changes in patients with major mental disorders such as manic-depressive psychosis, severe melancholia, and schizophrenia [14, 16].

Berger also noted that sensory stimuli (e.g., visual, auditory, tactile), voluntary movements, or even the intention to perform them, as well as mental activities, all elicited similar EEG changes. He attributed these effects to "directing attention toward a stimulus," emphasizing that "only a stimulus that enters consciousness and becomes a perception is capable of producing EEG changes" [12]. He observed that repeated application of sensory stimuli during a single session led to progressively less distinct EEG changes, with longer delays before these changes appeared. Berger speculated, "It seems as if the subject loses interest in the stimulus, and correspondingly, the changes occur more slowly and less distinctly. This could possibly be interpreted as indicating involvement of emotional processes, despite all precautions" [12]. Berger described this phenomenon as habituation, a concept originally identified in 1873 by the botanist Wilhelm Pfeffer (1845–1920) in the plant *Mimosa pudica* [17] and later studied in animals by Ivan Pavlov (1849–1936) [18]. Berger interpreted habituation as a conscious phenomenon linked to a loss of interest, rather than an unconscious process.

In his first publication, Berger described two types of brain waves, now known as alpha and beta waves, which he initially termed waves of the first and second order [1]. He wrote:

> The electroencephalogram represents a continuous curve with constant fluctuations, in which, as I have emphasized, larger first-order waves with an average

duration of 90 milliseconds can be differentiated from smaller second-order waves with an average duration of 35 milliseconds. The larger deflections measure at most 150 to 200 microvolts [1].

Berger recognized that these wave patterns corresponded to those identified in curarized animals by Vladimir Pravdich-Neminsky [19]. He noted that alpha waves appeared when subjects were awake and calm with their eyes closed, disappearing upon eye opening (Fig. 3.4). In contrast, beta waves, which were faster and of lower amplitude, became more prominent when the eyes were open. Contrary to expectations based on earlier animal studies by researchers such as Richard Caton, Ernst Fleischl von Marxow, Adolf Beck, and Napoleon Cybulski, Berger found that sensory stimulation did not produce larger EEG wave deflections but instead caused the disappearance of alpha waves.

Initially, Berger hypothesized that alpha waves reflected active mental processes. He suggested that sensory stimulation of specific cortical areas had an inhibitory effect on surrounding cortical regions, leading to the suppression of alpha waves. This observation supported his belief that "in terms of function, the brain is a unified whole" [12]. This idea echoed theories by Ivan Pavlov and Alexei Ukhtomsky (1875–1942), particularly the concept of dominance, which describes a stable focus of increased cortical excitability reinforced by incoming stimuli, while inhibition spreads across other regions [20, 21].

Over time, Berger revised his views, identifying beta waves as markers of cortical activation. He referred to them as "the material concomitants of mental processes" [22] and "the material concomitants of psychophysiological processes" [23]. In his final article on the EEG, Berger distinguished between a "passive" and an "active" EEG [23]. The passive EEG, dominated by alpha waves, represented "continuous automatic cortical activity undisturbed by external or internal stimuli." Psychophysiologically, this state corresponded to "undisturbed passive ideation," as seen in dreams or moments of restful wakefulness with closed eyes [23]. Conversely, the active EEG, characterized by beta waves, emerged during heightened attention and mental exertion. Berger described this as "an active conscious process (attentional focus, mental effort, thinking in the truest sense of the term)" [23]. Thus, Berger concluded: "I would like to reiterate once more that during mental exertion or what is termed active consciousness, i.e., during the transition from passive EEG to active EEG, there is always a significant reduction in the brain's potential fluctuations" [23].

The pronounced decrease in the posterior dominant alpha rhythm upon eye opening—now known as the *Berger effect*—remains a foundational observation in EEG research. This phenomenon, reflecting neuronal desynchronization, is intrinsically linked to the increased complexity of brain function during cognitive and attentional tasks.

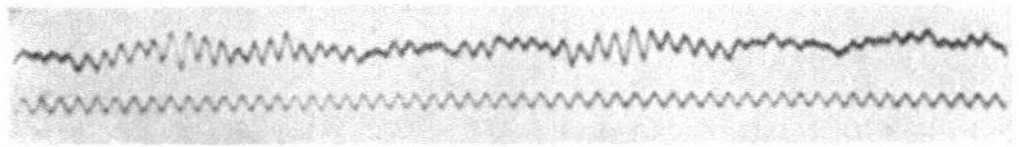

Fig. 3.4 Sample EEG recorded by Hans Berger from his 15-year-old son, Klaus. The recording was obtained using lead band electrodes positioned on the forehead and the back of the head. The upper trace displays scalp-derived electrical waves, later identified by Berger as alpha waves. The lower trace represents the timeline, depicted as a continuous 10 Hz sinusoidal wave. Reproduced from Berger, 1929 [1]

3.7 EEG in Neurological and Psychiatric Conditions

Throughout his career, Hans Berger conducted extensive EEG recordings across a range of neurological and psychiatric conditions, including schizophrenia, melancholia, intellectual impairment, senile dementia, *dementia paralytica* (a manifestation of tertiary syphilis), *dementia praecox* (schizophrenia), complete blindness, brain tumors, and epilepsy. He investigated various brain states, such as resting and active cognitive tasks, and explored the effects of numerous drugs—including scopolamine, morphine, atropine, cardiazol, veronal, cocaine, and caffeine—on brain activity [8]. Berger was also the first to document slow waves associ-

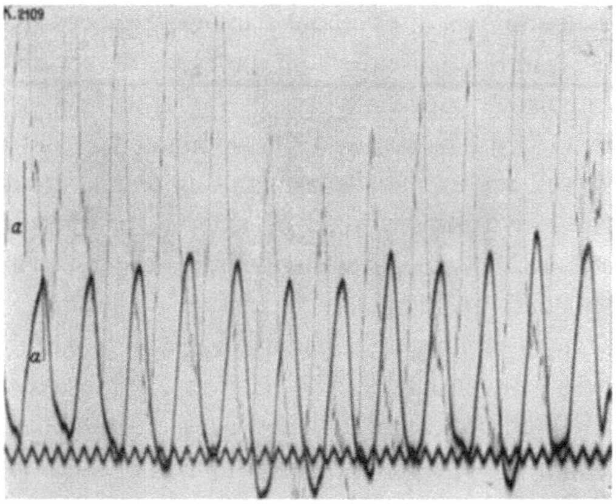

Fig. 3.5 Sample EEG recorded by Hans Berger from G.G., an 18-year-old patient with genuine epilepsy (generalized idiopathic epilepsy), showing high-amplitude 3 Hz discharges. The recording was derived from electrodes placed on the right frontal and left occipital sides, interconnected with oscillographs and a coil galvanometer. The upper trace represents the oscillograph curve, while the lower trace depicts the coil galvanometer curve. The timeline is displayed as a continuous 10 Hz sinusoidal wave, with the letter "a" indicating simultaneous points on the curves. Berger initially speculated that these discharges originated in the motor cortex and were localized, influenced by small twitching movements observed in the patient's left hand. This recording is likely the first documented EEG observation of a typical absence seizure. Adapted from Berger, 1932 [24]

ated with brain lesions. Initially, he misinterpreted these as slow alpha waves and rejected the term "delta waves" [8].

Berger made particularly significant contributions to the understanding of epilepsy through his EEG research. Although he never recorded the ictal EEG of a generalized tonic-clonic seizure, he documented several hallmark findings. These included marked amplitude reduction during post-ictal coma [14], EEG suppression during seizures with impaired consciousness [24], and interictal epileptiform abnormalities [25]. In 1930, Berger identified the characteristic 3-Hz slow-wave pattern associated with typical absence seizures (Fig. 3.5). However, he initially doubted the significance of these waves, considering them potential artifacts, and delayed publishing this observation [25, 26].

In 1933, Berger described the ictal EEG of a 45-year-old woman experiencing a partial motor seizure (characterized by clonic movements of the right hand). The EEG revealed high-amplitude sharp waves over the left central areas that preceded the clinical signs of the seizure [25]. This observation supported John Hughlings Jackson's (1835–1911) hypothesis that seizures result from abnormal and synchronized firing of cortical neurons in the brain region corresponding to the observed clinical symptoms [8].

Berger's EEG recordings solidified the concept of epilepsy as an "electrical disorder" and introduced the notion of epilepsy as a "paroxysmal brain dysrhythmia" [27]. This marked a departure from earlier theories that attributed epilepsy to chemical imbalances in the brain [28].

3.8 Electroencephalogram and the "Transfer of Thoughts"

As aforementioned, Hans Berger's interest in neurophysiology was initially inspired by his fascination with the "transfer of thoughts" and the biological mechanisms that might underpin it. He sought to understand how feelings or thoughts generated in one mind could be detected by another. However, his research with the EEG ultimately led him to disprove the plausibility of "telepathic communication" without physical intermediaries. In his final communication on the EEG, Berger conclusively stated:

> For the α and β-waves of my EEG, any long-distance effect is excluded; they cannot propagate through space. On the advice of experienced electrophysiologists, I refrained from attempting to establish any remote effects [23].

The cooperative activity of countless neurons in the brain generates electrical currents that are exceedingly small. For these currents to propagate, they require a conductor with low resistance, such as neural tissue. The propagation of such signals through air is physically impossible due to the absence of an adequate conductive medium.

3.9 Electroencephalogram: Writing the Brain Through Electricity

Regarding the naming of the method for recording brain electrical activity in humans, Hans Berger expressed his linguistic and scientific preferences:

> Because for linguistic reasons I consider the word 'electrocerebrogram' [the term proposed by Pravdich-Neminsky in 1925] to be a barbarism, compounded as it is of Greek and Latin components, I would like to propose, in analogy to the name 'electrocardiogram,' the name *'electroencephalogram' [Elektrenkephalogramm]* for the curve which here for the first time was demonstrated by me *in man* [italics by the author] [1].

Berger's proposed German term, *Elektrenkephalogramm*, adhered to the Greek root *enkephalo-* (meaning "brain") to ensure linguistic consistency and scientific precision.

From the outset, Berger was captivated by the synchronization of electrical waves between the two cerebral hemispheres and believed that the EEG reflected widespread cortical activity. Drawing on the unified concept of psychophysical function proposed by Austrian psychiatrist Josef Berze (1866–1957), Berger wrote:

> All of this led me to formulate the working hypothesis that the main waves of the EEG represent accompanying phenomena [or epiphenomena, *Begleiterscheinungen*] of those material cortical processes referred to as psychophysical, since they can, under certain circumstances, be associated with phenomena of consciousness. Therefore, based on the findings reported above, I consider the EEG as the electrical epiphenomena of a wave of activity that traverses in a specific direction across the human cerebral cortex, unifying both hemispheres into a cohesive whole [*einheitliches Ganzes*] that also responds in the same way [i.e., as a whole] [29].

Hans Berger believed the EEG revealed an intimate correspondence between psychophysical function and a unified physiological process distributed across the entire cerebral cortex. He proposed that the brain waves observed in the EEG represented "the electrical epiphenomena of a uniform process that spreads across the entire cortex, involving the nerve cells uniformly and independently of local structural differences in the cortical areas" [29].

Initially, the term *electroencephalogram* was often used interchangeably with the alpha rhythm. This association stemmed partly from Berger's early hypothesis that alpha waves reflected specific neural functions, while beta waves were linked to broader metabolic and nutritional processes. The hallmark disappearance of alpha waves upon eye opening and focused attention became a defining feature of the EEG.

In their groundbreaking 1934 article published in *Brain*, Edgar Douglas Adrian (1889–1977) and his engineer Bryan Harold Cabot Matthews (1906–1986) of the Cambridge Physiological Laboratory proposed the term *Berger rhythm* to describe "the remarkable electric effect which can be detected in the human subject by electrodes applied to the head." They wrote, "Berger calls it the electrencephalogram, but the shorter title avoids the suggestion that the rhythm is produced by the entire cortex" [30]. Their tribute to Berger acknowledged his discovery while reflecting a different interpretation of the EEG signal's origin. Adrian and Matthews claimed, "The potential waves are due to the spontaneous beat of an area in the occipital cortex which is normally occupied by activities connected with pattern vision" [30].

In response, Berger reaffirmed his perspective on the global origin of the EEG signal:

> The EEG originates *everywhere* in the cerebral cortex and constitutes a directed wave of activity of the cortex ... In the EEG, a *fundamental function* of the human cerebrum intimately connected with the psychophysical processes becomes visibly manifest [italics by the author]. This is clearly demonstrated by the above-discussed close relationship of the EEG with processes of attention, which represent the most important psychophysical function. Therefore, I also consider it more correct to retain the term EEG, which I coined as the discoverer of these potential oscillations in man, rather than to exchange it for the term chosen by Adrian and Matthews, by which I felt greatly honored! [31].

Adrian and Matthews' reluctance to adopt the term EEG likely reflected an unconscious resistance to Berger's broader concept of the brain as a unified entity, rather than a collection of distinct functional areas. Nevertheless, their 1934 replication of Berger's findings was instrumental in bringing international recognition to his work and boosting his reputation.

3.10 Conclusions

From his early experiments with plethysmography to his later focus on electrophysiology and the development of the EEG, Hans Berger consistently pursued the visualization of psychic activity through graphical representation. This approach reflected a broader trend among contemporaries who employed graphical methods in biology to illuminate the workings of organic structures, often likening them to machines [32, 33].

Berger's initial exploration of the relationship between neurophysiological and vascular processes in the brain ultimately contributed to the development of modern multimodal imaging techniques. The combination of functional magnetic resonance imaging with EEG recording represents the fulfillment of Berger's lifelong aspiration: to depict changes in neural activity during motor and cognitive tasks alongside their associated metabolic and electrical changes [34].

Although Berger initially envisioned the EEG as a type of "psychoscope" capable of identifying neurophysiological correlates of specific mental states [13], it became evident that the EEG's recording surface served as an opaque psychophysical interface between the mind and brain. Nevertheless, over time, the EEG proved to be an effective "cerebroscope," enabling the investigation of both normal and abnormal brain functions.

The EEG trace, in essence, acts as a canvas on which the brain "inscribes" its activity, with electrical signals serving as both medium and representation. Berger aptly referred to the EEG as the *Hirnspiegel* ("mirror of the brain"), a term that captures its ability to reflect the intricate workings of the human mind. Through this "mirror," we continue to unravel the mysteries of the brain, venturing ever deeper into its captivating and endless "wonderland."

References

1. Berger H. Über das Elektrenkephalogramm des Menschen. Arch Psychiatr Nervenkr. 1929;87: 527–70.
2. Berger H. Psyche. Jena: Gustav Fischer; 1940.
3. Millett D. Hans Berger: from psychic energy to the EEG. Perspect Biol Med. 2001;44(4):522–42.
4. Berger H. Vorlesung: Parallelprinzip und Wechselwirkung. – Psychische Energie und ihre Rückverwandlung in physische Energie. In: Berger H. Psychophysiologie in 12 Vorlesungen. Jena: Gustav Fischer, 1921.
5. Mosso A. Über den Kreislauf des Blutes im menschlichen Gehirn. Leipzig: Veit; 1881.
6. Berger H. Zur Lehre von der Blutzirkulation in der Schädelhöhle des Menschen namentlich unter dem Einfluss von Medikamenten (Experimentelle Untersuchungen). Jena: Gustav Fischer; 1901.
7. Caton R. The electric currents of the brain. Br Med J. 1875;278
8. Gloor P. Hans Berger and the discovery of the electroencephalogram. Electroencephalogr Clin Neurophysiol. 1969;Suppl 28:1–36.
9. Berger H. Untersuchungen über die Temperatur des Gehirns. Jena: Gustav Fischer; 1910.
10. Dube W, Besel R, Maier F. Nikolai Guleke–Begründer und Wegbereiter der Neurochirurgie in Thüringen [Nicolai Guleke--the founder and pioneer of neurosurgery in Thuringia]. Zentralbl Neurochir. 1989;50(3–4):210–2.
11. Borck C. Hans Berger's long path to the EEG. In: Borck C, editor. Brainwaves. A cultural history of electroencephalography. Translated by Ann M. Hentschel. Routledge; 2018.
12. Berger H. Über das Elektrenkephalogramm des Menschen. Zweite Mitteilung. J Psychol Neurol. 1930;40:160–79.
13. Borck C. Writing brains: tracing the psyche with the graphical method. Hist Psychol. 2005 Feb;8(1):79–94.
14. Berger H. Über das Elektrenkephalogramm des Menschen. Dritte Mitteilung. Arch Psychiatr Nervenkr. 1931;94:16–60.
15. Baumann T, Sparing F, Martin M, Fangerau H. Neurophysiologen im Nationalsozialismus – Hans Berger, Paul Hoffmann, Richard Jung und Alois E. Kornmüller Klin Neurophysiol. 2020;51:14–41.
16. Berger H. Über das Elektrenkephalogramm des Menschen. Zwölfte Mitteilung Arch Psychiat Nervenkr. 1937;106:165–87.
17. Pfeffer W. Physiologische Untersuchungen. Leipzig: W. Engelmann; 1873.

18. Pavlov IP. Conditioned reflexes. London: Oxford University Press; 1927.
19. Prawdicz-Neminski WW. Zur Kenntnis der elektrischen und der Innervationsvorgänge in den funktionellen Elementen und Geweben des tierischen Organismus. Elektrocerebrogramm der Säugetiere. Pflügers Archiv für die Gesamte. Physiologie. 1925;209:362–82.
20. Ukhtomsky A. The Dominant. Saint Petersburg: Leningrad State University; 1966.
21. Brenner JE, Igamberdiev AU. Philosophy in reality: scientific discovery and logical recovery. Philosophies. 2019;4(2):22.
22. Berger H. Über das Elektrenkephalogramm des Menschen. Dreizehnte Mitteilung. Arch Psychiatr Nervenkr. 1937;106:577–84.
23. Berger H. Über das Elektrenkephalogramm des Menschen. Vierzehnte Mitteilung. Arch Psychiatr Nervenkr. 1938;108:407–31.
24. Berger H. Über das Elektrenkephalogramm des Menschen. Fünfte Mitteilung. Arch Psychiatr Nervenkr. 1932;98:231–54.
25. Berger H. Über das Elektrenkephalogramm des Menschen. Siebente Mitteilung. Arch Psychiatr Nervenkr. 1933;100:301–20.
26. Jung R, Berger W. Fünfzig Jahre EEG. Hans Bergers Entdeckung des Elektrenkephalogramms und seine ersten Befunde 1924–1931. Arch Psychiatr Nervenkr (1970). 1979;227(4):279–300.
27. Gibbs FA, Gibbs EL, Lennox WG. Epilepsy: a paroxysmal cerebral dysrhythmia. Brain. 1937;60:377–88.
28. Brigo F. Study Group for the History of Neurology of the Italian Neurological Society; Epilepsy Study Group of the Italian Neurological Society. A century of EEG: unveiling the electrical nature of epilepsy. Neurol Sci. 2024; https://doi.org/10.1007/s10072-024-07919-5.
29. Berger H. Über das Elektrenkephalogramm des Menschen. Sechste Mitteilung. Arch Psychiatr Nervenkr. 1933;99:555–74.
30. Adrian ED, Matthews BHC. The Berger Rhythm: potential changes from the occipital lobes in man. Brain. 1934;57:355–85.
31. Berger H. Über das Elektrenkephalogramm des Menschen. Zehnte Mitteilung. Arch Psychiatr Nervenkr. 1935;103:444–54.
32. Brigo F, Gazzaniga V, Iorio S, Martini M. Mirroring the mind and brain: reflections on Hans Berger's EEG and the scientific perception of reality. Medicina Nei Secoli J Hist Med Med Humanit. 2024;36(3):171–4. https://doi.org/10.13133/2531-7288/3032.
33. Marey E-J. La méthode graphique dans le sciences expérimentales et principalement en physiologie et médécine. Paris: Masson; 1878.
34. Rossini PM, Cole J, Paulus W, Ziemann U, Chen R. 1924–2024: First centennial of EEG. Clin Neurophysiol. 2024;170:132–5. https://doi.org/10.1016/j.clinph.2024.11.021.

The Development of Most Important Schools of Clinical Electroencephalography and Epileptology in North America and Europe

4

Oriano Mecarelli

4.1 Introduction

Hans Berger (1873–1941) carried out the first human electroencephalogram (EEG) on July 6, 1924, during neurosurgery on a 17-year-old boy. Berger worked very closely with neurosurgeons and began his studies on the human EEG by placing electrodes directly above the cerebral cortex, also using the large bone windows produced by wartime head injuries. Subsequently, he dealt with EEG recordings from the scalp, with the application of two electrodes, one frontal and one occipital, using his own son as a volunteer. At the time, Berger only had a rudimentary EEG machine at his disposal, especially built for him by Engineer Siemens, equipped with a double-coil galvanometer which guaranteed a sensitivity of 130 μV/cm. The first tracings recorded by Berger, which were very short and on a roll of scrolling photographic paper, allowed only one EEG channel in fronto-occipital bipolar derivation to be visualized in addition to a channel for an electrocardiogram and a channel for a time marker. The numerous recordings made by Berger after his first experience in 1924 allowed him to accumulate important scientific material which was, however, not used for a scientific publication until 1929. In his first scientific report, Hans Berger described the characteristics of the normal occipital background rhythm (reported in an awake state, with eyes closed and with responsiveness to eye opening), identified with the Greek letter "*alpha,*" and the more rapid frequency activity called "*beta*" [1, 2].

A year later, Berger had accumulated 1133 records from 76 subjects, and from 1929 to 1937, this material allowed him to publish 14 scientific reports [3]. Unfortunately, however, the papers were all written in German and had little circulation, also because Berger received little credit at home due to the political events of the time and the opposition of some German scientists.

In spite of these difficulties, news of Berger's discovery soon spread around the world, and many other neuroscientists devoted themselves to recording human EEGs, even using increasingly advanced equipment. The rapid development of electroencephalography was, in fact, possible mainly due to the great collaboration that had already been established between bioengineers and neurophysiologists [4, 5].

At that time, in England, Lord Edgar Adrian (1889–1977) and Brian Matthews (1906–1986) of the Cambridge Physiological Laboratory confirmed Berger's research using copper gauze electrodes and introducing differential amplifiers with valves into the systems. In 1934, they then published their results in English in the journal *Brain*, making them more easily accessible to the Anglophone scientific community [6].

O. Mecarelli (✉)
Department of Human Neurosciences, Sapienza University, Rome, Italy

F. Brigo, O. Mecarelli (eds.), *EEG: The First 100 Years*,
https://doi.org/10.1007/978-3-031-86878-8_4

Always in England, from 1936, William Grey Walter (1910–1977) became the pioneer of clinical electroencephalography. Grey Walter built an oscilloscope capable of recording three EEG channels, and this method began to be used for diagnosing cerebral tumors and as a rudimentary monitoring system, both during surgery and to confirm the effects of anesthesia. It was Grey Walter who, following Berger's example, identified slow EEG activity with the Greek letters "delta" and "theta" and who, as early as 1943–1944, introduced the first methods for automatically analyzing the cerebral bioelectric signal [7, 8].

Fig. 4.1 Wilder G. Penfield (1891-1976)—photograph of Author's personal property

4.2 The Development of Electroencephalography and Epileptology Schools in North America

The most important pioneer of the EEG in North America was Herbert H. Jasper (1906–1999), who in 1933, set up an experimental and clinical EEG laboratory at Bradley Hospital in Providence (Rhode Island). In 1934, Jasper, in collaboration with psychologist Leonard Carmichael, recorded the first human EEG at his laboratory, confirming the existence of the alpha rhythm described by Berger under waking conditions with the eyes closed and the blocking reaction when the eyes were opened. The recording was made using high-gain direct current amplifiers designed and constructed by Howard Andrews, a physicist and electrical engineer from Brown University [9].

In 1935, Jasper traveled to Europe, first going to Paris to discuss his doctoral thesis on "*Electroencephalographie chez l'homme*" and then visiting Berger in Jena, and Grey Walter and Adrian in London and Cambridge, respectively. Back in the United States, Jasper also published several papers in 1935 on the human EEG and the technical means of recording it, confirming what had been published just an year earlier by Adrian and Matthews who in turn, had confirmed the data published by Berger. Furthermore, Jasper, at Brown University, devoted himself to improving the technical characteristics of the instrumentation required to record a good-quality EEG, especially in order to localize epileptogenic foci.

The work of Jasper and his collaborators actually formed the basis for the first EEG monitoring systems for epilepsy. In fact, Jasper collaborated early on with Wilder Penfield (1891–1976) (Fig. 4.1), a neurosurgeon at the Montreal Neurological Institute, and persuaded him to operate on the first two patients with epileptic seizures on the basis of focus localization using the EEG. Jasper soon moved from Brown University to Montreal, completing the organization of the EEG Laboratory together with Andrew Cipriani, physician and electronics engineer, in 1939 [10].

In just 1 year, approximately 1000 examinations were carried out at the Jasper Laboratory on over 500 patients, more than half of whom were considered candidates for eventual neurosurgery. The development of Jasper's EEG Laboratory was continuous, and in 1949, Jasper and John Hunter described the first method for simultaneously recording EEGs and clinical seizures using a movie camera. This was the first system described in the literature of Video-EEG, which was also based on the photographic method already described by Robert Schwab at Harvard in 1938.

In a few years, thanks to the work of Penfield and Jasper, Montreal became the most important center of reference for the neurosurgical treatment of focal epilepsies, and the publication of the book "*Epilepsy and the Functional Anatomy of the Human Brain*" in 1954 was, in a way, the crowning achievement of the fruitful 20-year collaboration between the two neuroscientists [11].

Immediately after Jasper, the second American electroencephalographer was Alfred L. Loomis (1887–1975) who set up a private laboratory in New York in 1938. Loomis' system was equipped with a switch box allowing 12 electrodes to be connected to six channels of amplification; the writing system was equipped with 13 ink pens that wrote using an ingenious paper-handling machine. Loomis also recorded EEGs during sleep and described the changes in the EEG during the night, laying the foundation for modern polysomnographic study.

The third group in America which confirmed Berger's observations on alpha rhythm and beta activity was that of Hallowell Davis (1896–1992) who worked at Harvard Medical School in Boston. In 1934, *Frederic A. Gibbs* (1903–1992), together with his wife Erna Leonhardt-Gibbs (1904–1987) [12]—who was one of the first self-trained EEG technicians in the world—started work at the Davis Laboratory, and a very important group was formed, not least because of the participation of William Gordon Lennox. Recordings on epilepsy patients were initially made using a single-channel ink writing recorder from Western Union, but the equipment soon became more advanced through collaboration with Albert M. Grass (1910–1992), an engineer working at the Massachusetts Institute of Technology in the Mechanical Engineering Department.

Beginning in 1935 at the Department of Physiology at Harvard Medical School, a multidisciplinary working group was formed consisting of the Gibbses and Albert Grass and Ellen H. Robinson, the latter a biologist who devoted herself primarily to the neurophysiological study of the auditory pathways and who in 1936 married Albert Grass (Fig. 4.2). Mr. and Mrs. Grass then started a commercial company (the Grass Instrument Company) that would provide equipment of high quality to neuroscientists for over half a century, and finally to the formation of The Grass Foundation, which has benefited the neuroscience community since 1955 [13].

Fig. 4.2 From left to right: Albert Grass, Frederic Gibbs, Ellen Grass, Robert Morrison, and Erna Gibbs at the third International Congress of Electroencephalography and Clinical Neurophysiology, held in Boston in 1953. (With permission by SJ Zottoli and the Grass Foundation)

Albert Grass built the first three-channel electroencephalograph in 1935; one unit was destined for Boston City Hospital (where Gibbs and Lennox worked), and the second was provided to the Davis Laboratory at Harvard Medical School. The commercial version of the Grass Model I was sold from 1936 onward; it had three channels of differential amplification and an ink-writing system on scrolling roll paper (Fig. 4.3). Subsequently, in 1938–1939, the Grass Model II was produced and sold in two versions with four or six channels, and in 1946, the Grass Model III, the first 8- and 16-channel EEG device, was introduced to the market (Figs. 4.4 and 4.5). Approximately 5000 units of the Grass Model III were produced and sold worldwide immediately after the Second World War, demonstrating that electroencephalography was now recognized to be a useful neurophysiological study method in both the clinical and the research fields [14].

In the summer of 1935, the Gibbs couple traveled to Russia for the International Congress of Physiologists; they took the opportunity to visit Berger in Jena and show him the EEG patterns of discharges of spike-and-wave complexes recorded in patients with absence seizures.

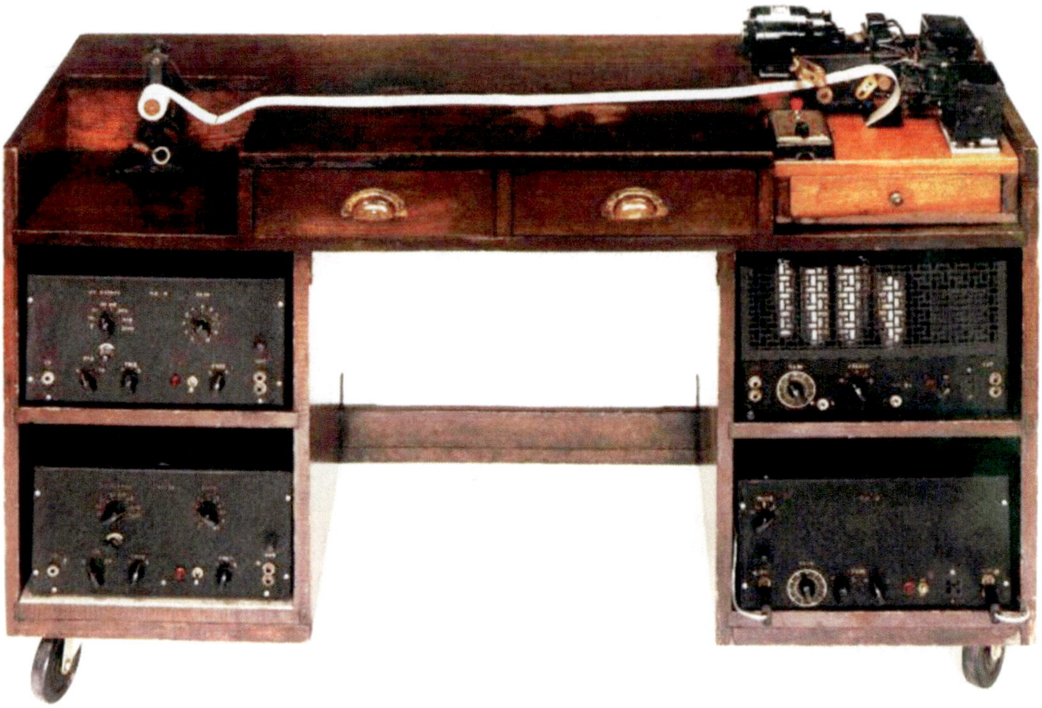

Fig. 4.3 The Grass Model I Electroencephalograph was built in 1935 and had only three channels, two for EEG and one for time marker. Hallowel Davis and his wife Pauline designed the "console." The principle of the design was to have a long table top to view the record, the ink writer on the right, and a roll paper take-up device on the left. Knee space was provided. Model I did more than establish basic parameters to record EEG. Being the only available high-sensitivity amplifier with a bandwidth of 1.0–10,000 Hz, it was used for such things as single nerve cell response with microelectrodes by Renshaw. (Image courtesy of Natus Medical Incorporated, natus.com)

Berger appeared very saddened at the time and told the Gibbs that his equipment had been confiscated by the state and he was encountering significant scientific and political opposition from Alois Kornmuller (1905–1968) at the Kaiser Wilhelm Institute.

In those years, a very important working group was formed in Boston by the Gibbses and William Gordon Lennox. WG Lennox (1884–1960) was always involved in clinical electroencephalography and epileptology and worked closely with Stanley Cobb, and Erna and Frederic Gibbs at Harvard Medical School. Lennox in 1944 became the founding Chief of the Seizure Unit at the Boston Children's Hospital which was the first comprehensive pediatric epilepsy unit in the United States. From the scientific point of view, Lennox's role in the international community was very important; he was president (from 1935 to 1949) and honorary president (from 1949 to 1953) of the International League Against Epilepsy (ILAE), and from 1941 to 1952, he was the editor of Epilepsia. In 1951, Lennox, together with Henri Gastaut, described the Lennox-Gastaut syndrome and contributed to spreading epileptological culture throughout the world with his numerous scientific publications.

The "*Gibbs-Gibbs-Lennox era*" proved to be one of the most productive periods in the history of clinical electroencephalography. In 1935, Gibbses and Lennox were the first to describe the characteristic spike-and-wave pattern in 12 patients with absence seizures, and successively they reported the EEG correlates of hyperventilation, tonic–clonic seizures, psychomotor seizures, and status epilepticus. The Atlas of Electroencephalography by Gibbs and Gibbs has allowed the dissemination of knowledge regarding electroencephalography worldwide, becoming the "bible" of electroencephalogra-

Fig. 4.4 The Grass Model II Electroencephalograph had four- and six-channel capability and was built and marketed by 1938–1939. In the picture, Donald Lindsley recorded a four-channel EEG at North-western University circa 1946. (Image courtesy of Natus Medical Incorporated, natus.com)

phers for many decades. The first volume was published in 1941 and was dedicated to Hans Berger. In it, the normal EEG of the child and adult and the pathological patterns in patients with epilepsy, brain tumors, brain injuries, and various other neurological and psychiatric disturbances were systematically illustrated for the first time [15]. Subsequently, the Atlas was republished in three volumes. The first volume, again with a dedication to Berger, saw the light of day in 1950 and dealt with methodology and normal EEG [16]. The second volume in 1952 was devoted entirely to the EEG patterns of epilepsy and was dedicated to Lennox [17], while the third was published in 1964 and showed the EEG patterns of neurological and psychiatric disorders. The dedication of the third volume was specifically for Albert Grass and I want to quote it here in its entirety because it shows how both then and now in electroencephalography is grounded in an effective collaborative relationship between clinicians, neurophysiologists, and bioengineers: "*to Albert Grass, scientist, friend of scientists and instrument-maker extraordinary, whose own characteristics exceeded the specifications for resolution, discrimination and faithfulness. His genius made it easy to see that shimmering submolecular world, where the infinitely small and extremely fast wave the stuff that dreams are made off*" [18].

In 1944, Mr. and Mrs. Gibbs left Boston and moved to the University of Illinois at Chicago. Here Frederic Gibbs devoted himself primarily to the use of EEG for localizing the epilepto-

Fig. 4.5 The Grass Model III EEG was the first eight-channel electroencephalograph and was put on the market in 1946. The photo shows four Grass Model III EEGs, eight channels each for 32 channels of simultaneous recording in the operating room. (Image courtesy of Natus Medical Incorporated, natus.com)

Fig. 4.6 Left, Frederic and Erna Gibbs; right, Fred Gibbs and his team recording the electrocorticography. (Photographs reproduced from Vannemreddy et al., Neurosurgery 70: 774–782,2012, Wolters Kluwer Health, Inc. permission)

genic focus in temporal epilepsy not necessarily due to structural brain lesions. In 1948, neurosurgeons Percival Bailey and Oscar Sugar performed the first anterior temporal lobectomy in a patient with refractory epilepsy, while Gibbs and his team recorded electrical activity by electrocorticography [19] (Fig. 4.6).

In 1958, Lennox retired, and he passed away on July 21, 1960, a few days after a ceremonial dinner at the Harvard Club in Boston. The latter had been organized to celebrate the publication of the textbook *Epilepsy and Related Disorders* which Lennox had written together with his daughter Margaret [20]. However, Lennox's personality was complex and controversial. On the one hand, he always fought to combat epilepsy-related stigma and to comprehensively improve the care of people with epilepsy; however, this contrasted with his closeness to eugenic theories, and his advocacy of voluntary euthanasia and sterilization for individuals with severe intellectual disabilities and severe psychiatric disorders [21].

Another important Laboratory in America concerning human electroencephalography was that established by Lee Edward Travis (1896–1987) at the University of Iowa in 1933. Lee Travis was a brilliant psychologist and speech pathologist, and at his Laboratory, he and his collaborators also devoted themselves to EEG recordings in stutterers and normal speakers. At the Travis Laboratory, Jasper himself and also Donald Lindsley, Charles Henry, and John Knott trained in electroencephalography, using equipment built by Engineer Theodore Hunter. The first published report on manual quantitative analysis of the voltage and frequencies of the EEG was published, edited by John Knott from the Travis Laboratory in 1936.

Also at the Mayo Clinic, in 1935, *Charles L. Yeager* began to work on electroencephalography in collaboration with a biophysicist, E.J. Baldes. From 1940, Reginald G. Bickford, in collaboration with A. Faulconer Jr., began to study EEG patterns under various conditions of anesthesia, and in 1953, the Mayo Clinic was one of the first medical centers to implement intracerebral electroencephalography in humans.

4.3 Henri Gastaut and the Marseille School's Contribution to the Development of Epileptology and Clinical Electroencephalography

In Europe, during the post-war period, the most prestigious epileptological and electroencephalographic school was that of Henri Gastaut and collaborators at the Centre Saint Paul in Marseille (France). Henri Gastaut was born in Munich on April 5, 1915, and passed away on July 15, 1995. In 1946, Gastaut went to Bristol (UK) to Grey Walter's Laboratory where he learned the basics of electroencephalography and discovered the activating effect of intermittent light stimulation. Back in Marseille, Gastaut bought a Grass four-pen electroencephalograph and set up a small EEG laboratory at the Timone Hospital, which very quickly became a point of reference for young French and foreign colleagues attracted by the new technique and the brilliant neurologist's personality. At the end of 1949, Gastaut went to the Montreal Neurological Institute in Canada for a short period and worked with Penfield and Jasper, increasing their knowledge of temporal epilepsy in humans and studying the genesis of generalized epileptic discharges in animals [22].

On his return to Marseille, he founded the Centre Saint Paul for epileptic children. Gastaut, together with his collaborators (in particular, Robert Naquet, Joseph Roger, Annette Beaumanoir, Michelle Bureau, Charlotte Dravet, and many others) represented the point of reference for neurologists and child neuropsychiatrists in Europe and beyond for many years, organizing the famous series of "Colloquia" at Marseille from 1950 to 1980 (Fig. 4.7). in 1973, in recognition of his great contribution to the knowledge of electroencephalography and related fields in 1973, the University of Marseille appointed him Professor of Clinical Neurophysiology, a position he held until his retirement.

The first Marseille Colloquium was held in 1950, while the 25th and last was organized in his honor in 1980 when he retired as President of the University of Aix-Marseille.

The Colloquia contributed a great deal to the training of specialists in epileptology and electroencephalography and were held nearly every year and nearly always in Marseille. From 1954, the Colloquia were sponsored by the International Federation of Societies for EEG and Clinical Neurophysiology (IFSECN) and by the International League Against Epilepsy (ILAE), then by the International Institute for Advanced Education of the North American Treaty Organization and, since

Fig. 4.7 Center of the photo Henri Gastaut, between Anne Beaumanoir (1923–2022), one of his collaborators in Marseille, and Raffaello Vizioli (1926–2006), a famous Italian neurologist. (Photograph of Author's personal property)

1976, by the World Health Organization (WHO). The number of participants increased rapidly from 34 in 1950 to 110 in 1954 and has been between 150 and 250 since then. A total of 15–20 countries were represented each time. One-third were French, one-third were Spanish or Italian, and the remaining one-third spoke Dutch, English, German, or Scandinavian [23].

Henri Gastaut was the Director of the Centre St. Paul for epileptic and neurologically handicapped children from 1960 to 1972 and was mainly concerned with Clinical Electroencephalography and Epileptology in both his clinical and his research work.

Electroencephalography, especially as a method for studying brain function and describing the various forms of epilepsy, was Gastaut's lifelong passion. Together with his wife Yvette Reynaud and his closest collaborators, Gastaut identified some of the most characteristic patterns of the waking human EEG, such as lambda waves, mu rhythm, rolandic spikes, and so on. He established several methods for activating abnormal EEG discharges, and he used EEG to study the various stages of altered consciousness and paroxysmal sleep disorders. In the field of epileptology, the Gastaut School has identified or helped to describe many forms of epilepsy and epileptic encephalopathy, such as photosensitive epilepsy, startle epilepsy, occipital focal epilepsy of childhood, temporal lobe epilepsy, hemiconvulsion-hemiplegia-epilepsy syndrome, West syndrome, and Lennox–Gastaut syndrome. Henri Gastaut published several hundred papers and book chapters as well as several important books starting with the classic *The Epilepsies: Electro-Clinical Correlations* in 1954. In the preface to this book, Wilder Penfield wrote: *Henri Gastaut has produced a lucid statement of present knowledge of the epilepsies, a statement much needed at this time. More than that, there is originality in his thinking and a brilliance in his writing that characterizes the best French literature* [24].

One of the most important works by Gastaut and his school was the definition and classification of epileptic seizures and epilepsy. His first classification of seizures was presented and discussed at the Marseille Colloquium in 1964. The classification was then discussed in the following years at various Congresses; it was approved at the 12th Congress of the ILAE in New York in 1967 and finally adopted in 1969 [25, 26]. The Classification of Epilepsies and Epileptic Syndromes which ILAE published in 1989 was also based on earlier proposals of the Gastaut School [27]. Gastaut believed that the use of standard terminology in epileptology was essential, and as early as 1964, proposed a *Dictionary of Epilepsy* to the WHO. The first version was discussed in two meetings in 1967 and 1968 and was only published in 1973 after 9 years of unremitting effort. The Dictionary was published by the WHO initially in four official languages—English, French, Spanish, and Russian, and it was subsequently translated into German, Chinese, Japanese, and so on [28].

Gastaut also gave great impetus to the study of the etiological and physiopathological aspects involved in the development of the various forms of epilepsy, emphasizing the importance of epidemiological, neuroradiological (when computed tomography of the brain was introduced), and neuropsychological correlations.

While dedicating his life to the study of epilepsy in its entirety, Gastaut did not neglect to address the psychological and social needs of people with epilepsy and their caregivers. He was involved in stimulating the formation of a working group within the WHO regarding the need to improve knowledge regarding epilepsy in society and, in the last years of his life, also emphasized the importance of dealing with epilepsy in developing countries.

In addition to being a great epileptologist and an excellent teacher, Gastaut made a considerable contribution to the development of both national and international scientific societies in the field of epileptology. He founded the French League Against Epilepsy in 1949 and served on it until 1973, while simultaneously devoting approximately a quarter of his life to the ILAE from 1953 to 1977. The ILAE was founded in Budapest in 1909 but suffered long periods of inactivity mainly due to the world wars. In 1947, Epilepsia, the official organ of ILAE, began to be published once a year as a review of the literature published in that year. In 1949, Gastaut proposed organizing the ILAE Congress jointly with IFSECN and the World Neurology Society to the ILAE President WG Lennox, and in 1953, he was commissioned to organize the ILAE Congress in Lisbon. This was the first truly scientific Congress of the ILAE and was devoted to the subject of temporal lobe epilepsy. From 1957 to 1969, Gastaut devoted all his energy to the ILAE as Secretary-General until he became its President in 1969 and Past President in 1973. During these years *Epilepsia* became an increasingly prestigious journal in the field of both clinical and experimental epileptology.

Henri Gastaut's great interest in clinical epileptology always complemented his interest in electroencephalography. After spending 6 months at the Burden Neurological Institute in order to learn electroencephalography at the laboratory directed by Grey Walter, in 1947, Henri Gastaut, together with his intellectual mentor, participated in the first International Congress of Electroencephalography in London during which the creation of the Federation EEG Societies and of the EEG Journal was proposed. On his return from London, Henri Gastaut devoted himself to the foundation of the French-speaking EEG

Society (1948) and played a major role in the preparation of the Second International EEG Congress (Paris, 1949) during which the IFSECN was officially created.

For approximately 20 years after 1949, Gastaut devoted himself to the IFSECN, serving it as Secretary (1949–1957) and then President (1957–1961). In those years, the International Congresses were held jointly with those of the World Federation of Neurology and the ILAE.

In remembering Henri Gastaut after his death in Marseille in 1995, two of his closest collaborators, Charlotte Dravet and Joseph Roger, wrote: "*He was one of the great pioneers who established epileptology as a respected discipline within neurology and whose contributions advanced the knowledge and treatment of epilepsy enormously. His intelligence was to keen and his personality so exceptional that one who met him could ever forget the encounter*" [29].

References

1. Berger H. Uber das Elektrenkephalogramm des Menschen. Arch Psychiatr Nervenkr. 1929;87:527–70.
2. Gloor P. Hans Berger and the discovery of the electroencephalogram. Electroenceph Clin Neurophysiol. 1969;Suppl 28:21–36.
3. Gloor P. Hans Berger on the electroencephalogram of man. The fourteen original reports on the human electroencephalogram. Amsterdam: Elsevier; 1969. p. 1–350.
4. Sutter R, Kaplan PW, Schomer DL. Historical aspects of Electroencephalography. In: Schomer DL, Lopes da Silva FH, editors. Niedermeyer's Electroencephalography. Basic principles, clinical applications, and related fields. 7th ed. Oxford University Press; 2018. p. 3–19.
5. Mecarelli O. Past, present and future of the EEG. In: Mecarelli O, editor. Clinical electroencephalography. Springer Nature Switzerland AG; 2019. p. 3–8.
6. Adrian ED, Matthews BHC. The Berger rhythm: potential changes from the occipital lobes in man. Brain. 1934;57:355–85.
7. Walter WG. The location of cerebral tumours by electroencephalography. Lancet. 1936;II:305–8.
8. Walter WG, Dovey VJ. Electroencephalography in cases of subcortical tumours. J Neurol Neurosurg Psychiatry. 1944;7:57–65.
9. Collura TF. History and evolution of electroencephalographic instruments and techniques. J Clin Neurophysiol. 1993;10:476–504.
10. Jasper HH. History of the early development of electroencephalography and clinical neurophysiology at the Montreal Neurological Institute: the first 25 years 1939–1964. J Can Sci Neurol. 1991;18(4):533–48.
11. Penfield WG, Jasper HH. Epilepsy and the functional anatomy of the human brain. Boston: Little, Brown and Company; 1954. p. 1–896.
12. Lennox M. In Memoriam: Erna Leonhardt Gibbs. Epilepsia. 1988;29(1):97.
13. Zottoli SJ. The origin of the Grass Foundation. Biol Bull. 2001;201:218–26.
14. Grass AM. The Electroencephalographic Heritage until 1960. Quincy: Grass Instrument Company; 1984. p. 1–44.
15. Gibbs FA, Gibbs EL. Atlas of electroencephalography. Cambridge, MA; 1941.
16. Gibbs FA, Gibbs EL. Atlas of electroencephalography. Volume one: methodology and controls. Second edition. Addison-Wesley Publishing Company, Inc; 1950.
17. Gibbs FA, Gibbs EL. Atlas of electroencephalography. Volume two: epilepsy. Addison-Wesley Publishing Company, Inc; 1952.
18. Gibbs FA, Gibbs EL. Atlas of electroencephalography. Vol. 3. Neurological and psychiatric disorders. Addison-Wesley Publishing Company, Inc; 1964.
19. Vannemreddy P, Stone JL, Slavin KV. Frederic Gibbs and his contributions to epilepsy surgery and electroencephalography. Neurosurgery. 2012;70:774–82.
20. Lennox WG, Lennox-Buchthal MA. Epilepsy and related disorders, vol. 1 and 2. Boston: Little, Brown; 1960. p. 1–1168.
21. Rosso M, Chu D, Santoro JD. William Gordon Lennox (1884–1960). J Neurol. 2021;268:3512–3.
22. Naquet R. In memoriam Henri Gastaut. Electroencephalogr Clin Neurophysiol. 1996;98:231–5.
23. Broughton RJ Editor. Henri Gastaut and the Marseille School's Contribution to the Neurosciences. Electroenceph clin Neurophysiol. 1982, Suppl 35:1–447.
24. Gastaut H. The epilepsies. Electro-clinical correlations. Springfield: Charles C Thomas Publisher; 1954. p. 1–149.
25. Commission on Terminology of the International League Against Epilepsy. A proposed international classification of epileptic seizures. Epilepsia. 1964;5:297–306.
26. Gastaut H. Clinical and electroencephalographic classification of epileptic seizures. Epilepsia. 1970;11:102–13.
27. Commission on Classification and Terminology of the International League Against Epilepsy. Proposal for revised classification of epilepsies and epileptic syndromes. Epilepsia. 1989;30(389–39):9.
28. Gastaut H. Dictionary of epilepsy. Geneva: World Health Organization; 1973.
29. Dravet C, Roger J. In Memoriam. Henri Gastaut 1915–1995. Epilepsia. 1996;37(4):410–5.

5 Understanding Brain Physiology Through the Electroencephalogram

Giuliano Avanzini

The history of neurophysiology is punctuated by controversies that testify to the vitality of our discipline: the first one that marks the very beginning of modern neurophysiology brought us much earlier than 1924 when Hans Berger provided the scientific community with the evidence that recording electrical signal of brain activity from the scalp was both feasible and meaningful. It brings us back to the turn of the eighteenth century, when the *Commentarius "De viribus electricitatis in motu musculari"* published in 1791 by Luigi Galvani [1] raised the famous Volta-Galvani controversy, with Alessandro Volta arguing against Galvani's claim about the evidence of "animal electricity." Volta was right in maintaining that the electricity responsible for frog muscle contraction was generated by the bimetallic arc used by Galvani and this conclusion led him to the invention of the pile named Voltaic after him. However, Galvani's intuition that nervous function depended on the movement of electric charges was the right one as Galvani himself and his nephew Giovanni Aldini subsequently demonstrated, thus establishing the basis of modern neurophysiology. More than a century later, Hans Berger had to face the skeptical attitude of the scientific community toward the evidence of his success in recording the electrical activity of the human brain through the scalp, a negative attitude that yielded only to the authority of the Nobel Prize Edgar Douglas Adrian, who spontaneously supported Berger's results ([2])—a sad demonstration that sometimes the authoritativeness of the presenter is more effective in granting the success of a discovery than its intrinsic value.

Today, there is no doubt about the importance of Berger's discovery nor is there any doubt that Berger deserves the full credit for solving the relevant technical problems and for envisaging the potential of the method to reveal the brain at work. The fact that his original goal, to capture the "psychic energy," remained elusive to his search does not reduce his merit because the subsequent work proved that the electroencephalogram (EEG) has indeed the potential to investigate higher brain functions, as the author will further elaborate in this chapter. Moreover, reading Berger's papers, one is impressed by the rigor and soundness of his scientific method and by his readiness to grasp the value of any scientific findings that he unexpectedly came across during his work, two attitudes that qualify a true scientist. The most clear example is provided by the epileptic discharges that he recorded between 1929 and 1933 and published after careful critical evaluation in 1933 [3] and in the monograph of 1938 [4]. Nowadays, after a century of experience, the clinical value of the EEG in clinical neurology is firmly established as it is exhaustively discussed throughout this book; in addition, it is clear that

G. Avanzini (✉)
Istituto Neurologico C. Besta, Milan, Italy
e-mail: giuliano.avanzini@istituto-besta.it

F. Brigo, O. Mecarelli (eds.), *EEG: The First 100 Years*,
https://doi.org/10.1007/978-3-031-86878-8_5

the EEG signal has great potential for investigating physiological brain functions, including higher brain functions.

This chapter reviews the cellular mechanisms underlying the EEG signal with particular regard to the excitation–inhibition interactions that generate field potential oscillations underlying the EEG rhythms. Examples of the value of frequency analysis and the derived methods to investigate cognitive activities such as language, visual and auditory perception, memory, and learning will be provided, and the value of EEG methods to define the involved neural networks will be discussed.

5.1 The EEG Signal

The enormous amount of work devoted to EEG after Hans Berger's discovery demonstrated that it directly reflects neural mass action and mainly synchronized postsynaptic potentials of aligned neurons, namely, pyramidal cells in the cortex. Postsynaptic potentials contribute much more than action potentials to the EEG signal because, due to the very short duration of the action potential, it is unlikely that many of them will occur in perfect synchrony. Only in special cases, such as an external powerful stimulation or an intrinsic excess of excitability, such as the one that characterizes the epilepsies, highly synchronized neuronal action potentials of a cell population occur perfectly in phase so that can summate, thus contributing to the generation of *EEG Spikes.*

Spatiotemporal synchronization of neural populations results in net polarization of extended brain regions, which may be transient, slow, or oscillatory.

It is worth remembering that after Oersted [5] we know that electricity and magnetism are interdependent phenomena as on one side any moving electric charge generates a magnetic field and on the other side any magnetic field induces electric charge movements. As soon as the technological advancements led to the development of suitable magnetometers (SQUID, acronym of *superconducting quantum Interference device),* the new method for recording brain-generated magnetic fields, magnetoencephalography (MEG), several laboratories started to test it. The experience accumulated so far allows us to say that although both EEG and MEG are well suited to measure neural activity with high temporal resolution, each one of them has advantages and drawbacks that have to be taken into account. MEG is superior to EEG for analyzing brain sources with tangentially oriented dipoles and capturing better short-range connectivity, whereas EEG gives a more precise evaluation of sources with radially oriented dipoles and better describes long-range connectivity with respect to MEG. Another difference is about the influence of volume conduction which affects the EEG signal more than the MEG one. Needless to say, to monitor situations that require freely moving subjects, MEG recording is useless at least with the present recording techniques.

Muthuraman et al. [6] carried out an EEG–MEG comparative analysis concluding that the combined approach EEG + MEG is superior to either EEG or MEG alone (see more details in the next paragraph on brain connectivity).

This chapter deals with the EEG signal, but most of the following considerations the author will apply as well to MEG.

The largest EEG potentials are generated by cortical sources which vary in depth and orientation, reflecting the cortical folding; using realistic models of the head and applying actual values of the skull to brain resistivity, it is possible to calculate the scalp electric field caused by neuronal sources, thus solving what is called the *forward problem.* Obviously, what is of much interest to the student of EEG is the *inverse problem,* that is, the estimation of the sources from the scalp fields, which without constraints has an unlimited number of solutions, which means no solution unless one makes specific assumptions about intracerebral sources [7]. To stress the cortical origin of EEG sources does not mean to deny the contribution of subcortical structures; indeed, it must be remembered that large populations of cortical neurons can be driven by subcortical structures (namely, thalamic nuclei) that substantially contribute in pacing their discharge. The most striking result of this cortico-subcortical interaction is oscillatory activities, the EEG rhythms, which are described according to their dominant frequency, amplitude, shape, and distribution over the scalp, and which are the main objectives of this chapter. It must be stressed that no oscillatory activity within a recognized EEG frequency band can be considered an EEG rhythm; to identify an EEG rhythm, it must demonstrate the existence of a spectral peak within the frequency band of interest.

In normal adults, the following frequency bands were originally identified: *Delta* (<4 Hz): is present only during deep sleep and is particularly prominent over the frontocentral regions; *Theta* (4–7 Hz): in drowsiness and light sleep is also prominent over the frontocentral regions and spreads toward the posterior regions. In wakefulness, theta waves are mainly seen over the temporal regions; *Alpha* (8–12 Hz): during wakeful relaxation with closed eyes predominates over the occipital regions, it is typically attenuated or suppressed by eye opening and mental engagement; *Beta* (13–30 Hz): predominates over the anterior regions; with eye opening, it spreads across other regions where it becomes predominant over alpha activity. Although the term *Gamma* waves to designate low amplitude beta-like waves at frequencies above 35–45 Hz had been employed since 1938 by Jasper and Andrews [8] and Das and Gastaut [9] suggested that a 40-Hz EEG activity could be associated to cognitive functions, systematic investigations of gamma frequency band and of its correlations with higher-level brain functions started only after Singer and Gray's seminal work of 1989. An important distinction must be made between *narrowband gamma activity* (30–80 Hz) and "*broadband/high gamma oscillations* (60–200 Hz)" that differ from each other in terms of neurophysiological mechanisms and significance as it will be discussed later.

In addition to the above "canonic" rhythms, some other EEG rhythms have been described that bear functional significance: *Mu rhythm* (8–13 Hz) first described as rhythm "en arceau" recordable over the Rolandic regions and typically blocked by voluntary movement and the *spindle rhythm* (also called *sigma waves*) observed during light sleep. Spindle frequency ranges between 12 and 14 Hz (9–16 Hz or 11–16 Hz in more inclusive instances).

Several lines of evidence support the role of oscillatory activities in brain function. Premises are that a neuronal population becomes a coordinate assembly when they work in synchrony [10] and that the most efficient way to establish synchrony in a neuronal population is by creating oscillations, which modulate membrane potentials collectively [7, 11]. Further data on mechanisms and functional significance of the EEG rhythms will be provided in the following paragraphs; here, the author wants to stress the concept that rhythmic oscillations of groups of neurons condition their functional activity promoting synergy both within the group and with other functionally related neuronal populations.

In discussing the relevance of brain oscillation, Lopes da Silva [7] states that slow oscillations are particularly suited to a functional bias throughout a large population of neurons, frequencies in alpha and theta bands are optimal to modulate or gate the transfer of information across specific neuronal populations (e.g., thalamus and cortex or hippocampal and associated cortical areas), and higher frequencies in beta and gamma range are particularly adequate to engage discrete neuronal populations in achieving transfer of specific information among neuronal assembly.

5.2 Neurophysiological Bases of EEG Rhythms

The results of many human and animal studies support the important role of thalamic reticular nucleus RTN in pacing rhythmic activities relevant to the EEG rhythms.

As shown by Fig. 5.1 left, RTN is a shell-shaped nucleus surrounding the thalamic mass, which includes all other thalamic nuclei projecting to the cortex. It is entirely composed of inhibitory GABAergic neurons that project exclusively to all other thalamocortical (TC) neurons and receive collaterals from both TC and corticothalamic projection (Fig. 5.1 right).

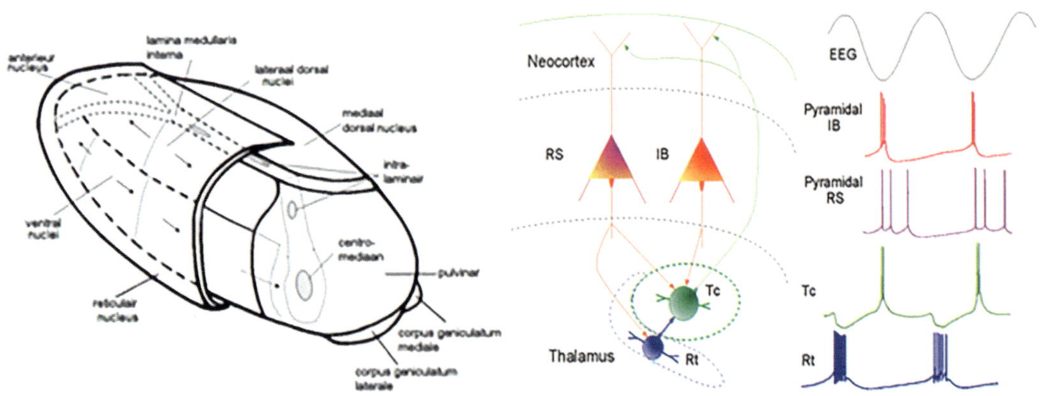

Fig. 5.1 *Left*: Schematic representation of thalamic nuclei showing the situation of RTN (reticular nucleus) with respect to TC nuclei (anterior nucleus, ventral nucleus, lateral dorsal nuclei, medial dorsal nuclei, pulvinar corpus genicolatum mediale, and corpus geniculatum laterale). The figure also shows the lamina medullaris interna within which the diffusely projecting intralaminar and centromedian nuclei are drawn. *Right*: Thalamocortical (TC) circuitry responsible for EEG rhythms with corresponding oscillatory cell discharges RS. Regular spiking pyramidal neurons, IB intrinsically bursting pyramidal neurons. Nuances of orange color indicate excitation and blue color indicates inhibition. The first trace depicts the resulting EEG activity. For further explanation, see the text

A key physiological feature of thalamic neurons is a low threshold Ca^{2+} conductance that is inactive at resting membrane potential and de-inactivates at membrane potentials more negative than −60 mV, thus allowing the low-threshold calcium current I_T to flow across the membrane generating a strong depolarization (*Calcium spike*) carrying a burst of sodium action potential. Every burst is followed by a profound potassium-dependent after-hypolarization (AHP) which de-inactivates I_T and starts a new burst/AHP sequence and so on, resulting in a regular oscillation of the membrane potential of thalamic neurons Every influence capable of setting membrane potential sufficiently above or below −60 mV potential can switch thalamic neurons from bursting to the tonic mode of firing or vice versa. The I_T is particularly expressed in reticular neurons (Rt), which plays a leading role in generating oscillatory activities that induce in Tc neurons inhibitory postsynaptic potentials that rhythmically de-inactivate I_T involving the Rt–Tc–cortex–Rt loop in a widespread rhythm revealed by the EEG.

A switch of the TC loop from firing to tonic mode results in the well-known finding of alpha suppression by eye opening, which was first observed by Hans Berger himself and is still routinely reproduced in the standard EEG recording procedure. This simple and easily reproducible reaction gave Hans Berger a demonstration that a correlation between the behavioral state of a subject and his EEG activity was indeed possible. Further studies culminating in the monumental work of Moruzzi and Magoun [12] led to consider alpha suppression as part of the arousal reaction induced by the activation of the ascending reticular system and the EEG as a primary tool to investigate vigilance. Reciprocally when the sleep inducer hypothalamic ventrolateral preoptic area inhibits the ascending reticular activating system, I_T de-inactivation is promoted that in turn leads to 12–14 Hz (more inclusively 9–16 or 11–16 Hz) *sigma activity* which is reflected by the EEG spindles. The I_T-dependent discharge mode of thalamic neurons and its role in generating spindles was first proposed by Jansen and Llinas [13, 14] and by Steriade and Deschenes [15] and subsequently confirmed and further elaborated by Mulle et al. [16], Steriade et al. [17], Spreafico et al. [18], and Avanzini et al. [19]. The role of oscillating thalamo-cortico-thalamic loops in generating theta and delta activities associated with lower levels of vigilance was actively investigated, leading to some hypotheses about physiological mechanisms accounting for the parallel course of slowing down EEG activities and the lowering level of vigilance.

Interestingly, the cropping up of REM sleep disrupts such a parallel course demonstrating that, due to activation of cholinergic projections from the basal forebrain [20], the cortex is very active, but its reactivity to external stimuli stays low because it is functionally disconnected. After the pioneering work of Frederic Bremer, Giuseppe Moruzzi, William Dement, Nataniel Kleimann, and Michel Jouvet, it became clear that EEG is a prominent method for sleep research, so the definition of the stages of sleep is mainly based on the corresponding EEG picture. For further details about EEG in sleep, the author refers the reader to the specific chapter in this book, suffice here to say that if there were no other demonstrations, the one provided by sleep would be enough to recognize the EEG as a powerful investigation tool to study higher brain functions such as attention memory and other cognitive and behavioral functions that are modulated by sleep.

The alpha suppression phenomenon is often referred to as an "EEG desynchronization," meaning that the synchronized regular alpha rhythm seen in quiet wakefulness is substituted during alert wakefulness by asynchronous fast activities. Actually, Gray and Singer in 1989 provided compelling evidence that neurons of the visual cortex respond to visual stimulation with synchronous, rhythmic activity at gamma frequency in phase with electrocortical gamma oscillations, which makes inappropriate the term "desynchronization." Gray et al. [21] found synchronous rhythmic cellular responses in the gamma range in neurons with different receptive fields but identical orientation tuning, suggesting that Gamma rhythm may reflect perceptual ongoing activities. These observations stimulated a new interest in fast EEG activities and led to the conclusion that cortical inhibitory GABAergic interneurons play a central role in their generation [22, 23]; in particular, parvalbumin-positive interneurons, namely, fast-spiking basket neurons, are thought to play a role in gamma oscillations, whereas somatostatin-expressing interneurons have a timescale in the beta range [24–26]. As far as gamma oscillations are concerned, the narrowband gamma activity (30–80 Hz) is to be considered to be driven by neuronal synchrony of basket interneurons and to be correlated with cognitive activities ([24]), whereas broadband gamma oscillations (60–200 Hz) correlate with asynchronous multiunit firing [27, 28] and has a less definite functional correlation (see [29]). The neurophysiological analysis of gamma oscillations took advantage of the fact that narrowband gamma oscillations can be elicited by visual stimulation [28, 30]; interestingly, the repetitive visual stimulation that was found to be effective in eliciting gamma response in the normal subject was also provocative of photoparoxysmal responses in patients with photosensitive epilepsies, suggesting that an enhancement of gamma synchrony is the mechanism for photoparoxysmal response [31–33]. The author will not further discuss this point, which is out of the scope of this chapter, and the author refers the reader to Hermes et al. [34] and Avanzini et al. [29] for further information. What is relevant here is that the sensory stimulus activates a circuit of fast recurrent excitation and slower feedback inhibition responsible for narrow-band gamma oscillations that have been correlated with perceptual activities.

This statement introduces the fascinating though controversial topic of the *role of gamma activity in higher cortical function*. Whereas there is a general consensus about gamma rhythm as the signature of excitation–inhibition interactions in the brain, whether it also provides a mechanism for information processing is still a matter of debate [28, 35]. Arguments in support of gamma rhythm's role in cognitive processes are the observations on visual [36], olfactory, [37], and somatosensory [38] perception in experimental animal and human studies that are discussed in the next paragraphs. The main objections deal with low and inconsistent power of the gamma signal, and with conduction delays that may affect the gamma phase synchrony in different areas [28]. These signal processing issues require further investigations to define the role played by gamma rhythm in signal processing relevant to the brain's higher functions and to design the best strategies by which gamma neurophysiology can contribute in highlighting cortical mechanisms responsible for them (Fig. 5.2).

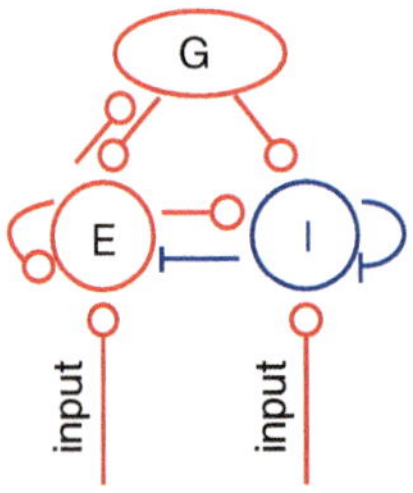

Fig. 5.2 Model of Gamma generating network developed by Jia et al. [39] to explain changes in firing and gamma oscillations with varying stimulus size and noise. The network includes excitatory (E) and inhibitory (I) input-driven neurons and a global excitatory component (G). (Reproduced from [29] (modified))

Mu Rhythm Although this EEG activity has long been familiar to every EEG student and practitioner, its interest in investigating brain functions has long been neglected. In recent years, the Parma group analyzed Mu reactivity during the observation of actions done by others to study mirror responses [40][1] finding a consistent decrease in Mu amplitude during the observation of a precision grip, due to desynchronization of mu rhythm generators in the sensorimotor cortex. The results show that Mu rhythm is sensitive to subtle changes in observed motor behavior and suggest its use as a non-invasive method to investigate mirror responses in humans.

5.3 Brain Connectivity Revealed by the EEG

The interest in analyzing the synchronization of oscillatory brain activity in various frequency bands depends on the idea that this may be one of the key mechanisms used by the brain to integrate information processing in multiple specialized local brain areas that are engaged in a common task. The idea was put forward by Gray and Singer [36] based on their study of the visual cortex and further elaborated by Singer [41], which stimulated a large bulk of investigations aimed at defining neural networks that are specifically active during different perception- or motion-related brain activities. The EEG signal is particularly suitable for analyses based on the *network science* that can be applied to different fields (e.g., internet telecommunication network, etc.) and which is based on *nodes* and *links,* talking about EEG the *nodes* correspond to the points from which the EEG signal is recorded (i.e., electrode location on the skull) and the *links* the measure of interaction between the *nodes.* Basically, the *functional connectivity* between two recording points is measured by statistical evaluation of the degree of interdependence of their activities. The most commonly used methods (*correlation and coherence*) give a reliable evaluation of the existence of *functional connectivity* between two nodes but cannot tell anything about the direction of the information flow. Other methods, based on *Granger causality,* which measures the capacity of a given series of observations, to predict the values of another series can add this information and are therefore defined as *directional.*

Figure 5.3 is reproduced from Muthuraman et al. [6], who compared the results of *directional* assessment of connectivity on EEG and MEG signal as well as EEG–MEG fusion in terms of sensitivity in the relaxed, awake state and concluded that the combined EEG–MEG approach is better than EEG or MEG alone. Note that the network has a similar conformation in the three different approaches both eye close and eye open conditions; note also that the combined EEG + MEG approach is superior to EEG or MEG alone. The dotted lines indicate weak connections (EEG alone and MEG alone), and the bold line indicates strong relations (EEG + MEG combination). The arrows indicate the direction of the information flow.

The hypothesis that coherent activation of different nodes in a neural network in a frequency band may reflect the engagement of the network in a specific brain function has been intensively tested. Rodriguez et al. [43] and Trujillo et al. [44] observed an enhancement of phase synchrony in the gamma band, between several EEG derivations, to be specifically associated with the recognition of one side of money. Many other

[1]The first author (Pietro Avanzini) whom I know personally and highly esteem for his scientific work is simply my homonymous and doesn't have any other connection with me.

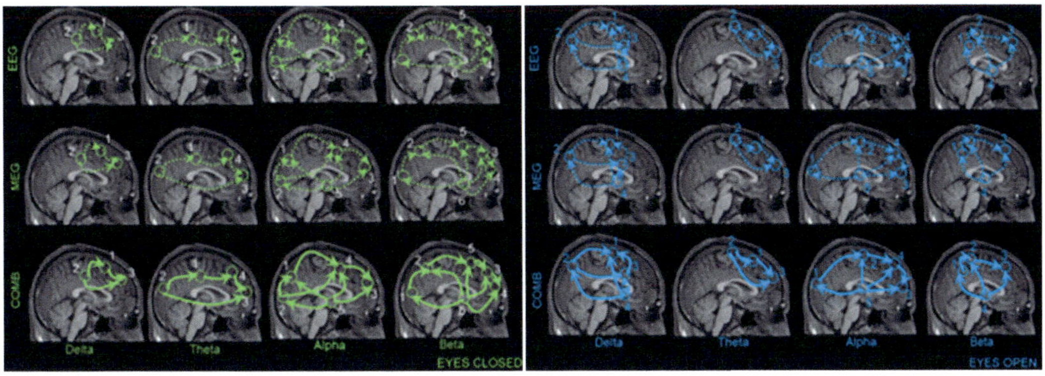

Fig. 5.3 Information flow between the coherent sources in resting state with eye closed (left) and eye open (right). First row: EEG recording; second row: MEG recording fusion of EEG–MEG recording. Each column refers to a single frequency band as indicated in the figure. The spatial filter approach dynamic imaging of coherent sources allows application of calculations within the scope of Granger causality to demonstrate the directionality of information flow. (From [6, 42], licensed under CC-BY 4.0)

subsequent observations supported the assertion that neurophysiological processes mediating conscious perception involve an increase in phase synchrony of gamma oscillations [7].

One of the brain systems that attracted the highest number of investigations using the EEG/MEG connectivity is language. In their systematic review of articles on functional brain connectivity of language function in pediatric age published between 1995 and 2018, Gaudet et al. [45] included 24 articles. Their interest in understanding language development is discussed in this excellent review. In general, brain connectivity proved to be a powerful method to characterize the topological properties of complex language networks. The method allows us to characterize the course of the shift from right to left dominance after 3 years of age and the role of the subcortical regions (*putamen and globus pallidus*) in expressive language [46].

The remarkable development of the methods to assess brain connectivity, namely, as their mathematical dimension is concerned, makes it now feasible to apply it to the most challenging problems such as the neurosciences of consciousness. Several studies of brain EEG connectivity in disorders of consciousness have been published in recent years (see the review in Szirmai et al. [47]) demonstrating the sensitivity of the method in evaluating the level of consciousness and in predicting its evolution. The efforts of many investigators are concentrated on measuring by EEG what Fingelkurts defines as the "operational synchrony" of different brain areas in "altered states of selfhood" [48].

5.4 Investigating Higher Brain Function by EEG

Since the time of my first training in neurophysiology (more than five decades ago!), it occurred to me repeatedly to listen to statements such as "oh yes Has Berger has the great merit of developing an outstanding method to investigate brain activities … however whereas EEG proved to be an irreplaceable method to describe brain dysfunctions it failed to fulfill Hans Berger's primary goal of disclosing us the higher brain functions." Well though the diagnostic and investigational power of EEG in the pathophysiology of brain dysfunction, namely, affecting excitability would *per sé* deserve our biggest gratitude, the EEG potential to study brain functions is by no means negligible, rather it proves increasingly valuable as testified by the information reported above, which are referred to and further extended in this last paragraph.

Arousal and Vigilance One of the first Hans Berger's observations was the striking effect of arousal on the EEG background activity: the regular 8.5–9.5 alpha rhythm seen in quiet relaxed wakefulness is suddenly substituted by low-amplitude beta and gamma rhythms when the subject opens his eyes [4] and subsequent investigations demonstrated that a similar type of reaction of the Mu rhythm is elicited by voluntary movements. Moreover, Pfurtscheller and Aranibar [49] demonstrated the role of attention in modulating alpha and Mu rhythms reactivity to warning signals or voluntary movements as *event-related desynchronization* and *event-related synchronization* can occur at the same time in different cortical areas depending on where subject's attention is directed [50].

The most important contribution to the understanding of the mechanisms of vigilance came from the studies of sleep EEG: the four stages of synchronous sleep are characterized by a progressive slowing of the EEG signal from the sigma to delta bands that parallels a progressive decrease of the level of vigilance evaluated as the strength of the awakening stimulus. The TC circuitry responsible for synchronous sleep EEG and the cholinergic projection, which is activated during REM sleep, have been discussed in the section on neurophysiological mechanisms of the EEG rhythms.

Cognitive Functions To studies of visual physiology, we owe the first evidence of the role of gamma activity in perception and the hypothesis that coherence of brain oscillations (namely, in the gamma band) in a neuronal population expresses a common engagement in a functional task [36, 51–53]. A number of EEG/MEG observations in humans confirmed the notion that phase synchronization within the gamma frequency band is associated with conscious perception [43, 44, 54, 55]. The specific role of different frequency bands in perception still needs to be better clarified: Kosem and van Wassenhove [56] reported a distinct contribution of low- and high-frequency neural oscillations to speech comprehension. The development of sound methods to analyze brain connectivity broadened the scope of coherence analysis of brain oscillations to different brain areas seen as nodes of neural networks. The author already quoted a group of papers investigating language networks in different populations of patients and the perspectives of this approach appear very promising in many other fields, including the neuroscience of consciousness which is probably the hardest problem of the neurosciences.

Memory and Learning The role of EEG oscillations in the recollection of contextual information has been investigated by Herweg et al. [57] by simultaneous EEG and fMRI recording, and their results demonstrated that theta-alpha power is linked to hippocampal connectivity with the striatum and prefrontal cortex, supporting a role of the hippocampus in coordinating neocortical activity via low-frequency oscillations. Moreover, they indicate that a combination of blood oxygenation level-dependent (BOLD)-derived connectivity with oscillatory synchrony may result in a powerful method to investigate neural mechanisms with spatial and temporal precision. More recently, Tang and Jadhav [58] reviewed the evidence that memory consolidation during sleep and memory retrieval in the waking state are associated with the same EEG pattern known as *sharp-wave ripples* [59, 60]. Tang and Jadhav suggest that plasticity driven by *sharp-wave ripples* plays a central role in learning.

5.4.1 EEG-Assisted Evaluation of Brain Transient Dysfunction

Besides providing the neuroscientist with an effective method to investigate directly several aspects of brain function, the EEG proved helpful in monitoring pathological ictal events affecting brain functions, thus allowing an indirect evaluation of the function through its reversible alteration. A typical example is transient brain dysfunction depending on the epileptic discharge whose onset, course, and cessation can be defined by EEG with high precision and correlated with the time course of function alteration. In this way, several hypotheses about the neural networks responsible for conscious perceptions are being tested by analyzing ictal changes whose time can be precisely defined by EEG monitoring [42, 61]. The power of such an approach can be fully appreciated in some studies such as the one of Heydrich et al. [62] where a detailed multimodal analysis of ictal experiential phenomenology [63, 64] led the authors to conclude that depersonalization (i.e., feeling of detachment from one's own body with lost control on it) and derealization (altered perception of one's surrounding that is perceived as unreal) are distinct phenomena and that the first one involves mechanisms of self-identification with one's own body taking place in premotor cortex.

5.5 Conclusions and Perspectives

One hundred years after its invention, the EEG technique and its younger sister MEG are still unsurpassed in showing us in real time the human brain at work. In this chapter, the EEG contribution to understanding the brain processes underlying higher brain functions such as arousal and vigilance, cognition, memory, and learning has been highlighted. Besides the pieces of information about many specific aspects of the above functions that the author tried to summarize, the author would like to stress some lessons that we learned from the experience of a group of dedicated people among whom any of us found his mentor, the first lesson being the school of methodological rigor that descends from the founder himself. People who met Hans Berger agree upon his extreme scrupulousness in evaluating every finding to exclude spurious results and artifacts, an attitude that every one of us has inherited from his mentor and transmitted to his pupils, as a fundamental premise of any study reliability. The importance of this attitude is increasingly evident with the technological advancements as every elaboration of the EEG signal, starting from its digital transformation, may obscure the presence of distortions and parasitic signals which would be obvious in the original record but difficult to detect in a manipulated signal. Along this line, the author stressed the need for cautiousness in dealing with correlating EEG with cognitive functions, a fascinating field that will advance our understanding of the brain as long as we are able to avoid shortcuts.

A second lesson we learned from Herbert Jasper and other pioneers of our discipline is the value of an active interaction with other branches of neurophysiology. The EEG is a neurophysiological method and it presupposes the specific knowledge and experience in biophysics, electrology, and informatics that qualify a neurophysiologist, this facilitates the transfer of information with reciprocal benefit. The contribution of basic neurophysiology to the interpretation of EEG patterns is the most obvious example of the importance of such an interaction, but the reciprocal contribution of EEG studies of oscillatory activities and connectivity is by no means negligible. As a consequence, EEG stands as a sound bridge between basic and applied neurophysiology.

This consideration introduces the next point about the technical innovations that are being made available in EEG equipment and elaborating systems. The most interesting methodological developments of the last 50 years deal with frequency analysis of oscillatory EEG activities and oscillatory interaction, which is assumed to reflect neural networks at work. This seems at present a harbinger of interesting developments, providing that interpretational caveats and common pitfalls reviewed by Bastos and Schoffelen [65] are duly taken into account. Many data concur in indicating that oscillatory activities revealed by EEG and MEG may contribute to coding and processing the information on which brain functions, including cognitive functions, are based. To what extent this I true is one of the forthcoming answers we expect from ongoing investigations.

References

1. Galvani L. De viribus electricitatis in motu musculari. Commentarius De Bononiensi Scientiarum et artium instituto atque Academia Commentarii. 1791;7:363–418.
2. Adrian ED, Mattews HO. The Berger rhythm: potential changes from the occipital lobes in man. Brain. 1934;57:355–85.

3. Berger H. Über das Elektrenkephalogramm des Menschen. Sechste Mitteilung. Arch Psychiatr Nervenkr. 1933;99:555–74.
4. Berger H. Das Elektrenkephalogramm des Menschen. Nova Acta Leopoldina, Halle (Saale). 1938;6:173–309.
5. Oersted HC. Experimenta circa effectum conflictus electrici in acum magneticum. JF Chem Phys. 1820;29:275–81.
6. Muthuraman M, Moliadze V, Mideksa KG, Anwar AR, Stephani U, Deuschl G, Freitag CM, Siniatchkin M. EEG-MEG integration enhances the characterization of functional and effective connectivity in the resting state network. PLoS One. 2015;10(10):e0140832.
7. Lopes da Silva F. EEG and MEG: relevance to neuroscience. Neuron. 2013;80(5):1112–28.
8. Jasper HH, Andrews HL. Electro-encephalography III. Normal differentiation of occipital and precentral regions in man. Arch Neurol Psychiatr. 1938;39:96–115.
9. Das N, Gastaut H. Variations de l'activite electrique du cerveau, du coeur et des muscles squelettiques au cours de la meditation et de l'extase yogique. Electroencephalogr Clin Neurophysiol. 1955;6:211.
10. Hebb DO. The organization of behavior. New York: Wiley; 1949.
11. Buzsáki G. Rhythms of the brain. New York: Oxford University Press; 2006.
12. Moruzzi G, Magoun HW. Brain stem reticular formation and activation of the EEG. Electroencephalogr Clin Neurophysiol. 1949;1(4):455–73.
13. Jahnsen H, Llinas R. Electrophysiological properties of Guinea-pig thalamic neurones: an in vitro study. J Physiol. 1984a;349:205–26.
14. Jahnsen H, Llinas R. Ionic basis for the electroresponsiveness and oscillatory properties of Guinea-pig thalamic neurones in vitro. J Physiol. 1984b;349:227–47.
15. Steriade M, Deschenes M. The thalamus as a neuronal oscillator. Brain Res Rev. 1984;8:1–62.
16. Mulle C, Madariaga A, Deschenes M. Morphology and electrophysiologicalproperties of reticularis thalami neurons in cat: in vivo study of a thalamic pacemaker. J Neurosci. 1986;6:2134–45.
17. Steriade M, Domich L, Oakson G. Reticularis thalami neurons revisited: activity changes during shifts in state of vigilance. J Neurosci. 1986;6:68–81.
18. Spreafico R, De Curtis M, Frassoni C, Avanzini G. Electrophysiological characteristics of morphologically identified reticular thalamic neurons from rat slices. Neuroscience. 1988;27:629–38.
19. Avanzini G, de Curtis M, Panzica F, Spreafico R. Intrinsic properties of nucleus reticularis thalami neurons of the rat studied in vitro. J Physiol. 1989;416:111–22.
20. Jouvet M. Research on the neural structures and responsible mechanisms in different phases of physiological sleep. Arch Ital Biol. 1962;100:125–206.
21. Gray CM, König P, Engel AK, Singer W. Oscillatory responses in cat visual cortex exhibit inter-columnar synchronization which reflects global stimulus properties. Nature. 1989;338:334–7.
22. Bartos M, Vida I, Jonas P. Synaptic mechanisms of synchronized gamma oscillations in inhibitory interneuron networks. Nat Rev Neurosci. 2007;8(1):45–56.
23. Buzsáki G, Leung LW, Vanderwolf CH. Cellular bases of hippocampal EEG in the behaving rat. Brain Res Oct. 1983;287(2):139–71.
24. Buzsáki G, Wang XJ. Mechanisms of gamma oscillations. Annu Rev Neurosci. 2012;35:203–25.
25. Chen G, Zhang Y, Li X, Zhao X, Ye Q, Lin Y, Tao HW, Rasch MJ, Zhang X. Distinct inhibitory circuits orchestrate cortical beta and gamma band oscillations. Neuron. 2017;96(6):1403–18. e1406.
26. Whittington MA, Traub RD, Kopell N, Ermentrout B, Buhl EH. Inhibition-based rhythms: experimental and mathematical observations on network dynamics. Int J Psychophysiol. 2000;38(3):315–36.
27. Manning JR, Jacobs J, Fried I, Kahana MJ. Broadband shifts in local field potential power spectra are correlated with single-neuron spiking in humans. J Neurosci. 2009;43:13613–20.
28. Ray S, Maunsell JH. Do gamma oscillations play a role in cerebral cortex? Trends Cognit Sci. 2015;19(2):78–85.
29. Avanzini G, Parra J, Hermes D. Gamma oscillations and photosensitive epilepsy. In: Kasteleijn-Nolst Trenite D, editor. The importance of photosensitivity for epilepsy. Springer; 2021.
30. Hermes D, Miller KJ, Wandell BA, Winawer J. Gamma oscillations in visual cortex: the stimulus matters. Trends Cognit Sci. 2015;19(2):57–8.
31. Parra J, Kalitzin SNJ, Iriarte BW, Velis DN, Lopes da Silva FH. Gamma-band phase clustering and photosensitivity: is there an underlying mechanism common to photosensitive epilepsy and visual perception? Brain. 2003;126(Pt 5):1164–72.
32. Varotto G, Visani E, Canafoglia L, Franceschetti S, Avanzini G, Panzica F. Enhanced frontocentral EEG connectivity in photosensitive generalized epilepsies: a partial directed coherence study. Epilepsia. 2012;53(2):359–67.
33. Visani E, Varotto G, Binelli S, Fratello L, Franceschetti S, Avanzini G, Panzica F. Photosensitive epilepsy: spectral and coherence analyses of EEG using 14Hz intermittent photic stimulation. Clin Neurophysiol. 2010;121(3):318–24.
34. Hermes D, Kasteleijn-Nolst Trenite DGA, Winawer J. Gamma oscillations and photosensitive epilepsy. Curr Biol. 2017;27(9):R336–8.
35. Fries P. Neuronal gamma-band synchronization as a fundamental processn in cortical computation. Annu Rev Neurosci. 2009;32:209–24.
36. Gray CM, Singer W. Stimulus-specific neuronal oscillations in orientation columns of cat visual cortex. Proc Natl Acad Sci USA. 1989;86(5):1689–702. 14.
37. Freeman WJ. Spatial properties of an EEG event in tthe olfactory bulb and cortex. Electroencephalogr Clin Neurophysiol. 1978;44:586–605.

38. Bouyer JJ, Rougeul A, Buser P. Somatosensory rhythms in the awake cat: a single unit exploration of their thalamic concomitant in nucleus ventralis posterior and vicinity. Arch Ital Biol. 1982;120:95–110.
39. Jia X, Xing D, Kohn A. No consistent relationship between gamma power and peak frequency in macaque primary visual cortex. J Neurosci. 2013;33(1):17–25.
40. Avanzini P, Fabbri-Destro M, Dalla Volta R, Daprati E, Rizzolatti G, Cantalupo G. The dynamics of sensorimotor cortical oscillations during the observation of hand movements: an EEG study. PLoS One. 2012;7(5):e37534.
41. Singer W. Neuronal synchrony: a versatile code for the definition of relations? Neuron. 1999;24(1):49–65, 111-25.
42. Blumenfeld H. Impaired consciousness in epilepsy. Lancet Neurol. 2012;11(9):814–26.
43. Rodriguez E, George N, Lachaux JP, Martinerie J, Renault B, Varela FJ. Perception's shadow: long-distance synchronization of human brain activity. Nature. 1999;397:430–3.
44. Trujillo LT, Peterson MA, Kaszniak AW, Allen JJ. EEG phase synchrony differences across visual perception conditions may depend on recording and analysis methods. Clin Neurophysiol. 2005;116:172–89.
45. Gaudet I, Hüsser A, Vannasing P, Gallagher A. Functional brain connectivity of language functions in children revealed by EEG and MEG: a systematic review. Front Hum Neurosci. 2020;14:62. https://doi.org/10.3389/fnhum.2020.00062. PMID: 32226367; PMCID: PMC7080982.
46. Youssofzadeh V, Williamson BJ, Kadis DS. Mapping critical language sites in children performing verb generation: whole-brain connectivity and graph theoretical analysis in MEG. Front Hum Neurosci. 2017;11:173.
47. Szirmai D, Zabihi A, Kói T, Hegyi P, Wenning AS, Engh MA, Molnár Z, Csukly G, Horváth AA. EEG connectivity and network analyses predict outcome in patients with disorders of consciousness - A systematic review and meta-analysis. Heliyon. 2024;10(10):e31277. https://doi.org/10.1016/j.heliyon.2024.e31277.
48. Fingelkurts AA, Fingelkurts AA, Kallio-Tamminen T. Me and I in the repertoire of spontaneously occurring altered states of selfhood: eight neurophenomenological case study reports. Cogn Neurodyn. 2022;16(2):255–82.
49. Pfurtscheller G, Aranibar A. Event-related cortical desynchronization detected by power measurements of scalp EEG. Electroencephalogr Clin Neurophysiol. 1977;42(6):817–26.
50. Pfurtscheller G, Lopes da Silva FH. Event-related EEG/MEG synchronization and desynchronization: basic principles. Clin Neurophysiol. 1999;110:1842–57.
51. Eckhorn R, Bauer R, Jordan W, Brosch M, Kruse W, Munk M, Reitboeck HJ. Coherent oscillations: a mechanism of feature linkingin the visual cortex? Multiple electrode and correlation analyses in the cat. Biol Cybern. 1988;60:121–30.
52. Fries P, Reynolds JH, Rorie AE, Desimone R. Modulation of oscillatory neuronal synchronization by selective visual attention. Science. 2001;291:1560–3.
53. Singer W, Gray CM. Visual feature integration and the temporal correlation hypothesis. Annu Rev Neurosci. 1995;18:555–86.
54. Doesburg SM, Green JJ, McDonald JJ, Ward LM. Rhythms of consciousness: binocular rivalry reveals large-scale oscillatory network dynamics mediating visual perception. PLoS One. 2009;4:e6142.
55. Fahrenfort JJ, Snijders TM, Heinen K, van Gaal S, Scholte HS, Lamme VA. Neuronal integration in visual cortex elevates face category tuning to conscious face perception. Proc Natl Acad Sci USA. 2012;109:21504–9.
56. Kosem A, van Wassenhove V. Distinct contribution of low- and high-frequency neural oscillations to speech comprehension. Lang Cognit Neeurosci. 2017;37:536–44.
57. Herweg NA, Apitz T, Leicht G, Mulert C, Fuentemilla L, Bunzeck N. Theta-alpha oscillations bind the hippocampus, prefrontal cortex, and striatum during recollection: evidence from simultaneous EEG-fMRI. J Neurosci. 2016;36(12):3579–87.
58. Tang W, Jadhav SP. Sharp-wave ripples as a signature of hippocampal-prefrontal reactivation for memory during sleep and waking states. Neurobiol Learn Mem. 2019;160:11–20.
59. Battaglia FP, Sutherland GR, McNaughton BL. Hippocampal sharp wave bursts coincide with neocortical "up-state" transitions. Learn Mem. 2004;11(6):697–704.
60. Jadhav SP, Kemere C, German PW, Frank LM. Awake hippocampal sharp-wave ripples support spatial memory. Science. 2012;336(6087):1454–8.
61. Danielson NB, Guo JN, Blumenfeld H. The default mode network and altered consciousness in epilepsy. Behav Neurol. 2011;24(1):55–65.
62. Heydrich L, Marillier G, Evans N, Seeck M, Blanke O. Depersonalization- and derealization-like phenomena of epileptic origin. Ann Clin Transl Neurol. 2019;6(9):1739–47.
63. Gloor P. The temporal lobe epilepsy and limbic system. Oxford University Press; 1997. p. 865.
64. Penfield W. Temporal lobe epilepsy. Br J Surg. 1954;41:1–7.
65. Bastos AM, Schoffelen JM. A tutorial review of functional connectivity analysis methods and their interpretational pitfalls. Front Syst Neurosci. 2016;9:175.

Preclinical Investigation to Understand Electroencephalogram Patterns

6

Marco de Curtis, Premysl Jiruska, and Christos Panagiotis Lisgaras

The recording of electrical signals in the mammalian brain precedes by 50 years the seminal study of alpha waves in humans [1]: in 1875, Richard Caton discovered electrical brain activity by firmly placing two electrodes on the surface of the cortex and on the skull of anesthetized cats and rabbits [2]. Since then, the electroencephalogram (EEG) has been studied and analyzed in normal and diseased brains of animals and humans [3, 4]. Nowadays, the EEG represents one of the main and most inexpensive tools to objectively explore the brain and its functions. The simplest EEG recording measures the changes in voltage between two sampling points (electrodes); these voltage field potentials can be recorded with electrode arrays with a high number of sensors arranged according to the clinical or experimental requirements and settings. The brain field potentials that generate EEG signals are very simple to capture but are extremely complex to interpret because the electrical brain activity recorded with the EEG recording electrodes is generated by diverse sources located at different distances from the recording site. The EEG signal retains a variety of averaged information that originates from neurons and other brain elements (such as glia, vessels, etc.) that can be located either close to (local field potentials) or away (volume-conducted potentials) from the recording electrode. Spontaneous network activity emerges from the interaction and convergence of several relatively simple elements responsible for localized extracellular potential changes that include chemical and electrical synapses, ion channels, and transporters. Moreover, EEG signals represent superimposed brain activities generated by electrical events that develop over multiple temporal scales—from the few milliseconds of an action potential to hundreds of seconds during the slow direct current (DC) potential shifts.

In this chapter, we will review the following aspects: (i) the experimental preparations and the neurophysiological techniques utilized to record brain activity in preclinical studies, (ii) the patterns and elements that characterize the physiological EEG analyzed with preclinical investigation tools, (iii) the network/tissue mechanisms underlying

M. de Curtis (✉)
Epilepsy Unit, Fondazione IRCCS Istituto Neurologico Carlo Besta, Milan, Italia
e-mail: marco.decurtis@istituto-besta.it

P. Jiruska
Department of Physiology, Second Faculty of Medicine, Charles University, Prague, Czech Republic
e-mail: premysl.jiruska@lfmotol.cuni.cz

C. P. Lisgaras
Department of Child & Adolescent Psychiatry, Neuroscience, Physiology, and Psychiatry, Neuroscience Institute New York University Langone Health, New York, NY, USA
e-mail: nchristos.lisgaras@nyulangone.org

F. Brigo, O. Mecarelli (eds.), *EEG: The First 100 Years*,
https://doi.org/10.1007/978-3-031-86878-8_6

EEG patterns and EEG fragment generated in pathological conditions, with special focus on epilepsy, and (iv) the integration of electrophysiology with new technologies to better understand brain signal synchronization. In the present report, we will not discuss the EEG changes that characterize wake and sleep states, because of the large literature available on this topic extensively reviewed in the last decades [5–7]. Moreover, rodents, felines, and primates (monkeys and humans) display different EEG patterns during sleep stages/phases that could possibly mediate diverse underlying networks [7, 8].

In this chapter, we assume the equivalence between EEG patterns observed in different animal species during wake and those recorded in humans. This postulation is based on the concept that patterns recorded in experimental model share common features and are considered as counterparts of human EEG patterns. The unavoidable bias derived by this assumption is the fact that we interpret EEG events on the basis of their electrographic features, through the "reading glasses" of the electrophysiologist. Nevertheless, we should be aware that different neurobiological mechanisms could be responsible for electrographically similar neurophysiological patterns, as demonstrated, for instance, by the demonstration that epileptiform interictal population spikes with similar electrographic features that can be generated by different underlying network mechanisms [9].

6.1 Electophysiological Tools to Study Brain Activity

Experimental neurophysiological recordings are commonly utilized to study brain events and activity at multiple spatial scales (ranging from micrometers to centimeters) and temporal scales. The use of EEG standard techniques in experimental studies has been recently reviewed in a series of manuscripts promoted by the joint Translational Task Force of the International League Against Epilepsy and the American Epilepsy Society [10–13]. Standard EEG recordings on animal models of neurological diseases are most commonly performed nowadays in rodents and lagomorphs: rats, mice, guinea pigs, and gerbils accounted for 83%, 20%, 4%, and 1%, respectively, of the total animal EEG studies [11]. As in the clinic, EEG recordings in animals can be realized with electrodes positioned either under the skin on the surface of the skull (comparable to human scalp EEG), or beneath the dura (as in human subdural grids and strips electrodes) or in deep brain structures, as for human intracerebral stereo-EEG recordings. The interpretation of EEG signals in animals is not as standardized as in human EEG diagnostics; EEG recordings can be implemented by varying the number and type of recording electrodes both on the brain surface and in deep brain structures, depending on the experimental setting and on the scope of the recording.

EEG electrode placement requires a surgically invasive procedure performed under anesthesia, followed by long-acting analgesic and antibiotic therapy to avoid postsurgical pain and infections. In preclinical studies, postoperative treatment and acclimation for at least 1 week usually precede the EEG signal acquisition session. EEG surface recording electrodes (typically metal micro-screws) can be either fixed to the skull under the skin or positioned in burr holes drilled in the skull for epidural recordings. Depth intracerebral recordings [10] are performed with a wide range of sharpened wires/electrodes made of metal (tungsten, stainless steel, silver, and platinum) used either as single leads or organized in diverse multiarray arrangements [14] (Fig. 6.1a and b); special recording arrays are the silicon probes that carry small platinum–iridium contacts linearly aligned on a thin silicate rod, utilizing the printed circuits technology [15] (Fig. 6.1c). Integrated circuit wiring has been largely utilized on new foldable and stretchable materials that enhance the resistance and the adherence of the recording sites to the brain surface, to develop multiarray electrodes for chronic and multisite population recording from large ensembles or from individual neurons [16].

Wire electrodes are used to record local field potentials, whereas tetrodes and thin wires are employed to record single units and multi-unit activity (see below). Tetrodes are constructed by

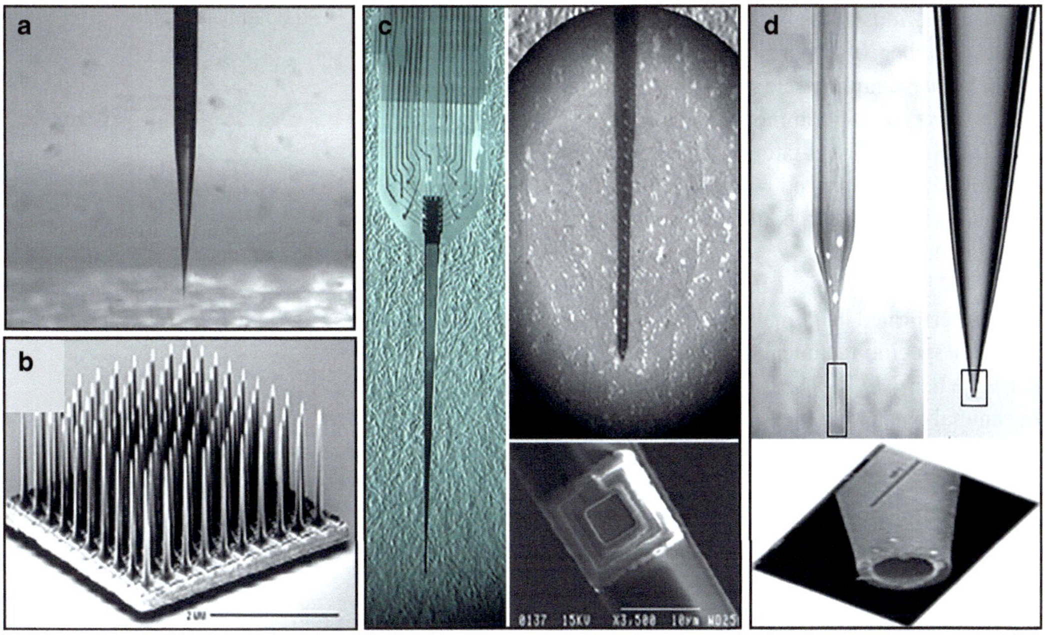

Fig. 6.1 Type of electrodes utilized to record brain activity in experimental animal models and preparations. (**a**) Single shaft metal electrode (tungsten) for intracerebral depth recordings in vivo and in vitro. (**b**) Multi-electrode *Utah* array featuring 96 stainless steel recording electrodes for in vivo depth recordings. (**c**) Silicon probe with 16 linearly arranged recording sites at the tip for in vitro and in vivo experiments, as illustrated in the upper right photograph. The detail of each platinum-iridium recording site is illustrated in the lower right microphotograph. (**d**) Borosilicate glass fire-pulled micropipette for intracellular recordings in vitro and in vivo. The final part of the electrode filled with a conductive saline solution is shown on the right, and the electrode tip is illustrated in the bottom microphotograph

bundling together several thin (<30 μm in diameter) insulated platinum/tungsten wires embedded in a single resin/glass shaft; each recording tip/element of a tetrode is very close to the other (>10 s of μm apart) to record from closeby neurons. Glass micropipettes heat-pulled from glass capillaries (Fig. 6.1d) are commonly used to record either extracellularly in different parts of the brain or intracellularly from single cells; glass recording electrodes can be used both on in vitro preparation and during acute and chronic in vivo experiments. Glass pipettes and multi-electrode arrays (MEAs) are the most widely utilized electrodes for in vitro experiments on brain tissue slices or in toto preparations (see the next paragraph; [17–18]). Long-term intracellular patch-clamp recordings with glass micropipettes are usually performed in anesthetized, head-restrained animals but are also achieved in anesthetized and unanesthetized, freely moving animals [19–21]; for these experiments, head-mountable recording devices or miniature microscopes have been developed that consent the recording of a single impaled neuron for several days in freely behaving rodents, often coupled with high-resolution, large-scale imaging of cortical brain activity ([22] for review).

With both epidural and depth EEG arrangements, the electrodes and a wire connector are secured to the skull surface using acrylic dental cement to form a mechanically firm headset. When wire-connected (tethered) videoEEG recordings are performed, the headset is connected to a cable attached to a swivel that allows reliable recordings in moving animals with little artifacts. Wireless recordings can be performed by using telemetric headstages that contain a pre-amplifier to improve EEG signals transmitted to a remote receiver. The absence of cables simplifies the recording of animal behavior. Tethered EEG is more laborious to the implant but consents to a higher number of recording electrodes and a much higher sampling rate compared to wireless EEG recordings—commonly limited to a 50–100 Hz sampling rate. In the last decade, technologically advanced wireless recording systems have been developed that consent to higher performances [23–25].

As for human clinical EEG diagnostic, EEG studies performed in animal models have to comply with specific technical and procedural standards. When EEG is performed to characterize the animal phenotype (i.e., to detect and quantify seizure occurrence or the sleep/wake cycles in an animal model), video recordings synchronized with EEG signals for electrical/behavioral correlations are performed with variable numbers of electrodes, based on the experimental quest. VideoEEG recording is a requirement for seizure/epilepsy phenotyping since seizures may be present with very subtle (if any) motor features, and EEG recording is essential to identify an ictal seizure pattern [26]. VideoEEG is routinely utilized to study physiological brain rhythms and patterns, to analyze sleep/wake cycles, to identify epileptiform and other pathological patterns, and more recently to study and understand circadian and multidien EEG pattern cycles [27–29]. Moreover, EEG signals have been included among the common data elements for preclinical epilepsy research [30], developed to assist epilepsy researchers in performing, collecting, and reporting EEG recordings in epilepsy models, for a reliable comparison between studies and different research centers.

When depth recordings are performed, local field potentials (LFP) generated within specific brain structures are the main target. The source of LFPs can be characterized with multichannel electrode arrays by performing laminar profiles in cortical or other laminated structures [31]. Laminar profiles (as recorded with a multichannel silicon probe in Fig. 6.2a) are essential to locate the current sink and sources (Fig. 6.2b) that generate the diverse components of the LFP within the depth of the laminar structure under analysis (usually a cortical structure). This technique has been applied to reconstruct local cortical networks evoked either by afferent stimulation (Fig. 6.2) or during the generation of population events such as interictal spikes.

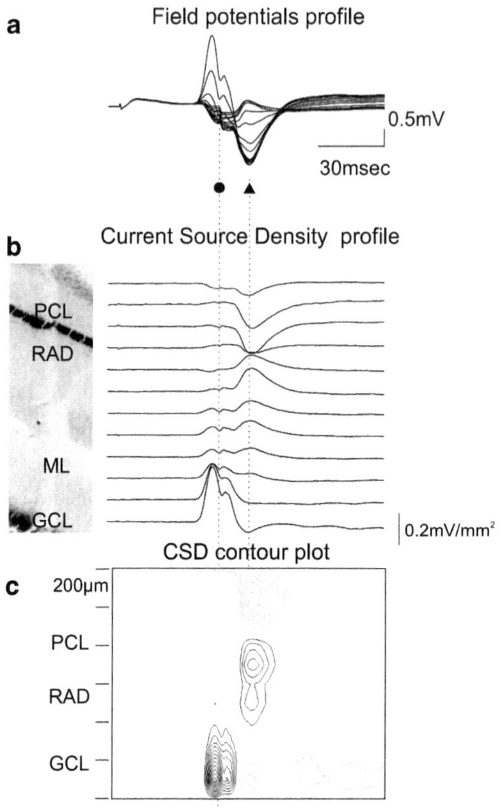

Fig. 6.2 Laminar profile recorded in the hippocampus of the isolated guinea pig brain preparation. In (**a**), the field potentials evoked by electrical stimulation of the lateral olfactory tract were simultaneously recorded with a 16-channel silicon probe similar to the electrode illustrated in Fig. 6.1c inserted in the ventral hippocampus of an in vitro isolated guinea pig brain. Each of the 16 recording sites was separated by 50 µm on a linear shaft. (**b**) Current source density analysis performed on the field potential laminar profile shown in **a**. The depth of each recording site is illustrated by showing a thionine-stained section of the hippocampal region that includes the CA1 and the dentate Gyrus (DG) layers. The recording tract of the electrode is evident in the stained section. PCL pyramidal call layer of CA1, RAD stratum radiatum, ML molecular layer, GCL granule cell layer of the DG. Active sinks (continuous concentric lines) and passive sources (dotted lines) of the CSD profile are illustrated in lower panel (**c**). The depth distribution of the sink and source contour plots describe the synaptic events generated in the CA1-DG region by the afferent stimulation

Several specific adaptations of LFP recordings were applied to acquire different neurophysiological patterns in animal models. As an example, the description of infraslow activity (so-called DC potentials; [32]), identified several decades ago [33] and largely neglected in EEG and LFP studies, proved to be relevant to understand activity-associated extracellular ions shifts. DC potentials were associated with physiological and pathological population events such as seizures and spreading depression [32]. DC shifts have been attributed to different mechanisms that include changes in intracranial transepithelial potentials associated with blood–brain barrier permeability changes [34–36], extracellular ion (potassium and proton) changes [9], and glial depolarization.

A further step in the analysis of EEG signals generation mechanisms is achieved by recording neuronal firing during physiological and pathological brain activity. Extracellular correlates of action potentials from individual or multiple cells can be recorded as a single or multiunit activity with glass micropipette, metal wire, or tetrode or with MEAs [37, 38]. With the latter technique, it is possible to identify the spatial distribution of neuronal clusters that sustain population events such as brain oscillatory activity. The recording of units and population activity is currently the optimal method to obtain information from hundreds of neurons simultaneously for hours and even days of continuous recording. Spike-sorting techniques are utilized to identify the activity of single neuronal subtypes—principal cells and inhibitory interneurons—based on specific features of extracellular or intracellular spike waveforms [37, 39]. The use of unit recording techniques has been exploited during presurgical EEG monitoring in patients with drug-resistant epilepsy (i) to analyze and reveal the cellular contributions of principal cells and interneurons to interictal and ictal epileptiform discharges [40, 41], (ii) to study cellular activities recorded during normal spontaneous activity (brain oscillatory patterns), (iii) to examine cellular activity during stimulus-evoked brain potentials in normal conditions [42, 43], and (iv) to understand the cellular correlates of higher brain functions, such as memory and visual recognition [44].

6.2 Experimental Preparations and Models to Study EEG Patterns and EEG Events

As illustrated in the previous paragraph, different EEG and LFP patterns can be recorded to analyze specific neurophysiological patterns in *naïve* subjects and in diseased animals that reproduce neurological humans dysfunctions. This level of in vivo recording (and analysis) allows us to describe the EEG patterns and their electrical sources and local generators with large-scale electrophysiology sensors and micro-electrode arrays. A more detailed understanding of the biological mechanisms that underlie the observed patterns at the cellular and subcellular level is achieved by reproducing macroscopic EEG patterns or fragments/parts of EEG signals in experimental in vitro preparations and models [17, 18, 45] that maintain physiological features in isolation for several hours. This can be achieved by analyzing brain signals with a wide range of technologies that include electrophysiological, sensor-specific, and imaging recordings.

Brain sections maintained in vitro (brain slices) represent the most widespread preparation utilized for neurophysiological studies. Brain slices are cut at a thickness that ranges between 300 and 500 μm and are maintained in vitro in incubation chambers, perfused with saline solutions that mimic the composition of the extracellular and cerebrospinal fluids [17]. Brain slices retain local tissue networks and generate EEG-like patterns when appropriately stimulated/activated by either pharmacological or electrophysiological manipulations. By altering the composition of the perfusion fluid in different types of brain slices, EEG components such as *theta, beta, gamma* oscillatory activities, epileptiform spikes, spindle sequences, high-frequency oscillations (HFOs), and seizure-like discharges and infraslow potential shifts have been reproduced. In vitro slices can include either specific portions of the brain, such as single brain areas (as the hippocampal slice depicted in Fig. 6.3a) or complex slices that include more than one brain region, such as hippocampal/parahippocampal and cortico/thalamic slices. Virtually, any

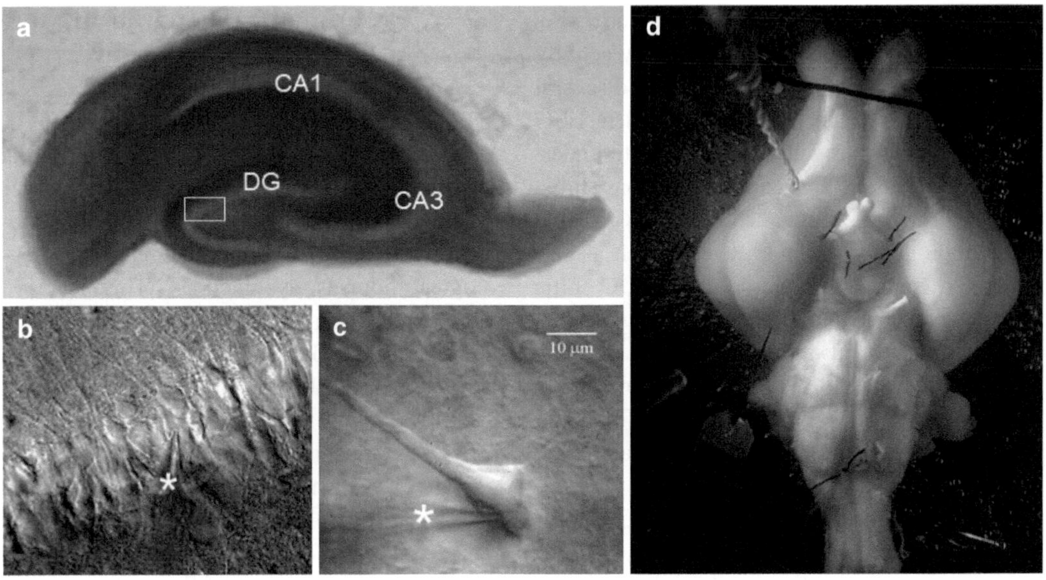

Fig. 6.3 In vitro preparations for neurophysiological studies. (**a**) In vitro maintained coronal hippocampal slice (350 µm thick) of a mouse. The CA1, CA3, and dentate gyrus subfields are marked. The area outlined by the rectangle in expanded in panel (**b**), where the position of the tip of a glass micropipette is marked by the double asterisks. (**c**) In vitro isolated pyramidal cell in culture. The recording patch clamp micropipette that records from the inside of the neuron is shown on the left side of the cell soma. (**d**) In vitro isolated guinea pig brain perfused through the basilar artery via the resident arterial system. A twisted silver wire utilized for electrical stimulation is positioned on the lateral olfactory tract on the left hemisphere. A micropipete and a silicon probe recording from the entorhinal cortex are shown in the lower left part of the figure

brain region has been sliced and maintained in vitro to study electrophysiological sources of specific brain signals [46].

Interestingly, brain slices can be maintained in culture with different methods to analyze events that occur during days or weeks. These organotypic slices can generate spontaneous interictal epileptiform activities that are not commonly observed in acute brain slices [18]. Human brain slices have been cut from postsurgical specimens obtained during brain surgery in patients with both pharmacoresistant epilepsy and brain tumors [47, 48] to directly analyze the determinants of EEG patterns and epileptiform activities directly in the human tissue.

In addition to in vitro slices, other in toto in vitro preparations have been developed to reproduce by pharmacological manipulation or by electrical stimulations EEG patterns observed in the normal and diseased brain. The generators of septohippocampal oscillatory and epileptiform patterns were analyzed in the in toto bilateral hippocampi maintained in vitro in double incubation chambers [49] that enables the application of different drugs separately in one of the hippocampi, to study the effect in the controlateral one. Another preparation is that the isolated in vitro whole brain of the guinea pig (Fig. 6.3d) has been originally prepared to analyze cerebellar physiology and was later exploited to study the neurophysiology of the olfactory/limbic system as well as the epileptiform activity generated in different cortical and subcortical brain regions [50, 51].

In vitro preparation is ideal to mechanistically identify membrane, cellular, and network constituents of EEG patterns, in particular of elements or fragments that form the EEG signals as a whole, such as (i) oscillatory activities from the Berger frequency bands, up to HFOs, (ii) infraslow potentials, (iii) interictal epileptiform patterns, (iv) seizure patterns, and (v) sleep-related patterns. In addition to the methods described in the previous paragraph (LFP and unit recordings), several technologies are available for the mechanistic analysis of EEG signals on in vitro preparations that cannot be utilized (at least in the large majority of neuroscience laboratories) during in vivo experiments. Ion-selective electrodes have been specifically employed in in vitro preparation to analyze the changes in extracellular ions (mainly potassium and protons/pH) that accompany brain activity and specific EEG elements [52]. This recording method has been extremely practical in understanding epileptiform activity and spreading depression phenomena [51, 53].

Signal analyses of multisite recordings (either electrical or imaging or both) increase the information yield for the interpretation of the network determinant of EEG patterns. Brain slices can be positioned on an MEA that features a large number of extracellular recording electrodes (up to several hundred in the so-called high-density MEA matrices; [54, 55]). These recording arrangements are ideal for analyzing propagation patterns and intrinsic network organization of specific EEG features—by achieving both LFP and single/multiple unit recordings. The increasing flexibility of MEA substrates has allowed us to surgically implant them in vivo for long-term recordings in animal models. An alternative in vitro method to analyze patterns at both single cell level and at the cell population scale is achieved by coupling neurophysiological recordings with high-resolution imaging signals, such as voltage-sensitive dyes or calcium indicators. This combined methodology allows the correlation of the electrical activity of single cells to the behavior of the network of neurons and nonneuronal cells, such as astrocytes and endothelial cells that surround it, during the expression of specific neurophysiological patterns. Currently available calcium and voltage indicators [56–58] allow recording of the activity of any brain cell type (principal neurons, interneurons, glial cells) with high cellular and subcellular spatial resolution. Selective dyes targeted to specific cell subtypes are utilized to analyze the contribution of these cells to the generation of brain activities. Calcium and voltage indicators facilitate the study of neurotransmission and neuromodulation during epileptiform activity. The use of high-resolution cellular imaging in vivo consents to repeated imaging sessions that can be performed chronically, allowing long-term monitoring of changes in neuronal or glial activity correlated with electrographic changes over disease progression when animal models of diseases are employed. Glutamate, GABA, acetylcholine, potassium, ATP, and other types of indicators allow us to study changes in neurotransmission, neuromodulation, and other molecules during epileptiform activity [59]. Complex emergent dynamics may be revealed by the application of statistical analysis methods to brain signals—obtained from both in vivo and in vivo multichannel recordings with MEA and high-resolution fast imaging. The study of synchronization, cross-frequency coupling, and nonlinear interactions by dedicated software are utilized for multichannel data postprocessing, to understand neuronal and network dynamics and to derive insight into the functional network organization.

Combining EEG with other neuroscience technologies enhances our understanding of epileptic activity. EEG paired with optogenetic stimulation or inhibition offers insights into the roles of different cell types during specific EEG events. Optogenetics uses light sources to control cells that express genetically engineered light-sensitive ion channels, in order to dissect the role of individual cells in the generation of specific EEG patterns [60]. Optogenetics have been exploited both in vitro and in vivo to identify the role of principal neurons and interneuron subtypes that selectively carry the optogenetic channels in their membrane in the circuits that control seizures, to understand epileptic networks and interfere with seizure generation [60, 61]. Closed-loop systems engineered to detect epileptic activity in real time and to trigger light stimulation to modulate specific cells show promise for on-demand seizure control activated by the fast detection of seizure onset. Optogenetics has advantages over electrical stimulation due to its cell-type specificity, though off-target effects must be managed. Chemogenetic methods also enhance research on EEG patterns by allowing long-term manipulation of neuronal populations to assess their contributions to EEG activity. This approach uses

Designer Receptors Activated Only by Designer Drugs [62] to control specific cell populations. DREADD chemogenetics enables prolonged interventions, with effects lasting for hours depending on the drug used as an activator.

6.2.1 Physiological EEG Activity Patterns

Early EEG studies in humans utilized the Greek letters *alpha, beta, gamma,* and *delta* to designate EEG oscillatory activity at 8–13, 12–30, 30–100, and below 4 Hz, respectively; the 3–8-Hz band was designated *theta* to stand for thalamus [63] because thalamic lesions in monkeys shifted cortical dynamics from alpha (8–12 Hz) to theta (4–7 Hz). *Alpha, beta,* and *gamma* brain oscillations represent electrographic correlates of highly orchestrated network activity sustained by precise interplays between different types of neurons during cognitive functions and during wake and sleep conditions. In the last decades, a massive amount of data on neuro-oscillatory dynamics was utilized to explain high cognitive functions spanning from attention, recognition, movement planning, speech or memory, and cognitive-affective processes [64, 65]. General features of different animal-model-derived knowledge on the main oscillatory patterns will be reviewed in the following paragraphs.

6.2.1.1 Theta Activity

In different animals, *theta* rhythms are generated in the hippocampus and entorhinal cortex, perirhinal cortex [66, 67], medial prefrontal cortex [68], medial mammillary body, supramammillary nuclei of the hypothalamus, anterior nuclei of the thalamus, amygdala, inferior colliculi, septal nuclei, and several brain stem nuclei [67, 69]. In rodents EEG, the term *theta activity* is used to describe 3–8 Hz oscillations that originate from the interaction of septal and limbic networks, observed in rodents during behavioral exploration and navigation, attentional processes, and memory encoding and altered in animal models of dementia [70]. The interaction between hippocampal pyramidal cells and septal nuclei, enhanced by the intrinsic resonant properties of neurons in these brain regions, regulates a *theta* cycle. The discovery of place cells in the entorhinal cortex and hippocampus [71] prompted the idea that during *theta* sequences, the activity in hippocampal neurons activated by a specific position of the animal in space (place cells) build a cognitive map used to construct spatial memory and facilitate memory-guided behaviors [72]. Specific changes in synaptic currents and inhibitory potentials occurring at different phases of *theta* rhythm could enhance either encoding or retrieval of memory that repeats on every *theta* cycle [73]. Moreover, *theta* rhythms in entorhinal cells and networks promote periodicity of neuronal firing that contributes to code space and time in models of episodic memory [74]. Finally, *theta* modulation of faster oscillation cycles in the gamma band could provide a substrate for ordering sequences of items within a memory recall [69, 75].

6.2.1.2 Gamma Activity

Fast oscillations in the *gamma* (30–90 Hz) and *beta* (15–30 Hz) frequency range are thought to be important for information processing (see also Avanzini chapter in this book). Gray and Singer [76] first demonstrated *gamma* oscillation triggered by visual stimuli in the cat's visual cortex. This stimulus-evoked pattern was utilized as a paradigm to synchronize local networks in different associative neocortical regions to modulate the multisensory perception during a visual–tactile stimuli matching task. Parvalbuminergic and somatostatinergic interneurons sustain *gamma* and *beta* activity in the cortex primarily by $GABA_A$ receptor-induced inhibitory postsynaptic potentials. *Gamma* oscillations are organized in short sequences of population events by the coordinated interaction of excitatory and inhibitory synapses modulated by coincident rhythmic activity of lower frequency (*theta* and *delta*; [65]). *Gamma* activity recorded in the hippocampal formation is associated with memory formation [75]. In the hippocampus in vitro and in vivo, $GABA_A$-receptor-mediated activity within networks of inhibitory interneurons has a primary role in *gamma* oscillation generation [77, 78]. Fast-spiking, parvalbumin-expressing basket cells extensively interconnected by mutual inhibition generate synchronized *gamma* activity when exposed to a tonic excitatory drive [78–80] and by cholinergic stimulation [81, 82]; *gamma* activity is stabilized through gap junctions that connect basket cells interneuron networks [83]. Interestingly, fast activity in the *beta/gamma* range has been observed at the onset of focal seizures, as part of the low-voltage fast oscillatory activity pattern described both in patients and in animal models (see below; [84, 85]).

6.2.1.3 HFOs

In the last two decades, EEG recordings and analyses have also focused on frequencies between 80 and 500 Hz the so-called HFO band. HFOs are currently thought to represent a promising epilepsy biomarker [86]. Based on frequency characteristics, HFOs have been typically classified into ripples (80–200 Hz) and fast ripples (200–500 Hz). The translational potential of HFOs was first introduced when fast ripples were recorded from epileptic foci (hippocampus) in epileptic rats and humans [87, 88]. However, ripples have also been shown to be of clinical utility in localizing the seizure onset zone in neocortical epilepsy, and importantly, these events can be recorded using both micro- and macro-electrodes [89]. In an animal model of temporal lobe epilepsy, fast ripples are present before the onset of epilepsy and are not observed in nonepileptic animals, supporting the idea that fast ripples may play a critical role in epileptogenesis [90, 91]. This notion is also supported by the finding that, with disease progression, there are increased rates of HFOs [92]. More recently, HFOs in the fast ripple band have been observed in three mouse lines of Alzheimer's disease independently of seizures, but their role in precipitating seizures in Alzheimer's models remains to be determined [93].

Mechanistically, ripples are thought to arise from inhibitory synaptic potentials onto the soma of pyramidal neurons [94]. In circumstances where GABAergic inhibition is conserved within cortical networks, ripples arise from patterns of inhibitory postsynaptic currents in pyramidal neurons [95]. It has been suggested that the fast ripples in epileptic hippocampus and cortex may reflect synchronous action potential discharges in pyramidal neurons [96]. Recent work has suggested that fast ripples arise at a network level from several groups of hypersynchronous pyramidal neurons that fire out of phase [97]. Synchronization of these groups of neurons would be required on a rapid timescale, and gap junction coupling provides an ideal nonsynaptic mechanism for this to occur [98].

6.3 Network/Tissue Mechanisms Underlying EEG Epileptiform Patterns

EEG studies brought mechanistic insights into the cellular and network processes that generate epileptiform activity. In this paragraph, we will summarize the main findings achieved by exploiting the techniques and methods described above to characterize the network/cellular/biological events underlying the generation of EEG phenomena in the field of epilepsy research. The number of available studies on the topic is vast (see a compendium in 45) and only a few general aspects will be covered in this paragraph, with specific reference to the use of the above-mentioned techniques and preparations to mechanistically explain the generation of focal seizures.

As mentioned in the introduction, videoEEG monitoring is mandatory for the characterization of the epileptic phenotype in animal models of epilepsy. This procedure derives from the use of videoEEG in a clinical setting. Even though videoEEG monitoring represents a crucial phenotyping step, it is still not extensively utilized when new models of epilepsy are developed and characterized. Epileptiform interictal and ictal discharges and the possible behavioral correlates are recorded with surface or depth electrodes, in particular to detect nonmotor and subtle seizure symptoms that cannot be identified by monitoring the animals exclusively with video recordings.

6.3.1 Interictal Abnormalities

Interictal activity is spontaneously generated by an epileptic tissue. The interictal epileptic activity takes different forms of brief and transient population discharges, the most frequently observed being sharp waves or spikes [99]. These have been hypothesized to be generated by the synchronous discharge of paroxysmal depolarizing potential or paroxysmal depolarization shift (PDS; [100, 101]) from a large group of principal, excitatory cells. PDS was largely studied in acute models of seizures induced by pharmacological or electrical manipulations in vivo and in vitro. In most in vitro preparations, PDSs were induced by tissue treatment with $GABA_A$ receptor antagonists and were blocked by antagonizing glutamatergic transmission. Based on these findings, it was proposed that the PDS and the associated population spike are generated by enhanced synaptic excitation and are due to the blockade of inhibitory synaptic activity [101]. Later studies demonstrated that interictal spikes and PDS also occur in interneurons during the focal interictal state; these interictal spikes were insensitive to glutamate receptor antagonists and were abolished by GABA receptor blockade (see review [9, 102]), suggesting that interictal spikes can be sustained by diverse neurotransmitters such as glutamate and GABA (Fig. 6.4).

A synchronous interictal pattern that is frequently observed in human focal epilepsies and in animal models of focal seizures as well as on in vitro cortical and hippocampal slices is the HFO pattern (see Sect. 6.2.1.3).

6.3.1.1 Focal Ictal Patterns

Human intracranial recordings performed in patients with pharmacoresistant seizures during presurgical intracerebral EEG monitoring demonstrated that focal seizures in the large majority of human focal epilepsies are sustained by low-voltage fast activity (LVFA; [85, 103–106]). In temporal lobe epilepsies, the onset of focal seizures is characterized by the abrupt cessation of the background interictal activity, substituted by low amplitude oscillatory activity in the *beta–gamma* range that precedes the later tonic and bursting recruitment of population spikes. In mesial temporal lobe seizures, ictal LVFA can be preceded by large amplitude *heading spikes*—also defined as pre-ictal spikes—and is usually associated with a large amplitude slow potential that lasts several seconds (Fig. 6.5). Moreover, the LVFA generated at seizure onset shows a gradual decline in frequency content (defined as *chirp)* that has been identified as a signature biomarker of the seizure onset zone in association with the very slow potential; these two biomarkers of ictogenesis were identified during intracerebral recordings performed in patients with different forms and locations of focal pharmacoresistant epilepsies [106, 107]. We have to mention here that the LVFA seizure onset is not the only one recorded in temporal lobe epilepsy; another pattern, defined as hypersynchronous because of the activation of large amplitude synchronous burst discharges at onset, has been described [88, 108]. In the following paragraph, we will illustrate how preclinical neurophysiological techniques and models described in the previous paragraphs proved to be critical to understanding the network mechanisms generated during the typical LVFA-at-onset seizure pattern recorded in the temporal lobe.

To analyze the neurobiological determinants of LVFA seizures, two levels of preclinical studies will be examined: (i) in vivo experiments that reproduce either temporal lobe epilepsies associated with hippocampal sclerosis or acutely induced temporal lobe seizures and (ii) in vitro studies in which LVFA seizures are evoked in limbic structures by acute pharmacological manipulations. Focal epilepsy with typical temporal lobe features can be induced in rodents either by intra-peritoneal or local intrahippocampal injection of excitotoxic agents, such as kainic acid [109] or pilocarpine [110], or by electrical stimulation of limbic structures (see [111] for review); these treatments induce a sustained status epilepticus followed within few days/weeks by hippocampal sclerosis and temporal lobe seizures. Acute temporal lobe seizures can be also generated by intrahippocampal injection of drugs such as 4-aminopyridine or pilocarpine [112–114]. In these acute and chronic models of

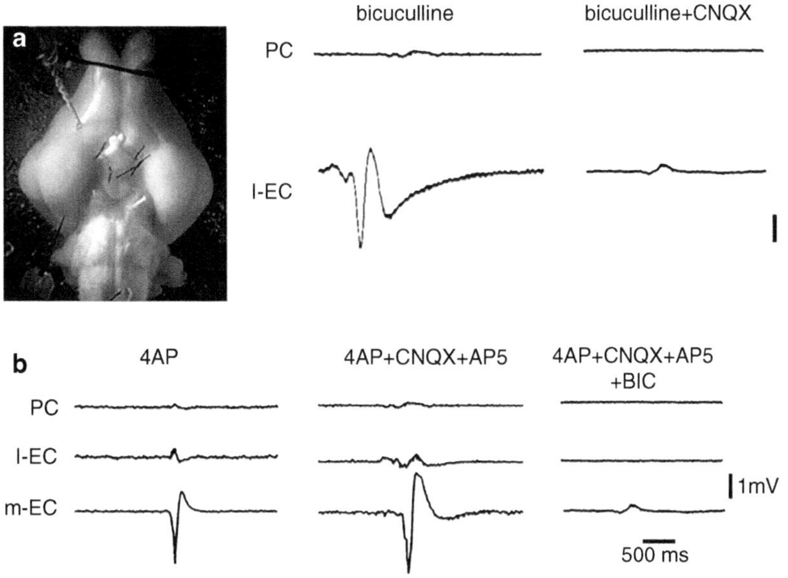

Fig. 6.4 Two types of interictal spikes recorded in the entorhinal cortex of the in vitro isolated guinea pig brain during perfusion with bicuculline (**a**) and 4-aminopyridine (**b**). Perfusion of the isolated brain preparation with 20 μmM bicuculline-induced epileptiform interictal spikes that were abolished by the perfusion of the isolated brain with the glutamate receptor blocker, CNQX 10 μM. Perfusion of 100 μM 4AP induced in the medial entorhinal cortex a population spike with electrographic features similar to the interictal spike illustrated in (**a**); this spike was not abolished by glutamate receptor blockers CNQX (10 μM) and APV (50 μM), but was abolished by 20 μM bicuculline. Bicuculline-induced spikes are sustained by excitatory transmission, whereas 4AP-induced spikes are sustained by inhibitory GABAergic networks

temporal lobe epilepsy and temporal lobe seizures, LFP and single/multiple unit recordings identified LVFA-onset seizures [115]. Moreover, LVFA-onset seizures could be generated during in vitro experiments performed on hippocampal and hippocampal-entorhinal cortex slices of rats and mice and on the in vitro isolated guinea pig brain preparation. Overall, the experiments demonstrated that the heading (preictal) spikes observed ahead of LVFA temporal lobe seizures are characterized by the reduction of principal neuron activity [116] associated with paradoxical bursting activity in GABAergic interneurons ([9, 117–120]; arrowhead in Fig. 6.5a). The onset of the LVA seizure is associated with a slow polarizing potential associated with massive neuronal depolarization and a large increase in extracellular potassium concentration coupled with the hyperactivity of interneuronal networks [117, 121] (Fig. 6.5a and b). Experiments performed in entorhinal cortex mouse slices demonstrated that all interneuron subtypes are active at LVA seizure onset, whereas principal excitatory cells are less active and are entrained in the LVFA seizure after interneurons ([122]; green and black neuronal spiking, respectively, in Fig. 6.5b). The involvement of interneurons and the reduction of principal cell activity during the LVFA were supported also by in vivo experiments on the temporal lobe epilepsy model [123, 124]; the in vivo and in vitro findings and the human studies on the type of neurons that are first and more intensely activated at seizure onset have been recently critically reviewed [115]. This initial phase of LVFA seizures is followed by a stereotyped sequence of events: the increase in extracellular potassium associated with the intense LVFA interneuronal discharge promotes neuronal depolarization and restores the neuronal firing of all neurons in the epileptic focus (Fig. 6.5b). When more neurons are activated, recurrent synaptic and nonsynaptic interactions reinforce and progressively recruit additional neuronal discharges that organize in the generation of population spikes (tonic phase

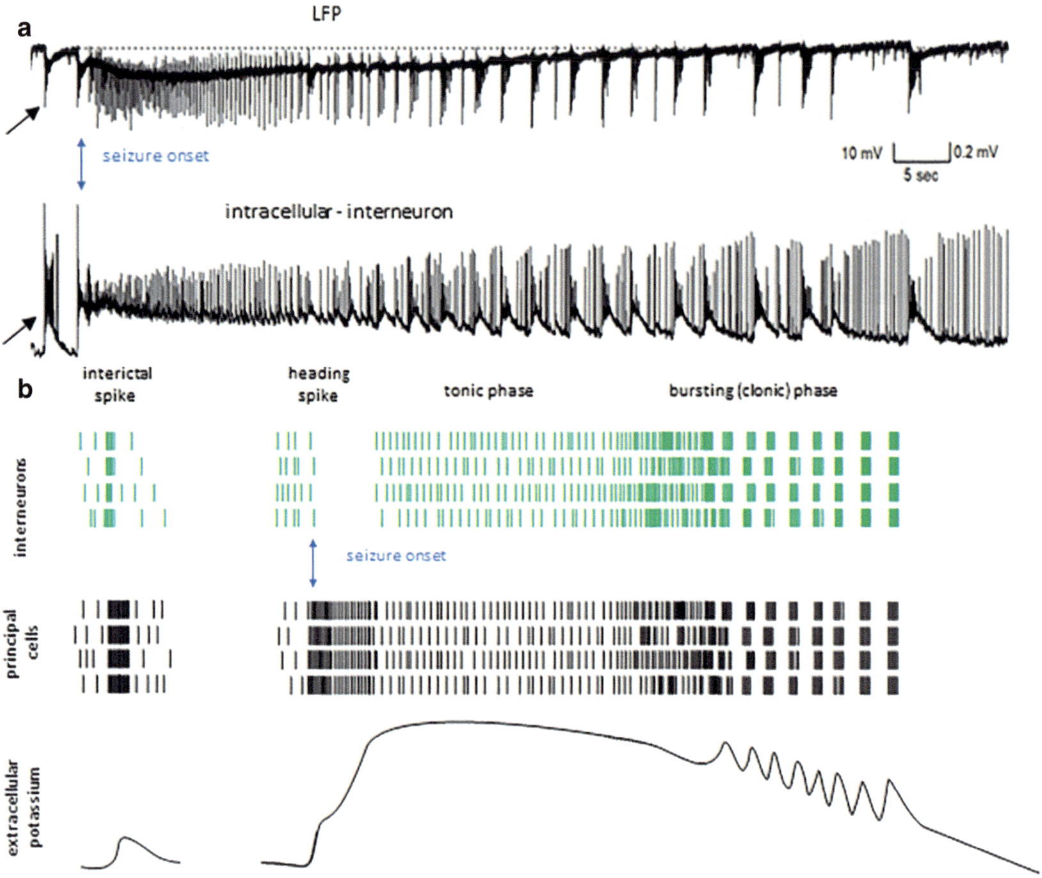

Fig. 6.5 Focal seizure with a low-voltage fast onset. (**a**) Local field potential (upper trace) and simultaneous intracellular recording (lower trace) from a GABAergic interneuron recorded in a hippocampal/entorhinal cortex in vitro mouse slice the entorhinal cortex during a focal seizure induced by 50 μM 4-aminopyridine. The arrowhead marks the preictal heading spike, associated with intense bursting discharge of the GABAergic interneuron. (**b**) Hypothesis on neuronal activity during a focal low-voltage fast seizure. Each line bar represents the firing of a neuron. In green, is the schematic firing of four interneurons and in black are four principal cells. The lower trace illustrated the changes in extracellular potassium associated with the above firing. Firing patterns during an interictal spike are dominated by principal cell firing followed by interneuronal discharge associated with the post-spike slow wave. During the heading spike, interneurons are active, and principal cells are silent. During the increase in extracellular potassium due to the interneuronal firing, all neurons are depolarized and start to fire with an irregular, tonic pattern, that progressively transforms into synchronized bursting that is prominent in all neurons during the final phase of the seizure. (Modified from de Curtis and Avoli [9])

in Fig. 6.5b; [9, 125]). Population discharges induce post-spike inhibition/depression, likely mediated by the activation of repolarizing potassium conductances and by the release of inhibitory neurotransmitters and neuromodulators (such as adenosine, pH changes, etc.; [126, 127]). As seizure activity progresses, neuronal bursting becomes more intense and more synchronous, and (as a consequence) the post-bursting depression larger (bursting phase in Fig. 6.5b). When the amplitude and spatial distribution of post-burst depression are large and synchronous enough, the activity within the epileptogenic area is damped and the seizure discharge is stopped [128, 129].

The preclinical findings that contributed to defining the above schematic hypothesis on LVFA seizure generation were verified in humans by recording single/multiunits with thin microwires during presurgical intracerebral videoEEG monitoring. Several groups confirmed that most units are silenced at the very onset of an LVFA seizure [40] when putative interneurons identified by single-unit spike sorting were active [41, 130]. The hypothesis that the enhanced interneuronal discharge represents the ultimate attempt of inhibitory neurons to stop the progression of a focal seizure has been put forward [131].

Other hypotheses on the initiation and propagation of hypersynchronous non-LVFA focal seizure patterns and LVFA nontemporal seizures have been proposed and the details on the possible underlying mechanisms have been reported [9, 115, 125, 132–134]. These mechanistic assumptions will be verified in humans when investigation technologies utilized in preclinical studies will be developed to the point of being used as diagnostic tools.

6.4 Limitations of Neurophysiology Techniques in Preclinical Research

As for clinical recordings in patients, the identification of artifacts is one of the major limitations of EEG and neurophysiological recordings in experimental research. Artifacts conceal cerebral activity, make EEG signal interpretation and analysis more difficult, and may lead to false results and misinterpretations. Animal movement and muscle activity represent the most common EEG artifacts; behaviors such as scratching, grooming, or behavior-related cable movements generate high-amplitude activity that could mimic rhythmic ictal discharges observed during seizures. Artifact detection can be improved by implementing synchronous video recording with EEG signals to identify nonpathological movement artifacts. Moreover, interictal discharges associated with myoclonic jerks can lead to EMG artifacts that interfere with HFOs that can be superimposed on interictal discharges. Bipolar recording, signal filtering, shielding, and pre-amplification can decrease artifact amplitude.

Nonphysiologic artifacts can be generated by electrodes, cables, connectors, and the various components of the recording device. Also in vitro setups can be a source of electrical artifacts. Electrode artifacts usually are the results of poor contact with the tissue, electrode degradation, damaged connectors, high electrode impedance, bad insulation, insufficient electrode position fixing, reference/ground electrodes, and so on. Most artifacts can be prevented by careful checklist controls and high-quality preamplifiers positioned as close as possible to the recording position. Power supply noise is one of the most common artifacts that can be induced by any of the situations described above. Applying filters can be an effective method to remove the power supply noise from the signal, but it also removes frequencies of cerebral origin. Telemetry recordings can minimize the artifacts emerging from the cable movement but can introduce different types of artifacts from the battery and transmitters.

All neurophysiological events are the manifestation of processes driven by different cell types. This brings us to one of the limitations of EEG because we cannot recognize whether the EEG waveform is due to excitatory or inhibitory transmission and at which part of the neuron it is generated. Using special multisite recording techniques and current source density analysis (see Fig. 6.3), it is possible to estimate which layer of the cortex or which compartment of the neuron generates the synaptic activity. By applying spike sorting techniques during single-unit recordings, action potential firing originating from principal cells and interneurons can be determined. Glial cells also influence EEG signals indirectly through their influence on the principal cell excitability, properties of extracellular space, and extracellular concentration of neurotransmitters, ions, and other metabolites. For these reasons, the simultaneous recording of neurophysiological signals with fast cellular imaging of selective fluorescent indicators specifically expressed in subpopulation of brain cells (neurons, glia, etc.) contributes to identifying the role of different cell types for the generation of a recorded electrical signal.

6.5 Conclusions

Preclinical neurobiological studies are indispensable to interpret and explain the biology of brain EEG activity in normal and pathological conditions. This chapter demonstrates that the complexity of the EEG signals can be reproduced in animal models and preparations that retain the essential features of the human EEG. The EEG can be fragmented into simplified patterns generated by well-defined networks, neuronal patterns, and neurobiological events that can be identified exclusively by using research technologies presently available for preclinical studies. In the future, the evolving preclinical neuroscience tools and devices will possibly be refined and standardized for a potential use in the human brain, to verify directly in patients the hypotheses that have been developed in the preclinical setting. This process will eventually contribute to improving clinical diagnostics, set the foundation for possible future treatments, and ultimately contribute to understanding how the human brain works.

References

1. Berger H. Über das Elektrenkephalogramm des Menschen. Archiv Psychiatrie. 1929;87:527–70.
2. Ormerod W. Richard Caton (1842–1926): pioneer electrophysiologist and cardiologist. J Med Biogr. 2006;14(1):30–5.
3. Adrian ED, Matthews BHC. The berger rhythm: potential changes from the occipital lobes in man. Brain. 1934;57:355–85.
4. Jasper HH, Carmichael L. Electrical potentials from the intact human brain. Science. 1935;81:51–3.
5. Brown RE, Basheer R, McKenna JT, Strecker RE, McCarley RW. Control of sleep and wakefulness. Physiol Rev. 2012;92(3):1087–187.
6. Zielinski MR, McKenna JT, McCarley RW. Functions and mechanisms of sleep. AIMS Neurosci. 2016;3:67–104.
7. Timofeev I, Schoch SF, LeBourgeois MK, Huber R, Riedner BA, Kurth S. Spatio-temporal properties of sleep slow waves and implications for development. Curr Opin Physiol. 2020;15:172–82.
8. Iotchev IB, Kubinyi E. Shared and unique features of mammalian sleep spindles—insights from new and old animal models. Biol Rev. 2021;96:1021–34.
9. de Curtis M, Avoli M. GABAergic networks jump-start focal seizures. Epilepsia. 2016;57:679–87.
10. Hernan A, Schevon C, Worrell G, Galanopoulou A, Kahane P, de Curtis M, Ikeda A, Quilichini P, Williamson, Garcia Cairasco N, Scott R, Timofeev I. Methodological standards and functional correlates of depth in vivo electrophysiological recordings in control rodents: a report of the TASK1-WG3 group of the AES/ILAE Translational Task Force of the ILAE. Epilepsia. 2017;58(S4):28–39.
11. Kadam SD, D'Ambrosio R, Duveau V, Roucard C, Garcia Cairasco N, Ikeda A, de Curtis M, Galanopoulou AS, Kelly KM. Methodological standards and interpretation of video-EEG studies in adult control rodents: a report of the TASK1-WG1 group of the AES/ILAE Translational Task Force of the ILAE. Epilepsia. 2017;58(S4):10–27.
12. Moyer JT, Gnatkovsky V, Ono T, Otáhal J, Wagenaar J, Stacey W, Noebels J, Ikeda A, Staley K, de Curtis M, Litt B, Galanopoulou AS. Standards for data acquisition and software-based analysis of in vivo electrophysiological brain recordings. A report of the TASK1-WG5 group of the AES/ILAE Translational Task Force of the ILAE. Epilepsia. 2017;58(S4):53–67.
13. Akman O, Raol YH, Auvin S, Cortez MA, Kubova H, de Curtis M, Ikeda A, Dudek E, Galanopoulou AS. Methodological recommendations and possible interpretations of video-EEG recordings in immature rodents used as experimental controls. A TASK1-WG2 report of the ILAE/AES Joint Translational Task Force. Epilepsia Open. 2018;3(4):437–59.
14. Nicolelis MAL, Ribeiro S. Multielectrode recordings: the next steps. Curr Opin Neurobiol. 2002;5:602–6.
15. Buzsáki G. Large-scale recording of neuronal ensembles. Nat Neurosci. 2004;7:446–51.
16. Vazquez-Guardado A, Yang Y, Bandodkar AJ, Rogers JA. Recent advances in neurotechnologies with broad potential for neuroscience research. Nat Neurosci. 2020;23:1522–36.
17. Raimondo J, Heinemann U, de Curtis M, Goodkin H, Dulla C, Janigro D, Ikeda A, Lin CC, Premysl J, Galanopoulou A, Bernard C. Methodological standards for in vitro models of epilepsy and epileptic seizures. a report of the TASK1-WG4 group of the AES/ILAE Translational Task Force of the ILAE. Epilepsia. 2017;58(S4):40–52.
18. Dulla C, Janigro D, Jiruska P, Raimondo J, Ikeda A, Lin CC, Goodkin H, Galanopoulou A, Bernard C, de Curtis M. How do we use in vitro models to understand epileptiform and ictal activity? A report of the TASK1-WG4 group of the ILAE/AES Joint Translational Task Force. Epilepsia Open. 2018;3(4):460–73.
19. Lee A, Epsztein J, Brecht M. Head-anchored whole-cell recordings in freely moving rats. Nat Protoc. 2009;4:385–92.
20. Harvey CD, Collman F, Dombeck DA, Tank DW. Intracellular dynamics of hippocampal

place cells during virtual navigation. Nature. 2009;461(7266):941–6.
21. Niell CM, Stryker MP. Modulation of visual responses by behavioral state in mouse visual cortex. Neuron. 2010;65(4):472–9.
22. Malvaut S, Constantinescu VS, Dehez H, Doric S, Saghatelyan A. Deciphering brain function by miniaturized fluorescence microscopy in freely behaving animals. Front Neurosci. 2020;14:819.
23. Musk E, Neuralink. An integrated brain-machine interface platform with thousands of channels. J Med Internet Res. 2019;21(10):e16194.
24. Erofeev A, Antifeev I, Vinokurov E, Bezprozvanny I, Vlasova O. An open-source wireless electrophysiology system for in vivo neuronal activity recording in the rodent brain: 2.0. Sensors. 2023;23(24):9735.
25. Sahasrabuddhe K, Khan AA, Singh AP, Stern TM, Ng Y, Tadić A, Orel P, LaReau C, Pouzzner D, Nishimura K, Boergens KM, Shivakumar S, Hopper MS, Kerr B, Hanna MS, Edgington RJ, McNamara I, Fell D, Gao P, Babaie-Fishani A, Veijalainen S, Klekachev AV, Stuckey AM, Luyssaert B, Kozai TDY, Xie C, Gilja V, Dierickx B, Kong Y, Straka M, Sohal HS, Angle MR. The Argo: a high channel count recording system for neural recording in vivo. J Neural Eng. 2021;18(1):015002.
26. Pitkänen A, Immonen RJ, Gröhn OH, Kharatishvili I. From traumatic brain injury to posttraumatic epilepsy: what animal models tell us about the process and treatment options. Epilepsia. 2009;50(S2):21–9.
27. Quigg M, Clayburn H, Straume M, Menaker M, Bertram EH 3rd. Effects of circadian regulation and rest-activity state on spontaneous seizures in a rat model of limbic epilepsy. Epilepsia. 2000;41(5):502–9.
28. Pitsch J, Becker AJ, Schoch S, Müller JA, de Curtis M, Gnatkovsky V. Circadian clustering of spontaneous epileptic seizures emerges after pilocarpine-induced status epilepticus. Epilepsia. 2017;58(7):1159–71.
29. Baud MO, Ghestem A, Benoliel JJ, Becker C, Bernard C. Endogenous multidien rhythm of epilepsy in rats. Exp Neurol. 2019;315:82–7.
30. Scharfman HE, Galanopoulou AS, French JA, Pitkänen A, Whittemore V, Harte-Hargrove LC. Preclinical common data elements (CDEs) for epilepsy: a joint ILAE/AES and NINDS translational initiative. Epilepsia Open. 2018;3(S1):9–12.
31. Mitzdorf U. Current source-density method and application in cat cerebral cortex: investigation of evoked potentials and EEG phenomena. Physiol Rev. 1985;65(1):37–100.
32. Kovac S, Speckmann EJ, Gorji A. Uncensored EEG: The role of DC potentials in neurobiology of the brain. Prog Neurobiol. 2018;165-167:51–65.
33. Aladjalova NA. Infra-slow rhythmic oscillations of the steady potential of the cerebral cortex. Nature. 1957;179(4567):957–9.
34. Vanhatalo S, Tallgren P, Becker C, Holmes MD, Miller JW, Kaila K, Voipio J. Scalp-recorded slow EEG responses generated in response to hemodynamic changes in the human brain. Clin Neurophysiol. 2003;114(9):1744–54.
35. Vanhatalo S, Palva JM, Holmes MD, Miller JW, Voipio J, Kaila K. Infraslow oscillations modulate excitability and interictal epileptic activity in the human cortex during sleep. Proc Natl Acad Sci USA. 2004;101(14):5053–7.
36. Voipio J, Tallgren P, Heinonen E, Vanhatalo S, Kaila K. Millivolt-scale DC shifts in the human scalp EEG: evidence for a nonneuronal generator. J Neurophysiol. 2003;89:2208–14.
37. Harris KD, Henze DA, Csicsvari J, Hirase H, Buzsáki G. Accuracy of tetrode spike separation as determined by simultaneous intracellular and extracellular measurements. J Neurophysiol. 2000;84:401–14.
38. Humphrey DR, Schmidt EM. Extracellular single-unit recording methods. In: Boulton AA, Baker GB, Vanderwolf CH, editors. Neurophysiological techniques. Neuromethods, vol. 15. Humana Press; 1990.
39. Barthó P, Hirase H, Monconduit L, Zugaro M, Harris KD, Buzsáki G. Characterization of neocortical principal cells and interneurons by network interactions and extracellular features. J Neurophysiol. 2004;92(1):600–8.
40. Truccolo W, Donoghue JA, Hochberg LR, Eskandar EN, Madsen JR, Anderson WS, Brown EN, Halgren E, Cash SS. Single-neuron dynamics in human focal epilepsy. Nat Neurosci. 2011;14:635–41.
41. Elahian B, Lado NE, Mankin E, Vangala S, Misra A, Moxon K, Fried I, Sharan A, Yeasin M, Staba R, Bragin A, Avoli M, Sperling MR, Engel J, Weiss SA. Low-voltage fast seizures in humans begin with increased interneuron firing. Ann Neurol. 2018;84:588–600.
42. Cash SS, Hochberg LR. The emergence of single neurons in clinical neurology. Neuron. 2015;86(1):79–91.
43. Quiroga QR, Kraskov A, Koch C, Fried I. Explicit encoding of multimodal percepts by single neurons in the human brain. Curr Biol. 2009;19(15):1308–13.
44. Fried I. Neurons as will and representation. Nat Rev Neurosci. 2022;23(2):104–14.
45. Pitkänen A, Buckmaster PS, Galanopoulou AS, Moshé SL. Models of seizures and epilepsy. 2nd ed. Elsevier; 2017.
46. Ballany K. Isolated central nervous system circuits. New York: Humana Press, Springer; 2012.
47. Kohling R, Vreugdenhil M, Bracci E, Jefferys JRG. Ictal epileptiform activity is facilitated by hippocampal GABAA receptor-mediated oscillations. J Neurosci. 2000;20:6820–9.
48. de la Prida LM, Huberfeld G. Inhibition and oscillations in the human brain tissue in vitro. Neurobiol Dis. 2019;125:198–210.
49. Khalilov I, Esclapez, Esclapez M, Medina I, Aggoun, Lamsa K, Leinekugel X, Khazipov R, Ben-Ari Y. A novel in vitro preparation: neurotechnique the intact hippocampal formation. Neuron. 1997;19(4):743–9.

50. Muhlethaler M, de Curtis M, Walton K, Llinas RR. The isolated and perfused brain of the Guinea pig in vitro. Eur J Neurosci. 1993;5:915–26.
51. de Curtis M, Librizzi L, Uva L. The in vitro isolated whole Guinea pig brain as a model to study epileptiform activity patterns. J Neurosci Meth. 2016;260:83–90.
52. de Curtis M, Uva L, Gnatkovsky V, Librizzi L. Potassium dynamics and seizures: why is potassium ictogenic? Epilepsy Res. 2018;143:50–9.
53. Raimondo JV, Burman RJ, Katz AA, Akerman CJ. Ion dynamics during seizures. Front Cell Neurosci. 2015;9:419.
54. Spira ME, Hai A. Multi-electrode array technologies for neuroscience and cardiology. Nat Nanotechnol. 2013;8(2):83–94.
55. Obien MEJ, Hierlemann A, Frey U. Accurate signal-source localization in brain slices by means of high-density microelectrode arrays. Sci Rep. 2019;9(1):788.
56. Abdelfattah AS, Zheng J, Singh A, Huang YC, Reep D, Tsegaye G, et al. Sensitivity optimization of a rhodopsin-based fluorescent voltage indicator. Neuron. 2023;111:1547–63.
57. Marvin JS, Shimoda Y, Magloire V, Leite M, Kawashima T, Jensen TP, et al. A genetically encoded fluorescent sensor for in vivo imaging of GABA. Nat Methods. 2019;16:763–70.
58. Zhang Y, Looger LL. Fast and sensitive GCaMP calcium indicators for neuronal imaging. J Physiol. 2024;602:1595–604.
59. Shimoda Y, Leite M, Graham RT, Marvin JS, Hasseman J, Kolb I. Extracellular glutamate and GABA transients at the transition from interictal spiking to seizures. Brain. 2024;147:1011–24.
60. Deisseroth K. Optogenetics. Nat Methods. 2011;8:26–9.
61. Walker MC, Kullmann DM. Optogenetic and chemogenetic therapies for epilepsy. Neuropharmacology. 2020;168:107751.
62. Roth BL. DREADDs for neuroscientists. Neuron. 2016;89:683–94.
63. Niedermeyer E. The normal EEG of the waking adult. Electroencephalography. Baltimore: Lippincott Williams Wilkins; 1999.
64. Karakaş S, Başar E. Preface: Quiet revolutions in neuroscience. Int J Psychophysiol. 2006;60(2):98–100.
65. Buzsaki G. Rhythms of the brain. New York: Oxford University Press; 2006.
66. Alonso A, Garcia-Austt E. Neuronal sources of theta rhythm in the entorhinal cortex of the rat. Exp Brain Res. 1987;67:493–501.
67. Buzsaki G. Theta oscillations in the hippocampus. Neuron. 2002;33(3):325–40.
68. Jones MW, Wilson MA. Theta rhythms coordinate hippocampal-prefrontal interactions in a spatial memory task. PLoS Biol. 2005;3(12):e402.
69. Losonczy A, Zemelman BV, Vaziri A, Magee JC. Network mechanisms of theta related neuronal activity in hippocampal CA1 pyramidal neurons. Nat Neurosci. 2010;13(8):967–72.
70. Mehak SF, Shivakumar AB, Kumari S, Muralidharan B, Gangadharan G. Theta and gamma oscillatory dynamics in mouse models of Alzheimer's disease: a path to prospective therapeutic intervention. Neurosci Biobehav Rev. 2022;136:104628.
71. O'Keefe J. Place units in the hippocampus of the freely moving rat. Exp Neurol. 1976;51:78–109.
72. Colgin LL. Five decades of hippocampal place cells and EEG rhythms in behaving rats. J Neurosci. 2020;40(1):54–60.
73. Hasselmo ME, Stern CE. Theta rhythm and the encoding and retrieval of space and time. NeuroImage. 2014;85 Pt 2(0 2):656–66.
74. Hafting T, Fyhn M, Molden S, Moser MB, Moser EI. Microstructure of a spatial map in the entorhinal cortex. Nature. 2005;436:801–6.
75. Lisman JE, Jensen O. The theta-gamma neural code. Neuron. 2013;77:1002–16.
76. Gray C, König P, Engel A, et al. Oscillatory responses in cat visual cortex exhibit inter-columnar synchronization which reflects global stimulus properties. Nature. 1989;338:334–7.
77. Whittington MA, Traub RD, Jefferys JG. Synchronized oscillations in interneuron networks driven by metabotropic glutamate receptor activation. Nature. 1995;373(6515):612–5.
78. Bartos M, Vida I, Jonas P. Synaptic mechanisms of synchronized gamma oscillations in inhibitory interneuron networks. Nat Rev Neurosci. 2007;8(1):45–56.
79. Mann EO, Suckling JM, Hajos N, Greenfield SA, Paulsen O. Perisomatic feedback inhibition underlies cholinergically induced fast network oscillations in the rat hippocampus in vitro. Neuron. 2005;45(1):105–17.
80. Buzsaki G, Wang XJ. Mechanism of gamma oscillation. Ann Rev Neurosci. 2012;35:203–25.
81. Fisahn A, Pike FG, Buhl EH, Paulsen O. Cholinergic induction of network oscillations at 40 Hz in the hippocampus in vitro. Nature. 1998;394(6689):186–9.
82. Dickson CT, Biella G, de Curtis M. Evidence for spatial modules mediated by temporal synchronization of carbachol-induced gamma rhythm in medial entorhinal cortex. J Neurosci. 2000;20(20):7846–54.
83. Buhl DL, Harris KD, Hormuzdi SG, Monyer H, Buzsáki G. Selective impairment of hippocampal gamma oscillations in connexin-36 knock-out mouse in vivo. J Neurosci. 2003;23(3):1013–8.
84. Avoli M, de Curtis M, Gnatkovsky V, Gotman J, Köhling R, Lévesque M, Manseau F, Shiri Z, Williams S. Specific imbalance of excitatory/inhibitory signaling establishes seizure onset pattern in temporal lobe epilepsy. J Neurophysiol. 2016;115:3229–37.
85. de Curtis M, Gnatkovsky V. Reevaluating the mechanisms of focal ictogenesis: the role of low-voltage fast activity. Epilepsia. 2009;50:2514–25.

86. Zijlmans M, Jiruska P, Zelmann R, Leijten FS, Jefferys JG, Gotman J. High-frequency oscillations as a new biomarker in epilepsy. Ann Neurol. 2012;71(2):169–78.
87. Bragin A, Engel J Jr, Wilson CL, Fried I, Buzsaki G. High-frequency oscillations in human brain. Hippocampus. 1999;9:137–42.
88. Bragin A, Engel J Jr, Wilson CL, Vizentin E, Mathern GW. Electrophysiologic analysis of a chronic seizure model after unilateral hippocampal KA injection. Epilepsia. 1999;40:1210–21.
89. Worrell GA, Gardner AB, Stead SM, Hu S, Goerss S, Cascino GJ, et al. High-frequency oscillations in human temporal lobe: simultaneous microwire and clinical macroelectrode recordings. Brain. 2008;131:928–37.
90. Bragin A, Wilson CL, Engel J Jr. Chronic epileptogenesis requires development of a network of pathologically interconnected neuron clusters: a hypothesis. Epilepsia. 2000;41 Suppl 6:S144–52.
91. Lévesque M, Avoli M. High-frequency oscillations and focal seizures in epileptic rodents. Neurobiol Dis. 2019;124:396–407.
92. Zijlmans M, Jacobs J, Zelmann R, Dubeau F, Gotman J. High frequency oscillations and seizure frequency in patients with focal epilepsy. Epilepsy Res. 2009;85(2–3):287–92.
93. Lisgaras CP, Scharfman HE. High-frequency oscillations (250–500 Hz) in animal models of Alzheimer's disease and two animal models of epilepsy. Epilepsia. 2023;64(1):231–46.
94. Ylinen A, Bragin A, Nádasdy Z, Jandó G, Szabó I, Sik A, Buzsáki G. Sharp wave-associated high-frequency oscillation (200 Hz) in the intact hippocampus: network and intracellular mechanisms. J Neurosci. 1995;15(1 Pt 1):30–46.
95. Menendez de la Prida L, Gal B. Synaptic contributions to focal and widespread spatiotemporal dynamics in the isolated rat subiculum in vitro. J Neurosci. 2004;24(24):5525–36.
96. Ibarz JM, Foffani G, Cid E, Inostroza M, de la Prida LM. Emergent dynamics of fast ripples in the epileptic hippocampus. J Neurosci. 2010;30:16249–61.
97. Foffani G, Uzcategui YG, Gal B, Menendez de la Prida L. Reduced spike-timing reliability correlates with the emergence of fast ripples in the rat epileptic hippocampus. Neuron. 2007;55(6):930–41.
98. Draguhn A, Traub RD, Schmitz D, Jefferys JG. Electrical coupling underlies high-frequency oscillations in the hippocampus in vitro. Nature. 1998;394(6689):189–92.
99. de Curtis M. Avanzini G Interictal spikes in focal epilepsies. Prog Neurobiol. 2001;63(5):541–67.
100. Matsumoto H, Marsan CA. Cortical cellular phenomena in experimental epilepsy: interictal manifestations. Exp Neurol. 1964;9:286–304.
101. Prince DA, Avoli M. The paroxysmal depolarizing shift (PDS): the first cellular marker of focal epileptogenesis. In: Noebels JL, Avoli M, Rogawski MA, Vezzani A, Delgado-Escueta AV, editors. Jasper's basic mechanisms of the epilepsies. 5th ed. Oxford University Press; 2024.
102. Avoli M, de Curtis M. GABAergic synchronization in the limbic system and its role in the generation of epileptiform activity. Progr Neurobiol. 2011;95:104–32.
103. Bartolomei F, Chauvel P, Wendling F. Epileptogenicity of brain structures in human temporal lobe epilepsy: a quantified study from intracerebral EEG. Brain. 2008;131:1818–30.
104. Perucca P, Dubeau F, Gotman J. Intracranial electroencephalographic seizure-onset patterns: effect of underlying pathology. Brain. 2014;137:183–96.
105. Singh S, Sandy S, Wiebe S. Ictal onset on intracranial EEG: do we know it when we see it? State of the evidence. Epilepsia. 2015;56:1629–38.
106. Gnatkovsky V, Pelliccia V, de Curtis M, Tassi L. Two main focal seizure patterns revealed by intracerebral electroencephalographic biomarker analysis. Epilepsia. 2019;60:96–106.
107. Di Giacomo R, Burini A, Chiarello D, Pelliccia V, Deleo F, Garbelli R, de Curtis M, Tassi L, Gnatkovsky V. Ictal fast activity chirps as markers of the epileptogenic zone. Epilepsia. 2024;65(6):e97–e103.
108. Ogren JA, Bragin A, Wilson CL, et al. Three-dimensional hippocampal atrophy maps distinguish two common temporal lobe seizure-onset patterns. Epilepsia. 2009;50:1361–70.
109. Lévesque M, Avoli M. The kainic acid model of temporal lobe epilepsy. Neurosci Biobehav Rev. 2013;37:2887–99.
110. Curia G, Longo D, Biagini G, Jones RSG, Avoli M. The pilocarpine model of temporal lobe epilepsy. J Neurosci Methods. 2008;172:143–57.
111. de Curtis M, Rossetti AO, Verde DV, van Vliet EA, Ekdahl CT. Brain pathology in focal status epilepticus: evidence from experimental models. Neurosci Biobehav Rev. 2021;131:834–46.
112. Lévesque M, Salami P, Gotman J, Avoli M. Two seizure-onset types reveal specific patterns of high-frequency oscillations in a model of temporal lobe epilepsy. J Neurosci. 2012;32:13264–72.
113. Lévesque M, Salami P, Shiri Z, Avoli M. Interictal oscillations and focal epileptic disorders. Eur J Neurosci. 2018;48:2915–27.
114. Shiri Z, Manseau F, Lévesque M, Williams S, Avoli M. Interneuron activity leads to initiation of low-voltage fast-onset seizures. Ann Neurol. 2015;77:541–6.
115. Wenzel M, Huberfeld G, Grayden DB, de Curtis M, Trevelyan AJ. A debate on the neuronal origin of focal seizures. Epilepsia. 2023;64 Suppl 3:S37–48.
116. Gnatkovsky V, Librizzi L, Trombin F, de Curtis M. Fast activity at seizure onset is mediated by inhibitory circuits in the entorhinal cortex in vitro. Ann Neurol. 2008;64:674–86.
117. Avoli M, Barbarosie M, Lücke A, Nagao T, Lopantsev V, Köhling R. Synchronous GABA-mediated potentials and epileptiform discharges in the rat limbic system in vitro. J Neurosci. 1996;16:3912–24.

118. Lopantsev V, Avoli M. Participation of GABAA-mediated inhibition in ictallike discharges in the rat entorhinal cortex. J Neurophysiol. 1998;79:352–60.
119. Uva L, Breschi GL, Gnatkovsky V, Taverna S, de Curtis M. Synchronous inhibitory potentials precede seizure-like events in acute models of focal limbic seizures. J Neurosci. 2015;35:3048–55.
120. Dossi E, Huberfeld G. GABAergic circuits drive focal seizures. Neurobiol Di. 2023;s 180:106102.
121. Librizzi L, Losi G, Marcon I, Sessolo M, Scalmani P, Carmignoto G, de Curtis M. Interneuronal network activity at the onset of seizure-like events in entorhinal cortex slices. J Neurosci. 2017;37(43):10398–407.
122. Scalmani P, Paterra R, Mantegazza M, Avoli M, de Curtis M. Involvement of GABAergic interneuron subtypes in 4-aminopyridine-induced seizure-like events in mouse entorhinal cortex in vitro. J Neurosci. 2023:RM-1190-22.
123. Grasse DW, Karunakaran S, Moxon KA. Neuronal synchrony and the transition to spontaneous seizures. Exp Neurol. 2013;248:72–84.
124. Toyoda I, Fujita S, Thamattoor AK, Buckmaster PS. Unitactivityofhip- pocampal interneurons before spontaneous seizures in an animal model of temporal lobe epilepsy. J Neurosci. 2015;35:6600–18.
125. Devinsky O, Vezzani A, O'Brien TJ, Jette N, Scheffer IE, de Curtis M, Perucca P. Epilepsy. Nat Rev Dis Primers. 2018;4:1–24.
126. Uva L, de Curtis M. Activity- and pH-dependent adenosine shifts at the end of a focal seizure in the entorhinal cortex. Epil Res. 2020;165:106401.
127. Lado FA, Moshérai SL. How do seizures stop? Epilepsia. 2008;49(10):1651–64.
128. Boido D, Gnatkovsky V, Uva L, Francione S, de Curtis M. Simultaneous enhancement of excitation and postburst inhibition at the end of focal seizures. Ann Neurol. 2014;76(6):826–36.
129. Aracri P, de Curtis M, Forcaia G, Uva L. Enhanced thalamo-hippocampal synchronization during focal limbic seizures. Epilepsia. 2018;59(9):1774–84.
130. Weiss SA, Staba R, Bragin A, Moxon K, Sperling M, Avoli M, Engel J Jr. Interneurons and principal cell firing in human limbic areas at focal seizure onset. Neurobiol Dis. 2019;124:183–8.
131. Schevon CA, Weiss SA, McKhann G Jr, Goodman RR, Yuste R, Emerson RG, Trevelyan AJ. Evidence of an inhibitory restraint of seizure activity in humans. Nat Commun. 2012;3:1060.
132. Avoli M. Inhibition, oscillations and focal seizures: an overview inspired by some historical notes. Neurobiol Dis. 2019;130:104478.
133. Uva L, Saccucci S, Chikhladze M, Tassi L, Gnatkovsky V, Milesi G, Morbin M, de Curtis M. A novel focal seizure pattern generated in superficial layers of the olfactory cortex. J Neurosci. 2017;37(13):3544–54.
134. de Curtis M, Garbelli R, Uva L. A hypothesis for the role of axon demyelination in seizure generation. Epilepsia. 2021;62(3):583–95.

7 EEG Interictal Spikes, Ictal Patterns, and Accepted Terminology

Stefano Meletti, Leandra Giunta, and Matteo Pugnaghi

7.1 Introduction

The electroencephalogram (EEG) is still the most useful diagnostic test when a seizure disorder is suspected, and often it is the only tool by which a diagnosis of epilepsy can be established; moreover, in some instances, EEG findings can aid in prognosis and determining the need for treatment.

This chapter deals with ictal EEG recorded by extracranial electrodes (scalp EEG), discussing ictal EEG patterns in generalized and focal seizures.

The following topics will be addressed. First, the definition of "spikes," that is, the graphoelements that have an epileptogenic meaning in the scalp EEG. If present in the interictal EEG, they represent the first electrographic biomarkers of a possible epilepsy. Although interictal EEG may often be sufficient to establish whether the subject is suffering from an epileptic disorder, or whether it is a focal or generalized epilepsy, in several other instances, ictal EEG recording may be essential. We will therefore describe in general terms the characteristics of ictal EEG patterns in focal and generalized seizures. The most characteristic ictal EEG patterns of certain types of focal seizures will be analyzed according to localization. Finally, we will briefly evaluate the importance of polygraphic recordings as an added value to EEG recordings through some examples.

We will not deal in detail with the usefulness of polygraphic recordings, which is the subject of a separate chapter. We will not deal here with ictal EEG patterns in status epilepticus for which we refer the reader to the specific chapter on EEG emergencies.

7.2 A Historical Perspective on Ictal EEG Patterns

The first documentation of an EEG recording of a focal epileptic seizure dates back to the 1930s, when the German psychiatrist Hans Berger reported the phenomenon. However, the first clear recording of epileptiform discharges (indicative of seizure activity) came from Gibbs, Lennox, and Davis in 1935–1936. This landmark paper presented detailed EEG patterns associated

S. Meletti (✉)
Department of Biomedical Metabolic Sciences and Neurosciences, University of Modena and Reggio Emilia, Modena, Italy

Neuroscience Department, Neurophysiology Unit and Epilepsy Centre, Azienda Ospedaliera-Universitaria di Modena, Modena, Italy
e-mail: stefano.meletti@unimore.it

L. Giunta · M. Pugnaghi
Neuroscience Department, Neurophysiology Unit and Epilepsy Centre, Azienda Ospedaliera-Universitaria di Modena, Modena, Italy
e-mail: giunta.leandra@aou.mo.it; pugnaghi.matteo@aou.mo.it

F. Brigo, O. Mecarelli (eds.), *EEG: The First 100 Years*,
https://doi.org/10.1007/978-3-031-86878-8_7

with epileptic seizures [1]. They described characteristic spike-and-wave discharges linked to epilepsy and showed how the EEG could be used to localize seizure activity in the brain. This was a monumental step in understanding and diagnosing epilepsy through EEG technology. Their work laid the foundation for the clinical use of EEG in epilepsy, making it a critical diagnostic tool in neurology.

Subsequently, the first study to report the EEG pattern associated with temporal lobe seizures is credited to William G. Lennox and F. A. Gibbs, published in the early 1940s. This seminal work identified specific localized EEG abnormalities linked to temporal lobe epilepsy (TLE), often characterized by rhythmic activity or sharp waves over the temporal region [2]. In this study, Gibbs and Lennox described the localization of epileptiform discharges over the temporal lobes as a critical advance in the recognition of TLE as a distinct clinical entity. This was a major step forward in understanding how different types of epilepsy could be linked to specific regions of the brain, influencing future diagnosis and treatment.

Their work contributed significantly to the development of the surgical treatment of TLE , which gained traction in the following decades starting from the seminal work of W. Panfield and H. Jasper [3], who described EEG patterns specific to TLE, particularly identifying interictal spikes and seizures originating from the temporal lobe. This was crucial in localizing the source of focal seizures in patients with epilepsy. Their research contributed significantly to the understanding of the relationship between abnormal electrical discharges in the brain and the clinical manifestations of TLE.

The same authors also reported the first studies on EEG patterns associated with frontal lobe seizures in the late 1940s and early 1950s. Their work on mapping the brain during epilepsy surgery led to the identification of specific frontal lobe seizure patterns [3]. Their book is one of the most comprehensive early sources on the functional anatomy of the brain, based on their groundbreaking studies of patients undergoing brain surgery for epilepsy. They recorded the electrical activity of different regions of the brain, including the frontal lobes, and were able to correlate specific EEG patterns with seizure activity originating in this region. Penfield and Jasper's work provided crucial insights into frontal lobe epilepsy, which can present with fast, rhythmic discharges and localized spikes in the frontal regions of the EEG.

These early studies laid the groundwork for future EEG classification and understanding of focal seizure patterns, contributing to the evolving knowledge of epilepsy as a disorder with distinct localizations and EEG signatures. Subsequently, fundamental work in the classification of epileptic seizures was done by the Marseilles School in the 1960s–1970s under the leadership of Henri Gastaut, bringing epileptology into modern times [4, 5].

7.3 Methods of Scalp EEG Recordings

Noninvasive ictal EEG is recorded by scalp electrodes, usually placed according to the 10–20 International system; other scalp electrode arrangements with reduced interelectrode distances (i.e., 10–10 system) may be used with the purpose of improving spatial localization. Actually, the International Federation of Clinical Neurophysiology (IFCN) suggests an extended montage with 25 leads covering the inferior temporal region [6].

Electrodes should be firmly applied to the scalp, possibly with collodion, to avoid their misplacement or detachment during seizures.

Standard EEG recording parameters, that is, low-filter settings of 1.0–1.6 Hz and high-frequency filters of 70 Hz, are usually employed. The digitization rate should be at least 200 Hz. Digital recording allows montage reformatting that can be useful for offline inspection of ictal discharges with different montages to obtain lateralizing or localizing information.

When seizures are extremely frequent, they can be recorded in routine EEG laboratories. Alternatively, ambulatory EEG may be also employed; however, limitations such as the absence of video data and the occurrence of technical artifacts masquerading EEG activity can significantly reduce its utility. The optimal setting for seizure recording requires an inpatient unit, where simultaneous video and EEG monitoring can be performed: indeed, long-term video–EEG monitoring for seizure recording is still the crucial test in the great majority of drug-resistant focal epilepsy evaluated for epilepsy surgery. In this setting, anti-seizure medication (ASM) tapering is often performed to facilitate seizure occurrence and reduce the duration of the admission in the inpatient monitoring unit; in this case, care has to be taken to avoid secondary generalization or even status epilepticus.

7.4 The Definition of Spike and Interictal EEG Abnormalities

EEG is the most commonly used investigative method in patients with epilepsy. Providing high-quality diagnostic testing and care for patients with suspected or established epilepsy requires accurate and reliable EEG interpretation.

Interictal epileptiform discharges (IEDs) confirm the diagnosis [7, 8], predict the recurrence of seizures [9], and help classify epilepsy [10]. Therefore, correctly classifying sharp transients on EEG as epileptiform or non-epileptiform carries high weight in the clinical decision-making process in many scenarios [11]. Both over- and under-calling IEDs can lead to epilepsy misdiagnosis and resultant harm to patients and healthcare systems [12, 13]. Notably, EEG misinterpretation is an issue not only when EEGs are read by neurologists without a specific EEG training—a common practice in the United States as well as in Italy—but also when studies are read by experts. The latter have been shown to have imperfect interrater reliability in identifying IEDs on routine EEG [14, 15].

Routine interictal EEG recording is one of the most abused investigations in clinical practice: Over-reading of EEG is the most frequent cause of epilepsy misdiagnosis. Approximately 30% of patients seen at epilepsy centers for drug-resistant seizures do not have epilepsy. Epilepsy misdiagnosis is detrimental to the patients, as it unnecessarily exposes them to ASMs, prevents them from driving, and limits their career choices. Therefore, "conservative" reading of the EEG has been recommended to achieve high specificity and avoid over-diagnosing epilepsy [12, 16].

Sharp transients are often misinterpreted as IEDs when presenting with sharp morphology and higher amplitude on EEG during early drowsiness. Several attempts have been made to define the characteristic morphology of IEDs [17–19]. In an effort to improve reliability in identifying IEDs, the IFCN, in 2017, proposed a new operational definition of IEDs consisting of six criteria [20] describing the morphological features of IEDs (Fig. 7.1):

1. Di- or tri-phasic waves with sharp or spiky morphology (i.e., pointed peak);
2. Different wave duration relative to ongoing background activity, either shorter or longer;
3. Asymmetry of the waveform—a sharply rising ascending phase and a more slowly decaying descending phase, or vice versa;
4. The transient is followed by an associated slow after-wave;
5. The background activity surrounding epileptiform discharges is disrupted by the presence of the epileptiform discharges;
6. Distribution of the negative and positive potentials on the scalp suggests a source of the signal in the brain, corresponding to a radial, oblique, or tangential orientation of the source [21].

By contrast with the older approach to IED identification consisting of a single step of recognition based on experience [22], the new IFCN criteria break down the evaluation of a candidate IED into a series of six binary decisions about the presence or absence of elementary features and recommends that a waveform be considered an IED if four or more of the six features are present. The IFCN operational criteria yielded high specificity (over 95%) and reasonable sensitivity (over 80%) when at least five of the IFCN criteria

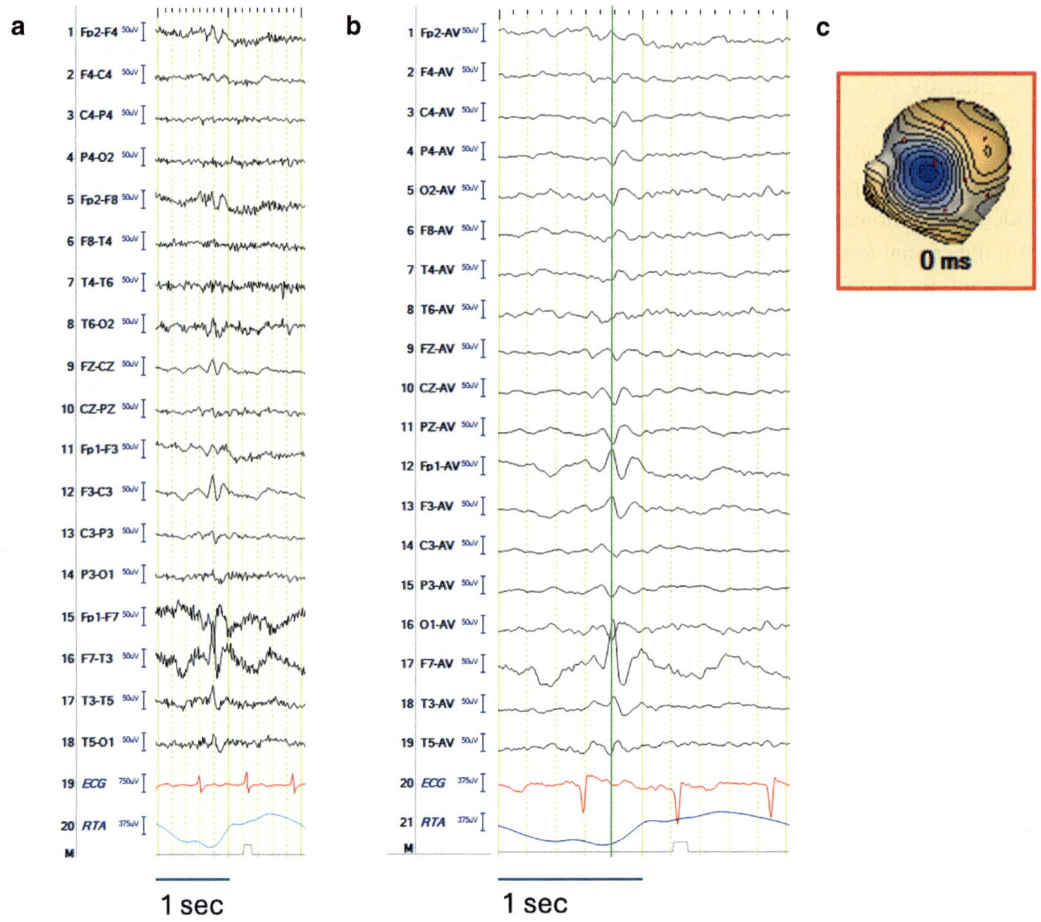

Fig. 7.1 An Interictal spike is shown in bipolar (**a**) and common average (**b**) montage. The spike fulfills the IFCN criteria that define an interictal epileptiform discharge (see text). In (**c**), the voltage spike field is shown (radial voltage map)

were present (in any combination). Validation studies suggest that this simple approach is useful for clinical practice and by electroencephalographers [23].

7.5 Ictal Pattern: The Term and Clinical Relevance

In EEG, an ictal pattern refers to the specific EEG activity that occurs during a seizure. The term "ictal" comes from the Latin word "ictus," which means "seizure" or "attack." An ictal pattern is a hallmark of epileptic activity and is critical for diagnosing and understanding seizures.

General characteristics of an ictal pattern in EEG:

1. Sudden Onset: Ictal patterns typically have a sudden onset, marking the beginning of the seizure activity.
2. Rhythmic Activity: The pattern often shows rhythmic, repetitive activity that is different from the background EEG. This could be in the form of spikes, sharp waves, or rhythmic slowing.
3. Evolution of Frequency and Amplitude: The frequency and amplitude of the EEG waves usually change during the ictal phase. Often, there is an increase in amplitude and a change

in frequency, such as from faster to slower waves as the seizure progresses.
4. Localization: Ictal patterns can be localized to a specific region of the brain (focal seizure) or generalized across both hemispheres (generalized seizure).
5. Duration: The ictal pattern lasts as long as the seizure itself, which can range from a few seconds to several minutes. Once the seizure ends, the EEG typically shows a postictal phase characterized by slowing or suppression of activity.

Clinical Importance

Identifying and analyzing ictal patterns is crucial for diagnosing epilepsy, determining the type of seizure, and localizing the seizure focus, which is especially important for surgical planning in epilepsy treatment:

- To establish whether the paroxysmal episodes are of epileptic origin or of different nature (for instance, syncope, functional/dissociative seizures, sleep disorders, and movement disorders).
- To classify the seizure type: For clinical purposes, the main goal of seizure classification is to establish whether it is a generalized or a focal seizure, with possible yields for the choice of ASM treatment.
- To localize the cerebral region(s) originating the ictal event(s), i.e., the epileptogenic zone, and the pathways of spread. Ictal EEG recording is an essential step in the presurgical evaluation of drug-resistant focal epilepsies.

In summary, an ictal pattern on an EEG is the characteristic electrical activity observed during a seizure, marked by changes in rhythm, amplitude, and frequency, and can be focal or generalized depending on the type of seizure.

7.6 EEG Ictal Patterns in Generalized Seizures

EEG ictal patterns in generalized seizures exhibit distinct features that reflect the widespread, bilateral involvement of the brain from the onset of the seizure. These patterns are essential for diagnosing and understanding generalized epilepsy. Generalized epileptic seizures originate suddenly within, and rapidly involve, bilaterally distributed networks that include subcortical as well as cortical structures, although not necessarily the entire cortex. It is assumed that they result from a generalized disturbance of the cortical function, either on a genetic background or from a diffuse brain sufferance. It may occur that individual seizures may display a lateralized onset; however, the location and lateralization may vary from seizure to seizure, indeed showing a degree of asymmetry.

Here are the main features of EEG ictal patterns in generalized seizures:

1. Bilateral, Symmetric Onset
   - Widespread Distribution: The ictal pattern typically begins simultaneously across both hemispheres of the brain, appearing on all or most EEG channels.
   - Symmetry: The activity is generally symmetric, meaning it looks similar on both sides of the brain, reflecting the global nature of the seizure.
2. Generalized Spike-and-Wave Discharges (Fig. 7.2)
   - Spike-and-Wave Complexes: One of the hallmark features of generalized seizures, particularly absence seizures, is the presence of generalized spike-and-wave discharges. These are characterized by a sharp spike followed by a slow wave, repeating at a frequency of about 3 Hz (3 cycles per second).
   - Polyspikes-and-Wave Complexes: In certain types of generalized seizures, such as myoclonic or tonic–clonic seizures, polyspikes-and-wave complexes (multiple spikes followed by a slow wave) may be seen.

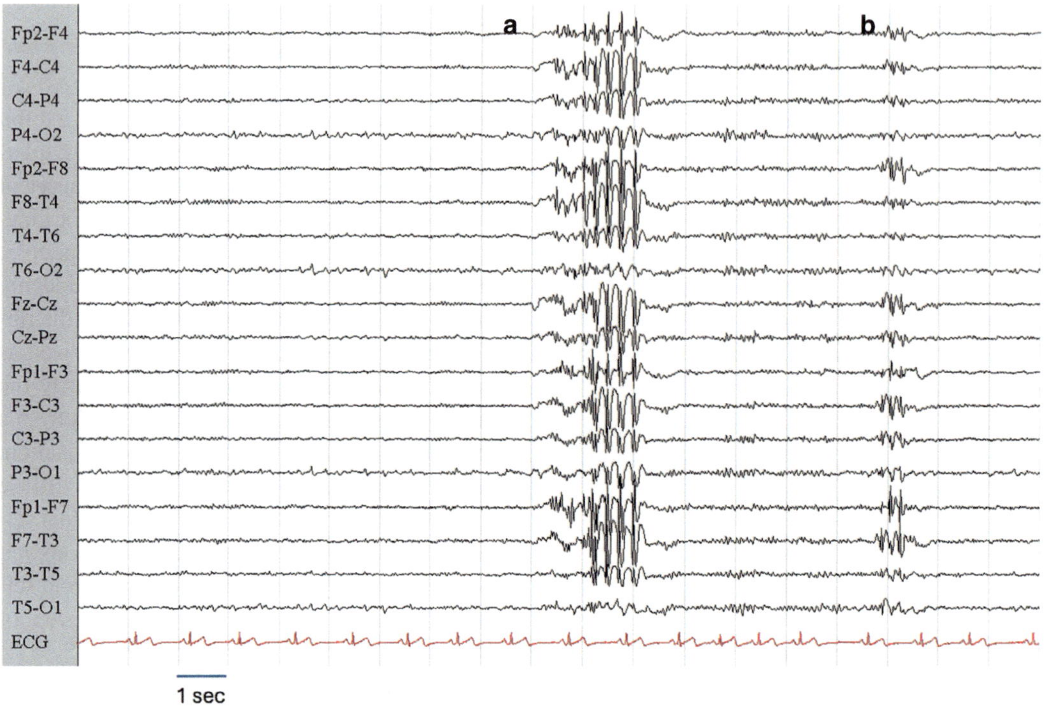

Fig. 7.2 Brief generalized spike-and-waves around 3 Hz are shown. In (**a**), the discharge starts with polyspike activity anteriorly predominant; in (**b**), the spike-and-wave discharge shows a clear anterior predominance

3. Rhythmic, Synchronous Activity
   - Rhythmic Discharges: The EEG during a generalized seizure often shows rhythmic, high-amplitude discharges that are synchronized across multiple regions of the brain.
   - Frequency Characteristics: Depending on the type of generalized seizure, the rhythmic activity may vary in frequency:
     - Absence Seizures: typically exhibit 3 Hz spike-and-wave complexes.
     - Myoclonic Seizures: characterized by fast, polyspikes-and-wave complexes or bursts of rapid spikes.
     - Tonic–Clonic Seizures: may start with low-voltage, fast activity, followed by generalized polyspikes-and-wave discharges or rhythmic slowing.
4. Lack of Focal Onset
   - No Localized Focus: unlike focal seizures, generalized seizures do not have a specific, localized onset. The abnormal EEG activity appears diffusely across the brain from the beginning, without a clear focal starting point.
5. Evolution of the Ictal Pattern
   - Progression: In tonic–clonic seizures, the ictal pattern may evolve, starting with a tonic phase of low-voltage, fast activity, followed by the clonic phase with rhythmic spike-and-wave or polyspikes-and-wave discharges.
   - Postictal State: After the seizure ends, the EEG often shows a period of generalized slowing or suppression of activity (postictal generalized EEG suppression), known as the postictal phase, as the brain recovers.
6. Generalized Background Disruptions
   - Disruption of Normal Background: The ictal pattern often disrupts the normal EEG background activity, which can include

alpha rhythms in awake adults or sleep spindles during sleep, replacing it with pathological discharges.

Clinical Relevance

Identifying these generalized ictal patterns on EEG is crucial for diagnosing generalized epilepsy and differentiating it from focal epilepsy. The patterns also help in selecting appropriate treatment strategies, which may include specific ASMs that are effective for generalized seizures.

In summary, EEG ictal patterns in generalized seizures are characterized by bilateral, symmetric, and synchronous discharges, such as generalized spike-and-wave or polyspikes-and-wave complexes, with no focal onset. These features reflect the global involvement of the brain during the seizure and are essential for accurate diagnosis and management of generalized epilepsy.

7.7 Specific Ictal Patterns in Generalized Seizures

7.7.1 Absence Seizures

Typical childhood absence seizures are usually characterized by a frontally predominant generalized 3–4 c/s spike-and-wave discharge, that progressively slows down [4, 5, 24]. The morphology of the spike-and-wave complex has been described in detail, consisting of a spike component, followed by a positive transient and then a high-amplitude slow wave [25]: the amplitude of this complex usually increases in the first 1–2 s of the discharge, whereas, at the end of the discharge, the spike component may be less evident. In some instances, the interictal EEG may show lateralized and even focal spikes intermixed with the generalized bursts. In juvenile absence epilepsy, the frequency of the generalized spike-and-wave discharge may be faster, around 3.5–4.5 Hz, as compared to childhood absence epilepsy (Fig. 7.3). On the contrary, in developmental and epileptic

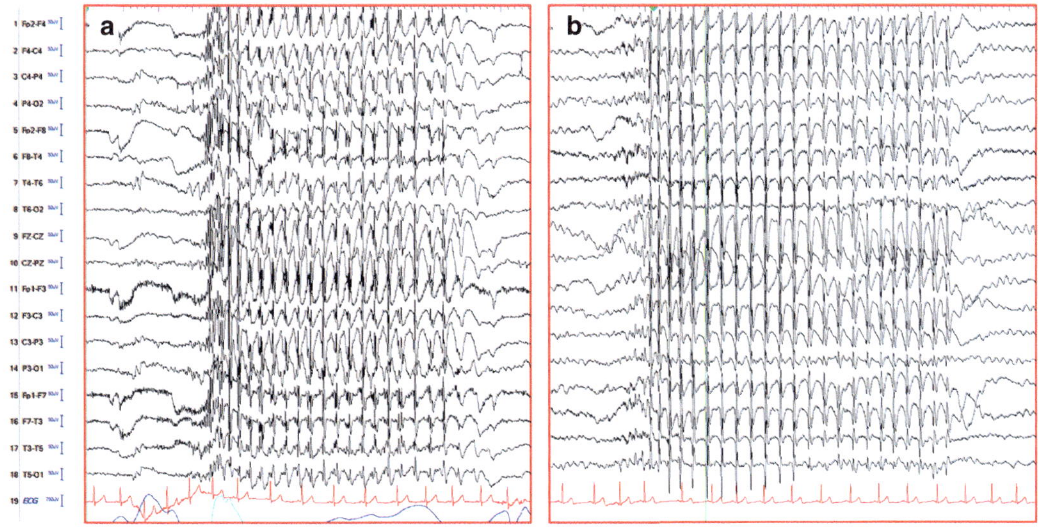

Fig. 7.3 Absence seizures: typical and atypical. In (**a**), an absence seizure is shown in a girl with Juvenile Absence Epilepsy: the discharge begins with polyspikes followed by a 3-Hz discharge. In (**b**), an absence seizure in a boy with intellectual disability: note the slow around 2 Hz frequency of the discharge

encephalopathies, absence seizures generally show a slow frequency of spike-and-wave complexes around 1.5–2 Hz or they can show a myoclonic component as in myoclonic absence seizures [4, 5, 26, 27].

7.7.2 Generalized Polyspikes Train

Refractory seizures in idiopathic generalized epilepsies may persist in adult life, associated with absence status and occasional tonic–clonic seizures; in these patients, distinctive EEG features are abundance of polyspikes discharges and fast paroxysmal activities during sleep (Fig. 7.4) in the interictal EEG, and ictal EEG showing runs of polyspikes preceding or intermixed with the absence related spike-wave discharge [4, 5]. Certainly, it has been repeatedly observed that the presence in patients diagnosed with IGE of polyspikes train discharges during sleep is a predictor of a worse response to drug therapy [28, 29].

7.7.3 Generalized Tonic–Clonic Seizures

The typical convulsive tonic–clonic or Grand Mal seizure in idiopathic generalized epilepsies is characterized by a generalized spike-and-wave discharge that is replaced by polyspike-wave activities and then the tonic phase intervenes whose EEG hallmark is a tonic discharge of continuous spikes; subsequently, this pattern starts to be fragmented by slow waves of progressively longer duration that interrupt the spike activity and correspond to the clonic phase. At the end of the seizure, the EEG shows a marked diffuse suppression of the background activity, followed by slowing in both hemispheres (Fig. 7.5).

7.7.4 Tonic Seizures

Tonic seizures are a hallmark of Lennox-Gastaut syndrome [30]. They are usually characterized by brief (5–20 s) tonic axial contraction that can display some degree of

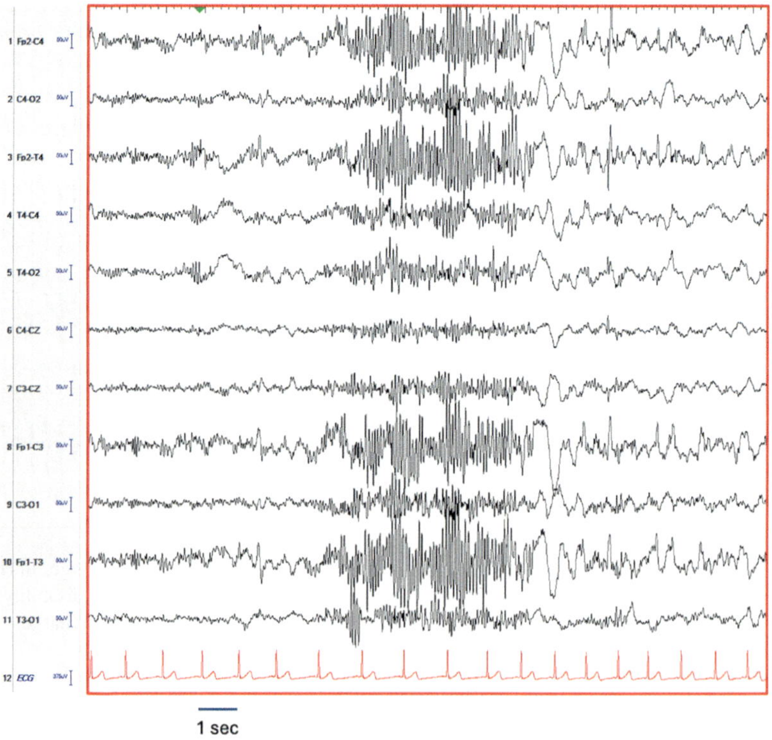

Fig. 7.4 Generalized paroxysmal fast activity in sleep in patients with drug-resistant generalized epilepsy. The EEG was recorded with a reduced montage. Sensitivity: 15 uV/mm high-pass filter: 0.1 s; low-pass filter: 35 Hz

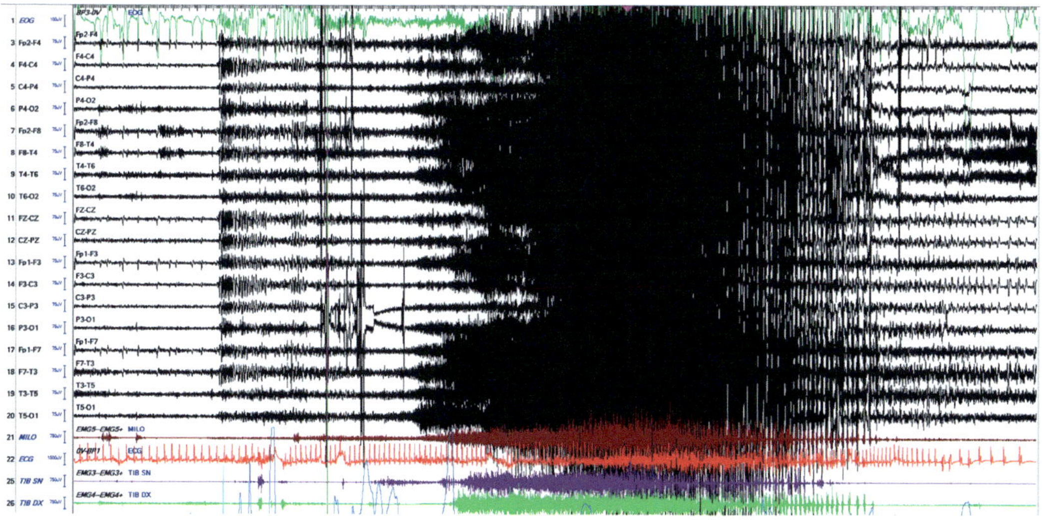

Fig. 7.5 A compressed view of 2-min EEG is shown displaying generalized tonic–clonic seizures in a man with idiopathic generalized epilepsy. The seizure begins with a generalized polyspike discharge followed by fast activity and then diffuses muscle artifacts during the tonic phase. Note the ECG trace that shows first a marked tachycardia followed by bradycardia in the post-ictal period

asymmetry; in some instances, they may terminate with a brief clonic phase. Vegetative phenomena (such as tachycardia, irregular tachypnea mixed with apneic periods, mydriasis, and vasomotor phenomena) may be prominent. Loss of consciousness is common, whereas postictal confusion may be occasional. The ictal EEG manifestations are always bilateral, symmetrical, and synchronous.

Rapid desynchronization with or without subsequent rapid synchronization, pure hypersynchronization at 10 c/s (the epileptic recruiting rhythm), followed by added slow-wave activity appearing as spike-waves or other patterns may be seen [4, 31] (Fig. 7.6). Since occurrence during sleep usually is common, EEG/polygraphic sleep recording may be crucial to detect this seizure type.

7.7.5 Atonic Seizures

In generalized epilepsies, atonic seizures due to pure loss of muscular tone are rare; they may be extremely brief, that is, drop seizures, or more prolonged (up to several minutes). Several ictal EEG patterns have been reported, such as generalized polyspikes-wave bursts, usually associated with drop seizures, polyspikes discharges followed by generalized spike-wave activity and low- or high-voltage fast activity [31].

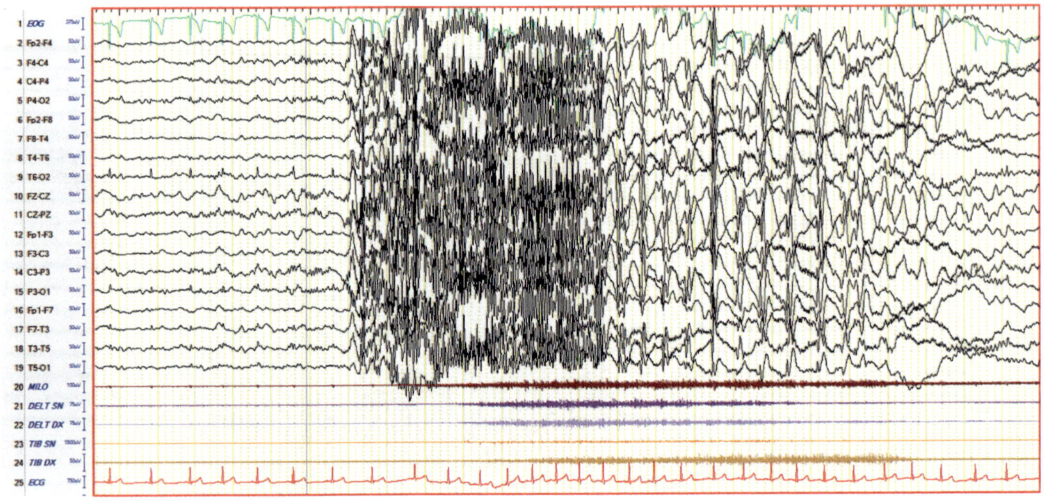

Fig. 7.6 A tonic seizure during sleep in a patient with Lennox-Gastaut syndrome is presented. Note the generalized polyspikes train/fast activity discharge that is followed by slight tonic contraction of the mylohyoid and deltoid muscles. The fast discharge is followed by slow spike-and-waves activity. Sensitivity: 15 uV/mm high-pass filter: 0.1 s; low-pass filter: 35 Hz

7.8 EEG Ictal Patterns in Focal Seizures

In focal seizures, the EEG ictal patterns exhibit specific features that reflect the localized origin of the seizure activity within a particular region of the brain. A wide variety of EEG seizure patterns can characterize focal seizures [32, 33]. The clinical and EEG expression of focal seizures may vary from one individual to another and to a lesser extent from one seizure to another in the same patient. The EEG onset of a seizure may involve a single or a few EEG leads, or the leads related to a whole brain lobe or either a whole hemisphere; finally, it can happen that the location and side of onset cannot be determined because the propagation of the ictal discharge may be so fast as to render difficult to detect the precise site of onset, the seizure pattern appearing as involving an entire hemisphere or both hemispheres. However, the features of the discharge usually allow its recognition as a focal and not a generalized seizure.

At their start, focal seizures may begin with any kind of EEG frequency (delta, theta, alpha, and beta); sometimes, they may be preceded by an increase or a sudden decrement till the complete disappearance of the interictal spikes [34] or by a diffuse high-amplitude slow spike–wave complex. Then, usually an organized rhythmic discharge, sometimes waxing and waning in amplitude and frequency, but clearly lateralized or localized to some electrodes (generally with a lobar or hemispheric topography) represents the more overt and common ictal EEG pattern of a focal seizure. This activity is usually followed by a postictal slowing that can be either diffuse or localized. Focal seizures may also evolve into a secondary generalized tonic–clonic convulsion.

The main features of EEG ictal patterns in focal seizures include the following:

1. Localized Onset
   - Focal Localization: The ictal pattern usually begins in a specific area of the brain. The onset is often restricted to one hemisphere or even a small region within a hemisphere.
   - Electrode Placement: The EEG changes may be visible only on a few electrodes that correspond to the brain region where the seizure originates.

2. Rhythmic Activity
    - Rhythmic Discharges: The ictal pattern typically consists of rhythmic discharges such as spikes, sharp waves, or rhythmic slow waves (theta or delta activity).
    - Repetitive Pattern: The activity tends to be repetitive and can increase in amplitude and change in frequency as the seizure progresses.
3. Evolution of the EEG Pattern
    - Change Over Time: During the seizure, the EEG pattern often evolves, showing changes in frequency, amplitude, and morphology. This can include a gradual increase in the frequency or a spread of rhythmic activity to adjacent brain regions.
    - Propagation: As the seizure progresses, the ictal pattern may spread from the focal region to other areas of the brain, leading to a broader distribution of abnormal activity.
4. Ictal Onset Zone
    - The point of onset is often marked by a sudden change from the normal background activity to a pathological rhythm. This ictal onset zone can be crucial for identifying the region of the brain responsible for seizure generation.
5. Temporal Dynamics
    - Post-Ictal Slowing: After the seizure ends, there may be a period of post-ictal slowing in the EEG, where the background activity is reduced in amplitude and frequency, reflecting the brain's recovery phase.
6. Frequency Characteristics
    - Initial Fast Activity: Some focal seizures may start with fast, low-amplitude activity (beta or gamma frequencies) before evolving into slower rhythmic discharges.
    - Slowing Down: The seizure might slow down in frequency over time, often shifting to theta or delta activity as it spreads or as the seizure progresses.
7. Asymmetry
    - Lateralization: The ictal pattern is often asymmetric, meaning that it is more prominent on the side of the brain where the seizure originates.
8. Clinical Correlation
    - Specific Symptoms: The EEG findings are often correlated with specific clinical symptoms, such as sensory disturbances, motor activity, or alterations in consciousness, depending on the location of the seizure focus.

Importance in Diagnosis

Recognizing these features is critical for diagnosing focal epilepsy, determining the exact location of the seizure focus, and planning appropriate treatment, which may include medication or surgery.

In summary, EEG ictal patterns in focal seizures are characterized by a localized onset, rhythmic discharges, evolution in frequency and amplitude, and clear temporal dynamics, with the potential for propagation to adjacent brain regions. These features are essential for the clinical assessment and management of epilepsy.

7.9 Limits in the Detection of Ictal Patterns from Scalp EEG

Scalp electrodes have significant limitations in exploring basal temporal, frontal, or occipital regions, and picking up activity originating from the mesial aspects of both hemispheres and from the depth of cerebral sulci. Additional electrodes, such as zygomatic electrodes, may be used to detect interictal EEG activity or seizure onset from basal temporal lobe structures. However, in drug-resistant focal epilepsies when extracranial EEG is unsatisfactory to determine the seizure onset zone and a more precise localization is necessary for surgical treatment, invasive recordings by means of intracranial electrodes are often mandatory.

Interpretation of ictal EEG data has to take into account inherent limitations of scalp EEG recording such as the possible effect of volume conduction, the spatial averaging properties of the scalp and the skull due to their electrical characteristics, the sensitivity of EEG recording almost exclusively to radially oriented electrical dipoles, as well as the inability to pick up activity originating from the cortex buried in the depth of sulci or from basal or interhemispheric cortical regions [21, 35].

Additional problems may be represented by the fact that ictal activity to be recorded requires a synchronous discharge in a relatively large amount of cortex. Moreover, seizure patterns may become evident only when they build up in a rhythmic high-amplitude discharge. This latter phenomenon may take place in regions far away from the seizure focus, particularly if this one is in medial regions or in severely damaged cortex. Focal seizures without impairment of consciousness may display very localized ictal seizure patterns. In other circumstances, such as in seizures with exclusively motor, sensory, or somatosensory phenomenology, or when the location of the seizure focus is restricted to portions of the cortex situated in cerebral regions insensitive to scalp EEG recording, it may be difficult to detect any seizure-related EEG modification.

Finally, artifacts related to muscle contractions or movements or to other ictal behaviors (such as chewing or blinking) can frequently obscure seizure patterns and render localization tentative a failure. Muscle artifacts can be reduced by digitally filtering the EEG signal reducing high-frequency filters; however, this procedure may modify muscle activity to resemble spikes or runs of fast activity, possibly leading to localization errors.

7.10 Ictal Patterns in Temporal Lobe Seizures

Temporal lobe seizures are the most frequently recorded in Epilepsy Monitoring Units; the onset can be mesial (thus deeper and less visible on scalp EEG) or lateral (closer to the recording electrodes from the scalp and thus better visible in its early part) [36]. This is followed by a phase of rhythmic activity that may be more or less prolonged (from a few seconds to a few minutes), often spreading into the peri-Sylvian leads of the onset hemisphere but still remaining localized in most cases within the affected hemisphere (Fig. 7.7).

In certain types of seizures, such as antero-mesial temporal lobe seizures, the seizure onset may be represented by a “flattening” of the EEG activity, localized in the anterior temporal leads, then a rhythmic 5–9 Hz discharge of spike activity, progressively slowing in frequency may follow, maintaining, although with some ipsi- and contralateral spread, a predominant temporal lobe location. Often at the onset or during an antero-mesial temporal lobe seizures, prominent muscular artifacts due to oro-alimentary automatisms may partially obscure the EEG activity [37]. Ictal seizure patterns in lateral temporal lobe seizure may appear as lateralized rhythmic theta activity that may be preceded or followed by a rhythmic spike discharge. This pattern may be localized in the temporal lobe channels, sometimes partially spreading to the ipsilateral hemisphere. Sometimes, more frequently in seizures of longer duration, the activity may spread to the contralateral hemisphere and result in a generalized tonic–clonic seizure.

Each phase of electrical activity has a correlate in clinical manifestations (subjectively: ascending epigastric sensation or déjà vu/déjà veçu or anxiety/fear in mesial temporal, auditory symptoms in temporal lateral. Objective clinical correlates of seizures are arrest and loss of con-

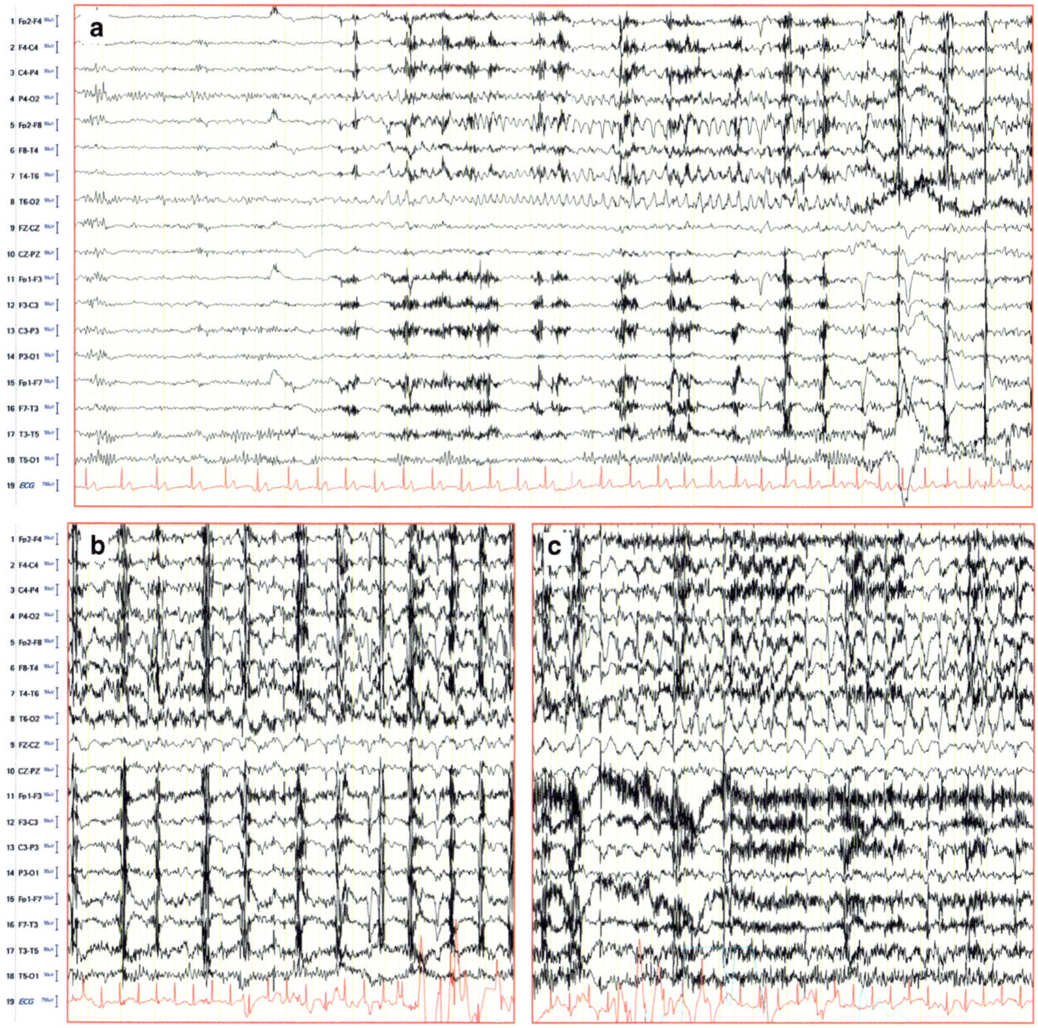

Fig. 7.7 Temporal lobe seizure in a 55-year-old man with focal epilepsy due to a disembrioplastic neuroepitelial tumor of the right temporo-mesial region. (**a**) Disappearance of the background rhythm in the right hemisphere followed by slow rhythmic activity on the right temporal region, mixed with swallowing artifacts, progressively increasing in amplitude and decreasing in frequency (**b**, **c**), with diffusion to the supra-Sylvian regions of the right hemisphere. From the clinical point of view, the patient presents with swallowing and oro-mandibular automatisms, which continue for about 2 min, followed by stiffening of the left upper limb, gaze deviation to the left, and finally clonic movements. Sensitivity: 15 uV/mm high-pass filter: 0.1 s; low-pass filter: 35 Hz

tact, oro-alimentary or gestural automatisms in mesial temporal, speech difficulty if dominant hemisphere in lateral temporal, or spreading of discharge) that can be observed by video-EEG recording.

Isolated auras frequently have no associated surface EEG signs due to localized and brief ictal discharges, and only intracerebral recordings may demonstrate an associated EEG modification.

7.11 Ictal Patterns in Frontal Lobe Seizures

Frontal seizures can vary in their clinical and EEG features depending on the part of the frontal lobe affected, given its wide extension [32, 33, 37, 38].

Seizures originating in the anterior region (fronto-polar or fronto-basal) often have absent or poor clinical correlates both objectively and subjectively (sometimes only reduction or suspension of contact may be observed), and an EEG representation in the anterior leads. If there is initial or later involvement of the more central regions, the clinical semeiology is enriched by motor manifestations (clonic movements or tonic stiffening) of the limbs contralateral to the involved hemisphere, and subsequently, there may be a spread of the discharge to the contralateral hemisphere, resulting in a generalized tonic–clonic manifestation.

Compared with temporal seizures, frontal seizures are generally of shorter duration, earlier contralateral spread, and more frequent occurrence in sleep (Fig. 7.8).

In frontal lobe seizures, either interictal or ictal EEG may be devoid of overt abnormalities, or they can be even misleading: this is particularly true for seizures originating from the mesial aspects or fronto-orbitary regions of the frontal lobes.

An interictal EEG pattern that can be observed in frontal lobe epilepsies with seizures originating from the antero-mesial areas is the secondary bisynchronism that is characterized by a bilateral, synchronous, and symmetrical epileptic activity, possibly depending on the rapid contralateral propagation of a focal discharge through callosal fibers. During sleep, the EEG may display runs of fast polyspikes, that, when prolonged, may correspond to tonic seizures; clinically, these patients often suffer from seizures complicated by sudden and abrupt falls to the ground. Sometimes, these EEE patterns can be difficult to distinguish from those found in generalized epilepsies, so the complete workup including clinical and imaging data remains crucial.

In addition, frontal seizures during sleep may have, more often than in seizures originating

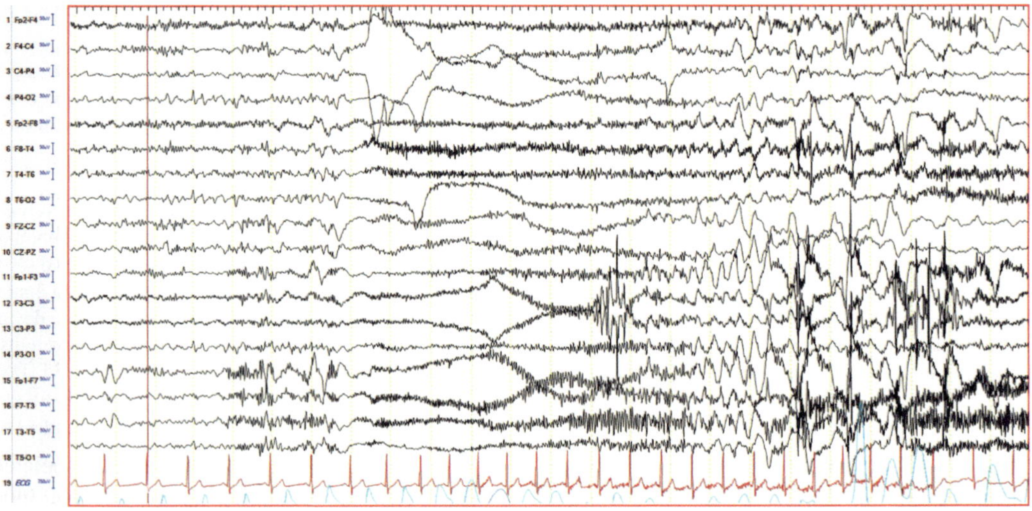

Fig. 7.8 Frontal lobe seizure in sleep in a 25-year-old female patient with sleep-related hypermotor epilepsy due to left frontobasal focal cortical dysplasia. The EEG pattern is characterized by a brief sequence of polyspikes followed by an abrupt flattening of the EEG trace maximal in the left fronto-central-temporal region, which evolves into a low-voltage fast activity, followed by a slow rhythmic activity in the same electrodes and then movement artifacts. Clinically, the patient wakes up and shows brief head flexion movement along with bilateral upper limbs proximal adduction followed by postural adjustment and asymmetrical tonic posture. Sensitivity: 10 uV/mm high-pass filter: 0.1 s; low-pass filter: 35 Hz

from other lobes, a hypermotor clinical correlate (pedaling, tilting, or violent, repetitive movements), which may be preceded on the EEG by few seconds of focal or diffuse EEG suppression or by low-voltage fast activity, more or less localized and visible depending on the depth of onset of the critical discharge and followed by movement artifacts. In patients presenting with hypermotor frontal lobe seizures, ictal EEG activity may be completely masqueraded by striking muscular and movement artifacts; indeed, in these patients, video-EEG recording of the seizures demonstrating the ictal clinical-EEG features, and their stereotypy may be necessary for a correct diagnosis [38].

In children, staring spells resembling absence seizures, often associated with some degree of cognitive impairment and refractory to treatment, may originate from frontal regions; in these cases, ictal EEG may display a rhythmic spike–wave discharge, with pronounced asymmetry of amplitude [37].

7.12 Ictal Patterns in Insular Lobe Seizures

Insular lobe seizures are among the rarest in terms of frequency and among the most difficult to record on scalp EEG in relation to the region from which they arise, located deep in the Sylvian fissure and covered by the temporal and frontal lobes. Due to its anatomical location and multiple connections with different cortical and subcortical regions, what we can see at scalp EEG is essentially the subsequent temporal and fronto-opercular spread of the discharge, often with slow/sharp rhythmic wave patterns, rather than the initial ictal discharge, which invariably remains hidden, and for its precise detection, invasive recording by stereo-EEG electrodes is often required.

For these reasons in the case of insular seizures, close correlation with clinical and imaging data is mandatory. In particular, the semeiology of insular seizures may include somatosensory symptoms (tingling, electric, or pain sensations involving the contralateral but sometimes also ipsilateral hemi-body, laryngeal spasm, in addition to auditory, speech or gustatory manifestation such as disgust) [39, 40], which may be followed by signs and symptoms typical of secondary temporal and/or fronto-opercular spread of the discharge.

Moreover, the semeiology of insular seizures can mimic that of other brain regions, and most of the insulo-opercular symptoms can be masked by those of extra-insular seizure propagation, making its detection even more difficult, leading to misdiagnosis and in some cases resulting in epilepsy surgery failure. In addition, sleep-related hypermotor seizures may also be a presentation of insular epilepsies, showing predominant hyperkinetic and tonic-dystonic features and scalp EEG patterns that are very often indistinguishable from seizures originating in the fronto-mesial regions [41].

7.13 Ictal Patterns in Parieto-Occipital Lobe Seizures

Occipital lobe seizures are usually characterized by symptoms due to the initial involvement of the occipital areas or due to the spread of the epileptic discharge to other cortical structures, superiorly or inferiorly the Sylvian scissure, or contralaterally through the splenium of the corpus callosum (Fig. 7.9). Elementary visual hallucinations (such as flashing or steady spots, simple geometric forms, that, when lateralized, depend on a contralateral occipital seizure discharge), eye blinking, tonic eye, and head deviation, either ipsilateral or contralateral to the discharge, vegetative phenomena (such as vomiting) may be observed at the onset of occipital lobe seizures. At seizure onset, ictal EEG may not display clear-cut abnormalities, particularly when the seizure focus is located in mesial or basal occipital lobe structures; during the evolution of the seizure, the ictal discharge may become more evident appearing as a low-voltage fast activity progressively followed by a rhythmic epileptiform discharge [42].

Often scalp EEG abnormalities may be seen in the posterior regions bilaterally, sometimes main-

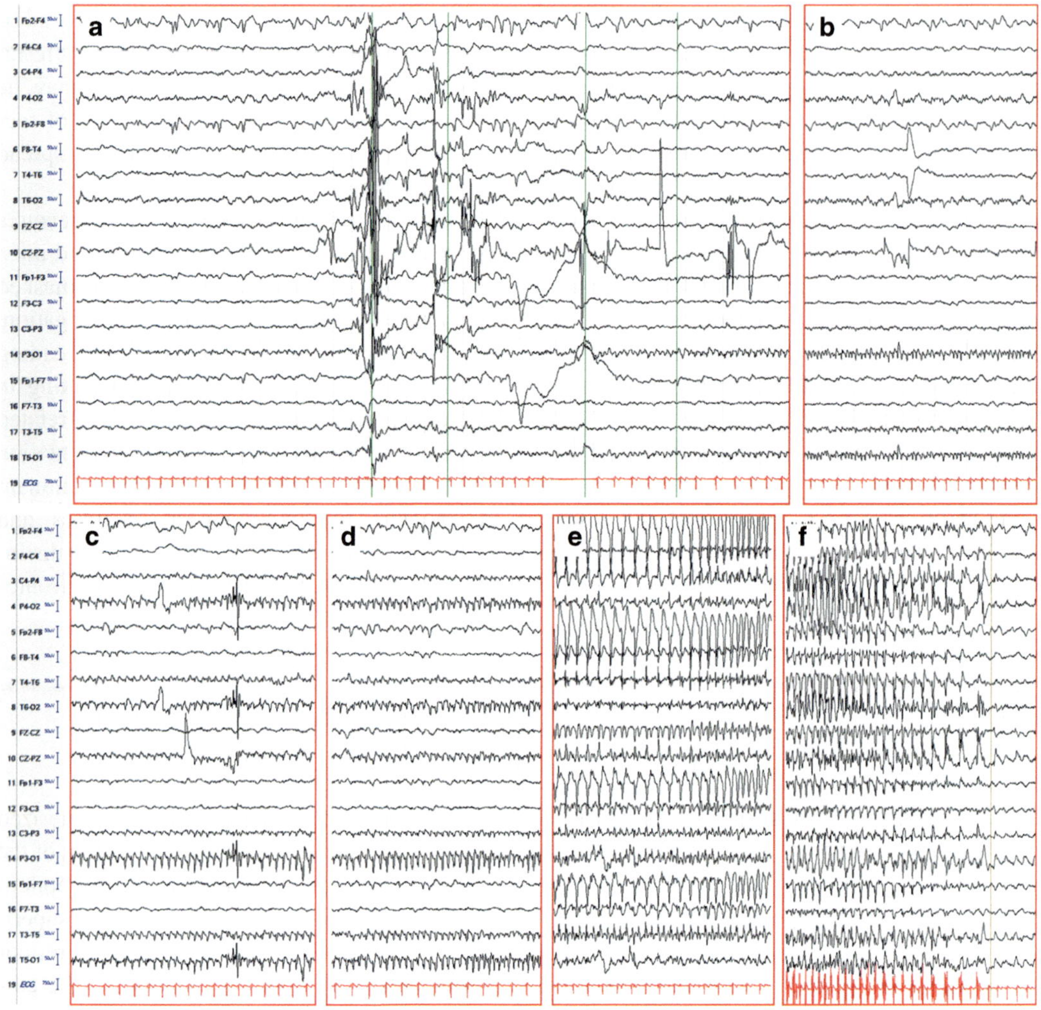

Fig. 7.9 Occipital lobe seizure. A 17-year-old patient with severe head trauma, multiple contusive-hemorrhagic lesions, and left-predominant occipital subdural hematoma. (**a–d**) Four consecutive EEG extracts; in **A,** we can see the appearance of low-voltage spikes on the left occipital leads (P3-O1 and T5-O1) which become more frequent and evolve into rhythmic spike waves and sharp waves. These epileptiform abnormalities remain localized in the left hemisphere for about 2 min (up to **b**), and then they involve the contralateral homologous regions (**c** and **d**). After a further 5 min (EEG trace not shown), the discharge progressively spreads anteriorly up to the bilateral frontal regions (**e**, **f**) increasing in amplitude and decreasing in frequency, until it becomes more diffuse and ends abruptly. Clinically, the patient presents in a coma state, and no objective changes are observed until the final stage when clonias are noted initially in the face and then in the four limbs. Given the duration and the EEG pattern, this electro-clinical event constitutes a non-convulsive focal status originating in the left occipital lobe and spreading anteriorly, finally culminating in clonic jerks. Sensitivity: 10 uV/mm high-pass filter: 0.1 s; low-pass filter: 35 Hz

taining a side prevalence, although the seizure and corresponding clinical manifestations (predominantly positive or negative visual symptoms) remain lateralized to the affected hemisphere.

In parietal lobe seizures, symptoms are of varying localizing and lateralizing value and seizure discharges may spread rapidly; indeed, perceived symptoms may reflect secondary spread rather than the primary site of seizure onset. The most common ictal clinical manifestations are paresthesia, painful, thermal, sexual, and apraxic disturbances of body image. Ictal EEG may be poorly informative, particularly in seizures without impairment of consciousness, or may show diffuse suppression or diffuse abnormalities at seizure onset. Then sharp waves may appear, spreading either anteriorly or posteriorly. The possible propagation of the ictal EEG discharge from parietotemporal regions to the frontal and parietal operculum as well as the spread from posterior parietal areas to medial frontal regions has been reported [42].

7.14 The Added Value of Polygraphy in the Evaluation of Ictal EEG Patterns

Finally, we emphasize the importance of considering in each patient undergoing video-EEG to add polygraphic channels depending on the clinical question and the type of recording made. In particular, it is advisable in sleep recordings to record cardio-respiratory polygraphic activity, as well as the mylohyoid muscle and the oculogram. In fact, polygraphic recordings may make it possible to detect phenomena related to seizure activity that are invisible to clinical observation. There are numerous examples of this in the literature, which are the subject of a dedicated chapter; we would like to focus here as a case-use example on peri-ictal respiratory changes, which have recently been the subject of much attention. Indeed, the study of peri-ictal respiratory disturbances has gained attention, mainly in relation to the study of factors associated with an increased risk of sudden unexplained death in epilepsy (SUDEP), which is the leading cause of death among people with epilepsy [43–45]. Ictal respiratory changes comprise central or obstructive apnea, tachypnea, bradypnea, apnea, and hypoxemia (see Fig. 7.10). Ictal central apnea has been widely described in association with both generalized tonic–clonic and focal seizures, especially of temporal origin [46–49]. To note, the first description of seizure-related apnea was made by Hughlings Jackson in 1899, many years before the advent of the EEG, even if this phenomenon has been studied in detail only in the twenty-first century [50].

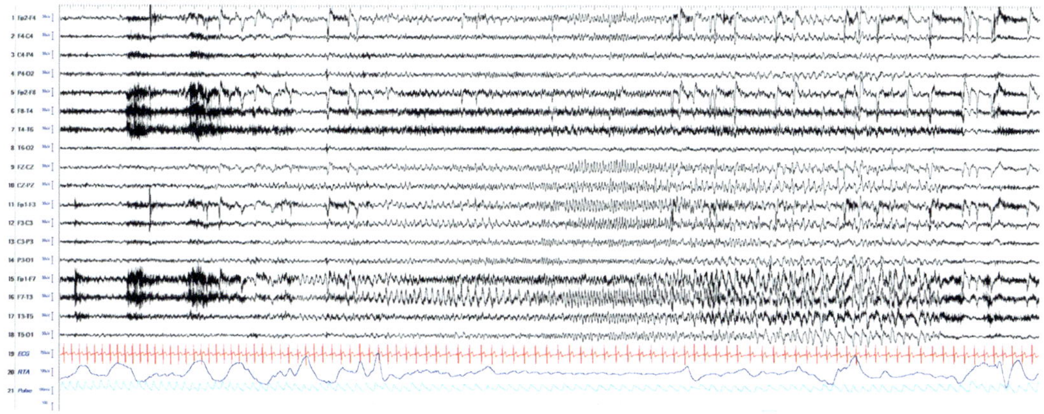

Fig. 7.10 Ictal apnea in temporal lobe seizure. 1 min of EEG trace is shown. A left temporal lobe seizures is recorded associated with a disruption and arrest of respiration (RTA: thoraco-abdominal breath) that persist for about 30 s. Note also the slight bradycardia during apnea. Clinically, no movement/automatisms were evident

References

1. Gibbs FA, Davis H, Lennox WG. The electroencephalogram in epilepsy and in conditions of impaired consciousness. Arch Neurol Psychiatr. 1935;34(6):1133–48.
2. Gibbs FA, Lennox WG. Electroencephalographic classification of epileptic patients and control subjects. Arch Neurol Psychiatr. 1941;46(6):1222–5.
3. Penfield W, Jasper HH. Epilepsy and the functional anatomy of the human brain. Little, Brown and Company; 1954.
4. Gastaut H, Roger J, Ouachi S, Timsit M, Broughton R. An electroclinical study of generalized epileptic seizure of tonic expression. Epilepsia. 1963;4:15–44.
5. Gastaut H, Tassinari CA. In: Handbook of electroencephalography and clinical neurophysiology, editor. Epilepsies, vol. 13. Amsterdam: Elsevier; 1975.
6. Seeck M, Koessler L, Bast T, Leijten F, Michel C, Baumgartner C, et al. The standardized EEG electrode array of the IFCN. Clin Neurophysiol. 2017;128:2070.
7. Fountain NB, Freeman JM. EEG is an essential clinical tool: pro and con. Epilepsia. 2006;47(Suppl 1):23–5.
8. Fisher RS, Cross JH, French JA, Higurashi N, Hirsch E, Jansen FE, et al. Operational classification of seizure types by the international league against Epilepsy: position paper of the ILAE Commission for Classification and Terminology. Epilepsia. 2017;58(4):522–30.
9. Van Donselaar CA, Schimsheimer RJ, Geerts AT, Declerck AC. Value of the electroencephalogram in adult patients with untreated idiopathic first seizures. Arch Neurol. 1992;49:231–7.
10. Pillai J, Sperling MR. Interictal EEG and the diagnosis of epilepsy. Epilepsia. 2006;47(Suppl 1):14–22.
11. Fisher RS, Acevedo C, Arzimanoglou A, Bogacz A, Cross JH, Elger CE, Engel J Jr, Forsgren L, French JA, Glynn M, Hesdorffer DC, Lee BI, Mathern GW, Moshe SL, Perucca E, Scheffer IE, Tomson T, Watanabe M, Wiebe S. ILAE official report: a practical clinical definition of epilepsy. Epilepsia. 2014;55(4):475–82.
12. Benbadis SR. Errors in EEGs and the misdiagnosis of epilepsy: importance, causes, consequences, and proposed remedies. Epilepsy Behav. 2007;11(3):257–62.
13. Benbadis SR, Tatum WO. Overintepretation of EEGs and misdiagnosis of epilepsy. J Clin Neurophysiol. 2003;20:42–4.
14. Jing J, Herlopian A, Karakis I, Ng M, Halford JJ, Lam A, Maus D, Chan F, Dolatshahi M, Muniz CF, Chu C, Sacca V, Pathmanathan J, Ge W, Sun H, Dauwels J, Cole AJ, Hoch DB, Cash SS, Westover MB. Interrater reliability of experts in identifying interictal epileptiform discharges in electroencephalograms. JAMA Neurol. 2020;77(1):49–57.
15. Nascimento FA, Jing J, Beniczky S, Benbadis SR, Gavvala JR, Yacubian EMT, Wiebe S, Rampp S, van Putten MJAM, Tripathi M, Cook MJ, Kaplan PW, Tatum WO, Trinka E, Cole AJ, Westover MB. One EEG, one read – a manifesto towards reducing interrater variability among experts. Clin Neurophysiol. 2022;133:68–70.
16. Engel J Jr. A practical guide for routine EEG studies in epilepsy. J Clin Neurophysiol. 1984;1:109–42.
17. Tatum WO, Husain AM, Benbadis SR, Kaplan PW. Normal adult EEG and patterns of uncertain significance. J Clin Neurophysiol. 2006;23:194–207.
18. Brazier MA. Preliminary proposal for an EEG terminology by the Terminology Committee of the International Federation for Electroencephalography and Clinical Neurophysiology. Electroencephalogr Clin Neurophysiol. 1961;13:646–50.
19. Maulsby RL. Some guidelines for assessment of spikes and sharp waves in EEG tracings. Am J EEG Technol. 1971;11:3–16.
20. Kane N, Acharya J, Beniczky S, Caboclo L, Finnigan S, Kaplan PW, et al. A revised glossary of terms most commonly used by clinical electroencephalographers and updated proposal for the report format of the EEG findings. Revision 2017. Clin Neurophysiol Pract. 2017;2:170–85.
21. Beniczky S, Schomer DL. Electroencephalography: basic biophysical and technological aspects important for clinical applications. Epileptic Disord. 2020;22(6):697–715.
22. Noachtar S, Binnie C, Ebersole J, Mauguiere F, Sakamoto A, Westmoreland B. A glossary of terms most commonly used by clinical electroencephalographers and proposal for the report form for the EEG findings. The International Federation of Clinical Neurophysiology. Electroencephalogr Clin Neurophysiol Suppl. 1999;52:21–41.
23. Kural MA, Aydemir ST, Levent HC, et al. The operational definition of epileptiform discharges significantly improves diagnostic accuracy and inter-rater agreement of trainees in EEG reading. Epileptic Disord. 2022;24(2):353–8.
24. Hirsch E, Panayiotopoulos CP. Childhood absence epilepsy and related syndromes. In: Roger J, BureauM DC, Genton P, Tassinari CA, Wolf P, editors. Epileptic syndromes in infancy, childhood and adolescence. 4th ed. Montrouge: John Libbey Eurotext Ltd; 2005. p. 315–35.
25. Weir P. The morphology of the spike-wave complex. Electroencephalogr Clin Neurophysiol. 1965;19:284–90.
26. Hirsch E, French J, Scheffer IE, Bogacz A, Alsaadi T, Sperling MR, Abdulla F, Zuberi SM, Trinka E, Specchio N, Somerville E, Samia P, Riney K, Nabbout R, Jain S, Wilmshurst JM, Auvin S, Wiebe S, Perucca E, Moshé SL, Tinuper P, Wirrell EC. ILAE definition of the idiopathic generalized Epilepsy syndromes: position statement by the ILAE task force on nosology and definitions. Epilepsia. 2022;63(6):1475–99.
27. Tassinari CA, Michelucci R, Gardella E, Rubboli G. Epilepsy with myoclonic absences. In: Engel J, Pedley TA, editors. Epilepsy: a comprehensive

textbook. 2nd ed. Philadelphia: Wolters Kluwer-Lippincott Williams & Wilkins; 2008. p. 2413–6.
28. Sun Y, Seneviratne U, Perucca P, Chen Z, Kee MT, O'Brien TJ, D'Souza W, Kwan P. Generalized polyspike train: an EEG biomarker of drug-resistant idiopathic generalized epilepsy. Neurology. 2018;91(19):e1822–30.
29. Cerulli Irelli E, Barone FA, Mari L, Morano A, Orlando B, Salamone EM, Marchi A, Fanella M, Fattouch J, Placidi F, Giallonardo AT, Izzi F, Di Bonaventura C. Generalized fast discharges along the genetic generalized Epilepsy Spectrum: clinical and prognostic significance. Front Neurol. 2022;10(13):844674. https://doi.org/10.3389/fneur.2022.844674.
30. Specchio N, Wirrell EC, Scheffer IE, Nabbout R, Riney K, Samia P, Guerreiro M, Gwer S, Zuberi SM, Wilmshurst JM, Yozawitz E, Pressler R, Hirsch E, Wiebe S, Cross HJ, Perucca E, Moshé SL, Tinuper P, Auvin S. International league against Epilepsy classification and definition of epilepsy syndromes with onset in childhood: position paper by the ILAE task force on nosology and definitions. Epilepsia. 2022;63(6):1398–442.
31. Tassinari CA, Michelucci R, Shigematsu H, Seino M. Atonic and myoclonic-atonic seizures. In: Engel J, Pedley TA, editors. Epilepsy: a comprehensive textbook. 2nd ed. Philadelphia: Wolters Kluwer-LippincottWilliams & Wilkins; 2008. p. 601–9.
32. Sperling MR, Morrell MJ. Atlas of electroencephalography. In: Pediatric and adult electroencephalography, vol. 2. Amsterdam: Elsevier; 1993.
33. Sperling MR, Clancy RR. Ictal electroencephalogram. In: Engel J, Pedley TA, editors. Epilepsy: a comprehensive textbook. 2nd ed. Philadelphia: Wolters Kluwer-Lippincott Williams & Wilkins; 2008. p. 825–54.
34. Williamson PD, Intensive monitoring of complex partial seizures: diagnosis and classification. In: Gumnit RJ, editor. Intensive neurodiagnostic monitoring. New York: Raven Press; 1987. p. 69–84.
35. Jayakar P, Duchowny M, Resnick TJ, Alvarez LA. Localization of seizure foci: pitfalls and caveats. J Clin Neurophysiol. 1991;8:414–31.
36. Engel J, Williamson PD, Wieser HG. Mesial temporal lobe epilepsy with hippocampal sclerosis. In: Engel J, Pedley TA, editors. Epilepsy: a comprehensive textbook. 2nd ed. Philadelphia: Wolters Kluwer-Lippincott Williams & Wilkins; 2008. p. 2479–86.
37. Lagae L, Pauwels J, Monte CP, et al. Frontal absences in children. Eur J Paediatr Neurol. 2011;5:243–51.
38. Delgado-Escueta A, Swartz B, Maldonado H, Complex partial seizures of frontal lobe origin. In: Wieser HG, Elger CE, editors. Presurgical evaluation of epileptics: basics, techniques, implications. New York: Springer-Verlag; 1987. p. 268–99.
39. Pugnaghi M, Meletti S, Castana L, Francione S, Nobili L, Mai R, Tassi L. Features of somatosensory manifestations induced by intracranial electrical stimulations of the human insula. Clin Neurophysiol. 2011;122(10):2049–58.
40. Mazzola L, Mauguière F, Isnard J. Electrical stimulations of the human insula: their contribution to the Ictal semiology of insular seizures. J Clin Neurophysiol. 2017;34(4):307–14.
41. Isnard J, Hagiwara K, Montavont A, Catenoix H, Mazzola L, Ostrowsky-Coste K, Guenot M, Rheims S. Semiology of insular lobe seizures. Rev Neurol (Paris). 2019;175(3):144–9.
42. Sveinbjornsdottir S, Duncan JS. Parietal and occipital lobe epilepsy: a review. Epilepsia. 1993;34:493–52.
43. Shorvon S, Tomson T. Sudden unexpected death in epilepsy. Lancet. 2011;378(9808):2028–38.
44. Ryvlin P, Nashef L, Lhatoo SD, et al. Incidence and mechanisms of cardiorespiratory arrests in epilepsy monitoring units (MORTEMUS): a retrospective study. Lancet Neurol. 2013;12(10):966–77.
45. Vilella L, et al. Postconvulsive central apnea as a biomarker for sudden unexpected death in epilepsy (SUDEP). Neurology. 2019;92(3):e171–82.
46. Lacuey N, Zonjy B, Hampson JP, et al. The incidence and significance of periictal apnea in epileptic seizures. Epilepsia. 2018;59(3):573–82. https://doi.org/10.1111/epi.14006.
47. Micalizzi E, Vaudano AE, Ballerini A, Talami F, Giovannini G, Turchi G, Cioclu MC, Giunta L, Meletti S. Ictal apnea: a prospective monocentric study in patients with epilepsy. Eur J Neurol. 2022;29(12):3701–10. https://doi.org/10.1111/ene.15547.
48. Tio E, Culler GW, Bachman EM, Schuele S. Ictal central apneas in temporal lobe epilepsies. Epilepsy Behav. 2020;112:107434. https://doi.org/10.1016/j.yebeh.2020.107434.
49. Meletti S, Duma GM, Burani M, Danieli A, Giovannini G, Osanni E, Micalizzi E, Mambretti F, Pugnaghi M, Vaudano AE, Bonanni P. Ictal and postictal central apnea in DEPDC5-related epilepsy. Neurol Genet. 2004; https://doi.org/10.1212/NXG.0000000000200183. Ahead of print.
50. Jackson JH. On asphyxia in slight epileptic paroxysms: on the symptomatology of slight epileptic fits supposed to depend on discharge-lesions of the Uncinate gyrus. Lancet. 1899;1:79–8.

8 The Development of Electroencephalography for Research and Clinical Purposes in Italy

Oriano Mecarelli, Flavio Villani, and Francesco Brigo

8.1 Introduction

In the field of medicine, physicians and clinical researchers who share a common approach or methodology often coalesce into schools—intellectual ecosystems where ideas flourish, methodologies converge, and legacies take root. Italy, too, witnessed the emergence of such schools as electroencephalography (EEG) gained prominence as a useful tool for diagnostic and research purposes. These schools of EEG often emerged around influential figures who had made significant contributions to clinical neurology and neurophysiology.

In this chapter, we have highlighted the role of the most influential schools of EEG in Italy, tracing their contributions from the 1930s onward, particularly those for which it was possible to retrieve ample historiographical sources. However, we acknowledge that there may be individuals or contributions we have inadvertently overlooked, and we apologize for any omissions.

O. Mecarelli (✉)
Department of Human Neurosciences, Sapienza University, Rome, Italy

F. Villani
Division of Clinical Neurophysiology – IRCCS Ospedale Policlinico San Martino, Genoa, Italy
e-mail: flavio.villani@hsanmartino.it

F. Brigo
Innovation, Research and Teaching Service; Teaching Hospital of the Paracelsus Medical Private University (PMU), Azienda Sanitaria dell'Alto Adige, Bolzano-Bozen, Italy
e-mail: Francesco.Brigo@sabes.it

8.2 Agostino Gemelli and the EEG Laboratory at Catholic University of the Sacred Heart in Milan

One of the pioneers of human EEG in Italy was Father Agostino Gemelli (1878–1959), a psychologist and philosopher who founded a Laboratory of Experimental Psychology at the Catholic University of the Sacred Heart in Milan in the early 1920s. This laboratory soon gained prestige, partly due to the connections Gemelli established with numerous neuropsychology research centers across Europe. In the late 1930s, two of his close collaborators, Trabattoni and Zunini, organized a highly efficient EEG laboratory at the Center for Experimental Psychology, directed by Gemelli. However, this laboratory was used exclusively for research purposes within the field of neuropsychology, focusing on EEG changes related to various psychic dynamics and sensory stimuli. The equipment used in the laboratory included linear power amplifiers connected to a kymograph with four writing pens (Fig. 8.1) [1, 2].

In 1947, during a conference on EEG advancements at Gemelli's Laboratory, the Italian Society

F. Brigo, O. Mecarelli (eds.), *EEG: The First 100 Years*,
https://doi.org/10.1007/978-3-031-86878-8_8

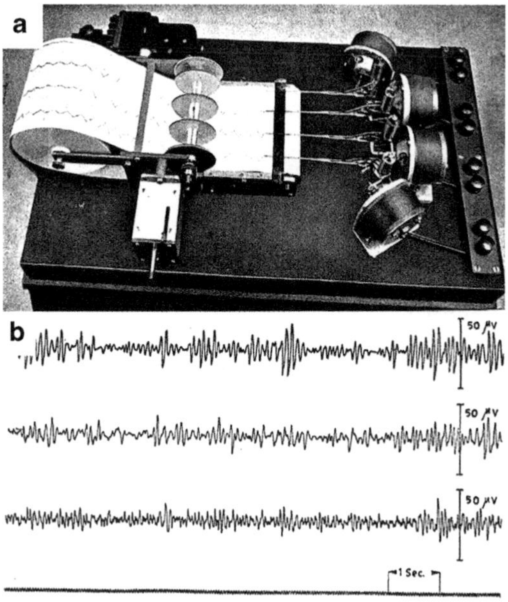

Fig. 8.1 (**a**) Kymograph with four writing pens, used to record EEGs in Agostino Gemelli's laboratory in 1939. (**b**) EEG derivations; from top to bottom: occipital, middle, and frontal. (Permission for reproduction granted by Vita e Pensiero Ed., Milan. www.vitaepensiero.it)

Fig. 8.2 Mario Gozzano in 1956

of Electroencephalography was founded. Gemelli was appointed President, with Giuseppe Moruzzi and Mario Gozzano serving as Vice-Presidents. However, historical sources are unclear about the exact year of the society's founding; some testimonies suggest that it was actually established in 1949 during the tenth Congress of the Italian Society of Neurology in Milan. The society was immediately recognized by the International Federation of Societies of Electroencephalography.

8.3 Mario Gozzano: His Pioneering Work and the Contribution of the Roman School

Mario Gozzano (1898–1986) was a true pioneer of clinical EEG and is often regarded as the founding father of the discipline in Italy [3] (Fig. 8.2). Born in the province of Cuneo in Piedmont, he graduated from the University of Torino with a degree in medicine and soon after went to Vienna to study experimental physiology techniques under Professor Otto Marburg. Upon his return to Italy, he became an assistant to Professor Giovanni Mingazzini in Rome. His academic career progressed with initial appointments as a professor of neurology at the universities of Napoli, Cagliari, Pisa, and Bologna, before finally settling at Rome's La Sapienza University in 1951.

Gozzano conducted significant experimental EEG research at the Berlin-Buch Brain Institute in the early 1930s, under the guidance of Oskar Vogt (1870–1959) and Alois E. Kornmüller (1905–1968). Kornmüller, who was familiar with the work of Hans Berger, operated as a sort of "competitor" to Berger, utilizing far more sophisticated apparatus in Berlin. Despite the unfavorable political climate—Oskar Vogt, a staunch anti-Nazi who protected his Jewish colleagues, was soon dismissed by government authorities—Gozzano continued his research, influenced by Kornmüller's "localizationist" theory of the brain as a mosaic of functionally different areas. Unfortunately, the results of Gozzano's groundbreaking studies, published in 1935 in Italian, had limited circulation internationally, although they were invaluable for Italian neurologists in understanding the fundamentals of EEG (Fig. 8.3) [4].

Gozzano's most important research findings in Germany and later in Italy included the description of strychnine-induced interictal and ictal patterns in the rabbit cortex and the study of reflex epilepsy, which explores the induction of epileptic manifestations in response to various stimuli. Gozzano was also among the first (pre-

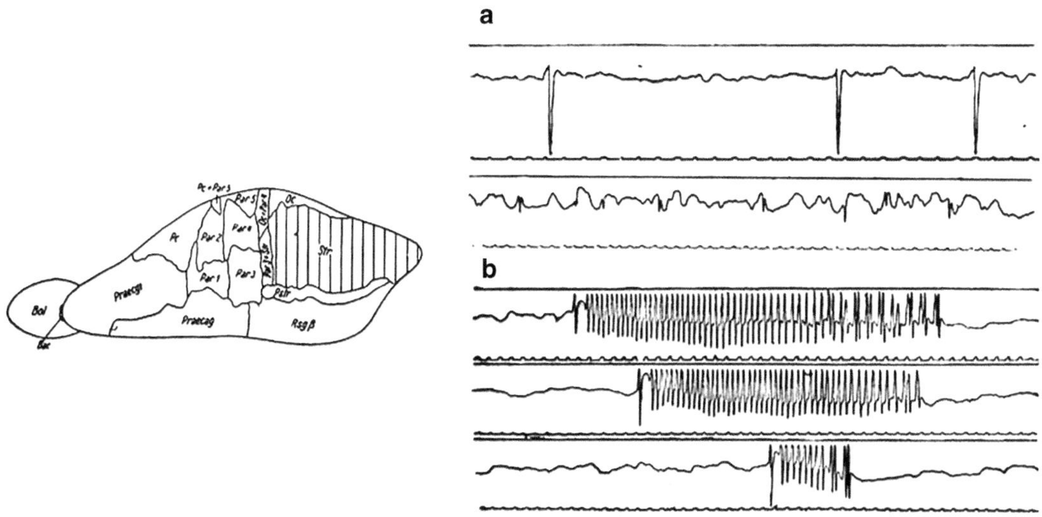

Fig. 8.3 Interictal (**a**) and ictal (**b**) epileptic discharges recorded by Gozzano on the cortex of a rabbit after the topical application of strychnine [3, 4]. (Courtesy of Salvatore Mazza)

ceded only by Berger) to identify rhythmic potential sequences at 11–12 Hz over the anterior cortical regions, though neither Berger nor Gozzano recognized this pattern as being characteristic of sleep. He was also the first to identify EEG frequencies exceeding the beta range following painful sensory stimulation [5].

In 1951, Gozzano, together with Spartaco Colombati, published the first Italian Handbook of Clinical Electroencephalography, which significantly advanced the knowledge of diagnostic techniques and EEG patterns characteristic of the known forms of epilepsy at the time [6].

Gozzano's contributions were not limited to academia and research; he also played a crucial role in the development of scientific societies in this field. He was one of the founders of the Italian EEG Society (which he presided over from 1950 to 1962, and which was renamed the Italian EEG and Neurophysiology Society in 1958). He organized the Fifth Congress of the International Federation of Societies for EEG and Clinical Neurophysiology in Rome in 1960. Additionally, he was the first president of the Italian League Against Epilepsy (LICE) in 1955, although the society remained inactive until it was reestablished in 1972.

Many young neurologists trained at the various universities where Gozzano was a professor, and we cannot mention all of them, so we will just provide brief biographical profiles of two of his collaborators who significantly contributed to EEG in both experimental and clinical settings.

Raffaello Vizioli (1926–2006) was one of Professor Gozzano's many collaborators. At just 25 years of age, he joined Gozzano in Bologna, where he managed the EEG Laboratory, and then to Rome, where he participated in research on experimental epilepsy in cats alongside his clinical work. His academic career later took him to Cagliari and then to Naples. Over the years, Vizioli became increasingly interested in psychopathology and medical anthropology, advocating for a holistic approach to patients with neurological disorders and fighting against the stigma associated with epilepsy. He emphasized that while EEG is a crucial diagnostic tool in this field, it should not be overused, and its limitations and frequent pitfalls of interpretation must be carefully considered. From 1960 to 1972, Vizioli served as secretary of LICE, an acronym for Lega Italiana Contro l'Epilessia, equivalent to Italian League Against Epilepsy [7–9].

Another of Gozzano's prominent collaborators was *Gianfranco Ricci* (1925–2000), who trained in experimental neurophysiology during the 1950s and 1960s, first in Pisa at Moruzzi's Laboratory and then at Jasper's Laboratory at the Montreal Neurological Institute. Ricci, along with his wife Bice Ballerini, a child neuropsychiatrist and pioneer of infantile EEG in Italy, went to Canada. Under Jasper's guidance, Ricci collaborated with Benjamin Doane on developing a technique for recording from single cells in monkeys restrained in a chair without anesthesia. This experimental setup allowed for the first time the recording of single neurons in different cortical areas of monkeys, both asleep and awake, during the formation of conditioned responses to various stimuli, and enabled the comparison of EEG surface recordings with intracellular recordings. The results of these experiments had a significant impact and were presented by Ricci and Doane in 1957 at the Fourth International EEG Congress in Brussels, and in 1958 at the Moscow Colloquium on Electroencephalography of Higher Nervous Activity (Figs. 8.4 and 8.5) [10].

Upon returning to Italy, Ricci continued his work in experimental neurophysiology, particularly in epileptology, while maintaining close connections with both Henri Gastaut and Robert Naquet, who had recently described the photosensitive epilepsy model in *Papio papio* baboons [11, 12]. Ricci eventually moved from the Neurological Clinic to the prestigious Child Neuropsychiatry Unit on Via dei Sabelli in Rome, directed by Professor Giovanni Bollea. Here, Ricci's focus gradually shifted to EEG and clinical epileptology, working closely with his wife, Bice Ballerini. Despite the challenges of maintaining the experimental laboratory while working in a clinical context, Ricci continued to pursue experimental epilepsy research for another decade, assisted by his brilliant collaborator Enrico Cherubini, who, after a year in Marseilles with Gastaut and Naquet, returned to Rome to conduct experimental research and study clinical neurophysiology in children [13–16]. However, in the early 1980s, Cherubini decided to focus entirely on research, eventually relocating first to Boston and then to INSERM in Paris, in the unit directed by Y. Ben-Ari. At this point, Gianfranco Ricci abandoned experimental epilepsy altogether and devoted himself entirely to developing Clinical Neurophysiology at the Sapienza University of Rome, where he was undoubtedly a pioneer.

Ricci's academic achievements were significant. In 1972, he founded the Special School for Technicians of Neuropathophysiology at Sapienza University, which, in the same year—thanks to Professor Elio Lugaresi—was also inaugurated at the University of Bologna. His entire academic career was subsequently dedicated to training neurophysiology technicians and physicians specializing in EEG and epilepsy. In 1990, he also succeeded in establishing the School of Specialization in Neurophysiopathology in Rome, although it was unfortunately discontinued by the Ministry of Education in 2010, creating a training gap in EEG and epileptology that has been difficult to fill.

Fig. 8.4 Left: Gianfranco Ricci with his wife Bice Ballerini (in the circle) in 1956 at the Montreal Neurological Institute. Right: Gianfranco Ricci in 1993 in his office at Sapienza University of Rome, while documenting an EEG on paper

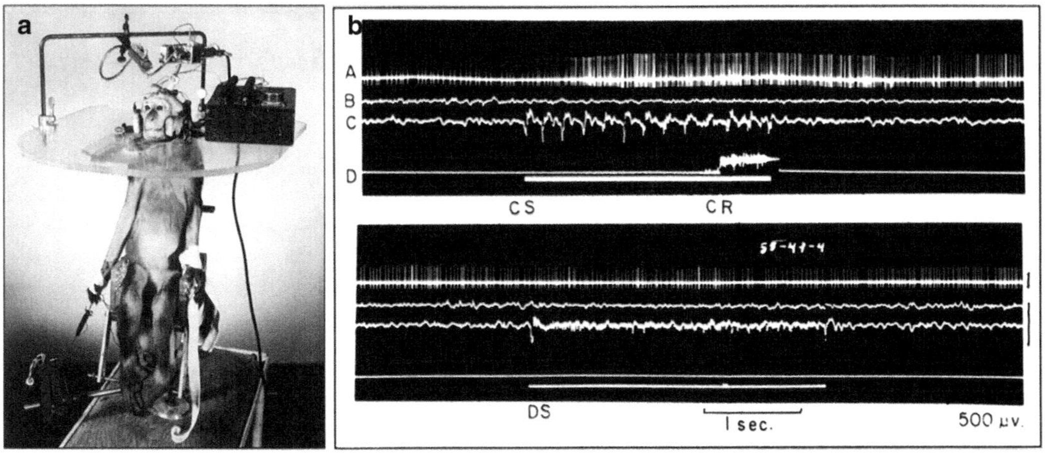

Fig. 8.5 Chronic Microelectrode Studies conducted at the Montreal Neurological Institute by Jasper, Ricci, and Doane (1956). (**a**) Monkey in a chair with a micromanipulator attached to its skull and hand attached to a switch that it pulls to avoid an electric shock. (**b**) Sample record of units from the motor cortex during conditioning; top line = units, second line = surface ECG, third line = surface evoked potentials from the occipital cortex in response to the conditioning stimulus (CS), fourth line = EMG and signal of conditioned response (CR), and fifth line = signal of CS. Lower records show a lack of response to the same intermittent photic stimulus at a higher frequency, not associated with shock (the differential stimulus, DS). (Reproduced with permission from Jasper [10], Cambridge University Press)

8.4 The School of Clinical Neurophysiology of Genoa

The first clinical-diagnostic use of EEG at the Neurological Clinic of the University of Genoa (directed by Lionello De Lisi, 1885–1957) began in the early 1950s, largely due to the experience and dedication of Carlo Loeb (1921–2005). Loeb had previously completed an internship at the Center for Neuroradiology and Neurophysiology at the Hôpital de la Pitié-Salpêtrière in Paris, under the direction of Hermann Fischgold. Fischgold, who collaborated with the renowned physiologist Giuseppe Moruzzi, conducted both experimental and clinical work on EEG.

Upon his return to Genoa, Loeb focused on the clinical application of EEG in various neurological disorders. He also demonstrated the EEG's specific utility in studying cerebral bioelectrical activity in animals under physiological conditions, particularly during the sleep–wake cycle, or following functional or structural brain insults (Fig. 8.6). The use of EEG for both experimental and clinical purposes was further encouraged by Cornelio Fazio (1910–2005), who succeeded De Lisi as the Clinic's director. Fazio supported Loeb in establishing and maintaining a highly productive electrophysiological research group, which produced a significant number of publications in prestigious international journals. Among the many young neurologists and neurophysiologists trained in Fazio's and later Loeb's schools, the most notable were Emilio Favale, Mario Manfredi, and Andrea Seitun.

From the late 1950s to the early 1970s, the experimental research conducted by the Genoese school focused primarily on the study of cortical-subcortical evoked potentials in felines during the sleep–wake cycle. Later, the researchers shifted their attention to the spontaneous activity of single cortical and subcortical motor neuron projections in freely moving animals during the sleep–wake cycle, utilizing both extracellular and intracellular recordings. This research period was followed by extensive investigations on in vivo and in vitro sections of the murine hippocampus, exploring the neuroprotective effects of drugs and energy precursors (such as phosphocreatine and its derivatives) on anoxia and post-anoxic recovery of nervous tissue.

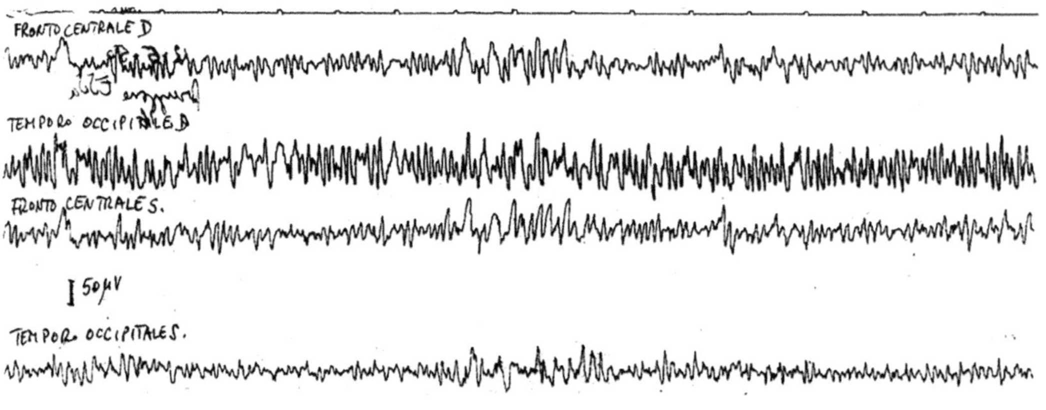

Fig. 8.6 EEG No. 4 recorded at the University of Genoa EEG Laboratory on June 9, 1950, in a 16-year-old boy with generalized tonic-clonic seizures, treated with phenobarbital. Four EEG traces (Top to bottom: right fronto-central, right temporo-occipital, left fronto-central, and left temporo-occipital). (Courtesy of Andrea Seitun)

Clinically, Loeb and his colleagues were among the pioneers in describing EEG changes in coma resulting from different etiologies. The Genoa research group published numerous international studies detailing EEG patterns in cerebrovascular, inflammatory, and tumor-related disorders. A notable contribution was the description of a specific EEG pattern in comatose patients, later defined "alpha-coma," which was indicative of an unfavorable outcome. This pattern was first described by Loeb and Poggio in 1953 following a ponto-mesencephalic hemorrhage [17, 18]. Loeb's 1958 paper, which analyzed recordings from 25 selected comatose patients, was one of the first attempts to establish an anatomo-electro-clinical correlation for prognostic purposes [19] (Fig. 8.7). Loeb was also among the first to describe EEG changes associated with cerebrovascular disorders. In 1967, in collaboration with Cesare Fieschi (1932–2016), he documented the relationship between EEG patterns and regional cerebral blood flow in 40 cases of cerebral infarction [20].

In subsequent decades, the EEG Laboratory at the Neurological Clinic of Genoa continued its significant diagnostic work under the leadership of Alberto Primavera. The lab became increasingly involved in prolonged EEG polygraphic recordings in coma and status epilepticus. In the early 1970s, Loeb founded the Center for the Diagnosis and Treatment of Epilepsy, which had dedicated and specifically trained medical and technical staff and was equipped with the first video-EEG and prolonged monitoring systems for wakefulness and sleep. This center, directed by Loeb for many years, also conducted pharmacological and neurophysiological research on experimental mouse models of epilepsy until its transformation in the late 1980s into an autonomous service under the direction of Andrea Seitun.

The Genoa School of Experimental and Clinical Neurophysiology was also instrumental in establishing another Neurophysiology Center in the late 1950s within the San Martino Hospital, at the Neurosurgical Clinic. The key figures in this Experimental and Clinical Neurophysiology Unit were Gianfranco Rossi (1927–2011) and Guido Rosadini (1928–2009). In 1969, Rossi was appointed to direct the Neurosurgical Clinic at the "Gemelli" Catholic University of Rome, while Rosadini remained in Genoa, where he established an important clinical and research group in the early 1970s. He secured the Chair of Neurophysiopathology from the University of Genoa and, in 1978, founded Italy's first School of Specialization in this discipline. The group organized by Rosadini, which included Walter Sannita, Franco Ferrillo, and Guido Rodriguez, focused primarily on qualitative and quantitative spectral analysis of the human EEG, particularly

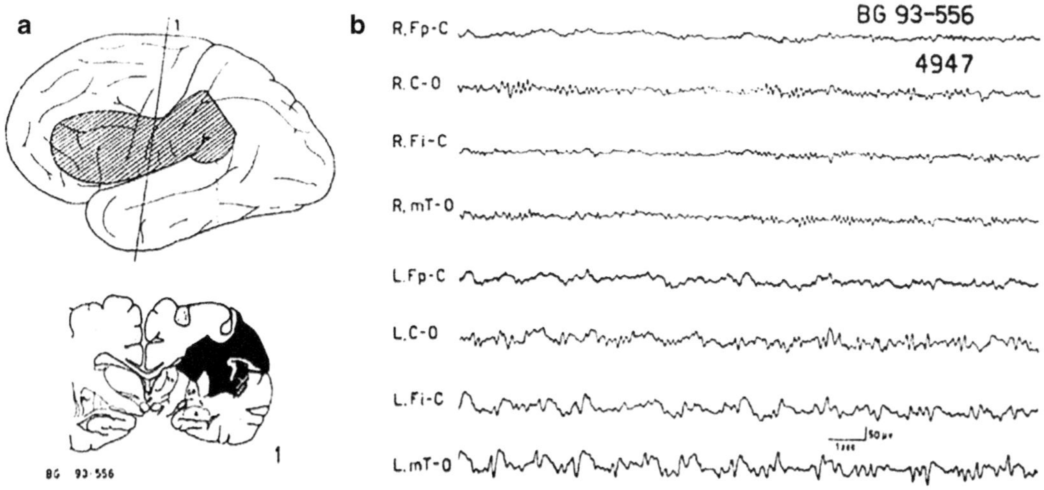

Fig. 8.7 EEG changes following a large infarct in the left hemisphere. (Original figure taken from Loeb, 1958, and reproduced with permission from Elsevier [19])

in relation to cerebral blood flow; the neurophysiological effects of drugs and psychoactive substances; and neurophysiological changes in sleep disorders, cerebrovascular diseases, and neurodegenerative conditions.

8.5 The School of Neurology and Electroencephalography at the University of Bologna

Clinical EEG in Italy began at the Neurological Clinic of the University of Bologna, led by Professor Mario Gozzano in the early 1950s. The development of EEG continued under Dr. Giovanni Maria Corsino, who, in 1956, published a widely circulated manual that facilitated the spread of EEG knowledge among generations of young physicians [21]. At that time, the Neurological Clinic was directed by Gozzano's successor, Professor Paolo Ottonello, who strongly supported the EEG Laboratory, where Dr. Elio Lugaresi was already working. In the introduction to his book, Corsino expressed gratitude to all his colleagues at the Clinic, particularly acknowledging Miss Rosa Cervi, the laboratory's indispensable technician and one of the first in Italy to dedicate herself to this profession (Fig. 8.8).

The Bologna school of EEG owes much of its development to the work of Elio Lugaresi (1926–2015), who dedicated his entire career to neurology, particularly the neurophysiological study of epilepsy and various sleep disorders. Shortly after graduating from medical school, Lugaresi became interested in epilepsy, inspired by a monograph by the renowned neurophysiologist Giuseppe Moruzzi. During the difficult years before and immediately after World War II, many brilliant Italian researchers—such as Pampiglione, Ajmone Marsan, Lombroso, and Chatrian—became passionate about the study of epilepsy and EEG but were compelled to emigrate abroad. However, in war-ravaged and impoverished Europe, Henri Gastaut succeeded in organizing what would become one of the world's most prestigious epilepsy centers in Marseille, attracting many Italian neurologists due to its geographic proximity. Like many of his Italian colleagues, Lugaresi spent a brief but pivotal period in 1956 at Gastaut's Center in Marseille, an experience that profoundly influenced his academic career. Gastaut became a mentor and friend to Lugaresi, and their relationship endured until Gastaut's death. They frequently met at scientific and organizational conferences. In 1967, in collaboration with Gastaut, Lugaresi organized the first International

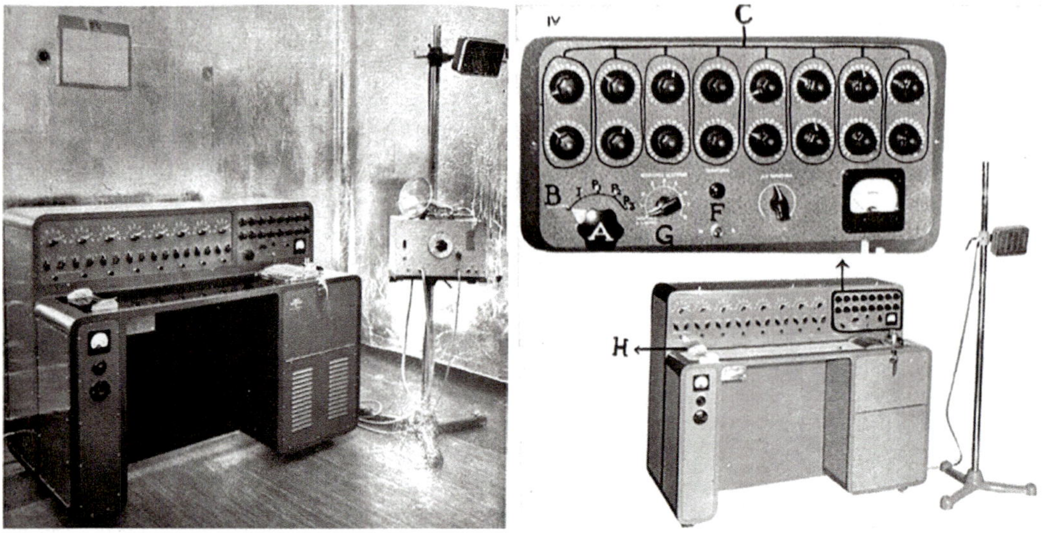

Fig. 8.8 ATE electroencephalograph with eight writing pens used at the EEG Laboratory in Bologna from 1951 to 1956, with which the tracings shown in G.M. Corsino's manual [21] were recorded. (Reproduced with permission from Cappelli Editore, Bologna)

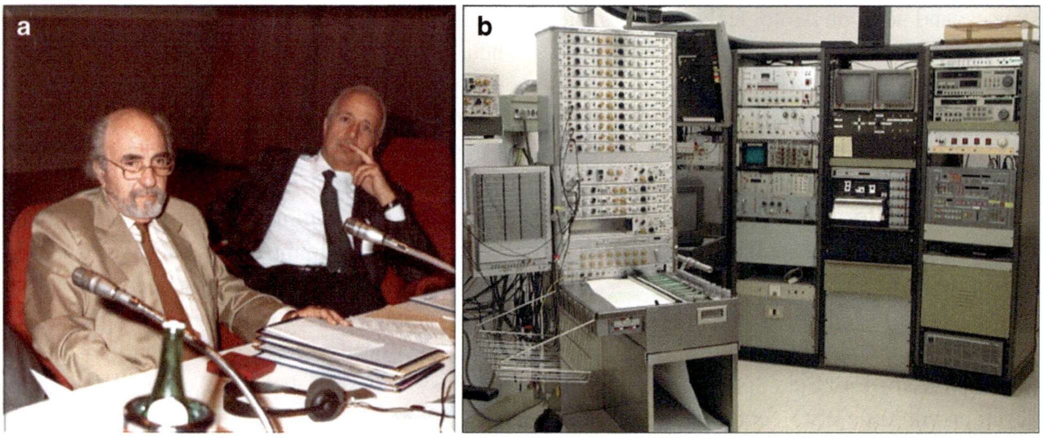

Fig. 8.9 (**a**) Henri Gastaut and Elio Lugaresi chairs at a conference in Bologna in the late 1980s. (**b**) Equipment used for Polysomnography in the 1980s–1990s at the University of Bologna Neurological Clinic. (Courtesy of Federica Provini)

Conference on Sleep Disorders (Fig. 8.9a), and in 1972, Gastaut participated in the International Colloquium on the Prognosis and Evolution of Epilepsy in Venice, where Lugaresi was elected President of the refounded LICE, with Raffaele Canger as Secretary. From 1981 to 1988, the Neurological Clinic of the University of Bologna hosted biennial Update Courses in Clinical Epileptology (later known as Polycentric Meetings in Epileptology, still organized annually by LICE). These courses were well attended by young neurologists who had the opportunity to discuss complex and unresolved clinical cases with leading epileptologists of the time, including non-Italians such as Gastaut, Roger, Janz, and Andermann and Italians such as Bergamini, Mutani, Angeleri, Baruzzi, Canger, Tassinari, Dalla Bernardina, and Munari.

Elio Lugaresi served as President of LICE from 1972 to 1976, and he also held the presidencies of the Italian EEG and Clinical Neurophysiology Society (1969–1972), the Italian Society of Neurology (1984–1987), and the Italian Society of Sleep Medicine (1990–1994). In addition to his work in epilepsy (he was awarded the title of Ambassador for Epilepsy by the International League Against Epilepsy [ILAE] in 1979), Lugaresi was a pioneer in the field of sleep disorders. His scientific achievements in this area remain his most enduring legacy, serving as a benchmark for all those involved in sleep research (Fig. 8.9b). Lugaresi conducted the first polysomnographic studies on what was initially called "nocturnal myoclonus," and he was the first to study the hemodynamic and ventilatory effects of obstructive sleep apnea [22, 23]. Lugaresi and his collaborators also first described a new form of epilepsy occurring during sleep, later called nocturnal frontal lobe epilepsy, and currently classified as sleep-related hypermotor epilepsy [24–26]. However, Lugaresi's most significant contribution to sleep medicine is the comprehensive clinical, neurophysiological, pathological, and genetic characterization of the prion disease he named Fatal Familial Insomnia [27].

Subsequent research by Lugaresi and his group revealed that an autoimmune encephalopathy (Morvan's syndrome) and an acute psychotic state triggered by alcohol withdrawal syndrome (delirium tremens) shared clinical and polysomnographic features with fatal familial insomnia. This led to the identification of a syndrome he termed agrypnia excitata, characterized by a loss of sleep accompanied by generalized motor and autonomic hyperactivation, likely due to an anatomic or functional disruption of the thalamolimbic circuits that regulate the sleep–wake cycle and body homeostasis [28].

Among Lugaresi's notable collaborators, those who have since passed away include Pasquale Montagna (1950–2010), Giorgio Coccagna (1936–2024), and Paolo Pazzaglia (1940–2019), who led the Epilepsy Centre at the Bologna Neurological Clinic for many years. Many neurologists trained at Lugaresi's School or collaborated with him, too numerous to list comprehensively. Notably, some of them, such as Carlo Alberto Tassinari and Paolo Tinuper, continue to make significant contributions to the field of epileptology.

8.6 The School of Clinical Neurophysiology at the University of Turin

Cosimo Ajmone Marsan (1918–2004) was among the first pioneers in EEG at the University of Turin. In 1948, he began recording the first EEG tracings using a six-channel Grass Model II, a device he purchased at his own expense. However, by 1950, Marsan's career took him abroad, first to the Montreal Neurological Institute under the direction of Wilder Penfield, and then to the National Institutes of Health in Bethesda, Maryland. There, until 1979, he directed the EEG and Clinical Neurophysiology Unit. He later moved to the University of Miami, where he continued his work until his death.

Another key figure in Turin's neurophysiology community was Lodovico Bergamini (1924–1996), whose interests spanned electromyography, cortical evoked potentials, and both experimental and clinical epileptology. From the 1950s onward, Bergamini was a strong advocate for the prognostic use of EEG in evaluating comatose patients [29, 30]. A dedicated educator, Bergamini was instrumental in training young neurologists. In 1974, he published the first edition of his *Handbook of Neurology*, which quickly became a textbook in many Italian universities. This work saw numerous editions over the years, often in collaboration with his close associates, Bruno Bergamasco (1937–2007) and Roberto Mutani, who served as the president of the LICE from 1983 to 1986. Bergamasco, Bergamini's first collaborator, was deeply involved in neurophysiological research, particularly in the diagnosis and prognosis of various stages of coma during the 1960s and 1970s. The Laboratory of Neurophysiology at the Neurological Clinic in

Turin was at the forefront of evoked potential recording in Italy during the 1960s. Their work led to the publication of *Cortical Evoked Potentials in Man* in 1967, which gained widespread recognition, with a foreword by Henri Gastaut [31] (Fig. 8.10).

Gian Emilio Chatrian (1926–2024), another notable figure associated with the Turin School, made significant contributions to clinical neurophysiology in the United States. Born in Turin, Chatrian pursued research in clinical neurophysiology with Henri Gastaut in Marseille and stereo-EEG with Jean Bancaud and Jean Talairach in Paris. Emigrating to the United States in 1954, he conducted research at Harvard Medical School and later at the Mayo Clinic in Rochester. Eventually, he founded and led the Division of EEG and Clinical Neurophysiology at the University of Washington School of Medicine in Seattle, where he worked until his retirement. Chatrian's research in EEG, epilepsy, evoked potentials, and intraoperative neurophysiologic monitoring was groundbreaking. He served as President of the American EEG Society (later the American Clinical Neurophysiology Society) in 1968 and the American Epilepsy Society in the 1970s. A member of the International Federation of Societies of EEG and Clinical Neurophysiology (IFCN), he authored several chapters in the *Handbook of EEG and Clinical Neurophysiology* and other textbooks. Chatrian is credited with describing some of the most characteristic EEG patterns, including periodic lateralized epileptiform discharges [32]. In 1974, he co-authored the *Glossary of EEG Terms*, the first international effort to establish a common language for EEG reporting. This glossary, which also included contributions from Bergamini, was officially translated into Italian in 1978 by the Italian Society of EEG and Neurophysiology [33, 34].

Fig. 8.10 Equipment used in the 1960s by Bergamini and Bergamasco at the University of Turin's Laboratory of Neurophysiology. (Figure taken from the book Cortical Evoked Potentials in Man, published in 1967 [31]; reproduced with permission from Charles C Thomas Publisher)

8.7 Raffaele Canger and the Residential Courses in Gargnano on Epileptology and Clinical Electroencephalography

Raffaele Canger (1938–2019), after training at the University of Heidelberg under Dieter Janz, became a professor of Clinical Neurophysiology at the University of Milan. He dedicated his life to the diagnosis and treatment of epilepsy at the Epilepsy Centre of San Paolo Hospital, where he led a team of epileptologists who cared for both children and adults with epilepsy. His work was pivotal in developing the Scientific Society LICE, where he served as the Secretary from its refounding in 1972 and later as President from 1990 to 1993. From 1973 to 2009, Canger edited the regular publication of the *LICE Bulletin*, which included proceedings from National Congresses, LICE recommendations and guidelines, and other resources useful for Italian epileptologists in clinical practice. For his contributions to LICE, he was made an Honorary Member in 2008. On the international stage, he was recognized as an Ambassador for Epilepsy by the ILAE in 1977 and served as a chairman for the ILAE European Congress in Florence in 2000.

One of Canger's most significant contributions to education was the establishment of the residential course in Epileptology in 1987 and in Clinical EEG in 1989 in Gargnano on Lago di Garda. Canger organized the Gargnano Courses with great enthusiasm until shortly before his death, inviting internationally renowned figures such as Dieter Janz, Peter Wolf, Annette Beaumanoir, Joseph Roger, Charlotte Dravet, Giuliano Avanzini, and Claudio Munari, among many others (Fig. 8.11). Virtually, all Italian epileptologists and electroencephalographists have participated in the Gargnano Courses at least once. The courses, held biannually, continue to attract young specialists and residents in child neuropsychiatry and neurology.

In clinical research, Canger made substantial contributions in areas such as epilepsy and pregnancy, the teratogenic effects of antiseizure medications, the treatment of pharmacoresistant epilepsies, and the cognitive and psychiatric comorbidities associated with epilepsy [35–38]. Beyond his clinical and research contributions, Canger was also a tireless advocate for improving public understanding of epilepsy and fighting stigma [39]. In the early 1980s, he co-founded the Lombardy Association Against Epilepsy for individuals with epilepsy and their families. This organization played a key role in organizing a significant meeting on the prevention of epilepsy in Milan in 1983, with Dieter Janz as a speaker. Additionally, from 1985 to 1989, Canger served as Vice President of the International Bureau for Epilepsy.

8.8 Other Leading Italian Schools of Clinical Neurophysiology

The Florence School of EEG is closely associated with Roberto Zappoli Thyrion (1926–2015), a student of Osvaldo Maleci, who moved from Padua to Florence. Zappoli had a deep interest in Clinical Neurophysiology, serving as President of the Italian Clinical Neurophysiology Society from 1978 to 1981. In 1971, at the Hospital of Careggi, he established an independent Clinical Neurophysiology Unit with an attached Epilepsy Centre, emphasizing the discipline's autonomy from the Neurological Clinic. Starting in the mid-1970s, Zappoli, along with colleagues like Rosadini in Genoa and Ricci in Rome, vigorously promoted Clinical Neurophysiology both academically and at the ministerial level. This advocacy led to the creation of the School of Specialization in Clinical Neurophysiology at various universities, including Genoa in 1978 and Florence in 1981. In Florence, Zappoli and his team were strong proponents of applying EEG and clinical neurophysiology in critical care settings, particularly for the prognostic assessment of consciousness disorders. This work fostered close collaboration between resuscitators and neurophysiologists, which significantly

Fig. 8.11 Left: Raffaele Canger with Annette Beaumanoir in 1995 at an EEG Course in Gargnano, and (right) in 2008 in Rome, where he was awarded the title of Honorary Member of the Italian League Against Epilepsy (LICE) by then-President Paolo Tinuper

advanced scientific research [40–42]. Throughout his career, Zappoli trained numerous expert electroencephalographers, contributing to the spread of advanced clinical neurophysiology practices throughout Italy.

At the University of Pavia's Neurological Clinic (C. Mondino Neurological Institute), the first attempts to record a human EEG were made by Paolo Pinelli (1921–2015) in 1948, using home-made equipment. Pinelli published a paper in Italian on alpha rhythm changes under various clinical conditions in *Minerva Medica* [43]. From 1949 to 1950, Pinelli studied at the Institute of Neurophysiology in Copenhagen under Fritz Buchtal, a pioneer in electromyography (EMG). There, he collaborated with Margareth A. Lennox in pediatric EEG, gaining comprehensive training in clinical neurophysiology. Pinelli's work in EMG did not diminish his collaboration with epileptologists. Notably, when he organized the first International Congress of EMG in Pavia in 1961, he invited Henri Gastaut to chair the event. Pinelli's academic career culminated in his transfer to the University of Milan in 1980. The Pavia School of Neurophysiology was notable for its excellence in both EEG and electromyography, with significant contributions from professors Antonio Arrigo (1929–1990) and Arrigo Moglia (1947–2020).

Among the EEG researchers at the University of Pavia, Faustino Savoldi (1930–2016) played a prominent role in studying EEG changes in migraine and other neurological disorders early in his career. Succeeding Pinelli in 1980 as the director of the Neurological Clinic and the Mondino Institute, Savoldi consistently supported both clinical and experimental EEG research. In 1983, he famously stated to a young resident, Carlo Andrea Galimberti, that "electroencephalography is the noblest thing a neurologist can devote himself to." Starting in the 1960s, Amelia Tartara (1936–2013) emerged as a leading figure in EEG research. In 1977, she was appointed director of the EEG Laboratory and the Center for the Diagnosis and Treatment of Epilepsy. Under her leadership, the EEG Laboratory also developed a focus on sleep medicine, a field that gained increasing interest in the 1980s. After Tartara's retirement in 2001, the EEG Center for Sleep Disorders was directed by Raffaele Manni, while Carlo Andrea Galimberti led the Epilepsy Center.

At the University of Parma's Neurological Clinic, experimental and clinical neurophysiology saw significant advancements in the 1950s, primarily due to Mirko Carreras (1927–2005), who had trained in Giuseppe Moruzzi's laboratory in Pisa and later in the United States. Carreras conducted important research on the organization of thalamic nuclei and their cortical connections, as well as the centrifugal control of sensory transmission, in collaboration with Giuliano Avanzini. Later, Domenico Mancia (1935–2004), known for his intellectual prowess and communication skills, advanced clinical neurophysiology in epileptology and disorders of consciousness, particularly toxic-metabolic and infectious encephalopathies. Another prominent figure from the Parma School, Franco Angeleri (1928–2008), moved to Perugia, where he established a leading Laboratory of Experimental and Clinical Neurophysiology. Angeleri served as LICE President from 1976 to 1979 and organized the XI Epilepsy International Congress in Florence in 1979. He played a key role in defining the criteria for recognizing Centers for the Diagnosis and Treatment of Epilepsy and organized the first scientific meetings of Neo-Latin language Leagues. Angeleri's career continued at the University of Ancona, where he worked until his retirement in 2000.

The School of Clinical Neurophysiology in Parma also contributed significantly to the international development of EEG, thanks to Mario Giovanni Terzano (1944–2020). Terzano, who trained at the Marseille School, focused his research on sleep EEG, in both physiological and pathological contexts. He made significant contributions by describing the microstructure of sleep and the characteristics of the cyclic alternating pattern [44, 45]. Terzano co-founded the Italian Society of Sleep Medicine and served as its President from 1997 to 2000.

The Verona School of Neurology developed under the guidance of Hrayr Terzian (1925–1988), who trained under Gianbattista Belloni at the University of Padua and Giuseppe Moruzzi at the Institute of Physiology in Pisa. Terzian's interests spanned the clinical definition of neurological syndromes to the development of biological and neurophysiological diagnostic techniques [46]. He equipped the Neurological Clinic at the University of Verona with a cutting-edge Laboratory of Neuropathology and Applied Neurobiology while maintaining a strong focus on advancements in psychiatry. Working alongside Carlo Agostino Terzuolo (1925–1999), Terzian explored the cortical mechanisms related to epilepsy induced by strychninization of the cortex in rabbits—an animal model developed by Antonino Clementi (1888–1968). In 1952, Terzian investigated the electroencephalographic changes induced by chlorpromazine, an antipsychotic medication developed a year earlier by Paul Charpentier in the French laboratories of Rhône-Poulenc [47]. Furthermore, in Marseille, alongside Henri Gastaut and his wife Yvette, Terzian provided the initial description of the rolandic arched rhythm, also known as the mu rhythm [48].

8.9 Conclusions

The history of EEG in Italy is marked by significant contributions from several pioneering figures and institutions that have advanced the field both nationally and internationally. Their legacy is characterized by their dedication to research, education, and clinical excellence. Their collective efforts have not only shaped this field in Italy but also contributed significantly to the global understanding and application of the EEG for clinical and research purposes.

Acknowledgments We gratefully thank the following colleagues for helping us with their testimonies in writing this chapter:

- Gabriella Colicchio and Salvatore Mazza for Gemelli's and Gozzano's School;
- Andrea Seitun, Mario Manfredi, Alberto Primavera, and Walter G. Sannita for the School of Genoa;
- Roberto Mutani and Paolo Benna for the School of Turin;
- Aldo Amantini for the School of Florence;
- Carlo Andrea Galimberti and Paolo Mazzarello for the School of Pavia.

References

1. Gemelli A, Trabattoni C. Un elettroencefalografo a penna scrivente a inchiostro per uso clinico. In: Vita e Pensiero, editor. Contributi del Laboratorio di Psicologia, vol. 1. Milano; 1944. p. 109–20.
2. Gemelli A, Ancona L. Electroencephalographic study of the psychic activity. Riv Neurol. 1951;21:358–9.
3. Mazza S, Pavone A, Niedermeyer E. Mario Gozzano: the work of an EEG pioneer. Clin Electroenceph. 2002;33:155–9.
4. Gozzano M. Ricerche sui fenomeni elettrici della corteccia cerebrale. Riv Neurol. 1935;8:212–61.
5. Trabka J. High frequency components in brain waves. Electroencephalogr Clin Neurophysiol. 1963;14:453–64.
6. Gozzano M, Colombati S. Elettroencefalografia Clinica. Torino: Rosenberg & Sellier; 1951. p. 1–137.
7. Vizioli R. A clinical and EEG survey on infantile and childhood epilepsy. Epilepsia. 1962;3:1–13.
8. Vizioli R. Tribute to Mario Gozzano (1898–1986). Ital J Neurol Sci. 1988;9:405–6.
9. Vizioli R. Musicogenic epilepsy. Int J Neurosci. 1989;47:159–64.
10. Jasper HH. History of the early development of Electroencephalography and Clinical Neurophysiology at the Montreal Neurological Institute: the first 25 years 1939–1964. Can J Neurol Sci. 1991;18(4):533–48.
11. Naquet R. Photosensitive epilepsy of the baboon Papio papio. Proc R Soc Med. 1972;65(2):180.
12. Naquet R, Catier J, Menini C. Neurophysiology of photically induced epilepsy in Papio papio. Adv Neurol. 1975;10:107–18.
13. Naquet R, Cherubini E, Carlier E, Gastaut H. Facial spasm and photosensitivity epilepsy in Papio papio. Rev Neurol. 1972;127:529–35.
14. Cherubini E, Gonella J, Mancia D, Mecarelli O, Tassinari CA. Effects of PTZ-induced generalized epilepsy on intestinal activity in cats. Epilepsia. 1981;22:309–14.
15. Ricci GF, Cherubini E, de Feo MR, Mecarelli O. Development of epileptic activity and its correlation with GABAergic and glutamergic systems during brain maturation in the rat. Prog Clin Biol Res. 1983;124:77–86.
16. Cherubini E, de Feo MR, Mecarelli O, Ricci GF. Behavioral and electrographic patterns induced by systemic administration of kainic acid in developing rats. Brain Res. 1983;285:69–77.
17. Loeb C, Poggio G. Electroencephalograms in a case with ponto-mesencephalic haemorrhage. Electroencephalogr Clin Neurophysiol. 1953;5(2):295–6.
18. Kaplan PW, Genoud D, Ho TW, Jallon P. Etiology, neurologic correlations and prognosis in alpha coma. Clin Neurophysiol. 1999;110:205–13.
19. Loeb C. Electroencephalographic changes during the state of coma. Electroencephalogr Clin Neurophysiol. 1958;10(4):589–606.

20. Loeb C, Fieschi C. Electroencephalograms and regional cerebral blood flow in cases of brain infarction. Electroenceph. Clin Neurophysiol. 1967;Suppl 25:111–8.
21. Corsino GM. Introduzione all'Elettroencefalografia Clinica. Bologna: Cappelli Editore; 1957. p. 1–195.
22. Lugaresi E, Coccagna G, Mantovani M, Berti Ceroni G, Pazzaglia P, Tassinari CA. The evolution of different types of myoclonus during sleep. A polygraphic study. Eur Neurol. 1970;4(6):321–31.
23. Lugaresi E, Cirignotta F, Coccagna G, Baruzzi A. Snoring and the obstructive apnea syndrome. Electroencephalogr Clin Neurophysiol Suppl. 1982;35:421–30.
24. Plazzi G, Montagna P, Tinuper P, Cerullo A, Provini F, Lugaresi E. Autosomal dominant nocturnal frontal lobe epilepsy (ADNFLE). Epilepsia. 1997;38(6):739–40.
25. Provini F, Plazzi G, Lugaresi E. From nocturnal paroxysmal dystonia to nocturnal frontal lobe epilepsy. Clin Neurophysiol. 2000;111(Suppl 2):2–8.
26. Tinuper P, Bisulli F, Provini F, Montagna P, Lugaresi E. Nocturnal frontal lobe epilepsy: new pathophysiological interpretations. Sleep Med. 2011;12(Suppl 2):39–42.
27. Lugaresi E, Medori R, Montagna P, Baruzzi A, Cortelli P, Lugaresi A, Tinuper P, Zucconi M, Gambetti P. Fatal familial insomnia and dysautonomia with selective degeneration of thalamic nuclei. N Engl J Med. 1986;315(16):997–1003.
28. Montagna P, Lugaresi E. Agrypnia Excitata: a generalized overactivity syndrome and a useful concept in the neurophysiopathology of sleep. Clin Neurophysiol. 2002;113(4):552–60.
29. Bergamasco B, Bergamini L, Doriguzzi T, Fabiani D. EEG sleep patterns as a prognostic criterion in post-traumatic coma. Electroencephalogr Clin Neurophysiol. 1968;24(4):374–7.
30. Bergamini L, Bergamasco B, Benna P, Gilli M. Acquired etiological factors in 1,785 epileptic subjects: clinical-anamnestic research. Epilepsia. 1977;18(4):437–44.
31. Bergamini L, Bergamasco B. Cortical evoked potentials in man. Springfield: Charles C Thomas Publ; 1967. p. 116.
32. Chatrian GE, Shaw CM, Leffman H. The significance of Periodic Lateralized Epileptiform Discharges in EEG; an electrographic, clinical and pathological study. Electroencephalogr Clin Neurophysiol. 1964;17:177–93.
33. Chatrian GE, Bergamini L, Dondey M, Klass DW, Lennox-Buchthal M, Petersen I. A glossary of terms most commonly used by clinical electroencephalographers. Electroencephalogr Clin Neurophysiol. 1974;37:538–48.
34. Chatrian GE, Bergamini L, Dondey M, Klass DW, Lennox-Buchthal M, Petersen I. Dizionario dei termini più usati in Elettroencefalografia Clinica. Versione italiana a cura della Società Italiana di Elettroencefalografia e Neurofisiologia. In: Gaggi A, editor. Riv It EEG e Neurofisiol, vol. 1 Fasc 1. Bologna; 1978. p. 3–20.
35. Canger R, Janz D. The EEG in primary and secondary psychomotor epilepsy. Riv Neurol. 1966;36:321–6.
36. Canger R, Cornaggia C, Bianchi M. Epilepsy and pregnancy. Prog Clin Biol Res. 1983;124:353–60.
37. Cornaggia C, Canevini MP, Giuccioli D, Pinelli P, Pruneri C, Canger R. Monotherapy and polytherapy for intractable epilepsies. Ital J Neurol Sci. 1985;6:201–5.
38. Canger R, Dalla Bernardina B, Avanzini G, Vigevano F, Tassinari CA. Genetic research in epilepsy: the Italian League Against Epilepsy contribution. Acta Neurol Scand Suppl. 1992;140:47–50.
39. Canger R, Cornaggia C. Public attitudes toward epilepsy in Italy: results of a survey and comparison with U.S.A. and West German data. Epilepsia. 1985;26:221–6.
40. Zappoli R. Effects of electrographic seizure discharges on conditioned avoidance responses in man. Electroencephalogr Clin Neurophysiol. 1970 Feb;28(2):213.
41. Denoth F, Zappoli R, Wenegrat B. Current methodological issues in ERP research: some remarks and proposals. A review. Electroencephalogr Clin Neurophysiol Suppl. 1987;40:24–30.
42. Zappoli R, Zappoli Thyrion GD. Cognitive potentials and corticocortical connections in humans. Electroencephalogr Clin Neurophysiol Suppl. 1999;50:74–84.
43. Pinelli P. Observations on the behavior of the alpha rhythm in various clinical conditions. Minerva Med. 1948;39(22):547.
44. Terzano MG, Mancia D, Salati MR, Costani G, Decembrino A, Parrino L. The cyclic alternating pattern as a physiologic component of normal NREM sleep. Sleep. 1985;8(2):137–45.
45. Terzano MG, Parrino L, Fioriti G, Spaggiari MC, Piroli A. Morphologic and functional features of cyclic alternating pattern (CAP) sequences in normal NREM sleep. Funct Neurol. 1986;1(1):29–41.
46. Brigo F. Hrayr Terzian (1925-1988): a life between experimental neurophysiology and clinical neurology. Neurol Sci. 2021;42(9):3939–42.
47. Terzian H. Studio elettroencefalografico dell'azione centrale del largactil (4560 RP) [Electroencephalographic study of the central action of largactil (4560 RP)]. Rass Neurol Veg. 1952;9:211–5.
48. Gastaut H, Terzian H, Gastaut Y. Etude d'une activité électroencéphalographique méconnue: le rythme rolandique en arceau [Study of a little electroencephalographic activity: rolandic arched rhythm]. Mars Med. 1952;89:296–310.

Development and Clinical Applications of Video-EEG

9

Lorenzo Ferri, Lidia Di Vito, Paolo Tinuper, and Francesca Bisulli

9.1 Video-EEG History

Video-electroencephalography (video-EEG) is a vital tool in clinical neurophysiology, enabling the simultaneous acquisition of EEG signals and corresponding video footage, to assess the clinical correlates of recorded EEG signals. The development of video-EEG has spanned several decades marked by numerous technological innovations and progress.

The initial EEG recordings, pioneered by Hans Berger in 1924, permitted only brief recordings of brain electrical activity. With the advent of the continuous paper chart recorder in 1930 and later advancements in computer technology, it became feasible to extend recordings and conduct long-term monitoring (LTM) [1].

Before modern video-EEG, seizures were documented using photography. At La Salpêtrière hospital in Paris, Charcot and his colleagues applied photography to epilepsy and related conditions, exploring hystero-epilepsy and challenging the prevailing beliefs about demonic possessions, ecstasy, hallucinations, and intense religious phenomena.

The diagnostic utility of these recordings was limited since they lacked visual synchronization with EEG, which hindered the ability to correlate paroxysmal activities with clinical manifestations both in real-time and retrospectively.

In 1938, at the 96th Annual Meeting of the American Psychiatric Association, neurologist Robert S. Schwab presented a method for simultaneously recording patients and their EEG using photography. He incorporated mirrors into his recordings and projected the patient's movements and EEG readings through a mirror reflection onto the same film [2, 3].

In 1949, Hunter and Jasper in Montreal reported a system for recording both the EEG and a frame of the patient on the same cinematographic film (Fig. 9.1).

The main drawback of this system was the film's limited capacity, allowing only run for 20 min of recordings before it had to be changed. Various video techniques were developed

L. Ferri (✉) · L. Di Vito
IRCCS Istituto delle Scienze Neurologiche di Bologna, ERN-EpiCARE full Member, Azienda USL di Bologna, Bologna, Italy
e-mail: lorenzo.ferri13@unibo.it; lidia.divito@ausl.bologna.it

P. Tinuper
Alma Mater Studiorum, University of Bologna, Bologna, Italy
e-mail: paolo.tinuper@unibo.it

F. Bisulli
IRCCS Istituto delle Scienze Neurologiche di Bologna, ERN-EpiCARE full Member, Azienda USL di Bologna, Bologna, Italy

Department of Biomedical and NeuroMotor Sciences, University of Bologna, Bologna, Italy
e-mail: francesca.bisulli@unibo.it

F. Brigo, O. Mecarelli (eds.), *EEG: The First 100 Years*,
https://doi.org/10.1007/978-3-031-86878-8_9

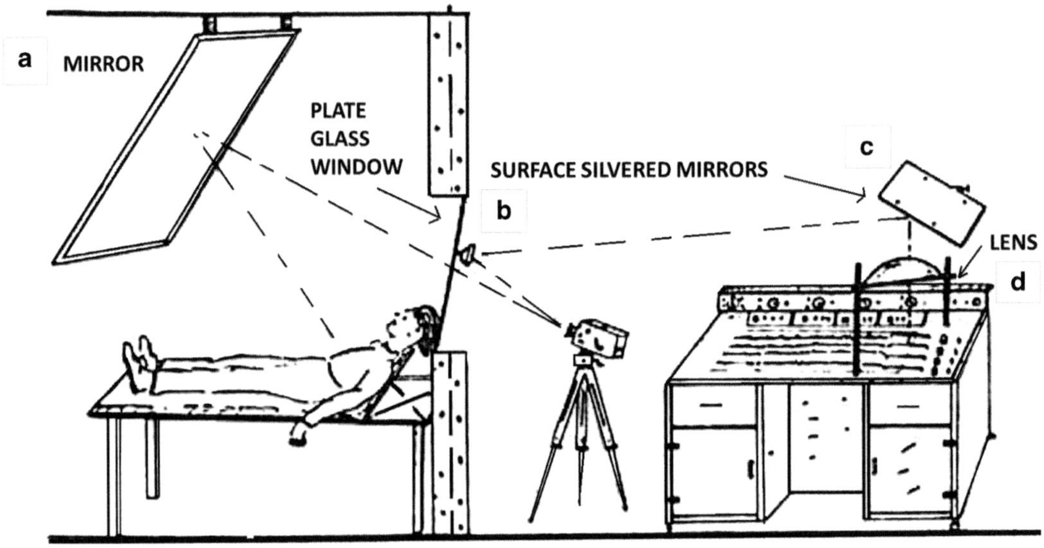

Fig. 9.1 Optical system used at the Montreal Neurological Institute in 1949 (modified with permission from Hunter and Jasper [4]. The patient and the recording system were in separate rooms between which was placed an observation window. The patient's image was visible, from the observation window, through a large mirror (**a**) positioned above the bed. The image of the polygraph pens and the EEG record was magnified by means of a plano-convex lens (**b**), reflected toward the silvered mirror (**c**) positioned above the lens and then conveyed to the small rectangular mirror (**d**) attached to the observation window [4]

subsequently and were commonly used to monitor patients with epilepsy.

In 1969 Penry and Dreifuss proposed recording both clinical and EEG data on a single video-monitor using a split-screen technique. If two or more cameras were used, the output was synchronized, and a composite picture of the patient and the simultaneously recorded EEG were produced. However, this method limited the number of clearly displayed EEG channels (Fig. 9.2).

Moreover, accurately observing both the semiological features of the clinical seizures and the concurrent EEG patterns simultaneously was often challenging, as only a brief and transient segment of the EEG was visible at any one time [1].

Ives and Gloor later developed an alternative method, involving separate recordings of clinical changes on videotape and EEG data on paper. Both the video and the EEG data had the same timecode enabling synchronization by the clock [1].

Early video-EEG using cine-film technology and EEG faced limitations due to the need for expensive equipment, short recording durations, and the inability to review recordings.

The development of videocassette recorder in the 1960s, which allowed data storage, and closed-circuit television (CCVT) technology enabled Ives and his colleagues to successfully create time-lapsed continuous video monitoring of patients without supervision [5].

In the 1970s, the first video-EEG monitoring units were developed in North America and Europe, allowing doctors to observe and record patients during epileptic seizures, linking electroencephalographic and polygraphic changes with visible clinical features.

By the end of the 1970s, a typical video-EEG system included three low-light monochrome cameras (whole-body, face close-up, and EEG), with reformatting to create a composite picture of these multiple views, enabling detailed off-line analysis of seizure semiology [6].

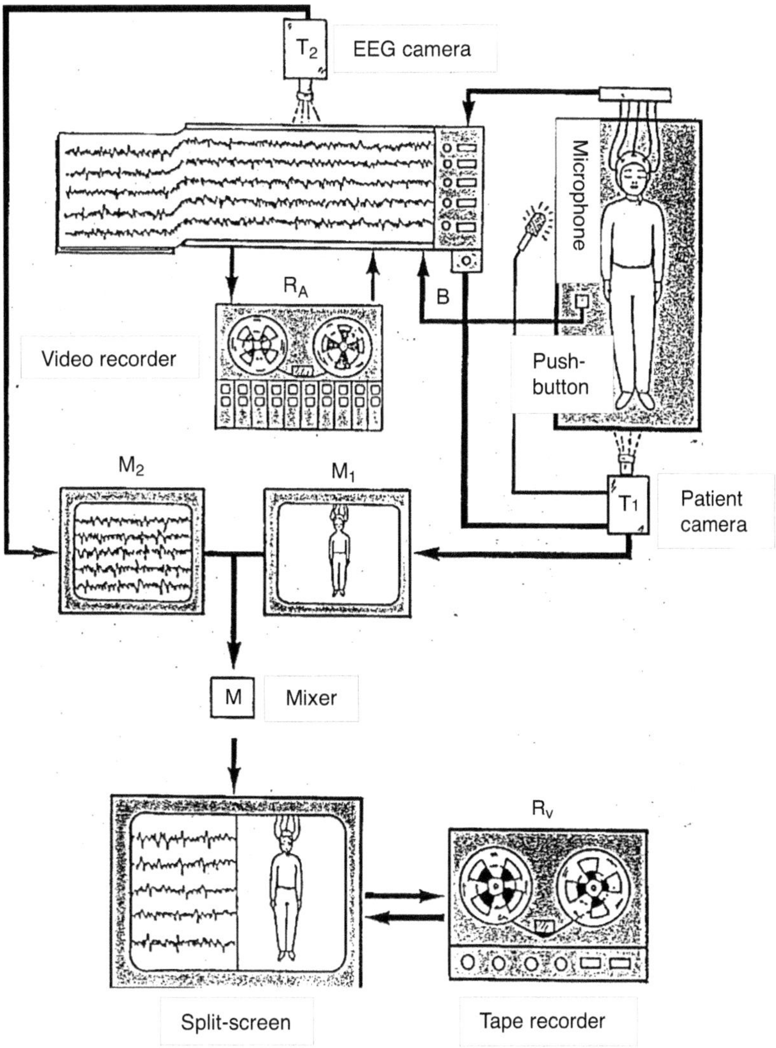

Fig. 9.2 Schematic representation of the split-screen technic used in EEG laboratory of Bologna in the 1980s. (Courtesy of Prof. Paolo Tinuper)

Further advancements were made possible by the development of new recording and transmission techniques such as radio and cable telemetry, camera-tracking mechanisms, and the miniaturization of recording equipment.

Until this period, the video-EEG was limited to recordings acquired in the patient's bed, requiring the patient to stay within the room and to be under constant direct supervision, which was labor-intensive and resource-dependent [1].

Cable telemetry consists of head-mounted EEG preamplifier and multiplexer unit connected by a long cable to a reading station. The multiplexer reduced the 16-signal traces to one for transmission, and a demultiplexer then reconstructed the original EEG signals. In radio telemetry, the multiplexer output modulates a radio transmission signal [ultra high frequency (UHF) or very high frequency (VHF)] allowing the patient to move within the hospital room, limited by the range of the cable or transmitter and the need for simultaneous video coverage [6].

With the miniaturization of components and recording apparatus, patients could move freely and even return home while being recorded with an ambulatory cassette recorder. Initially, this method allowed for only 4 channels to be recorded, but subsequent developments enabled 8- and 16-channel systems.

Continuous monitoring advancements improved patient assessment but also resulted in a significant increase in data. The development of computer technologies allowed data to be digitized, overcoming the issue of storage.

Between 1990 and 2000, advancements in modern cameras led to higher resolution and colored images for better patient recording. Additionally, analog EEG was gradually replaced by digital recording.

Nowadays, wearable devices and mobile apps have significantly advanced the field of electroencephalography and video-EEG telemetry, making it easier to monitor and diagnose epilepsy. These innovations allow for recording EEGs from patients' home, offering several advantages over traditional inpatient monitoring. Home video-EEG telemetry (HvET) has emerged as a cost-effective and patient-friendly alternative to inpatient monitoring [7].

HvET systems enable remote monitoring of patients by wirelessly transmitting EEG signals to central stations. There are three main types of HvET: (i) supervised HvET, with daily visits from a technologist to set up video-EEG at the patient's home, (ii) mobile HvET where patients and caregivers set up the equipment at home without a technologist, and (iii) cloud-based HvET, which continuously streams video-EEG data to healthcare providers, allowing real-time monitoring and remote troubleshooting. These systems aid in decision-making and timely intervention but require strong data security and privacy measures, such as robust encryption and access controls.

9.2 Technical Aspects

At the core of the video-EEG is the electroencephalograph (usually equipped with a standardized 10–20 montage system) connected to a video camera, ideally equipped with remote controls, zoom, and night vision filters (Fig. 9.3).

Other essential accessories for video-EEG studies are polygraphic recordings, such as the single-channel electrocardiogram (EKG) to record the heart rhythm, surface electromyography (EMG) channels to register muscle activity of clinical interest and/or for sleep studies, and thoracoabdominal tapes to record respiratory excursions.

Depending on the different clinical scenarios, center availability, and episode frequency, video-EEG can take different forms, including home monitoring, long-term studies, and real-time telemetry.

The main clinical video-EEG settings are the following:

(a) Short-term video-EEG monitoring (1–3 h) is typically conducted in a laboratory setting. This approach is particularly suited for patients who experience frequent clinical events or episodes that can be reliably induced or precipitated within the EEG laboratory. Such monitoring can often be integrated into an extended routine EEG recording session.
(b) Ambulatory video-EEG recording (<48 h) can be conducted in both inpatient and outpatient settings using ambulatory EEG recordings. This approach allows patients to maintain maximal mobility during the monitoring period. These studies facilitate the execution of routine activities, enabling patients to carry out their daily tasks even at home, ensuring comprehensive data collection in a naturalistic environment [8].
(c) Long-term inpatient video-EEG (vEEG) monitoring is a critical diagnostic procedure in tertiary epilepsy centers. Typically, this recording is conducted over a period of 2–5 days in hospitalized patients in the epilepsy monitoring unit (EMU). The duration can extend up to 2–3 weeks, depending on the clinical requirements [9]. Long-term monitoring in a dedicated environment allows better processing of clinical-instrumental data and reduction of artifacts, improving data quality and reliability. The

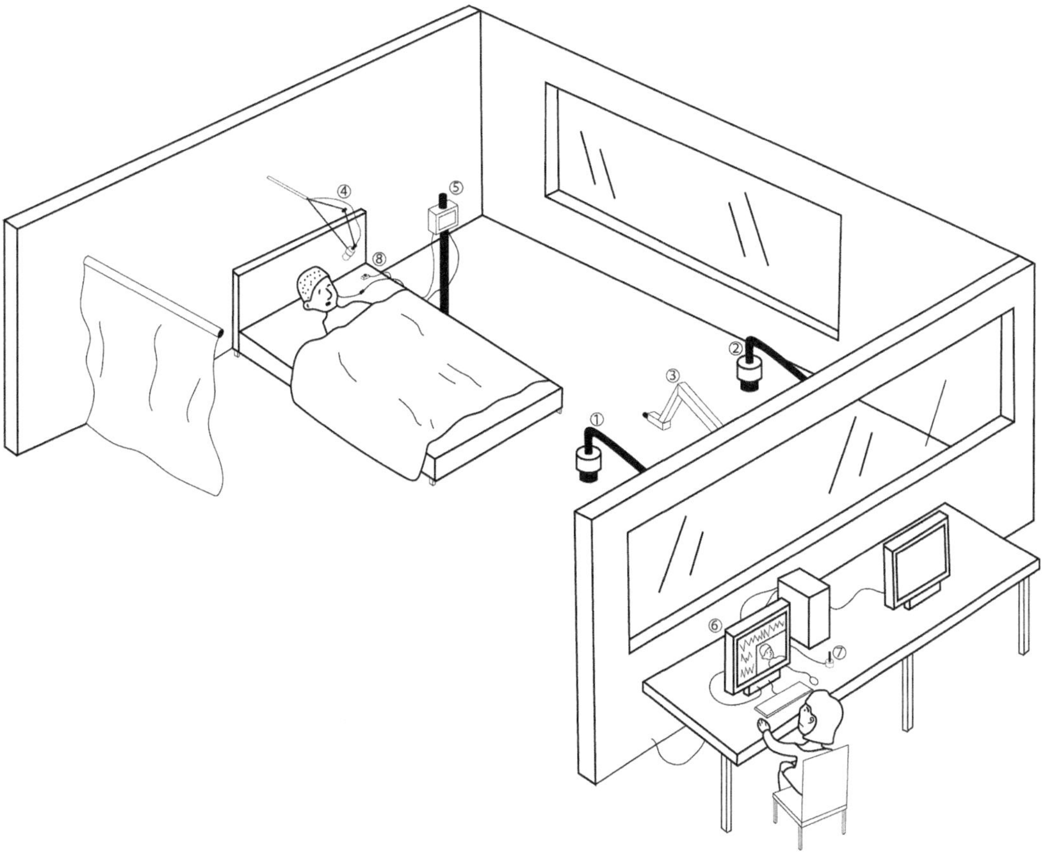

Fig. 9.3 Modern epilepsy monitoring unit with basic equipment: (1 and 2) high-definition cameras; (3) infrared cameras for night-time monitoring; (4) microphone; (5) EEG-machine junction box connect to scalp electrodes; (6) computers for real-time video-EEG monitoring and data storage; (7) cameras' joystick controller; (8) seizure alert button. (Designed by Davide Bolognini)

patient, admitted to a dedicated room with one or more patients, is monitored 24 h a day, while dedicated monitoring staff provides continuous clinical and neurophysiological supervision in an adjoining room, ensuring timely intervention during critical episodes.

In home video-EEG configurations, portability and ease of use are paramount. Patients are fitted with lightweight, battery-powered EEG recorders combined with compact cameras to provide continuous monitoring in a familiar environment. This is particularly useful, for example, for patients with physical and/or intellectual disabilities, to reduce the stress associated with outpatient/hospital care. While not as sophisticated as clinical-grade systems, these systems offer remarkable convenience and accessibility, allowing for prolonged monitoring without interrupting daily activities. However, ensuring data integrity and minimizing artifacts is more complex in the home environment and requires careful calibration and adequate patient education. A non-negligible limitation of this system is that it precludes, during a critical event, a real-time clinical assessment by an experienced examiner who can adequately test the patient.

Finally, as already mentioned, telemetry systems allow remote electroencephalographic monitoring and processing in nonhospital environments.

9.3 Clinical Application and Diagnostic Relevance of Video-EEG Methods

Video-EEG monitoring emerges from the clinical necessity to correlate paroxysmal events or clinical behaviors with EEG changes [10]. It was initially developed in specialized centers, where video-EEG was primarily reserved for the evaluation of patients considered for epilepsy surgery [11]. Over the years video-EEG has found increasing application; nowadays the primary indication for video-EEG is the differential diagnosis of paroxysmal events, particularly the distinction between epileptic seizures and their principal imitators, such as psychogenic nonepileptic seizures, syncope, and sleep or movement disorders [9]. Indeed, clinical symptoms alone often prove insufficient to distinguish nonepileptic behavioral events from seizures [12].

Improvements in digital technology and economies of scale have significantly increased the availability of video-EEG equipment together to the increasing demand for accurate diagnostic facilities [13]; as consequence the technique can now be performed in wide clinical setting [13].

Despite these advancements, video-EEG remains expensive and time-consuming due to the costs associated with equipment, hospitalization, and the need for specialized personnel. Assessing the cost-effectiveness of video-EEG is challenging; however, its clinical relevance is clear [12]. According to a large monocentric study at the Gowers unit over 612 patients, video-EEG led to a change in diagnosis in 58% of cases and a refinement in 13%, with only 29% retaining the same diagnosis after video-EEG. The most frequent diagnostic changes involved distinguishing epilepsy from nonepileptic attack disorders (55%) and differentiating focal from generalized epilepsies (35%). Additionally, video-EEG was especially useful in distinguishing frontal lobe seizures from generalized seizures and nonepileptic attacks. [14]. The clinical utility of video-EEG resulted independent from the main referral reason (diagnostic, seizure classification, and presurgical) according to other studies [15]. Multiple studies demonstrated that the differential diagnosis of psychogenic nonepileptic seizures (PNES) is the main indication and the principle diagnostic challenge in video-EEG [16, 17]. According to a previous study performed at Mayo Clinic, Rochester, over a series of 274 patients, 55% percent of patients were experiencing PNES; a combination of seizures and psychogenic spells occurred in 5.5% of patients [12]. According to a pediatric series, around 20% of the children monitoring in video-EEG have nonepileptic causes for their paroxysmal events [11]. For these reasons video-EEG is nowadays considered the gold standard for diagnosis of PNES; however, its accuracy in unknown because there is no definitive confirmatory gold standard [9]. Furthermore another study showed that interrater agreement in diagnosis of PNES has been proven to be only moderate (k . 0.57), highlighting the difficulties and subjective components related to the diagnosis of PNES [18]. Homemade video alone can be used for PNES diagnosis, but only in one-third of cases is it possible to establish a confident diagnosis [19]. Patients in which there are recurrent paroxysmal events in the context of normal interictal EEG findings, when the interictal EEG findings do not match the semiology of the clinical events, and when the history indicates a high clinical likelihood of an underlying nonepileptic disorder should be then referred for video-EEG monitoring [11].

Demonstrating its financial benefit is argued, since it requires analyzing the cost savings linked to therapeutic interventions informed by EEG study results [12].

Obviously, prolonged EEG studies are superior to the routine EEG in detecting seizures (50–70% compared with 2.5–7%) [12].

However, depending on clinical question and selecting patients with an high episodes frequency, various authors demonstrated that short-term video-EEG monitoring had high success rate in detecting and differentiating the nature of recurrent episode [20]; the diagnostic rate in selected patients of short-term video-EEG monitoring was comparable with LTM according to different studies [8, 21], with an evident optimization in terms of cost of the procedures.

Once recognized the epileptic nature of the episodes inpatient video-EEG could be used to:

1. Classification and Characterization of Seizure Types and Electroclinical Syndromes: Accurate classification is essential for determining an effective anti-seizure medication (ASM). For example, a patient with refractory generalized tonic-clonic seizures might have either generalized or partial epilepsy. In one study involving over 40 patients, long-term monitoring (LTM) modified/changed the diagnostic classification in 47.5% of cases [22]. Previously unrecognized seizure types were identified in 20% of patients with epilepsy. Improved seizure control was achieved in 60–70% of patients. Another series found that even brief outpatient video-EEG helped change the epilepsy syndrome diagnosis in 9 of 43 children, primarily due to the identification of additional seizure types that parents had not noticed [11].
2. Quantification of Seizures and Interictal Discharges: Accurate quantification of seizures, interictal subclinical epileptiform discharges, and precipitating factors can be challenging. This is particularly true for patients with absence seizures, especially atypical absence seizures, where the clinical manifestations are subtle and require repeated observations to confirm alterations in responsiveness. In patients with deteriorating function, precise determination of seizure frequency and interictal spikes is crucial to exclude epileptic encephalopathy [11]. Overnight studies are particularly useful in quantifying continuous spike-wave discharges in conditions like Landau-Kleffner syndrome (LKS) or other forms of suspected electrical status epilepticus during sleep (ESES).
3. Presurgical Evaluation in Medically Refractory Epilepsy: Identifying potential surgical candidates is a major indication for video-EEG monitoring in patients with drug-resistant seizures. video-EEG recording during presurgical aims to collect anatomo-electroclinical data to support the localization hypothesis. This includes representative sampling of interictal EEG discharges (IED) across all stages of the sleep-wake cycle. LTM can enhance sensitivity, detecting IEDs in 15% of temporal lobe epilepsy patients with serial negative routine EEG [23]. LTM may also reveal IEDs in regions previously undetected by video-EEG, providing more comprehensive information and identifying markers associated with poorer surgical outcomes [23]. The key aspect of LTM is documenting ictal EEG changes and their clinical correlations, which provide crucial information for hypothesizing the epileptogenic zone (EZ). If the EZ is not fully delineated, electroclinical data can be used to plan intracranial implantation for precise localization of the ictal-onset zone or functional mapping (see Chap. 15, The Use of the EEG for Epilepsy Surgery).

9.4 Implementation of Diagnostic Yield

The video-EEG is considered the gold standard in the diagnostic of paroxysmal event, yet it may not capture all correlated modifications or characterize all patient symptoms during seizures. Enhancing the diagnostic yield of video-EEG monitoring can be achieved through polygraphy, incorporating additional measures like electrocardiogram, blood pressure, and respiratory function. The use of extra electrodes, such as sub-temporal or zygomatic, can help "make the invisible visible," aiding also in the classification of physiologic nonepileptic disorders associated with paroxysmal events [12].

Continuous EKG monitoring and display from a single EKG channel during video-EEG is mandatory, as potentially serious cardiac abnormalities may occur peri-ictally [9]. Studies have identified various seizure-related cardiac arrhythmias occurring up to 3% of telemetered patients, and their recognition is necessary in order to identified cardiac life-treating symptoms and possible treatment (i.e., pacemaker in prolonged asystolia) [24]. On the other hand, the systematic literature review on seizure-related arrhythmias during video-EEG emphasizes the importance of postictal rather than ictal arrhythmias in understanding the pathophysiology of sudden unexpected death in epilepsy (SUDEP) [25]. While heart rate monitoring is common in epilepsy monitoring units (EMUs), automatic detection methods for ECG abnormalities are not widely implemented [26, 27].

In order to reduce the possibility that muscle artifact during an episode may obscure a single-lead EKG channel, it is suggested to use 3-lead EKG, which provided an excellent agreement as compared with the gold standard of conventional 12-lead ECG interpretation [13].

Pulse oximetry is another parameter highly recommended to be monitoring during video-EEG, especially at night. Desaturation (O2 saturation below 90%) and the duration of desaturation have been reported to correlate with certain seizure features, such as temporal or right hemispheric onset, contralateral spread, and male gender. Alarm thresholds of 80%–86% SpO2 were found to detect a significant proportion of generalized tonic-clonic seizures and focal seizures, with varying false alarm rates [28].

However, the role of technician and clinical testing remain crucial in detecting subtle clinical signs or identifyng negative symptoms (i.e., aphasia) that may not be visible on video. Several authors have reported the importance of standardized ictal and post-ictal testing among EMU in order to unify and compare practice among different centers [29, 30]. Also automatic testing such as Artie has been proposed but still are not widely used in clinical practice [31].

Finally, in order to improve the probability of recording seizures, provocation procedures should be routinely used. The most common used are ASM reduction, sleep deprivation, and, less commonly, exercise or the application of specific stimuli in reflex epilepsies that need to be individualized to each patient.

9.5 Safety Issues, Limitations, and Adverse Events

During the past 30 years, many studies and surveys have been performed in order to assess the EMUs safety [24, 32–38].

Due to the variability in how safety and quality data were reported in the included studies, it was not possible to analyze potential factors associated with adverse events in the EMU [39]. Sauro et al. performed a systematic review revealing that an EMU admission is reasonably safe: the pooled estimate of adverse events was 7% but with significant heterogeneity.

Adverse events in EMUs include mainly the following key categories: injuries and falls, cardiac arrhythmias, psychosis, status epilepticus, and, in rare instances, sudden unexpected death in epilepsy (SUDEP) [13].

Seizure-related falls are a common adverse event occurring in 1–20% of patients [40]. Most falls result in minor injuries, although some serious injuries have been reported such as acute epidural hematoma, vertebral fractures, cervical spine injury, and thoracic spine compression. All serious injuries were associated with generalized tonic-clonic seizures (GTCS) [33].

Among psychiatric adverse events, postictal psychoses are the most frequent; panic attacks and interictal psychosis are also reported. In a large series of 507 patients, Dobesberger et al. found that psychiatric adverse events occur in 5% of patients, half of them required hospitalization in a psychiatry department [33].

Status epilepticus (SE) is reported in 0–3.5% of patients in different studies [40]: most cases are nonconvulsive SE, while convulsive SE is rarer, except in children, where it is reported more frequently compared to adults [36].

Apart from the risk of aspiration during and/or immediately after a seizure, there is a serious risk of respiratory failure directly related to ictal activity: seizures in EMUs can also lead to cardiorespiratory arrest and death, including SUDEP. The International Mortality in EMUs

(MORTEMUS) study aimed to quantify this risk across all EMUs identified in Europe, Israel, Australia, and New Zealand. The incidence of SUDEP in adults in EMUs was 5.1 per 1000 patient-years (95% CI 2.6–9.2), with a risk of 1.2 per 10,000 video-EEG monitorings (95% CI 0.6–2.1). SUDEP in EMUs "primarily follows an early postictal, centrally mediated, severe alteration of respiratory and cardiac function induced by generalised tonic-clonic seizure, leading to immediate death or a short period of partly restored cardiorespiratory function followed by terminal apnoea then cardiac arrest." [41]. The authors found that the risk of SUDEP is likely increased by inadequate supervision and possibly by the withdrawal of antiepileptic drugs. Therefore, they emphasize the need for enhanced supervision in epilepsy monitoring units, particularly during nighttime [41].

Hamandi et al. conducted a survey and workshop of UK epilepsy monitoring units and found that although some variability in practice is inevitable due to different local arrangements and patient populations, there are clear areas of consensus, particularly regarding informed consent and patient safety [13]. Hamandi et al. conducted a survey and workshop on epilepsy monitoring units in the United Kingdom and found that, despite the inevitable variability in practice due to different local arrangements and patient populations, there are aspects that can be shared by all units, particularly concerning informed consent and patient safety. A crucial shared point is the need to provide all patients with written information before admission for video-EEG monitoring and to conduct a preliminary assessment including among others the type of seizures experienced by the patient, seizure frequency, occurrence of seizure clusters, history of generalized tonic-clonic seizures, and previous episodes of psychosis or status epilepticus. It has also been stressed the necessity to obtain signed consent for video-EEG and for ASM reduction with awareness of the possible associated risks including the recurrence of seizure clusters, status epilepticus, psychosis, or SUDEP.

Tapering off ASM is routinely performed in EMUs with the aim to promote the occurrence of seizures and decrease the length of hospital stay. However, ASM reduction might also increase the risk of seizure clusters, generalized tonic-clonic seizures (even in patients who usually do not suffer from this seizure type), and SE.

All SUDEP collected in MORTEMUS occurred following a generalized tonic-clonic seizure, most often in patients with more than 50–100% ASM s withdrawal [41].

While it is true that drug reduction and the sequence of medication tapering in polytherapies should be customized for each patient, taking into account various factors such as the inherent characteristics of the medications and the type of epilepsy the patient is experiencing. On the other hand, it has been suggested that the risk of seizure clusters can be reduced by adopting a specific protocol for managing drug tapering [42, 43].

The UK workshop reached a consensus that medication tapering should not be started before EMU admission, that provocation procedures and ASM reduction should be tailored to each patient, and that ASMs should be returned to their full dose at least 24 h before discharge [13].

Fostering a culture of safety within EMUs is paramount to reducing adverse events during LTM.

Identifying patients at heightened risk of adverse events based on clinical indicators is, therefore, critical.

This entails enhancing personnel training and establishing consensus guidelines on essential protocols to mitigate seizure-related risks. Continuous 24-hour surveillance by specially trained staff, including weekends, is important for ensuring patient safety and diagnostic precision.

EMUs should institute standard procedures for managing seizure clusters and status epilepticus.

Protective measures to prevent seizure-related injuries include implementing guard rails or cushioned cot sides, anti-suffocation pillows, low beds, or mattresses on the floor (in selected case such as patients with hypermotor seizures) and ensuring physical safeguards in monitoring and bathroom facilities [13]. For patients undergoing antiseizure medication (ASM) reduction, an intravenous cannula should be inserted, and ideally, the patient should remain in bed and within the camera's view at all times [13].

9.6 Future Perspectives

The future of video-EEG promises revolutionary advances in both device technology and clinical perspectives, ushering in an era of greater diagnostic accuracy, patient comfort, and personalized treatment strategies. As the technology continues to evolve, future video-EEG devices will incorporate innovative features to improve data quality, accessibility, and feasibility.

An important avenue of development is the ultra-miniaturization and integration of EEG recording hardware and the development of subscalp recording systems. Advances in microelectronics and wearable technology (such as smartwatch/wrist-wearable devices and smartphones) could enable the development of combined ultracompact EEG devices that can be seamlessly integrated into everyday clothing or accessories, improving continuous EEG recordings. These discreet and lightweight devices would offer unparalleled mobility and convenience, allowing continuous monitoring without significant disruption to the patient's lifestyle. In addition, wireless connectivity and cloud-based storage solutions would facilitate real-time data transmission and remote access, enabling seamless collaboration between healthcare professionals and patients regardless of geographical barriers, also extending the possibilities of developing seizure forecasting algorithms.

While technological developments in wearable devices promise to extend the capabilities of EEG signal acquisition in innovative ways, they are not yet compatible with the need for visual EEG monitoring, which ideally should follow the patient during daily activities. However, the integration of these devices with home recording methods may provide a medium-term opportunity to increase the tolerability of home video-EEG monitoring. A possibility to optimize or reduce the duration of video in home setting could be represented by acquisition of recordings only when the system of video-EEG is triggered by characteristic motion or by the support of artificial intelligence (AI)-driven decision systems, which can guide experts to review only those video segments most likely to contain seizures.

As already said, a frontier in the development of video-EEG devices is represented by incorporation of artificial intelligence (AI) and machine learning algorithms. Using large datasets of EEG recordings and associated clinical outcomes, AI-based analysis has the potential to revolutionize the detection, prediction, and characterization of epileptic seizures. These intelligent algorithms could learn to detect even the smallest changes in video-EEG data, enabling early identification of epileptiform activity and optimization of tailored treatment. In addition, AI-based decision support systems could assist in the interpretation of complex video-EEG results, improving diagnostic accuracy and reducing interpretation variability.

In addition, future video-EEG devices could be equipped with advanced biometric sensors and multimodal data fusion capabilities. The integration of physiological parameters related to heart rate variability, respiratory patterns, and autonomic responses into EEG recordings could provide a comprehensive view of the neurophysiological correlates of critical events and associated comorbidities.

9.7 Educational Issues

Beyond its recognized clinical utility, video-EEG recording is a paramount instrument to educate health care providers and students in recognize seizures and in elucidate the main symptoms and signs associated to epilepsy.

Its role is fully recognized by different scientific societies. The International League Against Epilepsy (ILAE) creates an online diagnostic manual where there are collected multiple videos. The site is principally designed for clinicians in primary and secondary with the aim to provide useful teaching aid (https://www.epilepsydiagnosis.org/). The Italian Chapter of International League Against Epilepsy (LICE) has developed an ad hoc task force (Videoteca LICE) to collect video-EEG recording of paroxysmal event of particular educational and scientific content, which could also be used to organize updating meetings and clinical case discussions with videos during international and the national congress. (https://www.lice.it/LICE_ita/commissioni/videoteca_new/doc_prog.php?idpadre=18).

Acknowledgments We express our gratitude to Paola Chessa and Riccardo Sgroi for their invaluable support in sourcing materials, drafting paragraphs, and producing images for this chapter has been instrumental. We also thank Davide Bolognini who conceived and designed Fig. 9.3.

References

1. Gloor P. Long-term monitoring in epilepsy. Electroencephalogr Clin Neurophysiol Suppl. 1985;37:xiii–xx.
2. Ho R, Carrazana EJ. The history of motion photography to video electroencephalography in the study of functional seizures and related seizure disorders: the first 100 years. Seizure. 2023;112:68–71.
3. Schwab RS, Schwab MW, Withee D, Chock YC. Synchronized moving pictures of patient and EEG. Electroencephalogr Clin Neurophysiol. 1954;6(4):684–6.
4. Hunter J, Jasper HH. A method of analysis of seizure pattern and electroencephalogram; a cinematographic technique. Electroencephalogr Clin Neurophysiol. 1949;1(1):113.
5. Ives JR, Gloor P. A long term time-lapse video system to document the patients spontaneous clinical seizure synchronized with the EEG. Electroencephalogr Clin Neurophysiol. 1978;45(3):412–6.
6. Shorvon SD. The idea of epilepsy: a medical and social history of epilepsy in the modern era. Cambridge, New York: Cambridge University Press; 2023. p. 760.
7. Brunnhuber F, Slater J, Goyal S, Amin D, Thorvardsson G, Freestone DR, et al. Past, present and future of home video-electroencephalographic telemetry: a review of the development of in-home video-electroencephalographic recordings. Epilepsia. 2020;61(Suppl 1):S3–10.
8. Kandler R, Ponnusamy A, Wragg C. Video ambulatory EEG: a good alternative to inpatient video telemetry? Seizure. 2017;47:66–70.
9. Baumgartner C, Pirker S. Video-EEG. In: Handbook of clinical neurology [Internet]. Elsevier; 2019. p. 171–83. [citato 28 May 2024]. Available at: https://linkinghub.elsevier.com/retrieve/pii/B9780444640321000114.
10. Connolly MB, Wong PK, Karim Y, Smith S, Farrell K. Outpatient video-EEG monitoring in children. Epilepsia. 1994;35(3):477–81.
11. Nordli DR. Usefulness of video-EEG monitoring. Epilepsia [Internet]. 2006;47(s1):26–30. [citato 28 maggio 2024]; Available at: https://onlinelibrary.wiley.com/doi/10.1111/j.1528-1167.2006.00656.x
12. Cascino GD. Video-EEG monitoring in adults. Epilepsia [Internet]. 2002;43(s3):80–93. [citato 28 maggio 2024]; Available at: https://onlinelibrary.wiley.com/doi/10.1046/j.1528-1157.43.s.3.14.x
13. Hamandi K. Current practice and recommendations in UK epilepsy monitoring units. Report of a National Survey and Workshop; 2017.
14. Yogarajah M, Powell HWR, Heaney D, Smith SJM, Duncan JS, Sisodiya SM. Long term monitoring in refractory epilepsy: the Gowers unit experience. J Neurol Neurosurg Psychiatry. 2009;80(3):305–10.
15. Alving J, Beniczky S. Diagnostic usefulness and duration of the inpatient long-term video-EEG monitoring: findings in patients extensively investigated before the monitoring. Seizure. 2009;18(7):470–3.
16. LaFrance WC, Baker GA, Duncan R, Goldstein LH, Reuber M. Minimum requirements for the diagnosis of psychogenic nonepileptic seizures: a staged approach: a report from the International League Against Epilepsy Nonepileptic Seizures Task Force. Epilepsia. 2013;54(11):2005–18.
17. Perez DL, LaFrance WC. Nonepileptic seizures: an updated review. CNS Spectr. 2016;21(3):239–46.
18. Benbadis SR, LaFrance WC, Papandonatos GD, Korabathina K, Lin K, Kraemer HC, et al. Interrater reliability of EEG-video monitoring. Neurology. 2009;73(11):843–6.
19. Erba G, Giussani G, Juersivich A, Magaudda A, Chiesa V, Laganà A, et al. The semiology of psychogenic nonepileptic seizures revisited: can video alone predict the diagnosis? Preliminary data from a prospective feasibility study. Epilepsia. 2016;57(5):777–85.
20. Chen LS, Mitchell WG, Horton EJ, Carter Snead O. Clinical utility of video-EEG monitoring. Pediatr Neurol [Internet]. 1995;12(3):220–4. [citato 27 maggio 2024]; Available at: https://linkinghub.elsevier.com/retrieve/pii/0887899495000217
21. Jamal Omidi S, Hampson JP, Lhatoo SD. Long-term home video EEG for recording clinical events. J Clin Neurophysiol Off Publ Am Electroencephalogr Soc. 2021;38(2):92–100.
22. Sutula TP, Sackellares JC, Miller JQ, Dreifuss FE. Intensive monitoring in refractory epilepsy. Neurology. 1981;31(3):243–7.

23. Cascino GD, Trenerry MR, So EL, Sharbrough FW, Shin C, Lagerlund TD, et al. Routine EEG and temporal lobe epilepsy: relation to long-term EEG monitoring, quantitative MRI, and operative outcome. Epilepsia. 1996;37(7):651–6.
24. Noe KH, Drazkowski JF. Safety of long-term video-electroencephalographic monitoring for evaluation of epilepsy. Mayo Clin Proc. 2009;84(6):495–500.
25. van der Lende M, Surges R, Sander JW, Thijs RD. Cardiac arrhythmias during or after epileptic seizures. J Neurol Neurosurg Psychiatry. 2016;87(1):69–74.
26. Kobulashvili T, Höfler J, Dobesberger J, Ernst F, Ryvlin P, Cross JH, et al. Current practices in long-term video-EEG monitoring services: a survey among partners of the E-PILEPSY pilot network of reference for refractory epilepsy and epilepsy surgery. Seizure. 2016;38:38–45.
27. Rubboli G, Beniczky S, Claus S, Canevini MP, Kahane P, Stefan H, et al. A European survey on current practices in epilepsy monitoring units and implications for patients' safety. Epilepsy Behav EB. 2015;44:179–84.
28. Bateman LM, Li C-S, Seyal M. Ictal hypoxemia in localization-related epilepsy: analysis of incidence, severity and risk factors. Brain J Neurol. 2008;131(Pt 12):3239–45.
29. Beniczky S, Neufeld M, Diehl B, Dobesberger J, Trinka E, Mameniskiene R, et al. Testing patients during seizures: a European consensus procedure developed by a joint taskforce of the ILAE—Commission on European Affairs and the European Epilepsy Monitoring Unit Association. Epilepsia. 2016;57(9):1363–8.
30. Ferri L, Vignatelli L, Alvisi L, Fabbri M, Boscarato S, Zenesini C, et al. If seizures left speechless: CA-P-S C-A-R-E, a proposal of a new ictal language evaluation protocol. Neurol Sci Off J Ital Neurol Soc Ital Soc Clin Neurophysiol. 2021;42(8):3249–55.
31. Touloumes G, Morse E, Chen WC, Gober L, Dente J, Lilenbaum R, et al. Human bedside evaluation versus automatic responsiveness testing in epilepsy (ARTiE). Epilepsia. 2016;57(1):e28–32.
32. Atkinson M, Hari K, Schaefer K, Shah A. Improving safety outcomes in the epilepsy monitoring unit. Seizure. 2012;21(2):124–7.
33. Dobesberger J, Walser G, Unterberger I, Seppi K, Kuchukhidze G, Larch J, et al. Video-EEG monitoring: safety and adverse events in 507 consecutive patients. Epilepsia. 2011;52(3):443–52.
34. Ley M, Vivanco R, Massot A, Jiménez J, Roquer J, Rocamora R. Safety study of long-term video-electroencephalogram monitoring. Neurol Barc Spain. 2014;29(1):21–6.
35. Shafer PO, Buelow J, Ficker DM, Pugh MJ, Kanner AM, Dean P, et al. Risk of adverse events on epilepsy monitoring units: a survey of epilepsy professionals. Epilepsy Behav EB. 2011;20(3):502–5.
36. Arrington DK, Ng Y-T, Troester MM, Kerrigan JF, Chapman KE. Utility and safety of prolonged video-EEG monitoring in a tertiary pediatric epilepsy monitoring unit. Epilepsy Behav EB. 2013;27(2):346–50.
37. Spanaki MV, McCloskey C, Remedio V, Budzyn D, Guanio J, Monroe T, et al. Developing a culture of safety in the epilepsy monitoring unit: a retrospective study of safety outcomes. Epilepsy Behav EB. 2012;25(2):185–8.
38. Kandler R, Lai M, Ponnusamy A, Bland J, Pang C. The safety of UK video telemetry units: results of a national service evaluation. Seizure. 2013;22(10):872–6.
39. Sauro KM, Wiebe N, Macrodimitris S, Wiebe S, Lukmanji S, Jetté N. Quality and safety in adult epilepsy monitoring units: a systematic review and meta-analysis. Epilepsia. 2016;57(11):1754–70.
40. Rheims S, Ryvlin P. Patients' safety in the epilepsy monitoring unit: time for revising practices. Curr Opin Neurol. 2014;27(2):213–8.
41. Ryvlin P, Nashef L, Lhatoo SD, Bateman LM, Bird J, Bleasel A, et al. Incidence and mechanisms of cardiorespiratory arrests in epilepsy monitoring units (MORTEMUS): a retrospective study. Lancet Neurol. 2013;12(10):966–77.
42. Di Gennaro G, Picardi A, Sparano A, Mascia A, Meldolesi GN, Grammaldo LG, et al. Seizure clusters and adverse events during pre-surgical video-EEG monitoring with a slow anti-epileptic drug (AED) taper. Clin Neurophysiol Off J Int Fed Clin Neurophysiol. 2012;123(3):486–8.
43. Rose AB, McCabe PH, Gilliam FG, Smith BJ, Boggs JG, Ficker DM, et al. Occurrence of seizure clusters and status epilepticus during inpatient video-EEG monitoring. Neurology. 2003. 2003;60(6):975–8.

The Role of the EEG Technicians

10

Lara Alvisi, Lorenzo Bevacqua,
and Angelantonio Rocco Mastrillo

10.1 Introduction

Richard Caton first recorded the electrical activity of the brain in animals approximately 150 years ago. Later, in 1924, H. Berger documented the spontaneous electrical activity of the human brain from the scalp. Today, electroencephalography (EEG) holds a prominent position in research and clinical neurophysiology, thanks to advancements in amplification and recording techniques. EEG is a graphical representation of the brain's bioelectric activity, recorded from electrodes placed on the scalp. It reflects the difference in electrical potential between various scalp areas, providing insights into the electrical functioning of neuronal populations in the cerebral cortex and the subcortical mechanisms that synchronize neuronal activity [1].

The EEG is recorded by a neuropathophysiology technician (NT), who provides critical information on the characteristics of both critical and intercritical brain activity. This data helps in defining seizure types and distinguishing between epileptic and non-epileptic seizures. Before each recording session, the NT must ensure that all equipment is fully operational. The laboratory should be organized, safe, and welcoming. To ensure accurate and reliable results, the NT needs comprehensive information about the patient, including biographical data, medical history, and, if available, previous EEG recordings. Understanding the purpose of the examination, the patient's profile, and the type of recording will help the NT select the most appropriate electrodes for the procedure.

A clear and simple explanation of the procedure to the patient is essential. This approach aids in easing the patient's anxiety, promoting relaxation, and ensuring better cooperation. During the recording, the NT must continuously monitor the patient's behavior, including alertness, response to stimuli, depth of sleep, and contact during any seizures. Depending on the recording's progress, the NT may need to apply additional electrodes and polygraph leads to provide valuable information for the neurophysiologist's interpretation of the EEG. The NT is responsible for the entire recording process and will oversee the patient until they leave the laboratory [1].

L. Alvisi (✉)
Department of Biomedical and Neuromotor Sciences
University of Bologna, Bologna, Italy

IRCCS Institute of Neurological Sciences of
Bologna, Bologna, Italy

Epilepsy Center (full member of the European
Reference Network EpiCARE), Bologna, Italy
e-mail: lara.alvisi@unibo.it

L. Bevacqua · A. R. Mastrillo
IRCCS Institute of Neurological Sciences of
Bologna, Bologna, Italy
e-mail: lorenzo.bevacqua2@unibo.it;
angelo.mastrillo@unibo.it

F. Brigo, O. Mecarelli (eds.), *EEG: The First 100 Years*,
https://doi.org/10.1007/978-3-031-86878-8_10

In various, sometimes complex, settings such as intensive care units or inpatient wards, the NT must adapt to challenging environments and interact with patients who may not cooperate easily. When working outside routine outpatient departments, the NT must coordinate with multidisciplinary teams. The NT should also be able to offer preliminary interpretations of the EEG signal, identify and document any electrical interference, and, when possible, address these issues to ensure the clarity of the recording.

10.2 The Organization of Societies for Electrophysiological Technology (OSET)

The Organization of Societies for Electrophysiological Technology (OSET) was founded in 1977. The founding societies included those from Italy, Australia, Canada, Germany, Great Britain, Finland, France, the Netherlands, Norway, Switzerland, and the United States.

The founding process spanned over 15 years, with initial activities dating back to the 1960s when physicians and EEG technicians from around the world convened at the first Refresher Courses. The inaugural course was held in Marseille in 1961, followed by the second in Salzburg in 1965. Key figures in this early initiative included A. Rémond (France) and L. Ponsen (the Netherlands).

In 1968, the idea of establishing an international society for EEG technicians was proposed through the International Federation of EEG and Clinical Neurophysiology Societies (IFSECN). This initiative aimed to address the generally low levels of medical and technical expertise by focusing on enhancing the preparation and competence of practitioners.

Globally, opportunities for medical and technical training were limited, with only a few centers of excellence available. During this period, L. Ponsen was particularly instrumental in advocating for the creation of a European group of national EEG technician associations. After the meeting in Marseille in 1973, the first European Conference of EEG Technicians took place in Arnhem (the Netherlands) in 1975, where a preliminary committee was formed, including L. Ponsen, M. Dopping (Germany), and D. Nelligan (Great Britain).

By the end of 1975, a second committee was established to draft the statutes. This committee included L. Ponsen, M. Esslinger (Germany), G. Klem (USA), D. Nelligan, C. Brayshaw (Great Britain), K. Parviainen (Finland), and C. Raess (France). Following the approval of the statutes in Freiburg in 1976, OSET was officially founded in Amsterdam in 1977. Shortly thereafter, R. Naquet (France) and A. Marsan (USA) initiated a survey to evaluate the training levels of EEG technicians across all IFSECN member countries.

This survey was conducted in 1981 by OSET, with C. Brayshaw and G. Klem presenting the findings at the IFSECN Congress in Kyoto (Japan). The survey revealed that only the Netherlands, Great Britain, and the United States had qualified training programs certified by qualifying examinations. This concerning situation led OSET to prioritize the development of guidelines for both basic and advanced training for EEG technicians. To address this need, special commissions (Education Commission) were established to create standards for vocational training. Between 1981 and 2020, several documents were produced, providing a foundation for consistent and high-quality professional training for technicians at an international level. These documents have been published in the official journals of various technician associations, including ANS (formerly EPTA Great Britain), ASET (USA), ANSA (Australia), and AITN (Italy) [2, 3].

10.3 The Role of the Neuropathophysiology Technician Worldwide

In the European healthcare landscape, the NT plays a crucial role in supporting physicians with the diagnosis of various neurological disorders. The NT's specialized skills and areas of expertise necessitate ongoing training that evolves with scientific advancements and professional development.

In Europe, NTs are employed in a range of settings including public and private healthcare services, research centers, and university hospitals. Although training requirements vary by country, generally, becoming an NT involves completing secondary education followed by a degree or professional diploma course lasting between 3 and 4 years. This is the case in countries like Italy, Portugal, the Netherlands, Germany, and the United Kingdom. In contrast, in some countries, such as France, the qualification is achieved through a specific vocational training program or a 1-year university diploma in neurophysiology for professionals already working in the healthcare field, such as nurses or other technical staff.

Outside Europe, NTs are also present in regions like the United States, Canada, Australia, Argentina, and other countries, though training pathways differ. These may include university diplomas or technical and vocational training programs with varying durations.

NTs globally are active members of national and international professional organizations. Through participation in conferences, workshops, and refresher courses, NTs stay updated with the latest technological and scientific advancements in clinical neurophysiology and research.

In Italy, recent efforts have been made to promote internationalization of NT degree programs, though challenges remain due to differences in educational systems and professional recognition. For instance, Italian degree programs in neurophysiopathology are typically housed within the Faculty of Medicine and Surgery, while in other European countries, they are often part of faculties related to science, technology, or applied sciences. Specifically, the Degree Course in Neurophysiopathology Techniques at the University of Bologna is working to enhance students' educational and professional experiences through international opportunities. In the 2022/2023 academic year, the university implemented an internship project abroad for two third-year students, each spending 2 months in Switzerland (University of Zurich) and Germany (Charité, University of Berlin) through a competitive application process and related scholarships. In 2024, the University of Bologna achieved a milestone by developing a 3-month Erasmus project with the Dutch vocational university of Zuyd, making it the first and only institution in Italy to establish such a partnership.

10.4 The Birth of the Neuropathophysiology Technician Profession in Italy: A Brief History

The profession of NT in Italy was first formalized by regulations in 1969, with the establishment of cardiology, respiratory physiopathology (Art. 32), and neurophysiopathology (Art. 33) services in hospitals. According to Article 31, "regulations on staffing must ensure an adequate number of NTs." However, since there was no officially recognized training or specialization for this role at that time, Presidential Decree 128/69 stated in Article 51 that "service in the same discipline for at least three years in hospitals, clinics, university institutes, or recognized scientific institutes would be considered equivalent to specialization" [4].

In 1972, to provide adequate training for technical personnel, the State established the first two university programs specifically for NTs at the universities of Rome and Bologna. The initiative was spearheaded by the respective directors, Gianfranco Ricci in Rome and Elio Lugaresi in Bologna. Initially, these programs were 2-year courses but were extended to 3 years in 1988 and 1991. Between 1977 and 1981, additional programs were established at the universities of Ancona, Pavia, Genoa, Milan, and Naples, with Pisa joining in 1988 and Siena in 1990.

On July 16, 1975, the Associazione Italiana Tecnici di Neurofisiopatologia (AITN) was founded by some of the first graduates of the specialized programs in Rome and Bologna. The AITN aimed to formally recognize the role of the NT. The association has maintained a close relationship with the history of the profession, securing formal recognition through ongoing collaboration with medical scientific societies, health profession federations, regional authorities, Parliament, and various ministries, including health, university, and other technical-scientific associations.

In 1990, the NT profession was officially included among the professions within the National Health System. By 1994, the Ministry of Health confirmed this recognition by including NTs in the "Regulation Containing the Modalities for the Ascertaining and Certification of Death" and assigned them the task of performing EEGs. The professional profile of the NT was officially defined in 1995, classifying NTs as technical health personnel. This profile outlines that:

The NT is a health professional specializing in the diagnosis of nervous system pathologies. They apply specific diagnostic methods in neurology and neurosurgery, such as electroencephalography, electroneuromyography, polygraphy, evoked potentials, and ultrasound, based on medical prescriptions.

The NT selects appropriate methods for recording bioelectrical phenomena and interacts directly with patients and equipment to execute diagnostic or neurophysiological research programs in close collaboration with specialist physicians. They are responsible for managing diagnostic methods, creating technical reports, and ensuring the accuracy of results. NTs employ diagnostic tools to assess cerebral activity for clinical or legal purposes, manage and maintain equipment, and perform their duties in both public and private health facilities, either as employees or freelancers [4].

In 1996, university degree for NTs were introduced at 18 universities nationwide and were upgraded to a 3-year Bachelor's Degree in 2001. The NT profession is categorized under the third class "Technical Healthcare," subgroup "diagnostic," alongside audiometry technicians, laboratory technicians, and radiology technicians. In 2008, following a period of unclear competencies, the Ministry of Health confirmed that the performance of EEGs for brain death assessment is an exclusive competence of NTs.

Today, practicing as an NT requires a 3-year Bachelor's degree in "Neurophysiopathology Techniques." This program is offered at the Faculty of Medicine in 14 Italian universities, training 120–130 technicians annually, with a total of approximately 2000 NTs in Italy. The Bachelor's degree includes compulsory practical training, and postgraduation opportunities include 2-year master's programs, additional master's degrees, and postgraduate courses. Additionally, NTs must participate in continuing medical education (CME) programs for ongoing professional development.

In 2017, the Ministry of Health established technical-scientific associations (ATS) tasked with defining specific guidelines for health professions. AITN was included in this list, which now comprises up to 293 societies and associations. Finally, in 2018, the Professional Register of NTs, part of the National Federation of Medical Radiology Health Technicians and Technical, Rehabilitation, and Prevention Health Professions (TSRM and PSTRP), was created [4].

10.5 The Evolution of the Method and New Skills of the Neurophysiopathology Technician

The EEG has long been valued for its rapid preparation, noninvasiveness, low cost, immediate results, and high temporal resolution. However, the role of the NT has transformed significantly, especially since the shift from analog to digital bioelectrical signals. The capability to process data post-acquisition has opened new avenues, particularly since the 1990s. Advances in survey instruments and technologies are creating new research and application sectors in neurophysiology, necessitating increasingly sophisticated skills from NTs in collaboration with neurophysiologists, biomedical engineers, and physicists.

Below, we provide a brief overview of the various contexts in which an EEG technician operates:

1. *Conventional Electroencephalography with Synchronized Video Recording (VEEG)*
   VEEG is the gold standard for evaluating and monitoring brain function associated with observable clinical changes, particularly for diagnosing and monitoring epileptic seizures (Fig. 10.1). The NT's role includes optimizing the video system integration and managing patient interventions in response to electroencephalographic changes or clinical manifestations. Critical tasks involve assessing contact, altered awareness, or impairment of certain functions [5], (Fig. 10.2).
2. *EEG Associated with Polysomnography*
   This method allows for the simultaneous recording of EEG and other physiological parameters over short or long periods. It is essential for evaluating sleep disorders, movement disorders, or suspected epilepsy episodes occurring during wakefulness or sleep. The NT is responsible for correct sensor placement, filter settings, and sampling rates, as well as conducting sleep staging and preparing the technical examination report [6–8], (Fig. 10.3).
3. *EEG with Electromyographic (EMG) Activity*
   When EEG is recorded alongside EMG, it helps detect brain electrical potentials associated with involuntary movements and study temporal correlations on the scalp. The NT must carefully sample EMG parameters, ensure accurate electrode placement, and prevent signal contamination by artifacts. If an automatic analysis system is not used, the NT may manually select muscle events through visual inspection, a process that, although time-consuming, yields clinically significant results [9].
4. *Ambulatory EEG Recording*
   Since the late 1970s, ambulatory EEG recording has allowed for prolonged monitoring while patients engage in daily activities. This method is crucial for managing epilepsy, recording critical episodes, quantifying abnormalities, and assessing drug therapy effects. The NT's responsibilities include ensuring stable electrode placement for extended periods, setting up the recording equipment properly, and recognizing artifacts, especially during active waking phases [10].
5. *Neurophysiological Monitoring During Carotid Endarterectomy*
   During carotid endarterectomy, EEG combined with somesthetic evoked potentials of the upper extremities is used to detect ischemic CNS injury and other cerebral circulation pathologies. The NT must possess

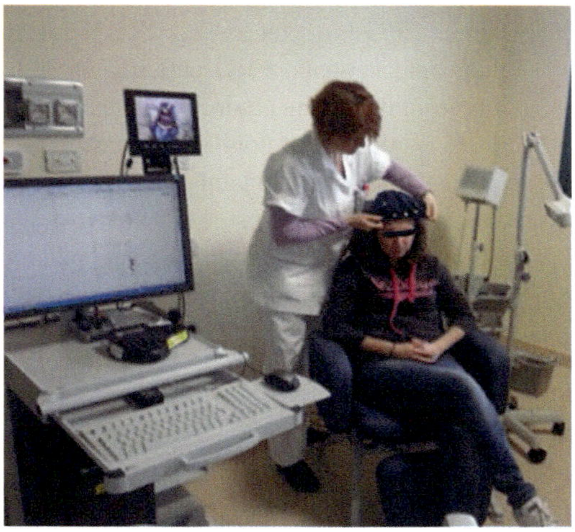

Fig. 10.1 A neurophysiopathology technician places a pre-wired EEG headset on the patient's scalp to conduct a routine video-EEG examination

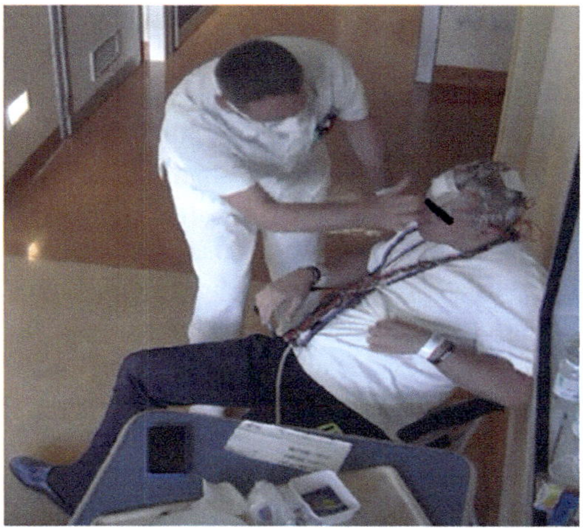

Fig. 10.2 A neurophysiopathology technician assists a patient during an epileptic seizure and assesses his responsiveness with simple questions

Fig. 10.3 A neurophysiopathology technician monitors the EEG recording from the control room

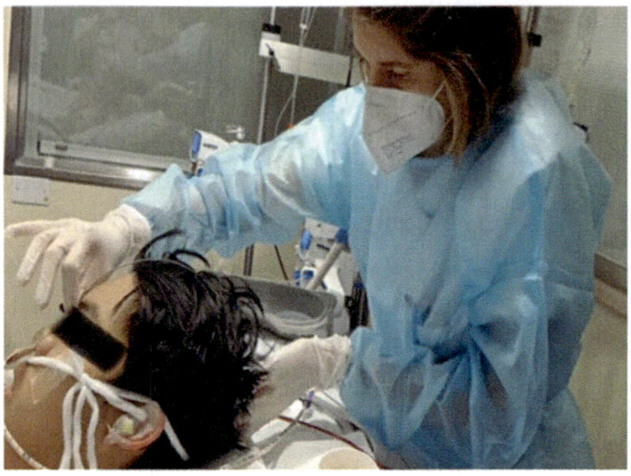

Fig. 10.4 A neurophysiopathology technician prepares the equipment for extended video-EEG monitoring of a patient in the intensive care unit

professionalism and experience for operating room recordings and collaborate closely with the surgical team [11].

6. *Continuous EEG Monitoring in the ICU*
   Continuous EEG monitoring in the ICU is essential for documenting neurological conditions and integrating with other monitoring systems. The NT must contextualize EEG patterns with clinical conditions, document and eliminate artifacts, and adhere to specific legislation for brain death assessment. Effective interaction with the entire healthcare team is crucial in this high-intensity care setting [12, 13], (Fig. 10.4).
7. *EEG-fMRI*
   EEG-fMRI combines EEG and functional magnetic resonance imaging (fMRI) to study brain activity and pathologies such as epilepsy. The NT must prepare the patient, manage artifacts specific to the MRI environment, and assist the neurophysiologist in data interpretation. This technique involves simultaneous EEG and BOLD (blood oxygen level dependent) imaging, which requires careful handling and post-analysis support [14].
8. *High-Resolution EEG (HREEG)*
   HREEG overcomes conventional EEG limitations by providing greater spatial resolution through a higher number of electrodes (60–256). This allows for detailed mapping of cortical areas involved in sensory and motor events. The NT must ensure precise electrode placement, maintain patient cooperation, and assist in data interpretation [15].
9. *Amplitude-Integrated EEG (aEEG)*
   Commonly used in neonatology, aEEG is simpler and less sensitive than VEEG but has improved with digital techniques and simultaneous EEG trace recording. The NT must understand neonatal EEG patterns and artifacts from recording environments and work closely with neonatologists and other specialists to enhance neonatal care [16, 17].
10. *Invasive EEG Recording*
    Invasive EEG recording involves placing devices directly on the brain, with the neurosurgeon responsible for device placement. The NT handles signal recording, equipment setup, and monitoring, requiring expertise in managing critical episodes and ensuring safety [18–21], (Figs. 10.5, 10.6, and 10.7).
11. *Role During the COVID-19 Pandemic*
    During the COVID-19 pandemic, NTs were essential in performing EEGs to identify

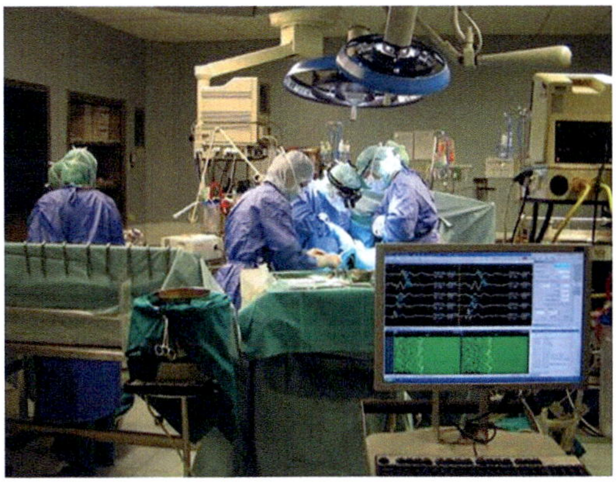

Fig. 10.5 A neurophysiopathology technician performs intraoperative monitoring during a surgical procedure

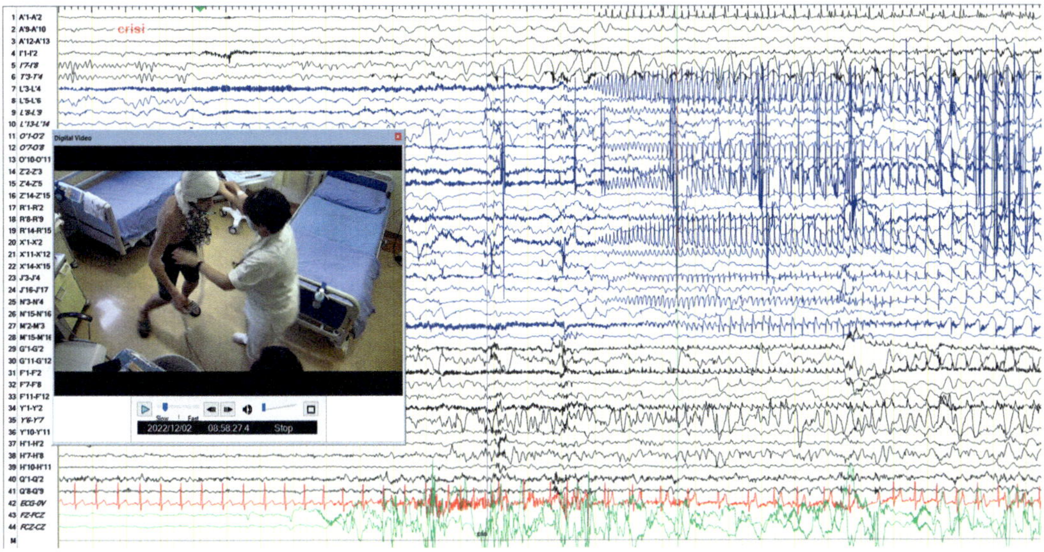

Fig. 10.6 A neurophysiopathology technician helps restrain a patient during an epileptic seizure to prevent self-injury while conducting a stereo-EEG recording

altered alertness or consciousness changes and in conducting electroneuromyography for acute neuromyopathies. This period saw the adoption of new protocols, use of Wi-Fi technology, and telemedicine solutions to manage patient care effectively [22], (Fig. 10.8).

Modern clinical neurophysiology increasingly relies on computer systems for recording, processing, and evaluating bioelectrical signals. The rise of telemedicine has facilitated remote service delivery, allowing NTs to conduct exams and share data through secure networks, enhancing the scope and efficiency of neurophysiological care.

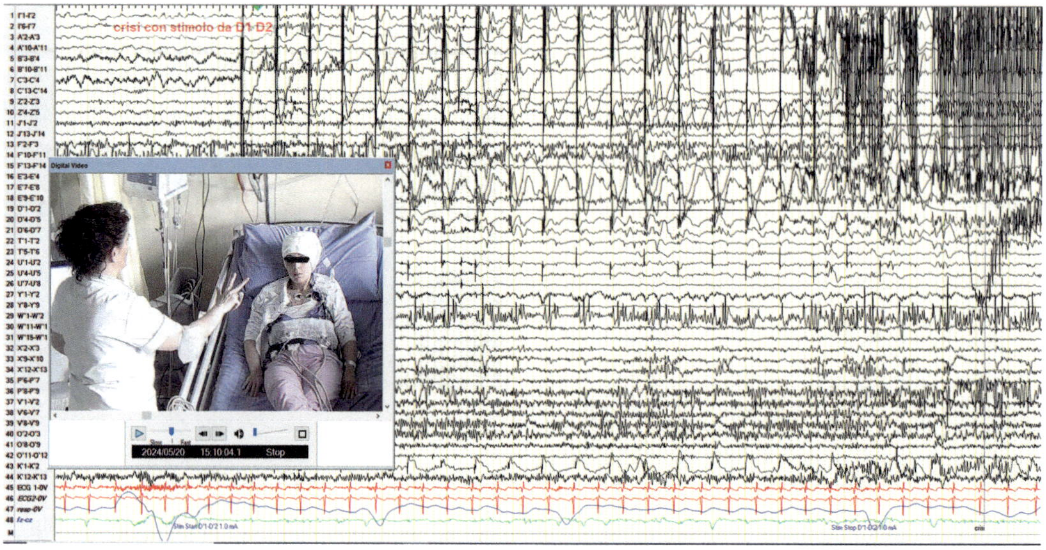

Fig. 10.7 A neurophysiopathology technician evaluates a patient during an epileptic seizure induced by electrical stimulation in a stereo-EEG recording

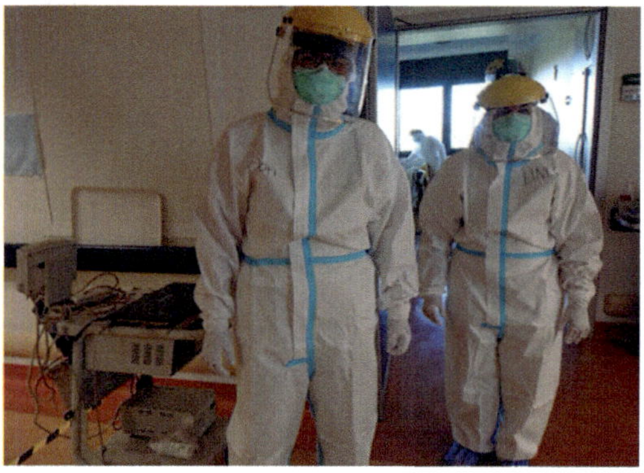

Fig. 10.8 Two neurophysiopathology technicians working in a COVID-19 department

10.6 Conclusions

The evolution of the NT over time has been marked by significant advancements and expanding responsibilities. Initially focused on basic EEG and its clinical applications, the NT's role has broadened significantly. The transition from analog to digital technologies has transformed the practice, enabling more sophisticated data processing and analysis. This technological evolution has introduced new research and diagnostic capabilities, requiring NTs to acquire a diverse set of skills and work closely with neurophysiologists, biomedical engineers, and physicists.

Acknowledgments We are grateful to Professor Paolo Tinuper for sharing his profound passion and expertise in electroencephalography with us. Our thanks also go to Professor Oriano Mecarelli for his unwavering support and belief in our professional role. Additionally, we are deeply appreciative of Dr. Lorenzo Ferri for his invaluable contribution to translating this chapter.

References

1. Mecarelli O. EEG laboratory: patient care and the role of the EEG technicians. In: Mecarelli O, editor. Clinical electroencephalography. Springer; 2019. p. 91–107.
2. Associazione Italiana Tecnici di Neurofisiopatologia (AITN). Newsletter, n. 4. ott, dic; 1999. pp. 12–21
3. Stone JL, Hughes JR. Invited review. early history of electroencephalography and establishment of the American Clinical Neurophysiology Society. J Clin Neurophysiol. 2013;30:28–44.
4. Mastrillo A. Istituzione e regolamentazione del Profilo del Tecnico di Neurofisiopatologia. Associazione Italiana Tecnici di Neurofisiopatologia (AITN), Newsletter; 2005. n. 1, pp. 8–15
5. Tassi L, Mariani V, Pelliccia V, Mai R. Video electroencephalography (Video-EEG). In: Mecarelli O, editor. Clinical electroencephalography. Springer; 2019. p. 305–17.
6. Alvisi L, Bisulli F, Licchetta L, Tinuper P. Polygraphic tecniques. In: Mecarelli O, editor. Clinical electroencephalography. Springer; 2019. p. 259–79.
7. Chokroverty S, Radtke R, Mullington J. Polisomnography: tecnical and clinical aspect. In: Schomer DL, Lopes da Silva FH, editors.

Niedermeyer's electroencephalography basic principles, clinical application, and related fields. 6th ed. Wolters Kluwer, Lippincott Williams & Wilkins; 2011. p. 817–65.
8. Radtke RA. Sleep disorders: laboratory evalutation. In: Current practice of clinical electroencephalography, IV edition, ed. Ebersole-associated editor Husain AM, Nordii DR. Philadelphia: Lippincott Williams & Wilkins; 2014. p. 599–630.
9. Rubboli R, Tassinari CA, Gardella E. Polygraphic investigation and back-averaging techniques in the study of epileptic motor phenomena. In: Mecarelli O, editor. Clinical electroencephalography. Springer; 2019. p. 281–96.
10. Brienza M, Davassi C, Mecarelli O. Ambulatory EEG. In: Mecarelli O, editor. Clinical electroencephalography. Springer; 2019. p. 297–304.
11. Yamada T, Yeh M. Monitoring EEG during carotid surgery. In: Schomer DL, Lopes da Silva FH, editors. Niedermeyer's electroencephalography basic principles, clinical application, and related fields. 6th ed. Wolters Kluwer, Lippincott Williams & Wilkins; 2011. p. 787–808.
12. Mecarelli O, Vicenzini E. Brain death. In: Mecarelli O, editor. Clinical electroencephalography. Springer; 2019. p. 767–87.
13. Saurabh RS, Hirsch LJ. Continuos EEG monitoring in the intensive care unit. In: Current practice of clinical electroencephalography, IV edition, ed. Ebersole-associated editor. Husain AM, Nordii RD. Philadelphia: Lippincott Williams & Wilkins; 2014. p. 543–98.
14. Meletti S. Simultaneous recording EEG in fMRI. In: Mecarelli O, editor. Clinical electroencephalography. Springer; 2019. p. 345–57.
15. Beniczky S, Sharma P. Electromagnetic source imaging, high-density and MEG. In: Mecarelli O, editor. Clinical electroencephalography. Springer; 2019. p. 329–43.
16. Dilena R, Raviglione F, Cantalupo G, INNESCO Group, et al. Consensus protocol for EEG and amplitude-integrated EEG assessment and monitoring in neonates. Clin Neurophysiol. 2021;123:886–903.
17. Mizrahi EM, Moshè SL, Hrachovy RA. Normal EEG and sleep: preterm and term neonates. In: Niedermeyer's electroencephalography basic principles, clinical application, and related fields. 6th ed. Wolters Kluwer, Lippincott Williams & Wilkins; 2011. p. 153–62.
18. Tassi L. Invasive EEG. In: Mecarelli O, editor. Clinical electroencephalography. Springer; 2019. p. 319–28.
19. Seeck M, Schomer DL, Niedermeyer E. Intracranical monitoring: depth, subdural, and foramen ovale electrodes. In: Niedermeyer's electroencephalography basic principles, clinical application, and related fields. 6th ed. Wolters Kluwer, Lippincott Williams & Wilkins; 2011. p. 677–714.
20. Gotman J, Nuwer M, Emerson RG. Principles and techniques for long term EEG recording (EMU, ICU, Ambulatory). In: Niedermeyer's electroencephalography basic principles, clinical application, and related fields. 6th ed. Wolters Kluwer, Lippincott Williams & Wilkins; 2011. p. 724–40.
21. Kahane P, Dubeau F. Intracerebral depth electrode. In: Current practice of clinical electroencephalography, IV edition, ed. Ebersole-associated editor Husain AM, Nordii DR. Philadelphia: Lippincott Williams & Wilkins; 2014. p. 393–441.
22. Stipa G, Gabbrielli F, Rabbito C, Telemedicine Working Group. The Italian technical/administrative recommendations for telemedicine in clinical neurophysiology. Neurol Sci. 2021;42:1923–31.

11 The Impact of EEG on the Concept of Epilepsy

Simon Shorvon

The idea of what a disease is, its meaning and its signification, evolves over time [1]. This is perhaps surprising as the disease itself does not change. Epilepsy is a good example. Its nature and manifestations have not altered at all and were presumably the same in 2024 as they were in 400 BC when the Hippocratic collection titled *On the Sacred Disease* was written. A seizure had the same features and pathophysiology then as now, yet the *concept* of epilepsy has undergone radical change. Epilepsy in 2024 would be completely unrecognisable to any citizen, doctor or patient of 1860 or 1900 and barely similar even in 1940. Electroencephalography (EEG) has played a fundamental role in this transformation after its introduction into clinical practice in the later 1930s and has indelibly changed the way epilepsy is conceived both in medicine but also in society.

11.1 Epilepsy Prior to EEG

The era of modern epilepsy can perhaps be said to have been ushered in by the work of Hughlings Jackson and others in the 1860s. Prior to this time, epilepsy was viewed predominately as a disease entangled with magic, possession, religious belief and superstition. Although rationalists, from Hippocrates onwards, had sought to demolish these, now seemingly prehistoric, beliefs, they had persisted and prevailed. Jackson, with remarkable intuition, proposed that the physical manifestations of an epileptic seizure were caused by *a sudden and excessive discharge* of neurons in a *localised* area of the *cerebral cortex* (the superficial grey matter of the brain) [2]. He saw the physical manifestations of a seizure simply to be a reflection of this disturbed physiology[1] and in doing so established a new theory of epilepsy, but it was a theory that had to wait for some seven decades before it could be definitively validated, and this validation came from EEG.

Jackson, and others, thus over the next three decades managed to demolish the stranglehold of religion and superstition in which epilepsy was encircled, but these beliefs were then replaced by the theories of inherited degeneration. *Degeneration* became a predominant social and medical belief from the later nineteenth century reaching its zenith around 1910. It was believed to be caused by defective germplasm[2] and this defective germplasm resulted not only in seizures but also in a whole range of inherent behavioural

S. Shorvon (✉)
UCL Queen Square Institute of Neurology, London, UK
e-mail: s.shorvon@ucl.ac.uk

[1] Jackson believed the 'excessive discharge' was chemical and due to abnormal cellular nutrition. Today the discharge is due to 'abnormal excitation' and explained in terms of molecular biochemistry and physiology.

[2] The 'gene' is the modern equivalent to the concept of the 'germplasm'.

F. Brigo, O. Mecarelli (eds.), *EEG: The First 100 Years*,
https://doi.org/10.1007/978-3-031-86878-8_11

and psychological failings such as mental deficiency, personality abnormalities, criminality, amorality and sexual deviance. In the early twentieth century, there was a massive public interest and anxiety over the idea that such inherited defects were weakening the population and fabric of society, and this led to a powerful eugenic movement. Epilepsy was at the centre of this concern, and the eugenicists focused on the personality abnormalities in idiopathic epilepsy and the associated psychological defects and social worthlessness. From 1910 to 1940s, this led in some countries to the segregation of some individuals with epilepsy, their institutionalisation and then to their sterilisation and during the Nazi period, their state-sanctioned murder.

In the 1930s, other changes in concept were also underway. This was in part due to the rise of neurology as a specialty in medicine. Prior to this, epilepsy had been conceptualised as a mental disorder and largely clinically managed by psychiatrists. But as neurology became established, epilepsy was increasingly seen as part of its purview. The divide between neurology and psychiatry was widening, as psychoanalytical theories predominated in psychiatry, and the neurologists (and neurosurgeons) grew generally impatient with both eugenical and psychoanalytical concepts of epilepsy. To the advanced neurologists of the period, epilepsy was seen not a mental disorder, as had hitherto been universally accepted, but a medical condition. They largely rejected the concept of inherent personality defects or behavioural features. The focus of neurology began to move to causation. Prior to this, the term epilepsy had been restricted to the 'idiopathic' condition, but given the increasing number of causes of seizures being identified, the illogicality of this became self-evident, and it was increasingly held that the term epilepsy should either be replaced by a plural form—i.e. *the epilepsies*—or abandoned altogether. As an example, in the definitive pre-EEG textbook of neurology, Kinnier Wilson deliberately entitled the chapter on epilepsy 'The Epilepsies' and placed his chapter within the section titled 'Disease Conditions of Uncertain Nature' [3].

11.2 The Introduction of EEG in Epilepsy

Physiology had been a favoured science since the early twentieth century (surely reflected by the fact that the Nobel Prize awards, founded in 1901, were awarded for 'Physiology or Medicine'). Charles Scott Sherrington was the key figure in neurophysiology in this early period. He had coined the word *synapse* and demonstrated one-way transmission across synapses. He viewed neural action as in essence an interaction between reflexes and recognised that there was therein a balance between excitatory and inhibitory effects. He also produced a nuanced approach to cerebral localisation and in the next two decades, studies of neural and synaptic function made great advances. Curiously, however, neither Sherrington nor any of the basic physiologists of the period made much mention of epilepsy, despite the fact that their work had clear implications for the condition and the obvious relevance of these processes to the mechanisms of seizures.

This brings us to Hans Berger and his prolonged series of experiments, published in a series of 14 papers all titled *Über das Elektrenkephalogramm des Menschen*[3] [4]. Berger was an eccentric and considered by many to be a crank, with flimsy scientific credentials and method, and although authorities disagree, there is some justification for these views. He became interested in recording brain function to see if threw light on telepathy and what he called 'psychic energy' and spent many years failing to show this. His first human EEG recording was probably made on 6 July 1924, when he observed 'tremulous movements' on his galvanometer due to brain currents in a patient with a skull defect [5].

Perhaps because of Berger's lack of physiological method, as well as his intellectual isola-

[3] All, bar the second report, were published in the journal *Archiv für Psychiatrie und Nervenkrankheiten*. Peter Gloor published English translations in 1969 and provided an excellent commentary. It is this which is used as the source of the information in this chapter.

tion, his publications did not initially gain much international recognition. This situation was to change when, in 1934, Edgar Douglas (later Lord) Adrian, a physiologist with impeccable credentials, and his colleague Bryan Matthews, a gifted electrical engineer, published a landmark paper in *Brain*. With careful experimentation, they confirmed the existence of oscillating rhythms, including the alpha rhythm, with Adrian generously termed the 'Berger rhythm' and identified its occipital origin [6]. This was the publication which moved EEG into the mainstream. Berger instantly became an international celebrity. In 1937 he attended the International Congress of Psychology in Paris, where he was greatly feted. Nonetheless, in September 1938 he left his chair in Jena and his laboratory was dismantled. His scientific work came to an abrupt end, and Berger fell into a deep depression. In 1941, he took his own life.

In pursuing his work on psychic energy, in 1932 Berger had become the first to record the EEG (in one or two channels) during epileptic seizures and noticed that absence attacks were associated with high-voltage 3 Hz discharges (without spikes). He described high-voltage potential oscillations in the interictal EEG of patients with frequent seizures, which he interpreted, correctly, as illustrating a predisposition to epilepsy [7]. He recorded changes in minor (partial) seizures [8]. He also demonstrated focal high-voltage sharp waves over the left central region in a patient with clonic jerking of the right hand in the aftermath of a 'severe epileptiform paralytic attack' [9]. He correctly interpreted this activity as a demonstration that the jerks originated from discharges in the contralateral central cortex.

Epilepsy though was not the main focus of Berger's work, and he had failed to appreciate the immense value his new discovery would have on this condition. This realisation was made by the neurologist and clinician William Lennox and his colleagues, in Boston. In December 1934, Lennox, with Hallowell Davis and Frederic and Erna Gibbs, began recording human EEG in Boston. In 1935, Gibbs, Davis and Lennox published their first paper on the topic, describing interictal spike and wave and the 3/second pattern of clinical absence seizures [10] (the interictal spike, as a hallmark of focal epilepsy, had been described at roughly the same time by Max Fischer and Hans Lowenbach in Berlin) [11]. They then went on to describe various aspects of clinical electroencephalography, including the electrographic changes of grand mal seizures [12–15], and in 1936, only 2 years after Adrian and Matthew's publication, the first clinical EEG unit was opened at Massachusetts General Hospital, directed by Robert Schwab. The work in Boston in effect marked the beginning of clinical electroencephalography, with epilepsy at its centre, and the introduction of EEG that was to prove crucial in a rethinking of the concept of epilepsy. In 1937, Lennox opened his Brain paper of 1937 in characteristic style:

> Diseases change their names with increase, not of age (like Chinese children), but with increase of medical knowledge...However, some diseases never have outgrown their baby names. For example, ever since the days of Hippocrates, recurring and sudden loss of consciousness and of muscle control has been called 'the falling sickness' or (in Greek) a seizure, 'epilepsy'. Thanks to the pioneer work of Berger in developing the electroencephalograph, we can now make good the lack of thousands of years and adopt for this condition a name based on the underlying pathological physiology. We now know that epilepsy is due to the development of abnormal rhythms in the cerebral cortex; it is a *paroxysmal cerebral dysrhythmia*. This discovery places the study and understanding of epilepsy on a different and deeper level and requires a reorientation of our thinking. [16]

In this paper, Lennox and his colleagues reported that by then they had 400 epileptic patients, of whom 120 had seizures whilst being recorded, and they outlined what they called 'seven new facts about epilepsy': (1) '[A] clinically observed seizure is but the outward manifestation of a disordered rhythm of brain potentials'. These were seen in all but 2–3% of seizures. (2) In grand mal, petit mal and psychomotor seizures, the patterns were different (as Lennox later put it: 'The writhings of the electrographic pen. name the sort of seizure the patient is having' [17]). (3) The fundamental basis of the disorder was not the abnormal rhythm but a defective control of rhythm. (4) There were interictal changes (Lennox called these 'subclinical seizures') taking the form of spike-wave or spike discharges. (5) EEG changes could predict when a grand mal seizure would occur. (6) 'The point of origin of an abnormal rhythm' can be in one area (now referred to as a focus) or occur 'simultaneously in all leads' (now referred to as generalised).

11.3 The Impact of EEG on Epilepsy

Lennox's view, that epilepsy was a *paroxysmal cerebral dysrhythmia,* transformed the conceptualisation of epilepsy. No longer was it considered a psychiatric disorder, or a disorder due to defective inheritance associated with other social and medical degeneration, nor a condition with many causes, but it was redefined in physiological terms, with the emphasis on the seizure not other features. Printed out onto paper, the EEG changes were seen to be an actual visualisation of Jackson's previously hidden excessive discharges. The impact was enormous, and EEG was one of the first examples of modern basic science discovery translated into the clinic. The changes also moved the clinical gaze from aetiology to physiology, and seizures of all causes were grouped into a single disease entity 'epilepsy'.

EEG engulfed the entire subject of clinical epilepsy. By the 1950s, the squiggly lines of EEG were acknowledged to be in effect the signature of epilepsy, and it is difficult to overstate the stimulus this gave to epilepsy and epilepsy research. It was indeed this single investigation that placed epilepsy again in the vanguard of neurology. A huge epilepsy literature on EEG emerged, and there was hardly a paper on the topic which did not mention EEG. In 1941, Erna and Frederic Gibbs produced their renowned *Atlas of Electroencephalography* [18], dedicated to Hans Berger, and in 1950, Denis Hill and William Cobb produced the first edition of a book become known as the Bible of EEG [19]. Many of the subsequent developments during this time were technical and directed to efforts to record actual seizures and not just the interictal EEG but to link the ictal EEG to the filmed clinical manifestations of a seizure [20], and in 1947, the first radio-telemetered EEG was recorded [21]. Perhaps more than any other person, in the 1950s and 1960s, Henri Gastaut from Marselles took over the baton of EEG and became a champion of the technology. He had been an indifferent student, but then in 1946 spent 6 months with brilliant and innovative Grey Walter in Bristol, who introduced him to EEG. Following this and later after a 3-month sabbatical with Penfield and Jasper at the Montreal Neurological Institute, he started a lifelong research programme heavily based in EEG which was to alter the concept of epilepsy. Here can be mentioned just three striking examples of the many ideas that followed in the decade or so after the introduction of EEG and which depended on EEG; and in all three Gastaut was a central figure.

11.3.1 Temporal Lobe Epilepsy

EEG played a fundamental role in the birth of the concept of temporal lobe epilepsy. The idea that epilepsy could arise in the temporal lobe was, like so much else in the field, first recognised by Hughlings Jackson in his case report of 'Dr Z', and Jackson surmised that seizures with dreaminess, absence and cognitive symptoms had a temporal lobe origin [22]. No further work on this topic was carried out until Karl Heinz Stauder reported in 1935 that a high proportion of patients had a similar characteristic constellation of symptoms which were commonly associated with Ammon's horn sclerosis (i.e. a temporal lobe defect) [23]. In 1941 Herbert Jasper and John Kershman recognised that psychomotor seizures were associated with temporal spiking and proposed that the origin of the spikes and seizures was in the temporal lobe or deeper structures [24]. Then in 1948 Frederic and Erna Gibbs and Bartolomé Fuster published a landmark paper in which the 'psychomotor type of EEG discharge' was correlated with clinical findings in 300 patients of whom 90% were found to exhibit temporal lobe seizures [15].

In 1952 Gastaut devoted the 5th Colloque de Marseille to temporal lobe epilepsy [25]. He produced a manuscript for the conference, which opened with: 'An extremely important development has recently occurred in the field of epileptology, viz. the identification of a new form of epilepsy, which has come to occupy a prominent place in nosology.' He noted that diagnosis was due to the EEG signs which are 'sufficiently simple and constant to ensure a correct diagnosis by themselves: a focus of negative spikes occupying the anterior temporal region of the scalp, readily demonstrable under the conditions of a standard examination or, at the worst, during activation by sleep or metrazol.' The discovery of temporal lobe epilepsy of course also led to a resurgence of epilepsy surgery, greatly aided by EEG, first in Montreal and then rapidly in the USA and in countries in Europe.

11.3.2 Epilepsy Syndromes

EEG also facilitated the identification of epilepsy suyndromes. The West syndrome was the first condition to be defined by its EEG appearance, when the specific EEG appearance of hypsarrhythmia was identified [26]. The Lennox-Gastaut syndrome was a second syndrome defined by its EEG signature [27].

11.3.3 Classification in Epilepsy

The introduction of EEG into epilepsy practice had one further far-reaching effect—the rethinking of the classification first of *epileptic seizures* and then of *the epilepsies*. Since the time of Jackson, classification in epilepsy had been based on the clinical appearance of seizures ('seizure type') and had not changed in any fundamental way for 80 years. EEG opened a Pandora's box and raised the possibility of introducing physiology into what was previously only a clinical classification. This approach proved irresistible and stimulated considerable activity with a variety of schemes incorporating EEG by Charles Symonds, Francis McNaughton, Richard Masland and jointly by Penfield and Jasper [28–32].

It was however Gastaut who was to create 'an international classification', as he put it to 'provide a uniform system and to standardise communications worldwide'. He gathered together 120 leading figures to a meeting in Marseille on 1–2 April 1964 and presented a draft scheme which he bulldozed through (with very little modification despite days of discussion with the participants of the meeting). He then floated his draft to a wider audience, but despite his protestations of wide consultation, Gastaut accepted few amendments to his 1964 draft, and the final 1969/1970 version proved to be very similar in form and content [33, 34]. At the quadrennial International League Against Epilepsy (ILAE) Congress in New York in 1969, the classification was discussed but does not seem to have been formally approved, but despite this it became widely accepted and is still in large part used. The tagging of the classification with the ILAE name was a publicity coup for the organisation, and by this single act, the ILAE became synonymous with professional authority in epilepsy. This action, more than any other, moved the organisation into the top position in the world of epilepsy, and for this Gastaut (and perhaps also EEG) must take the lion's share of the credit.

11.4 EEG in Other Areas of Medicine

Fairly rapidly, after its public introduction, the value of EEG was explored in a whole range of neurological and psychiatric diseases and also in normal human behaviours. Its scientific credentials appealed to the personality of many neurologists who viewed it as a stringent science, unlike psychology (or even worse, psychiatry), and with the introduction of EEG, suddenly there was a physiological tool which scientific clinicians could utilise. In the words of historian Cornelius Borck, 'With scientific recognition of the EEG as a valid parameter of brain function, the entire life of a human being, from conception until death, could count as the subject of electroencephalographic examinations. In fact, hardly any human activity was omitted from representation in the form of an EEG curve' [35].

EEG had caught the public imagination and popular media was full of images of electrographic squiggles and brain waves (a term indeed derived from EEG). EEG was used to study such facets as intelligence, personality, psychopathy, psychosis, moral deviancy (e.g. homosexuality) and even psychosomatic illness (e.g. peptic ulcer). The technology achieved particular traction in forensic medicine, much as epilepsy had itself done in the nineteenth century by Lombroso, and EEG abnormalities were claimed to be found in more than half of psychopathic criminals [36].

Most of this work came to nought, and even by 1947, as Adrian put it: 'Some of the dreams of the adolescent have been abandoned: the EEG is not the sure guide to the state of the brain that we may have hoped it would become, but we now know far better what it can show and what it cannot and the physician and surgeon have far better reason to trust the advice you can give them' [37]. Even more ferocious were the criticisms of Francis Walshe a decade later when discussing the future of neurology considered that the rise of 'electronic recording' and electrophysiology consumed resources and manpower but were abstractions which had contributed nothing to 'those age-old burning problems of the etiology, pathogenesis, and treatment of the many chronic, disabling, and killing affections of the nervous system' [38]. To Walshe 'Is it not surprising, then, that we should find distinguished scientists who believe that there is nothing in the mind and intellect of man that cannot be described in terms of the … bloodless dance of action potentials' [39]. He was reflecting a growing disillusionment about the value of EEG in neurology and psychiatry in general, and indeed now it occupies a minor place only in general neurology and almost no place at all in psychiatry. However, in epilepsy and in disorders of disturbed consciousness, including sleep and anaesthesia, EEG has maintained its leading role, not a bad record a century after its introduction into human investigation.

11.5 The Impact of EEG on Epilepsy: A Balance Sheet

EEG, neuroimaging, laboratory science and genetics are the major investigatory technologies that have transformed epilepsy in the modern era—and EEG was the first and in some ways has had the greatest impact. That clinical EEG has been of benefit to epilepsy can be in no doubt, and it resulted in many advances and discoveries. It was a visualisation of Jackson's' 'excessive discharge, and epilepsy evolved into being seen as a physiological condition and not a mental disorder nor a brain degeneration, nor an inherited defect or a condition with associated moral and personality defects. It led to new thinking about epilepsy, new classifications and the recognition of new syndromes. It stimulated epilepsy surgery and became a vital methodology in experimental medicine. It initially was also used as a localisational tool, but its value is localising structural abnormalities in the brain was then entirely superseded by neuroimaging.

Without diminishing its importance, it should also be recognised that like all technologies, its impact has been somewhat of a curate's egg—and there have been drawbacks as well as benefits. EEG ended the emphasis on the aetiology of seizures, a focus of attention which had been gathering force in the 1930s, and led to the reinstatement of epilepsy as a unitary disease, which it is clearly not. Because of EEG, the epileptic seizure became to be seen not as a symptom of the many disparate conditions but as a symptom of epilepsy itself. EEG induced a rash of new classifications which may have straightjacketed research to a large extent, and one wonders what directions the world of epilepsy research might have taken in its absence. The focal abnormalities of EEG created the mirage of focal seizures, when in fact most seizures are based on widespread neural networks. Walshe considered that EEG has contributed nothing to the 'age-old burning questions of aetiology, pathogenesis or treatment', but this was perhaps rather unfair, for it did contribute to pathogenesis and also to diagnosis. But in many ways Walshe was correct. Few other areas of medicine produced charlatanism on such a wide scale (and still does) and the rapid dissemination of EEG units in psychiatric hospitals, and the extravagant claims and widespread misuse of EEG in the 1950s and 1960s were a massive waste of resources.

The course of discovery of EEG was typical of all new technologies (and exactly the same pattern has been followed by neuroimaging and epilepsy genetics). The technical discovery was succeeded by swift adoption into clinical practice, rapid advances ensued, but this was then followed by over-emphasis of its value and by over-promotion. Reams of EEG nonsense occupy the pages of medical journals. Borck pointed out that at the time of its introduction, the world was awash with 'currents'—electric lighting and telephonic communication, and this influenced how the EEG was conceived. It is important to see it as a technology of its time, and as times change so has its relevance.

With the coming of EEG (and then of neuroimaging and genetics), the course of epilepsy was altered, and the concept of epilepsy changed. In assessing the value of EEG disentangling what is important and what is irrelevant, what is reiterative or simply wrong has to wait for the passage of time. Throughout the history of epilepsy, many theories and treatments, once universally accepted, are now viewed as entirely wrong and some even harmful. Now as EEG in epilepsy reaches its centenary, such a reappraisal is timely.

Acknowledgement Much of this chapter is based on notions from my book *The Idea of Epilepsy: a Medical and Social History of Epilepsy in the Modern Era (1860-2020).* (1860–2020) [1].

References

1. Shorvon S. The idea of epilepsy: a medical and social history of epilepsy in the modern era (1860–2020). Cambridge: Cambridge University Press; 2023.
2. Jackson JH. A study of convulsions. Trans St Andrews Med Graduates' Assoc. 1869;3(3):162–204.
3. Wilson SAK (finished posthumously by Bruce AN), Neurology. Vol 3. Baltimore: Williams & Wilkins; 1940. p. 1469.
4. Gloor P. Hans Berger on the electroencephalogram of man: the fourteen original reports on the human electroencephalogram. Electroencephalogr Clin Neurophysiol. 1969;28:1–350.
5. Gloor P. Hans Berger on the electroencephalogram of man: the fourteen original reports on the human electroencephalogram. Electroencephalogr Clin Neurophysiol. 1969;28:8.
6. Adrian ED, Matthews BHC. The Berger rhythm: potential changes from the occipital lobes in man. Brain. 1934;57:355–85.
7. Gloor P. Hans Berger on the electroencephalogram of man: the fourteen original reports on the human electroencephalogram. Electroencephalogr Clin Neurophysiol. 1969;28:191–207.
8. Gloor P. Hans Berger on the electroencephalogram of man: the fourteen original reports on the human electroencephalogram. Electroencephalogr Clin Neurophysiol. 1969;28:151–71, 5th report.
9. Gloor P. Hans Berger on the electroencephalogram of man: the fourteen original reports on the human electroencephalogram. Electroencephalogr Clin Neurophysiol. 1969;28:197.
10. Gibbs FAH, Davis H, Lennox WG. The electroencephalogram in epilepsy and in conditions of impaired consciousness. Arch Neurol Psychiatr. 1935;34:1133–48.
11. Fischer MH, Löwenbach H. Aktionsströme des Zentralnervensystems unter der Einwirkung von Krampfgiften. I. Mitteilung: Strychnin und Pickrotoxin. Arch Exp Pathol Pharmakol. 1934;174:357–82.
12. Gibbs FA, Lennox WG, Gibbs EL. The electroencephalogram in diagnosis and in localization of epileptic seizures. Arch Neurol Psychiatr. 1936;36:1225–35.
13. Gibbs FA, Davis H, Lennox WG. The electroencephalogram in epilepsy and in conditions of impaired consciousness. Arch Neurol Psychiatr. 1935;34:1133–48.
14. Gibbs FA, Gibbs EL, Lennox WG. Cerebral dysrhythmias of epilepsy: measures for their control. Arch Neurol Psychiatr. 1938;39:298–314.
15. Gibbs E, Gibbs F, Fuster B. Psychomotor epilepsy. Arch Neurol Psychiatr. 1948;60:331–9.
16. Gibbs FA, Gibbs EL, Lennox WG. Epilepsy: a paroxysmal cerebral dysrhythmia. Brain. 1937;60:377–88.
17. Lennox WG, Lennox MA. Epilepsy and related disorders. Boston: Little Brown; 1960. p. 33.
18. Gibbs FA, Gibbs EL. Atlas of electroencephalography. Cambridge MA: Addison-Wesley; 1941.
19. Hill D, Cobb WA, editors. Electroencephalography: a symposium on its various aspects. London: Macdonald; 1950.
20. Schwab RS, Schwab MW, Withee D, Chock YC. Synchronized moving pictures of patient and EEG. Electroencephalogr Clin Neurophysiol. 1954;6:684–6.
21. Parker C, Breakell C. The radio-electrophysiologogram: radio transmission of electrophysiological data from the ambulant and active patient. Lancet. 1953:1285–8.

22. Jackson JH. On a particular variety of epilepsy ("Intellectual Aura"), one case with symptoms of organic brain disease. Brain. 1888;11:179–207.
23. Stauder KH. Epilepsie und schläfenlappen. Arch Psychiatr. 1935;104:181–212.
24. Jasper H, Kershman J. Electroencephalographic classification of the epilepsies. Arch Neurol Psychiatr. 1941;45:903–43.
25. Shorvon SD. An episode in the history of temporal lobe epilepsy: the quadrennial meeting of the ILAE in 1953. Epilepsia. 2006;47:1288–91.
26. Gastaut H, Rémond A. Études électroencéphalographiques des myoclonies. Rev Neurol. 1952;86: 596–609.
27. Gastaut H, Roger J, Soulayrol R, Tassinari CA, Régis H, Dravet C, Bernard R, Pinsard N, Saint-Jean M. Childhood epileptic encephalopathy with diffuse slow spike-waves (otherwise known as "petit mal variant") or Lennox syndrome. Epilepsia. 1966;7:139–79.
28. Symonds CP. Classification of the epilepsies. BMJ. 1955;1:1235–8.
29. McNaughton FL. The classification of the epilepsies. Epilepsia. 1952;3:7–16.
30. Masland RL. Classification of the epilepsies. Epilepsia. 1959/60;1:512–20.
31. Masland RL. Comments on the classification of epilepsy. Epilepsia. 1969;10(Suppl):22–7.
32. Penfield W, Jasper H. Epilepsy and the functional anatomy of the human brain. Boston: Little Brown; 1954.
33. Gastaut H. Clinical and electroencephalographical classification of epileptic seizures. Epilepsia. 1969;10(Suppl):2–13.
34. Gastaut H. Clinical and electroencephalographical classification of epileptic seizures. Epilepsia. 1970;11:102–13.
35. Borck C. Brainwaves: a cultural history of electroencephalography. London: Routledge; 2019. p. 197.
36. Silverman D. Clinical and electroencephalographic studies on criminal psychopaths. Arch Neurol Psychiatr. 1945;50:18–33.
37. Nuwer MR, Lücking CH. Wave length and action potentials: history of the IFCN. Clin Neurophysiol. 2010;51(Suppl. 1):3–10.
38. Walshe FMR. The present and future of neurology. Arch Neurol. 1960;2:93–8.
39. Walshe FMR. Mind and brain. Blackfriars. 1960;41: 246–56.

12 The Role of the EEG in the Diagnosis of Epileptic Encephalopathies

Bernardo Dalla Bernardina, Gaetano Cantalupo, and Carlo Alberto Tassinari

12.1 Introduction

The term "epileptic encephalopathy" (EE) was coined by Gastaut to define both West syndrome and Lennox-Gastaut syndrome [1–3] as a condition in which the appearance of seizures and significant EEG interictal paroxysmal activities are accompanied by a severe impairment of psychomotor development. Furthermore, the concept of a detrimental effect on cognition determined by epileptiform abnormalities was hypothesized also in relation to another condition, characterized by "status epilepticus during sleep" (SES) by Tassinari, Dravet and Roger in 1977 [4], asserting that "the condition of a protracted SES can be the factor leading to severe mental deterioration and psychic disturbances". Only many years later, in 2001, the term "epileptic encephalopathy" entered the International League Against Epilepsy (ILAE) Classification, as a "new concept", defined as "a condition in which the epileptiform abnormalities themselves are believed to contribute to the progressive disturbances in cerebral functions" [5]. More recently, acknowledging cases where EE occurs in subjects suffering from a pre-existing developmental encephalopathy, the term of "developmental and epileptic encephalopathy" (DEE) was introduced, alongside that of EE [6]. The DDE condition embodies the notion that the underlying cause (mostly genetic) itself results in developmental delay, with the superimposed epileptic encephalopathy further adversely affecting development. The obvious consequence on the therapeutic and prognostic level is that, while in EE an effective antiseizure treatment can lead to the recovery of the subject neurodevelopmental impairment, in the case of DEE the same treatment, even if capable of limiting the aggravation of neurodevelopmental impairment induced by the epileptiform activities, cannot modify the pre-existent impairment directly related to the underlying pathology. Given the high aetiologi-

B. Dalla Bernardina (✉)
Center for Research on Epilepsy in Pediatric age (CREP), University Hospital of Verona, Verona, Italy

Innovation Biomedicine section, Department of Engineering for Innovation Medicine, University of Verona, Verona, Italy
e-mail: bernardo.dallabernardina@univr.it

G. Cantalupo
Center for Research on Epilepsy in Pediatric age (CREP), University Hospital of Verona, Verona, Italy

Innovation Biomedicine section, Department of Engineering for Innovation Medicine, University of Verona, Verona, Italy

Child Neuropsychiatry Unit, University Hospital of Verona (Full Member of the European Reference Network EpiCARE), Verona, Italy
e-mail: gaetano.cantalupo@univr.it

C. A. Tassinari
Center for Research on Epilepsy in Pediatric age (CREP), University Hospital of Verona, Verona, Italy

University of Bologna, Bologna, Italy

F. Brigo, O. Mecarelli (eds.), *EEG: The First 100 Years*,
https://doi.org/10.1007/978-3-031-86878-8_12

cal and phenomenological polymorphism, that characterizes EE and, in particular, DEE, the diagnostic process is very demanding and requires a multifaceted approach, including clinical, instrumental, neuropsychological, and laboratory assessments that cannot be limited to electroclinical evaluation alone. However, in many cases adequate EEG-polygraphic recordings prove to be of crucial in both diagnosis and management.

12.2 West Syndrome: The Prototype of Epileptic Encephalopathy

The first EEG-polygraphic well-defined EE [1] is the West syndrome, classically characterized by the association of epileptic spasms in series, firstly described many years ago by West [7], hypsarrhythmic EEG pattern, and concomitant impairment of the acquired psychomotor skills. According to Gibbs and Gibbs [8] hypsarrhythmia is described as: "[...] random high voltage slow waves and spikes. These spikes vary from moment to moment, both in duration and location. At times they appear to be focal, and a few seconds later they will be seen to originate from multiple foci. Occasionally, the spike discharge becomes generalized, but it never appears as a rhythmically repetitive and highly organized pattern that could be confused with a discharge of the petit mal variant type. The abnormality is almost continuous and in most cases it shows as clearly in the waking as in the sleeping record. It is referred to as Hypsarrhythmia".

Over the following years, several authors, using long-lasting EEG-polygraphic recordings, highlighted that West syndrome can present with a very polymorphic electroclinical picture, as exhaustively summarized by Roger and Dulac [9]. The ictal pattern can be characterized either by a pseudoperiodic diffuse slow wave, a very brief discharge of fast activity, or a slow wave with preceding or superimposed very brief fast discharge [10, 11]. The pattern can vary during the series; frequently the slow wave corresponds to the more clinically obvious spams, while burst of pure fast activity tends to occur toward the beginning or end of the cluster, corresponding clinically to decreased contact or motionless periods [12] (Fig. 12.1). The clinical correlate can be an overt spasm or a more subtle or subclinical manifestation, often manifesting as an isolated and brief lateral eye deviation. The spasms can be axial, with simultaneous involvement of the neck and the limbs of both sides, but they can also be bilateral asynchronous, unilateral, or focal. They can be preceded by a focal seizure or be intermixed with the focal seizure itself [12–14]. Also the pathognomonic EEG pattern of West syndrome, hypsarrhythmia can vary greatly, presenting—according to Gastaut and co-workers [15]—with fragmented awake, with excessive slowing, and with fast activity, asymmetrical or associated with a focus, or, according to Harachovy et al. [16], with increased interhemispheric synchronization, asymmetrical, with a consistent focus, with voltage attenuation, or with high voltage slow activity. Hypsarrhythmia can persist or disappear between the spams [17]. Moreover the epileptic spams in series can also occur in the absence of hypsarrhythmia [18–20]. Finally an hypsarrhythmic pattern can persist following the spasms disappearance, or it can appear in the absences of spasms [17]. Given this high electroclinical polymorphism, only long-lasting and systematic EEG-polygraphic recordings can help in the diagnostic assessment and therapeutic approach as illustrated by the few following examples.

In cases of epileptic spams in series, often preceded by a focal seizure, occurring in a context of asymmetrical or unilateral hypsarrhythmia, eventually combined with unusually high amplitude (>150 mV), fast (>15 Hz) activity, and the disappearance of interictal activity between consecutive spasms, a cortical dysplasia can be suspected [17, 21]. In some cases the electroclinical features can even suggest a specific aetiology [22]. A continuous sequence of unusually ample and fast polyspikes followed by a slow wave, recurring with sustained rhythmicity both when awake and during sleep, in a newborn, can suggest a holoprosencephaly [17, 21]. A diffuse and high-

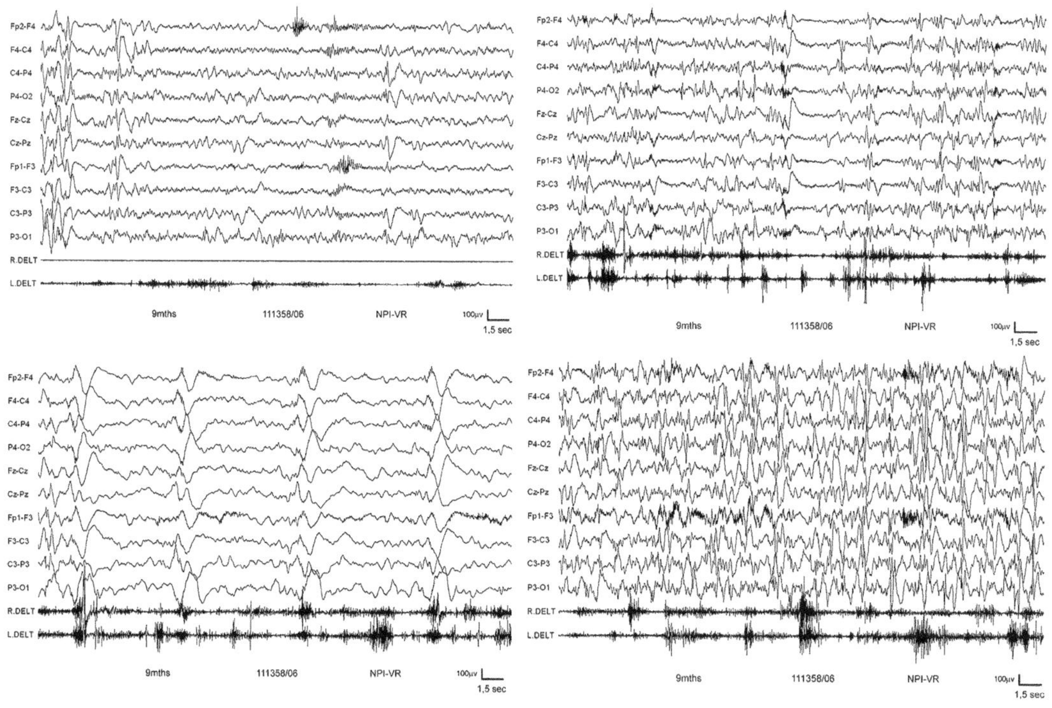

Fig. 12.1 A series of epileptic spasms in a 9 months old girl. As soon as the cluster of spasms starts, there is a great attenuation of the hypsarrhythmic pattern, and then the first events occur, in which the EEG feature is represented by bursts of pure fast activity without any observable deltoid muscle activation (upper left panel). Later on, a slow wave appears in association to the fast rhythm, gradually becoming higher in amplitude (upper right panel) and corresponding to the more clinically and polygraphically obvious spams (see deltoid muscle activations in the lower left panel); at the end of the cluster, hypsarrhythmia reemerges (lower right panel)

voltage hypsarrhythmic pattern intermingled by an unusually ample fast activity, appearing between 4 and 6 months of life, suggests a lissencephaly [17, 21, 23]. The finding in the first days or weeks of life of a unilateral or clearly asymmetric burst suppression or hypsarrhythmic pattern, in the presence of focal seizures and unilateral or even bilateral spasms (Fig. 12.2), strongly suggests a hemimegalencephaly or a posterior multilobar dysplasia [21, 24, 25]. An electroclinical picture characterized by early appearance of focal seizures that often initially are subclinical, followed by the appearance of spasm in series, presenting synchronous or asynchronous on the two hemisomes, in a context of a clear asynchrony of both interictal epileptiform paroxysmal activities and physiological sleep graphoelements, accompanied by high-voltage fast activities (unusually for the age), prompts to suspect an Aicardi syndrome [17, 26, 27].

However, it is worth taking in mind that at any age the presence of diffuse or focal unusually fast activities should induce to suspect a brain malformation [21].

Considering the West syndrome as a whole, an adequate analysis of the electroclinical pattern at onset and its modifications, according to the treatment and age progression, can give some useful suggestions. In the case of symptomatic aetiology, there is a tendency of the spasms to evolve in more prolonged tonic manifestations in series, accompanied by persistence of the hypsarrhythmic EEG pattern and predicting the evolution into a lifelong EE with a bad prognosis. In cases appearing in infants with normal development and unknown aetiology, the prognosis tends to be more favourable in the presence of the following features: (1) typical diffuse bilateral hypsarrhythmic pattern continuous when awake, more or less fragmented during slow sleep and

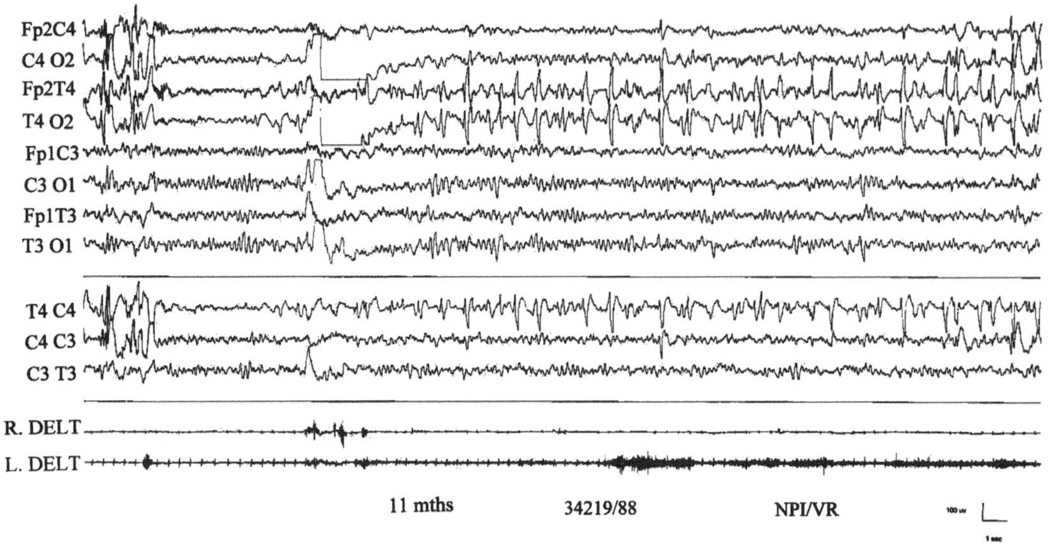

Fig. 12.2 Spasms in a 11 months old infant affected by hemimegalencephaly. In the first few seconds of the picture, it is recognizable the clearly asymmetric right-predominant hypsarrhythmic pattern that attenuates few seconds before the appearance of a bilateral spasm, superimposed to a sub-continuous focal epileptiform discharge on the right temporal region

disappearing during REM sleep, (2) persistent recognition of physiological rhythms during sleep, (3) reappearance of hypsarrhythmia between series of spasms, (4) absence of focal seizures, and (5) absence of residual focal abnormalities after disappearance hypsarrhythmic pattern following intravenous benzodiazepine and/or steroid treatment [17, 28].

Systematic EEG recordings are also essential for monitoring the progress of the hypsarrhythmic pattern after the clinical disappearance of spasms. In fact, hypsarrhythmia probably is not strictly an interictal EEG pattern but a component of a unique continuous ictal status, modified by the recurrence of—more or less easily detectable—motor events. Many authors furnished several elements supporting this hypothesis, summarized in Dalla Bernardina et al. [29] and here listed:

- The brief behavioural phenomena called "psychic awakening" [30, 31] concomitant with the transitory interruption of hypsarrhythmia induced by the spasm is the first evidence.
- The psychomotor deterioration and behavioural disturbances are milder in cases without hypsarrhythmia [32].
- Higher hypsarrhythmia scores correlate with worse prognosis [33] and electroclinical abnormalities [34–37].
- Neurosensory and cognitive impairment is generally short lasting in cases with transient electroclinical abnormalities.
- A prolonged duration of hypsarrhythmia appears to be a negative prognostic factor [28, 38–40].
- The risk of cognitive impairment increases if hypsarrhythmia lasts more than 3 weeks, independently from the aetiology [41].

Following these evidences, according to the opinion of West Delphi consensus group [42], the resolution of hypsarrhythmia itself must be considered an important therapeutic target.

Obviously, in the rare cases showing hypsarrhythmia without spasms and in all cases in which it persists after the spasms disappearance, the detrimental effect of hypsarrhythmia on neurodevelopmental trajectory can be detected only if the EEG study is associated with a neuropsychological assessment, adequate for age and adapted to clinical condition, as for patients with the EEG pattern of spike and wave activation during sleep with subtle or no apparent seizures (see below).

12.3 EEG Features Guiding to Monogenic DEE

It is well-known that a significant number of DEE are genetically determined [43, 44]. In several cases, adequate video-EEG-polygraphic recordings allow to suspect a specific monogenic DEE. For instance, the appearance in the first days of life of tonic seizures recurring often in clusters in a context of suppression-burst pattern, frequently responding to sodium channel blockers, strongly evokes a KCNQ2 DEE [45, 46]. Further, the early onset of peculiar seizures, characterized by an initial hypermotor manifestation related to a diffuse discharge of fast activity followed by a series of spasms and often associated to spontaneous or reflex massive jerks, evolving into a complex electroclinical picture associating generalized motor seizures spasms in series, myoclonic manifestations, and hypsarrhythmic pattern, is strongly suggestive a CDKL5-related DEE [47–50]. Moreover, the early appearance, mainly but not exclusively in females, of long-lasting focal seizures or generalized motor seizures in clusters, often realizing an epileptic status or akinetic seizures followed by an "obtundation status", occurring in a febrile context, suggests to suspect a PCDH19-DEE. This diagnostic hypothesis is further confirmed with the appearance of peculiar focal seizures with affective symptoms recurring always during febrile events [51–53]. Again, only adequate video-EEG-polygraphic records allow to recognize some peculiar conditions in whose both "fixation-off" and "eye-closure" can trigger seizures, namely, atypical absences with or without myoclonias, often realizing long-lasting "obtundation status". This condition, when associated to a peculiar EEG pattern characterized by the occurrence of parieto-occipital rhythmic delta waves induced by eye closure, should lead to strongly suspect a SYNGAP1-DEE [54].

12.4 Other DEEs with Pathognomonic EEG Findings

The occurrence in the first weeks or months of life of drug-resistant motor or non-motor seizures with autonomic symptoms manifesting on EEG as a focal flattening or a fast activity followed by a sequence of fast polyspikes recurring in clusters and involving alternatively—in a migrating manner—both hemispheres, imposes to suspect the diagnosis of an epilepsy of infancy with migrating focal seizures (EIMFS) (Fig. 12.3) [55, 56], whose most frequent cause is KCNT1 gene abnormality [57].

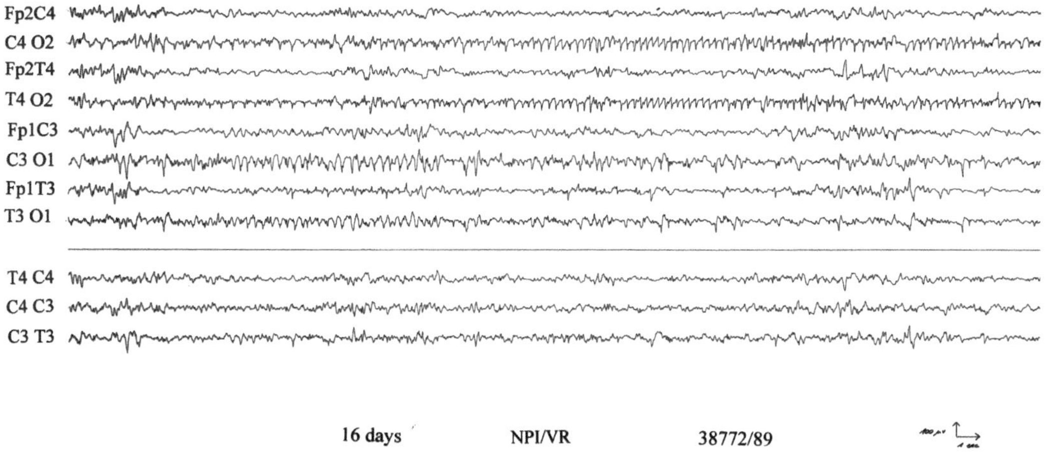

Fig. 12.3 Epilepsy of infancy with migrating focal seizures (EIMFS). A short segment of the EEG recording in a 16 days old newborn with EIMFS in which it can be appreciated the imbricated and asynchronous ictal discharges on the posterior regions of the two hemispheres

A further electroclinical entity for whose early diagnosis and EEG-polygraphic recordings are essential is the myoclonic status in nonprogressive encephalopathies [58–60]. This condition, mainly appearing during the first 2 years of life in children showing an obvious neurodevelopmental impairment, manifests with atypical absences with or without myoclonias and/or focal motor clonic seizures. The electroclinical picture evolves—often subtly—into a non-convulsive status epilepticus (NCSE) characterized by a sub-continuous, more or less diffuse, slow spike-and-wave (SW) discharge with superimposed spiky rhythmic theta on centroparietal regions, related with sub-continuous rhythmic or arrhythmic myoclonias and synchronous and asynchronous on different limbs (Fig. 12.4). Due to the pre-existence of severe intellectual disability and abnormal movements in those children, this condition is very difficult to recognize and may remain undiagnosed if well-performed EEG-polygraphic recordings are lacking. Further, only an accurate longitudinal monitoring with serial EEG-polygraphic recordings allows to detect the eventual progressive appearance of a chronic cortical myoclonus, forecasting a poorer evolution. The aetiology can be structural but is mainly genetic; more frequently appears in subjects with Angelman syndrome but can be observed in several other genetic disorders such as

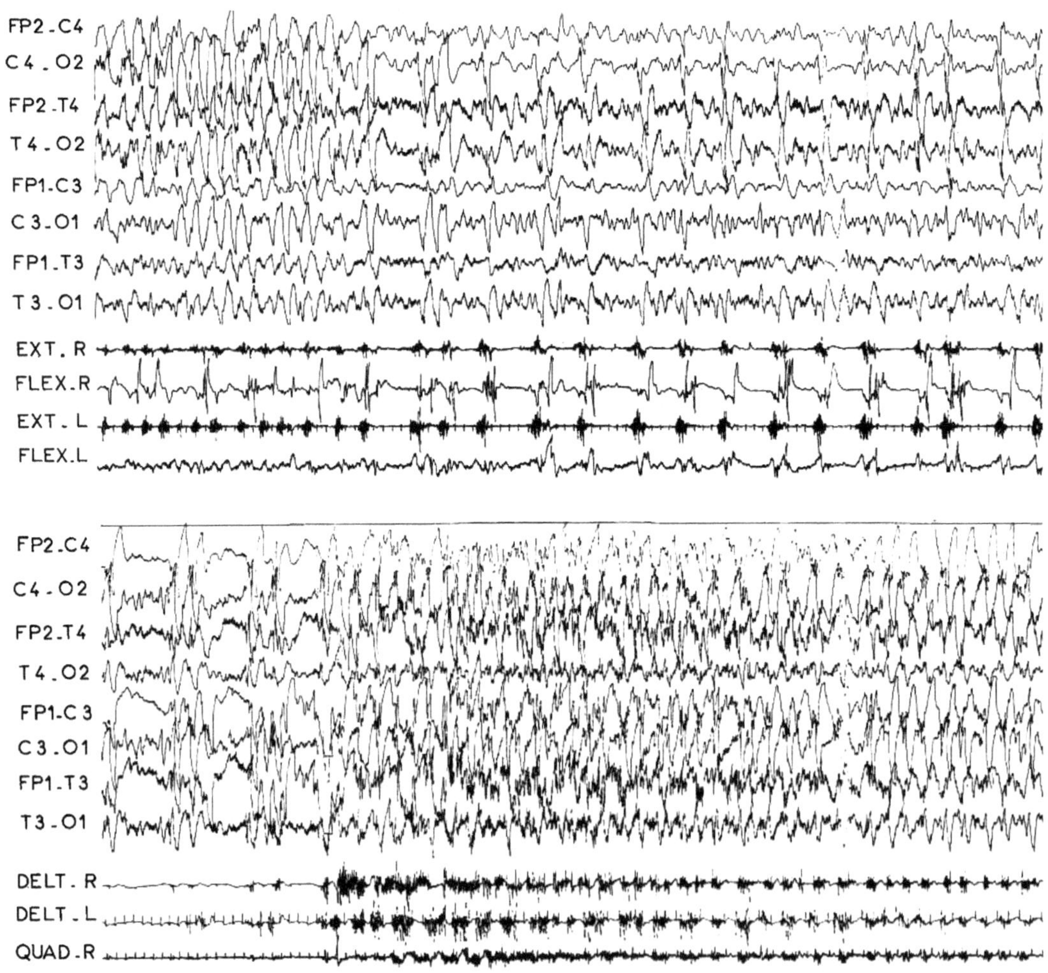

Fig. 12.4 Myoclonic status in nonprogressive encephalopathy. The EEG polygraphic recording of a 18 months old child demonstrates the presence of sub-continuous diffuse, slow spike-and-wave discharge with continuous rhythmic myoclonias, synchronous and asynchronous on different muscles

4p- syndrome, SCN8A-DEE, Rett syndrome, CDKL5-DEE, and ARV1-DEE [61–63].

Appearing at a later age and sustained by a great variability of aetiologies, Lennox-Gastaut syndrome is another exemplar DEE for whose diagnosis and management EEG-polygraphic recordings are mandatory. Typically it is characterized by the association of atypical absences with diffuse slow SW during wakefulness that become fast polyspikes and waves during slow sleep, tonic-atonic seizures inducing a sudden drop while awake and tonic seizures during sleep [2, 64]. At onset, the electroclinical picture can be incomplete, and some ictal events can be of ambiguous interpretation. For example, the ictal fall can evoke an epilepsy with myoclonic-atonic seizures; thus only the EEG-polygraphic recordings at awake and during sleep allow a correct differential diagnosis. In other cases, the absence seizures, at onset, can be transiently similar to those of childhood absence epilepsy. The tonic seizures during sleep may not be present in the first phases of the disease, or they can be difficult to recognize in the absence of nocturnal EEG recordings. Similarly, tonic seizures during sleep can be very subtle and so undetectable without sleep EEG at later ages, as in adolescents and adults.

Another DEE, appearing mainly in childhood, needing awake and sleep EEG-polygraphic recordings for early diagnostic suspicion is ring chromosome 20 syndrome. The initial presentation is often the abrupt appearance of multiple per day hypermotor seizures with terrifying hallucinations, often seep-related, frequently evolving into an obtundation status with loss of speech. The EEG correlate of the hypermotor seizure is a bilateral fast discharge followed by a generalized discharge of fast polyspikes and waves; the following non-convulsive status—that can last from some minutes to several days—is characterized by a sub-continuous sequence of high-voltage and diffuse slow SW [65, 66]. The recognition of a peculiar sub-continuous rhythmic theta-delta (3–7 Hz) activity on the interictal EEG is a further finding to support the diagnosis [67].

Last, but not least, as already mentioned in the introduction, another important electroclinical entity, in this context is the encephalopathy related to status epilepticus during slow sleep (ESES). This entity was initially recognized based exclusively on EEG by Patry, Lyagoubi, and Tassinari in 1971 in six children in which a peculiar EEG pattern was found, characterized by SW discharges occurring almost uninterruptedly while asleep [68]. They termed this condition as "subclinical electrical status epilepticus induced by sleep" as it was apparently not linked to any observable clinical event. A few years later, in 1977, expanding the initial cohort, Tassinari and co-workers acknowledged that the "electrical status epilepticus" was not completely subclinical and suggested that the extreme activation of sleep-related epileptiform abnormalities can have detrimental effect on cognitive/behavioural performances [4]. ESES is diagnosed in a patient presenting with epilepsy, with focal and apparently generalized seizures, in which a neurological deterioration occurs (involving cognitive, behavioural, and/or other domains) coupled with a worsening of the EEG abnormalities during sleep. This worsening is characterized by the appearance at sleep onset of diffuse spike-and-waves that can become continuous during sleep, leading to the term of "continuous spike-and-wave during sleep"(CSWS) to describe this EEG pattern. However, there is an extreme variability of the pattern, which can be also sub-continuous or more fragmented; it can also be asymmetric, hemispheric, or even focal. Since its first description, a numerical measure of epileptiform discharges abundance during sleep was used as a diagnostic criterion. The amount of epileptiform discharges was expressed as the percentage of the duration of slow sleep occupied by rhythmic spike-and-wave complexes and referred as "spike and wave index" (SWI), ranging from 85 to 100% in the seminal paper by Patry et al. [68]. However, we now know that this cut-off of 85% for diagnosis is too stringent and that probably there is a need to better define the features that will allow the clinician to foresee the evolution into ESES in patients at risk, besides the abundance only [69, 70]. Since the earlier descriptions, it was evident that—albeit fundamental—EEG cannot be the only parameter to use for diagnosis. In fact, diagnosis can be done only when a comprehensive clinical evaluation demonstrates the appearance or worsening of neurological, cognitive, and/or behavioural issues concomitantly to the presence of the pathologic EEG pattern. ESES can occur in patients with pre-existing neurological impairment, constituting a DEE, but it can also represent a complicated evolution of an otherwise idiopathic self-limited focal epilepsy of childhood [71]. The latter case constitutes a good example of an "epileptic encephalopathy" in which the subject has no pre-existing neurodevelopmental abnormality, and thus it is of crucial importance to recognize the possible worsening of the EEG (Fig. 12.5) and, in turn, strictly monitor the neurodevelopmental tra-

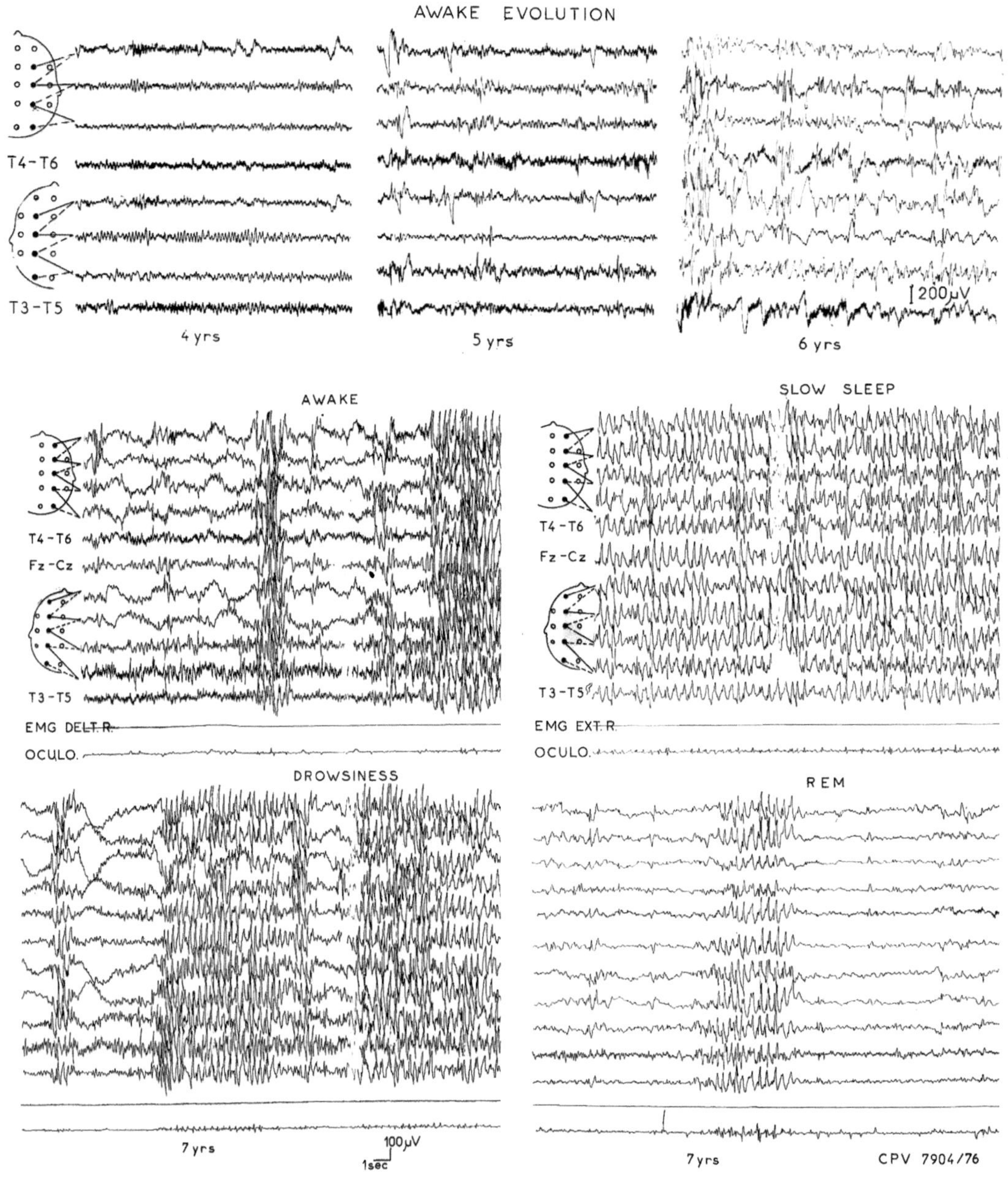

Fig. 12.5 Encephalopathy related to status epilepticus during slow sleep (ESES). In this case the patient was diagnosed with self-limited epilepsy with centrotemporal spikes (SeLECTS) at 4 years of age. During the follow-up an increasing activation of the epileptiform abnormalities was documented across the following years (upper line, samples of wake EEG at ages 4 years, 5 years, and 6 years, respectively). In the lower part of the figure, the worsening of the EEG both in wakefulness (upper left), drowsiness (lower left), and during slow sleep (upper right), with a reduction of paroxysmal discharges during REM sleep

jectory, to early diagnose and treat the patient, minimizing sequelae. Recently, the ILAE has proposed to replace the terms ESES and CSWS with the term "developmental and/or epileptic encephalopathy with spike-wave activation during sleep" (EE-SWAS/DEE-SWAS); however, there is a great number of conditions in which we can observe a spike-wave activation during sleep (SWAS), without implying a direct relationship between SWAS and encephalopathy, so it will be crucial to limit the use of EE and DEE to the situations in which we can demonstrate that seizures and/or epileptiform abnormalities themselves contribute to the progressive disturbances of brain functions, besides what is expected for the natural history of the disease.

12.5 Conclusion

As previously outlined, the diagnostic and therapeutic processes in DEE/EE need a multidisciplinary approach in which the EEG maintains an essential role even today. In fact, as outlined in this chapter, in several cases, the electroclinical phenotype represents the first and most important cue to the most probable aetiology, broader defined (structural, genetic, metabolic, etc.), and in some instances can suggest even a more specific diagnosis (e.g. gene involved, type of brain malformation). Furthermore, when facing a definite syndrome, EEG besides its diagnostic role is of great importance in prognosis, allowing detecting different variations in electroclinical phenotypes that have different predictive meaning. Finally, in cases in which obvious seizures are lacking or very rare (as in cases of hypsarrhythmia without spasms or in presence of the EEG pattern of CSWS), only the meticulous coupling of serial EEG evaluations (often by means of prolonged wake and sleep polygraphic recordings) and neuropsychological assessments can allow EE/DEE diagnosis and prompt a timely and adequate management.

References

1. Gastaut H, Roger J, Soulayrol R, Salamon G, Regis H, Lob H. Encéphalopathie myoclonique infantile avec hypsarythmie (syndrome de West) et sclérose tubéreuse de Bourneville [Infantile myoclonic encephalopathy with hypsarrhythmia (West's syndrome) and Bourneville's tuberous sclerosis]. J Neurol Sci. 1965;2(2):140–60.
2. Gastaut H, Roger J, Soulayrol R, Tassinari CA, Régis H, Dravet C, Bernard R, Pinsard N, Saint-Jean M. Childhood epileptic encephalopathy with diffuse slow spike-waves (otherwise known as "petit mal variant") or Lennox syndrome. Epilepsia. 1966;7(2):139–79.
3. Gastaut H. Dictionnaire de l'Epilepsie. Partie I: Definitions Organization Mondiale de la Santé; 1973.
4. Tassinari CA, Dravet C, Roger J. ESES: encephalopathy related to electrical status epilepticus during slow sleep. In: Proceedings of the ninth congress international federation of EEG and clinical neurophysiology. Amsterdam: Elsevier Science; 1977. p. 529–30.
5. Engel J Jr. International League Against Epilepsy (ILAE). A proposed diagnostic scheme for people with epileptic seizures and with epilepsy: report of the ILAE Task Force on classification and terminology. Epilepsia. 2001;42(6):796–803.
6. Scheffer IE, French J, Hirsch E, Jain S, Mathern GW, Moshé SL, Perucca E, Tomson T, Wiebe S, Zhang YH, Zuberi SM. Classification of the epilepsies: new concepts for discussion and debate-special report of the ILAE Classification Task Force of the Commission for Classification and Terminology. Epilepsia Open. 2016;1(1–2):37–44.
7. West W. On a peculiar form of infantile convulsions. Lancet. 1841;I:724–5.
8. Gibbs FA, Gibbs EL. Atlas of electroenchephalography: epilepsy, vol. II. Cambridge, MA: Addison-Wesley; 1952. p. 2.
9. Roger J, Dulac O. West syndrome: history and nosology. In: Dulac O, Chugani HT, Dalla Bernardina B, editors. Infantile spasms and West syndrome. London: W.B. Saunders; 1994. p. 6–11.
10. Kellaway P, Hrachovy RA, Frost JD Jr, Zion T. Precise characterization and quantification of infantile spasms. Ann Neurol. 1979;6(3):214–8.
11. King DW, Dyken PR, Spinks IL Jr, Murvin AJ. Infantile spasms: ictal phenomena. Pediatr Neurol. 1985;1(4):213–8.
12. Fusco L, Vigevano F. Ictal clinical electroencephalographic findings of spasms in West syndrome. Epilepsia. 1993;34(4):671–8.
13. Bour F, Chiron C, Dulac O, Plouin P. Charctères électrocliniques des crises dans le syndrome d'Aicardi [Electroclinical characteristics of seizures

in the Aicardi syndrome]. Rev Electroencephalogr Neurophysiol Clin. 1986;16(4):341–53.
14. Shewmon DA. Ictal aspects with emphasis on unusual variants. In: Dulac O, Chugani HT, Dalla Bernardina B, editors. Infantile spasms and West syndrome. London: W.B. Saunders; 1994. p. 36–51.
15. Gastaut H, Soulayrol R, Roger J, Pinsard N, editors. L'Encéphalopathie Myoclonique Infantile avec Hypsarythmie (Syndrome de West). Paris: Masson; 1964.
16. Hrachovy RA, Frost JD Jr, Kellaway P. Hypsarrhythmia: variations on the theme. Epilepsia. 1984;25(3):317–25.
17. Dalla Bernardina B, Watanabe K. Interictal EEG: variations and pitfalls. In: Dulac O, Chugani HT, Dalla Bernardina B, editors. Infantile spasms and West syndrome. London: W.B. Saunders; 1994. p. 63–81.
18. Gobbi G, Bruno L, Pini A, Giovanardi Rossi P, Tassinari CA. Periodic spasms: an unclassified type of epileptic seizure in childhood. Dev Med Child Neurol. 1987;29(6):766–75.
19. Caraballo RH, Ruggieri V, Gonzalez G, Cersósimo R, Gamboni B, Rey A, Poveda JC, Dalla BB. Infantile spams without hypsarrhythmia: a study of 16 cases. Seizure. 2011;20(3):197–202.
20. Caraballo RH, Gallo A, Reyes G, Flores G, Martín E, Intriago L, Ballesta D. Epileptic spasms in clusters without hypsarrhythmia in infancy and childhood: a single age-dependent type of epilepsy or well-defined epileptic syndrome? Epilepsy Res. 2024;202:107354.
21. Dalla Bernardina B, Perez-Jimenez A, Fontana E, Colamaria V, Piardi F, Avesani E, Santorum E, Grimau-Merino R, Tassinari CA. Electroencephalographic findings associated with cortical dysplasias. In: Guerrini R, et al., editors. Dysplasias of cerebral cortex and epilepsy. Philadelphia: Lippincott-Raven Publishers; 1996. p. 233–43.
22. Dalla Bernardina B, Dulac O. Introduction to etiology. In: Dulac O, Chugani HT, Dalla Bernardina B, editors. Infantile spasms and West syndrome. London: W.B. Saunders; 1994. p. 166–71.
23. Gastaut H, Pinsard N, Raybaud C, Aicardi J, Zifkin B. Lissencephaly (agyria-pachygyria): clinical findings and serial EEG studies. Dev Med Child Neurol. 1987;29(2):167–80.
24. Tjiam AT, Stefanko S, Schenk VW, de Vlieger M. Infantile spasms associated with hemihypsarrhythmia and hemimegalencephaly. Dev Med Child Neurol. 1978;20(6):779–98.
25. Vigevano F, Bertini E, Boldrini R, Bosman C, Claps D, di Capua M, di Rocco C, Rossi GF. Hemimegalencephaly and intractable epilepsy: benefits of hemispherectomy. Epilepsia. 1989;30(6):833–43.
26. Fariello RG, Chun RW, Doro JM, Buncic JR, Prichard JS. EEG recognition of Aicardi's syndrome. Arch Neurol. 1977;34(9):563–6.
27. Vigevano F, Maccagnani F, Bureau M, Dravet C, Roger J, Dalla Bernardina B, Daniele O, Tassinari CA. La sindrome di Aicardi. Boll Lega Ital Epil. 1981;32–33:29–35.
28. Dulac O, Plouin P, Jambaqué I. Predicting favorable outcome in idiopathic West syndrome. Epilepsia. 1993;34(4):747–56.
29. Dalla Bernardina B, Fontana E, Osanni E, Opri R, Darra F. Epileptic spasms: Interictal patterns. In: Guzzetta F, Dalla Bernardina B, Guerrini R, editors. Progress in epileptic spasms and West syndrome. Progress in epileptic disorders Vol. 4. John Libbey Eurotext (France); 2009. p. 43–60.
30. Launay C, Blanc C, Rebufatdeschamps M. Spasmes en fe flexion et hypsarythmia; mise au point à propos de six observations personnelles [Flexion spasms & hypsarhythmia; demonstration by 6 personal case reports]. Presse Med (1893). 1959;67(22):887–90.
31. Jeavons PM, Bower BD. Infantile spasms. In: Vinken PJ, Bruyn GW, editors. Handbook of clinical neurology, vol. 15. The Epilepsies. Amsterdam: Elsevier (North Holland); 1974. p. 219–34.
32. Jambaqué I, Chiron C, Dumas C, Mumford J, Dulac O. Mental and behavioural outcome of infantile epilepsy treated by vigabatrin in tuberous sclerosis patients. Epilepsy Res. 2000;38(2–3):151–60.
33. Kramer U, Sue WC, Mikati MA. Hypsarrhythmia: frequency of variant patterns and correlation with etiology and outcome. Neurology. 1997;48(1):197–203.
34. Matsumoto A, Watanabe K, Negoro T, Sugiura M, Iwase K, Hara K, Miyazaki S. Infantile spasms: etiological factors, clinical aspects, and long term prognosis in 200 cases. Eur J Pediatr. 1981;135(3):239–44.
35. Guzzetta F, Frisone MF, Ricci D, Randò T, Guzzetta A. Development of visual attention in West syndrome. Epilepsia. 2002;43(7):757–63.
36. Randò T, Baranello G, Ricci D, Guzzetta A, Tinelli F, Biagioni E, La Torre G, Epifanio R, Signorini S, Fazzi E, Mercuri E, Cioni G, Guzzetta F. Cognitive competence at the onset of West syndrome: correlation with EEG patterns and visual function. Dev Med Child Neurol. 2005;47(11):760–5.
37. Baranello G, Randò T, Bancale A, D'Acunto MG, Epifanio R, Frisone MF, Guzzetta A, La Torre G, Mannocci A, Ricci D, Signorini S, Tinelli F, Biagioni E, Veggiotti P, Fazzi E, Mercuri E, Cioni G, Guzzetta F. Auditory attention at the onset of West syndrome: correlation with EEG patterns and visual function. Brain Dev. 2006;28(5):293–9.
38. Lombroso CT. A prospective study of infantile spasms: clinical and therapeutic correlations. Epilepsia. 1983;24(2):135–58.
39. Kivity S, Lerman P, Ariel R, Danziger Y, Mimouni M, Shinnar S. Long-term cognitive outcomes of a cohort of children with cryptogenic infantile spasms treated with high-dose adrenocorticotropic hormone. Epilepsia. 2004;45(3):255–62.
40. Conti M, Matricardi S, Piscitello LM, Auconi M, Cursio I, Terracciano A, Vigevano F, Specchio N,

Marini C, Fusco L. Infantile epileptic spasms syndrome: when spasms come out of the blue. Epilepsy Behav. 2024;163:110180.
41. Primec ZR, Stare J, Neubauer D. The risk of lower mental outcome in infantile spasms increases after three weeks of hypsarrhythmia duration. Epilepsia. 2006;47(12):2202–5.
42. Lux AL, Osborne JP. A proposal for case definitions and outcome measures in studies of infantile spasms and West syndrome: consensus statement of the West Delphi group. Epilepsia. 2004;45(11):1416–28.
43. Guerrini R, Conti V, Mantegazza M, Balestrini S, Galanopoulou AS, Benfenati F. Developmental and epileptic encephalopathies: from genetic heterogeneity to phenotypic continuum. Physiol Rev. 2023;103(1):433–513.
44. Mei D, Balestrini S, Parrini E, Gambardella A, Annesi G, De Giorgis V, Gana S, Bassi MT, Zucca C, Elia M, Vetri L, Castellotti B, Ragona F, Mastrangelo M, Pisani F, d'Orsi G, Carella M, Pruna D, Giglio S, Marini C, Cesaroni E, Riva A, Scala M, Licchetta L, Minardi R, Contaldo I, Gambardella ML, Cossu A, Proietti J, Cantalupo G, LICE Collaborative Group, Trivisano M, De Dominicis A, Specchio N, Tassi L, Guerrini R. National survey on the prevalence of single-gene aetiologies for genetic developmental and epileptic encephalopathies in Italy. J Med Genet. 2024;62:jmg-2024-110328.
45. Weckhuysen S, Mandelstam S, Suls A, Audenaert D, Deconinck T, Claes LR, Deprez L, Smets K, Hristova D, Yordanova I, Jordanova A, Ceulemans B, Jansen A, Hasaerts D, Roelens F, Lagae L, Yendle S, Stanley T, Heron SE, Mulley JC, Berkovic SF, Scheffer IE, de Jonghe P. KCNQ2 encephalopathy: emerging phenotype of a neonatal epileptic encephalopathy. Ann Neurol. 2012;71(1):15–25.
46. Kato M, Yamagata T, Kubota M, Arai H, Yamashita S, Nakagawa T, Fujii T, Sugai K, Imai K, Uster T, Chitayat D, Weiss S, Kashii H, Kusano R, Matsumoto A, Nakamura K, Oyazato Y, Maeno M, Nishiyama K, Kodera H, Nakashima M, Tsurusaki Y, Miyake N, Saito K, Hayasaka K, Matsumoto N, Saitsu H. Clinical spectrum of early onset epileptic encephalopathies caused by KCNQ2 mutation. Epilepsia. 2013;54(7):1282–7.
47. Grosso S, Brogna A, Bazzotti S, Renieri A, Morgese G, Balestri P. Seizures and electroencephalographic findings in CDKL5 mutations: case report and review. Brain Dev. 2007;29(4):239–42.
48. Melani F, Mei D, Pisano T, Savasta S, Franzoni E, Ferrari AR, Marini C, Guerrini R. CDKL5 gene-related epileptic encephalopathy: electroclinical findings in the first year of life. Dev Med Child Neurol. 2011;53(4):354–60.
49. Solazzi R, Fiorini E, Parrini E, Darra F, Dalla Bernardina B, Cantalupo G. Diaper changing-induced reflex seizures in CDKL5-related epilepsy. Epileptic Disord. 2018;20(5):428–33.
50. Darra F, Monchelato M, Loos M, Juanes M, Bernardina BD, Valenzuela GR, Gallo A, Caraballo R. CDKL5-associated developmental and epileptic encephalopathy: a long-term, longitudinal electroclinical study of 22 cases. Epilepsy Res. 2023;190:107098.
51. Specchio N, Marini C, Terracciano A, Mei D, Trivisano M, Sicca F, Fusco L, Cusmai R, Darra F, Dalla Bernardina B, Bertini E, Guerrini R, Vigevano F. Spectrum of phenotypes in female patients with epilepsy due to protocadherin 19 mutations. Epilepsia. 2011;52(7):1251–7.
52. Marini C, Darra F, Specchio N, Mei D, Terracciano A, Parmeggiani L, Ferrari A, Sicca F, Mastrangelo M, Spaccini L, Canopoli ML, Cesaroni E, Zamponi N, Caffi L, Ricciardelli P, Grosso S, Pisano T, Canevini MP, Granata T, Accorsi P, Battaglia D, Cusmai R, Vigevano F, Dalla Bernardina B, Guerrini R. Focal seizures with affective symptoms are a major feature of PCDH19 gene-related epilepsy. Epilepsia. 2012;53(12):2111–9.
53. Trivisano M, Pietrafusa N, Terracciano A, Marini C, Mei D, Darra F, Accorsi P, Battaglia D, Caffi L, Canevini MP, Cappelletti S, Cesaroni E, de Palma L, Costa P, Cusmai R, Giordano L, Ferrari A, Freri E, Fusco L, Granata T, Martino T, Mastrangelo M, Bova SM, Parmeggiani L, Ragona F, Sicca F, Striano P, Specchio LM, Tondo I, Zambrelli E, Zamponi N, Zanus C, Boniver C, Vecchi M, Avolio C, Dalla Bernardina B, Bertini E, Guerrini R, Vigevano F, Specchio N. Defining the electroclinical phenotype and outcome of PCDH19-related epilepsy: a multicenter study. Epilepsia. 2018;59(12):2260–71.
54. Lo Barco T, De Gaetano L, Santangelo E, Bravi T, Proietti J, Cantalupo G, Brambilla I, Darra F. SYNGAP1-related developmental and epileptic encephalopathy: the impact on daily life. Epilepsy Behav. 2022;127:108500.
55. Coppola G, Plouin P, Chiron C, Robain O, Dulac O. Migrating partial seizures in infancy: a malignant disorder with developmental arrest. Epilepsia. 1995;36(10):1017–24.
56. Caraballo RH, Fontana E, Darra F, Cassar L, Negrini F, Fiorini E, Arroyo H, Ferraro S, Fejerman N, Dalla BB. Migrating focal seizures in infancy: analysis of the electroclinical patterns in 17 patients. J Child Neurol. 2008;23(5):497–506.
57. Burgess R, Wang S, McTague A, Boysen KE, Yang X, Zeng Q, Myers KA, Rochtus A, Trivisano M, Gill D, EIMFS Consortium, Sadleir LG, Specchio N, Guerrini R, Marini C, Zhang YH, Mefford HC, Kurian MA, Poduri AH, Scheffer IE. The genetic landscape of epilepsy of infancy with migrating focal seizures. Ann Neurol. 2019;86(6):821–31.
58. Dalla Bernardina B, Trevisan E, Colamaria V, Magaudda A. Myoclonic epilepsy ("Myoclonic Status") in non- progressive encephalopathies. In: Roger J, Dravet C, Bureau M, Dreifuss FE, Wolf P, editors. Epileptic syndromes in infancy, childhood and adolescence. John Libbey Eurotext Ltd; 1985. p. 68–72.

59. Dalla Bernardina B, Fontana E, Darra F. Myoclonic status in nonprogressive encephalopathies. Adv Neurol. 2005;95:59–70.
60. Caraballo RH, Cersósimo RO, Espeche A, Arroyo HA, Fejerman N. Myoclonic status in nonprogressive encephalopathies: study of 29 cases. Epilepsia. 2007;48(1):107–13.
61. Dalla Bernardina B, Fontana E, Darra F. Myclonic status. In: Bureau B, Genton P, Dravet C, Delgado-Escueta A, Tassinari CA, Thomas P, Wolf P, editors. Epileptic syndromes in infancy, childhood and adolescence. 5 th ed. John Libbey Eurotext Ltd; 2012. p. 431–9.
62. Darra F, Fontana E, Dalla BB. Myoclonic status in nonprogressive encephalopathies. In: Pellock JM, Nordli Jr DR, Sankar R, Wheless JW, editors. Pediatric Epilepsy. 4th ed. New York: DEMOS Medical Publishing; 2016. p. 373–82.
63. Caraballo R, Darra F, Reyes G, Armeno M, Cresta A, Mestre G, Dalla BB. The ketogenic diet in patients with myoclonic status in non-progressive encephalopathy. Seizure. 2017;51:1–5.
64. Specchio N, Wirrell EC, Scheffer IE, Nabbout R, Riney K, Samia P, Guerreiro M, Gwer S, Zuberi SM, Wilmshurst JM, Yozawitz E, Pressler R, Hirsch E, Wiebe S, Cross HJ, Perucca E, Moshé SL, Tinuper P, Auvin S. International league against epilepsy classification and definition of epilepsy syndromes with onset in childhood: position paper by the ILAE Task Force on nosology and definitions. Epilepsia. 2022;63(6):1398–442.
65. Vignoli A, Bisulli F, Darra F, Mastrangelo M, Barba C, Giordano L, Turner K, Zambrelli E, Chiesa V, Bova S, Fiocchi I, Peron A, Naldi I, Milito G, Licchetta L, Tinuper P, Guerrini R, Dalla Bernardina B, Canevini MP. Epilepsy in ring chromosome 20 syndrome. Epilepsy Res. 2016;128:83–93.
66. Gago-Veiga AB, Toledano R, García-Morales I, Pérez-Jiménez MA, Bernar J, Gil-Nagel A. Specificity of electroclinical features in the diagnosis of ring chromosome 20. Epilepsy Behav. 2018;80:215–20.
67. Avanzini P, Vaudano AE, Vignoli A, Ruggieri A, Benuzzi F, Darra F, Mastrangelo M, Dalla Bernardina B, Nichelli PF, Canevini MP, Meletti S. Low frequency mu-like activity characterizes cortical rhythms in epilepsy due to ring chromosome 20. Clin Neurophysiol. 2014;125(2):239–49.
68. Patry G, Lyagoubi S, Tassinari CA. Subclinical "electrical status epilepticus" induced by sleep in children. A clinical and electroencephalographic study of six cases. Arch Neurol. 1971;24(3):242–52.
69. Gardella E, Cantalupo G, Larsson PG, Fontana E, Dalla Bernardina B, Rubboli G, Darra F. EEG features in encephalopathy related to status epilepticus during slow sleep. Epileptic Disord. 2019;21(S1):22–30.
70. Cantalupo G, Pavlidis E, Beniczky S, Avanzini P, Gardella E, Larsson PG. Quantitative EEG analysis in encephalopathy related to status epilepticus during slow sleep. Epileptic Disord. 2019;21(S1):31–40.
71. Dalla BB, Fontana E, Michelizza B, Colamaria V, Capovilla G, Tassinari CA. Partial epilepsies of childhood, bilateral synchronization, continuous spike-waves during slow sleep. In: Manelis S, Bental E, Loeber JN, Dreifuss FE, editors. Advances in epileptology. XVIIth epilepsy international symposium. New York: Raven Press; 1989. p. 295–302.

13 Electroencephalography: Beyond the Waves

Carlo Alberto Tassinari and Gaetano Cantalupo

13.1 Introduction

The development of the electroencephalogram (EEG) began with Richard Caton (1842–1926), a physician from Liverpool, who in 1875 reported on the electrical activity observed in the exposed cerebral hemispheres of experimental animals. Later, Adolf Beck, a Polish physiologist, investigated the spontaneous electrical activity of the brains of rabbits and dogs, noting rhythmic oscillations that were affected by light, establishing the first documented observation of "brain waves." In 1912, Ukrainian physiologist Vladimir Pravdich-Neminsky published the first EEG recordings and evoked potentials in mammals [1]. By 1914, Napoleon Cybulski and Jelenska-Macieszyna had captured EEG traces of induced seizures. Indeed, the three seeds of the neurophysiology were sown: recording spontaneous waves, evoked responses, and abnormal brain activity. The modern era of EEG began with Hans Berger (1873–1941), a German physiologist and psychiatrist, who recorded the first human EEG in 1924 and coined the term electroencephalogram [2]. In 1934, British scientists Edgar Douglas Adrian and Bryan Harold Cabot Matthews confirmed Berger's findings, advancing the study of EEG [3]. Subsequent years saw a surge in EEG discoveries, including Fisher and Lowenback's 1934 demonstration of epileptiform spikes [4]. In 1935, Gibbs, Davis, and Lennox described the interictal spike-and-wave complexes associated with epileptic "absence seizures," marking the foundation of clinical electroencephalography [5]. The field continued to grow in the subsequent decade, leading to the foundation of the American EEG Society in 1947 and the setup of the first International EEG Congress, the first of a long series of national and international scientific societies and congresses focused on EEG and EEG-derived techniques (such as evoked potentials, event-related potentials, and event-related synchronization/desynchronization) to study normal and pathologic processes of the human brain. The development of the EEG faced many challenges from its inception, but of course the scientist worth of celebration is Hans Berger, whose work is celebrated in this book, on the centenary of his first recording of a human EEG. In fact, as Jasper and Penfield

C. A. Tassinari
University of Bologna, Bologna, Italy

Center for Research on Epilepsy in Pediatric age (CREP), University Hospital of Verona, Verona, Italy

G. Cantalupo (✉)
Center for Research on Epilepsy in Pediatric age (CREP), University Hospital of Verona, Verona, Italy

Innovation Biomedicine section, Department of Engineering for Innovation Medicine, University of Verona, Verona, Italy

Child Neuropsychiatry Unit, University Hospital of Verona (Full Member of the European Reference Network EpiCARE), Verona, Italy
e-mail: gaetano.cantalupo@univr.it

F. Brigo, O. Mecarelli (eds.), *EEG: The First 100 Years*,
https://doi.org/10.1007/978-3-031-86878-8_13

[6] noted in their tribute to Hans Berger: "his sustained vision which carried him through on exemplary series of rigorously controlled systematic experiments on human subjects finally convinced skeptical scientists throughout the world of the validity and value of records of the electrical activity of the human brain" [6]. The proof (if needed) that the advent of EEG should be considered a scientific paradigm shift or the emergence of a new scientific field is supported by the parallel growth of a new linguistic framework and the associated vocabulary [7, 8]. In fact, as expected, soon after the advent of EEG, neologisms and descriptive terms were proposed, culminating in the need for a shared standardized language, reflected in the glossary of commonly used EEG terms developed by Chatrian [9]. Due to the oscillatory nature of the EEG signal, most of the terms described rhythmic activities and waves, that often carry complex meanings.

13.2 Alpha Waves and the Berger Rhythm

Berger firstly described "a rhythmic oscillation of potential at a frequency of 10 cycles per second [...] present when the subject lies quietly with eyes closed [...]" [2]. Berger's work on this near-sinusoidal rhythmic activity was confirmed by Adrian and Matthews in 1934 [3], as detailed by Compston [10]. They demonstrated that the rhythm was not an artifact caused by pulsating blood vessels or head tremors, but indeed was the expression of spontaneous rhythmic activity of "cortical cells in [...] the occipital lobe [that] beat synchronously when [...] undisturbed"; furthermore they confirmed the relationship between the "Berger rhythm," mainly recorded in the posterior aspect of the brain, and vision, with visual stimuli easily interrupting it. Being the first building block of a new language, this rhythm was termed "alpha rhythm," after the first letter of the Greek alphabet, and the frequency band typical of the "alpha rhythm" in healthy adults was named, accordingly, alpha band (8–13 Hz). Topography, reactivity, and significance of this rhythm are similar also in children, in them; however, the frequency is lower, so to avoid misunderstandings between "alpha rhythm" and "alpha band," it may be more appropriate to use the term "posterior dominant rhythm" to designate the "Berger rhythm." This is also true, because there is a large body of evidence that the posterior alpha rhythm itself is composed by many alpha rhythms, with different reactivity to stimuli, as documented by means of intracerebral recordings since 1950 by Grey Walter: "[...] What one can see on the scalp is a spatial average of a large number of components [...]" [11]. Furthermore, rhythmic activities within the alpha band can be recorded in different parts of the brain and do not necessarily reflect an idling state of the brain region, being the prototype of spontaneous brain rhythm reflecting a dynamic process, which governs a large ensemble of integrative brain functions [12]. Besides being desynchronized by visual afferents and certain mental activities, a light flickering at frequencies within the alpha band—but also in lower and higher frequency ranges—could entrain the posterior alpha rhythm, synchronizing it at the driving frequency. Flickering lights, in predisposed subjects, can induce an excessive activation of the brain network, in which the visual areas represent a node (or an entry point), leading to the complex phenomenon of photosensitivity and, in the context of epilepsy, to photic-induced seizures. These are triggered by a variety of visual stimuli and involve multiple causes and clinical manifestations, ranging from myoclonus to consciousness disturbances, as observed in both baboons and humans.

13.3 Beta Waves and the Mu Rhythm

In 1949, Jasper and Penfield described a fast EEG activity, recorded intracranially, named the beta rhythm: "the beta rhythm (25 per sec) is characteristic of the electrical activity from the central region- is not affected by weak visual stimuli which blocked the alpha rhythm, [...] It was blocked by tactile stimulation to the contralateral side of the body [...] Blocking of the beta rhythm occurs, however only upon the initiation of movement and upon voluntary termination of a posture which the patient has assumed. It is not sustained during a maintained contraction such as clenching the fist, but blocking occurs at the beginning and cessation of the act. Sustained blocking occurs only with continuous consecutive movement or 'readiness to move' before the movement actually begins. These findings would be consistent with the suggestion that impulses reach the precentral gyrus at the beginning and at termination of a posture that is assumed voluntarily" [6]. Few years later, Gastaut identified a distinctive "arch-shaped" rhythm over the central/rolandic regions ("le rhythme rolandique en arceau"), later known as the mu rhythm [13, 14]. Although the frequency of this rolandic rhythm was in the alpha range, the comb-like morphology of the oscillation (at variance with the near-sinusoidal Berger rhythm) suggested that it would be the result of a "dédoublement" (halving) of the beta activity described by Jasper and Penfield, with which it shared the topography and reactivity: "The somatotopic reactivity was precise: the mid rolandic arceau was blocked by hand mobilization, while a 'superior – rolandic' (as recorded on vertex area) was blocked mainly by foot mobilization. Most interesting, thinking of performing a voluntary movement was effective to produce a reactivity of both beta and rhythm 'en arceau'". Going more in-depth on the significance of this rolandic rhythm, in an article describing a series of EEG changes observable in subjects during cinematographic presentation, Gastaut and Bert highlighted the reactivity (desynchronization) of the mu rhythm to the observation of others' action [15]. This rolandic rhythm reactivity—as stated by Avanzini et al. [16]—"remained for many years neglected" until, half a century later, interest in the mu rhythm resurged, particularly in relation to the discovery of "mirror neurons" by Giacomo Rizzolatti's research group [17], which are involved in both the execution and observation of actions. The relationship between mu rhythm and mirror neuron system activity, first suggested by Altschuler et al. [18], has since been confirmed by various studies, establishing a strong connection between rolandic rhythm reactivity and the observation of others' movements. Observation of motor acts determines a modulation of the rolandic rhythm similar to the one occurring during the execution of the same motor act. In particular the reactivity strictly tracks the kinematics of the observed movements [16], providing evidence of the presence in humans of a mechanism that allows a subject to map an observed action on its own action execution motor system. This mechanism, named "mirror mechanism" [19], is fundamental in understanding implicitly the action of others. Going further, it has been proposed that a mirror mechanism is also involved in empathy, allowing the observer's identification with the intention of the one who acts. In this regard it is very interesting to read the foreseeing statements of Gastaut and Bert from a 1954 paper in which they described that the rhythm "en arceau" "[...] also disappears when the subject identifies himself with an active person represented on the screen. This phenomenon is particularly interesting to study during a sequence of film showing a boxing match. A few seconds and at times, less than a second after the appearance of the boxers all type of rolandic activity disappears in spite of the fact that the subject seems completely relaxed and that there is no noticeable change of posture. The relation between the blocking of the 'arceau' rhythm and the image of boxers in action is unquestionable. In the middle of this particular film strip, the camera is suddenly turned from the ring to the spectators in the hall for a few seconds. In many subjects the rhythm 'en arceau' reappears during this short period and vanishes again as the boxers reappear on the screen. The motor characteristic of this identification is made obvious in a subject showing a bilateral rhythm 'en arceau' which is blocked only on the left hemisphere during the boxing match (it is with the right fist that one hits)" [15]. These observations have become a foundational element in modern neuroscience research, particularly in the development of brain–computer interface (BCI) technologies. BCIs enable communication between the human brain and external devices through analysis of EEG signals, allowing users to control aspects of their environment through the modulation of brainwave activity. By interpreting these brain signals, BCI systems can facilitate interaction with a wide range of tools and devices. For instance, BCI applications have already enabled individuals to manipulate robotic arms solely through thought processes [20] and, more recently, to control complex systems, such as a full-body exoskeleton, which allows a user to initiate and control movements as though it were a natural extension of their own body [21]. These innovations underscore the transformative potential of BCIs in augmenting human capabilities and providing new avenues for independence and mobility to those with severe physical impairments.

13.4 Delta Waves, Sleep, and Consciousness

The third component of our new "alphabet" was well described in 1949 by Brazier as an active component of the brain processes occurring during sleep: "[…] that slow potentials appear in the record when the subject is asleep was one of the first observations to be established in electroencephalography: and since Berger's original demonstration of this change, it has been plentifully confirmed by subsequent workers… The electrical activity of the brain during sleep is not the electrical concomitant of 'resting' cells, or of a quiescent brain, but the signal of activity in certain sub-cortical networks" [22]. These early studies revealed, in fact, that brain activity during sleep is far from quiescent, reflecting activity in specific sub-cortical networks. Today, sleep is known that EEG activity during normal sleep exhibits a highly complex architecture, structured physiologically into macro- and micro-dynamics, including the cyclic alternation of different sleep stages and a complex balance between stability and instability of the sleep process, reflected in a cyclic organization of fast rhythms and slow wave activity (SWA), better described as a cyclic alternating pattern (CAP), by Terzano and co-workers [23]. At the cellular level, SWA corresponds to slow oscillations in the membrane potentials of cortical neurons [24]. The regulation of SWA is homeostatic, with its amount increasing during wakefulness and gradually returning to baseline levels during sleep [25]. Slow wave sleep (SWS) has long been associated with restorative and recovery processes, though the exact nature of these functions remains an open question. One compelling explanation is the synaptic homeostasis hypothesis [26]. According to this hypothesis, wakefulness leads to a progressive increase in synaptic strength across many brain circuits due to the continuous encoding of new information. During sleep, particularly SWS, synaptic homeostasis is achieved by selectively weakening or pruning synapses, counteracting the over-strengthening that occurs while awake. SWS and its hallmark SWA are intricately linked to the brain's ability to disconnect from the external environment and enter a state of reduced consciousness. The relationship between slow waves and consciousness can be examined through the lens of the Integrated Information Theory (IIT) of Consciousness, which posits that consciousness arises from the brain's ability to integrate and differentiate information [27]. Slow wave activity (SWA) during sleep, particularly during slow wave sleep (SWS), offers a unique window into the relationship between neural dynamics and the levels or states of consciousness. Slow oscillations are thought to represent a state of global synchronization in cortical activity, reflecting the diminished integration of information across different brain regions. This phenomenon might underlie the characteristic "offline" state of consciousness during SWS, wherein sensory input and cognitive processing are markedly reduced. In this view, it is of particular interest to observe that the reactivity of the EEG rhythms, shifting brain functioning from a synchronized state (sleep) to a desynchronized one (wakefulness) relies on a functional interplay that connects various rhythmogenic areas and networks. This interplay is regulated by the brain stem reticular formation (RF), as originally described by Moruzzi and Magoun in the first volume of the first scientific journal dedicated to EEG (Electroencephalography and Clinical Neurophysiology) in 1949 [28]. The RF can be conceptualized as a transactional center that acts as an integrator and modulator of other systems [29]. Its function is influenced by its intrinsic excitability, which depends on peripheral inputs, humoral and metabolic signals, and reciprocal interactions with the cortex. Moruzzi and Magoun reported that the RF stimulation "[...] evokes changes in the EEG consisting in the abolition of synchronized discharges and introduction of low voltage fast activity in its place. The effect of Reticular Stimulation is strikingly like Berger's alpha wave blockade or any arousal reaction [...] The EEG modifications reflect the transition from sleep to wakefulness or from relaxation to alertness" [28]. It is very interesting to note the relevance that these observations have in understanding and formulating modern theories of consciousness.

13.5 Conclusions

If Hans Berger were alive today to evaluate the scientific progress made since his pioneering work on the EEG, he would undoubtedly be pleased but perhaps still searching for answers to the mind–body connection. In his last paper, Berger expressed his deep interest in psychophysiology, seeking to understand how brain processes translate into psychic manifestations [30]. While the EEG has become an invaluable tool in neuroscience, the quest to fully understand the brain's inner workings continues. Modern advancements, including the study of brain connectivity and network dynamics, are paving the way for new insights into both normal and pathological brain functions. On this line, Berger (as reported by Fischgold [31]) cited the statement of Angelo Mosso (1883): "As far as we apply instruments to evaluate the human brain functions, we should hope to understand the physical basis of consciousness. And even if do not reach satisfactory results we are on the right track" [32]. Even if EEG and EEG-derived techniques fall short of fully explaining consciousness, they keep us on the right path toward understanding to unravel how brain processes give rise to the mind.

References

1. Haas LF. Hans Berger (1873–1941), Richard Caton (1842–1926), and electroencephalography. J Neurol Neurosurg Psychiatry. 2003;74(1):9.
2. Berger H. Über das Elektrenkephalogramm des Menschen. Arch Psychiatr Nervenkr. 1929;87:527–70.
3. Adrian ED, Matthews BHC. The Berger rhythm: potential changes from the occipital lobes in man. Brain. 1934;57:355–85.
4. Fischer MH, Lowenbach H. Aktionsstrome des zentralnervensystems unter der einwikung von krampfgiften. I. Miteilung strychnine und pickrotoxin. Arch Exp Pathol Pharmakol. 1934;174:357–82.
5. Gibbs FA, Davis H, Lennox WG. The electroencephalogram in epilepsy and in conditions of impaired consciousness. Arch Neurol Psychiatr. 1935;34:1133–48.
6. Jasper H, Penfield W. Electrocorticograms in man: Effect of voluntary movement upon the electrical activity of the precentral gyrus. Archiv Fur Psychiat Zeitsh Neurol. 1949;183:163–74.
7. Fleck L. Entstehung und Entwicklung einer wissenschaftlichen Tatsache: Einführung in die Lehre vom Denkstil und Denkkollektiv. Basel: Benno Schwabe; 1935.
8. Kuhn TS. The structure of scientific revolutions. 2nd ed. Chicago: University of Chicago Press; 1970.
9. Chatrian GE, Bergamini L, Dondey M, Klass DW, Lennox-Buchtal M, Petersen I. A glossary of terms most commonly used by clinical electroencephalographers. Electroencephalogr Clin Neurophysiol. 1974;37:538–48.
10. Compston A. The Berger rhythm: potential changes from the occipital lobes in man. Brain. 2010 Jan;133(Pt 1):3–6.
11. Walter WG. Normal rhythms: their development, distribution and significance. In: Hill D, Parr G, editors. Electroencephalography. London: McDonald; 1950.
12. Başar E, Schürmann M, Başar-Eroglu C, Karakaş S. Alpha oscillations in brain functioning: an integrative theory. Int J Psychophysiol. 1997;26(1–3):5–29.
13. Gastaut H. Etude électrocorticographique de la réactivité des rhythmes rolandiques. Rev Neurol. 1952;37(2):176–82.
14. Gastaut H, Terzian H, Gastaut Y. Etude d'une activité électroencéphalographique méconnue: "le rhythme rolandique en arceau". Mars Med. 1952;89(6):296–310.
15. Gastaut H, Bert J. EEG changes during cinemotographic presentation. Electroencephalogr Clin Neurophysiol. 1954;6:443–4.
16. Avanzini P, Fabbri-Destro M, Dalla Volta R, Daprati E, Rizzolatti G, Cantalupo G. The dynamics of sensorimotor cortical oscillations during the observation of hand movements. An EEG study. PLos One. 2012;7(6):e37534.

17. di Pellegrino G, Fadiga L, Fogassi L, Gallese V, Rizzolatti G. Understanding motor events: a neurophysiological study. Exp Brain Res. 1992;91(1):176–80.
18. Altschuler EL, Vankov A, Wang V, Ramachandran VS, Pineda JA. Person see, person do: human cortical electrophysiological correlates of monkey see monkey do cells. In: Poster session presented at the 27th annual meeting of the society for neuroscience; New Orleans: LA; 1997.
19. Rizzolatti G, Sinigaglia C. The mirror mechanism. A basic principle of brain function. Nat Rev Neurosci. 2016;17(12):757–65.
20. Babiloni C, Carducci F, Cincotti F, Rossini PM, Neuper C, Pfurtscheller G, Babiloni F. Human movement-related potentials vs desynchronization of EEG alpha rhythm: a high-resolution EEG study. NeuroImage. 1999;10(6):658–65.
21. Benabid AL, Costecalde T, Eliseyev A, Charvet G, Verney A, Karakas S, Foerster M, Lambert A, Morinière B, Abroug N, Schaeffer MC, Moly A, Sauter-Starace F, Ratel D, Moro C, Torres-Martinez N, Langar L, Oddoux M, Polosan M, Pezzani S, Auboiroux V, Aksenova T, Mestais C, Chabardes S. An exoskeleton controlled by an epidural wireless brain-machine interface in a tetraplegic patient: a proof-of-concept demonstration. Lancet Neurol. 2019;18(12):1112–22.
22. Brazier MA. The electrical fields at the surface of the head during sleep. Electroencephalogr Clin Neurophysiol. 1949;1(2):195–204.
23. Terzano MG, Mancia D, Salati MR, Costani G, Decembrino A, Parrino L. The cyclic alternating pattern as a physiologic component of normal NREM sleep. Sleep. 1985;8(2):137–45.
24. Amzika F, Steriade M. Electrophysiological correlates of sleep delta waves. Electroencephalogr Clin Neurophysiol. 1998;107:69–83.
25. Borbely AA, Acherman P. Sleep homeostasis and models of sleep regulation. In: Kryger MH, et al., editors. Principles and practice of sleep medicine. Elsevier Saunders; 2005. p. 405–17.
26. Tononi G, Cirelli C. Sleep and the price of plasticity from synaptic and cellular homeostasis to memory consolidation and integration. Neuron. 2014;81:12–34.
27. Tononi G. An information integration theory of consciousness. BMC Neurosci. 2004;5:42.
28. Moruzzi G, Magoun HW. Brain stem reticular formation and activation of the EEG. Electroencephalogr Clin Neurophysiol. 1949 Nov;1(4):455–73.
29. Magoun HW. The waking brain. Springfield: Thomas; 1958.
30. Berger H. Das Elektrenkephalogramm des Menschen. Nova Acta Leopoldina, Halle (Saale). 1938;6:173–309.
31. Fischgold H, Dreyfuss- Brisac G, Pruvot P. Problémes de Base en Electroencephalographie. Masson; 1963. p. 82.
32. Mosso A. La paura. Milano: A. Treves; 1883.

Role of Scalp and Invasive EEG in Epilepsy Surgery

14

Veronica Pelliccia and Laura Tassi

14.1 Introduction

The study and knowledge about epilepsies began before the end of 1900, initially based mainly on clinical features. Since 1928, thanks to Hans Berger who invented the EEG, neurologists and neurophysiologists have been able to learn about cerebral neurophysiology, recording directly the electrical activity of the brain [1].

Since 1937 the pathological EEG modifications associated with epilepsy were described [2–5].

From the 1950s, 8 up to 16 channel hard-wired recordings became possible, and the EEG turned out to be an integral part of clinical activity. Only in the 1970s the first use of closed-circuit TV-EEG (CCTV) linked to analogic EEG appeared, and in the 1980s the first long-term monitoring (LTM) was introduced. Between 1990 and 2005 analogic was gradually replaced by digital EEG recordings, subsequently digital EEG with simultaneous video, for both standard recordings and telemetry/ambulatory monitoring, was widely used [6–8].

Over the years video-electroencephalography (Video-EEG) has allowed an enormous improvement in the knowledge of clinical modifications during seizures and their correlation and cortical localization with the contextual EEG modifications. This electrophysiological study also changed the way patients with focal epilepsy were studied, giving them the opportunity to be considered for surgical treatment [9].

By definition, Video-EEG is the combination of video recording (providing data related to clinical semiology of the seizures) and EEG (information about the type of interictal abnormalities and ictal discharges with corresponding anatomical localization), and it allows simultaneous (second-by-second) analysis and correlation of the EEG findings and clinical data [9].

Video-EEG is defined as an inpatient's long-term recording (more than 24 h) and provides different and more accurate information than standard EEG [10]. As because the manifestations of epilepsy are brief and intermittent, a standard 20- to 30-min EEG very often fails to show the ictal activity and it can also provide not significant information related to interictal abnormalities. Comparing the quality and usefulness of information obtained from Video-EEG monitoring with that from routine EEG, the intensive and prolonged monitoring was found to be far superior to the routine EEG examination, in directing to precise diagnosis of seizure disorders. The routine EEG recorded seizures in only 7% of patients, while with Video-EEG, following the withdrawal of anti-seizure medications

V. Pelliccia · L. Tassi (✉)
"Claudio Munari" Epilepsy Surgery Centre, Niguarda Hospital, Piazza Ospedale Maggiore 3, Milan, Italy
e-mail: veronica.pelliccia@ospedaleniguarda.it; laura.tassi@ospedaleniguarda.it

F. Brigo, O. Mecarelli (eds.), *EEG: The First 100 Years*,
https://doi.org/10.1007/978-3-031-86878-8_14

(ASMs), seizures were recorded in 70% of the patients [6, 11, 12].

Therefore, the International League Against Epilepsy (ILAE) recommends long-term Video-EEG when there is diagnostic uncertainty to classify seizure type or epilepsy syndrome, quantify seizures, and evaluate electro-clinical seizure characteristics when necessary for differential diagnosis, classification, seizure quantification, and seizure characterization for surgical management [10, 13].

14.2 Video-EEG: Definition

Video-electroencephalography is the combined and contemporary recording of the EEG and the clinical features during the seizures. It allows verifying the clinical signs and symptoms, already acquired in the clinical history of the patient [14], correlating it with any contemporary or preceding EEG modifications. In the clinical practice, Video-EEG is particularly useful not only to analyze and visualize the seizures, but also the repetitive or isolated paroxysmal activities and any other paroxysmal events.

The opportunity to record both clinical and EEG findings allows to provide the correct interpretation to each recorded event, classifying it as epileptic or other etiology, possibly recommending specific investigations. The comprehensive characterization of the clinical manifestations of the seizures and the analysis of associated EEG modifications is essential to diagnose and characterize the epilepsies, and it is crucial in the clinical presurgical work-up of the patients with more complicated types of epilepsy, in which is necessary to verify and check seizure semiology and corresponding EEG modifications. The digital EEG and the computed-assisted technologies make it possible to record for a long time, increasing exponentially the possibility to catch one or more seizures. The further advantage of 24 h recordings is to observe and analyze seizures occurring exclusively during nocturnal sleep or at the awakening, identify interictal abnormalities with sleep prevalence, and discriminate epileptic seizures from any sleep disorders.

Video-EEG monitoring is mainly used for:

1. Attain diagnosis of epilepsy or other non-epileptic paroxysmal disorders (e.g., psychogenic/paroxysmal non epileptic seizures (PNES), sleep disorders, and metabolic or cardiological disorders; Fig. 14.1) [10, 15–20].

   Diagnosis and classification of seizures and epilepsy syndromes are essential for appropriate therapeutic decision (Fig. 14.2). On the basis of the diagnosis, the necessary exams to better delineate it and the essential therapies and the timing of the follow-up can be indicated [12, 21–23].
2. Quantifying seizures or interictal epileptiform abnormalities by assessing any responses to therapy.

   In specific type of epilepsy, could be indicated to perform Video-EEG monitoring to verify the reduction or increase of interictal abnormalities related to ASM modifications. Similarly, in surgical cases treatment, the Video-EEG monitoring allows to verify the actual disappearance of seizures (e.g., in the case of seizures misrecognized by the patient or caregivers or poor patient reliability).
3. Achieve a correct diagnosis of the specific type of epilepsy, and in focal epilepsy, collect useful data to hypothesize the localization and extension of the Epileptogenic Zone (EZ).

   Therefore, the Video-EEG monitoring could be useful in the diagnosis of epilepsy and in discriminating focal versus generalized epilepsy, when the routine EEG and the clinical findings are not sufficient. In patients with drug-resistant focal epilepsy, it may be indicated to perform Video-EEG monitoring to identify and localize the EZ as essential part of the presurgical work-up (e.g. patients with negative brain MRI or extensive cerebral malformations, when it is necessary to establish the concordance between anatomo-electro-clinical findings or patients in which it is necessary to verify the seizures semiology).

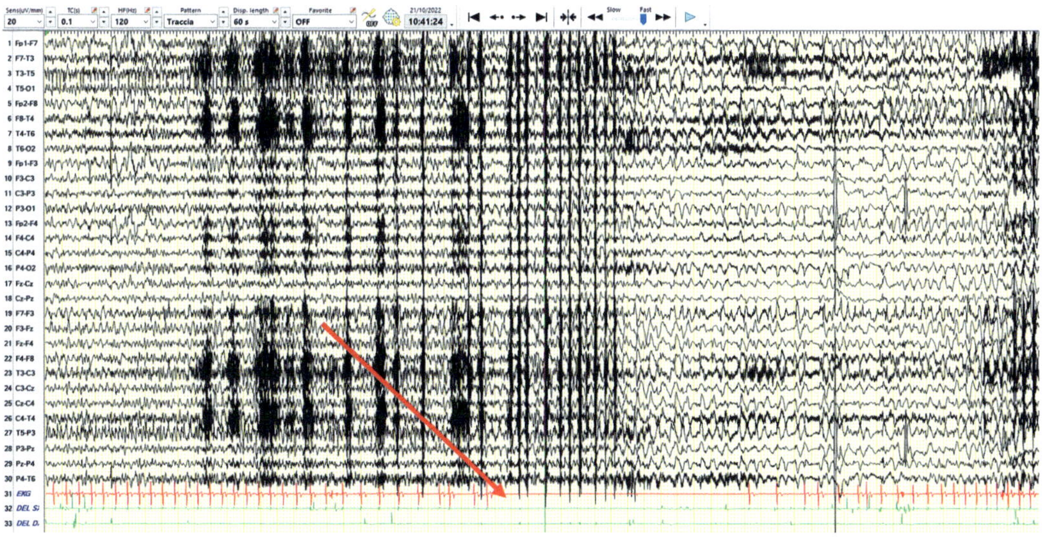

Fig. 14.1 Longitudinal and transverse montage showing (60-second page) a bitemporal ictal discharge inducing, visible on the EKG trace, a 13-second asystole (red arrow)

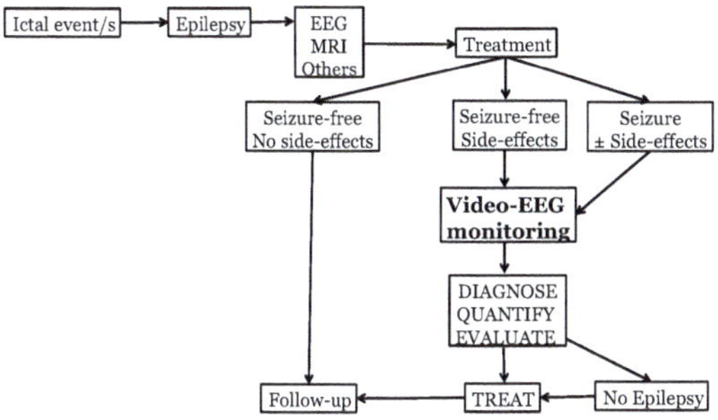

Fig. 14.2 Flow-chart concerning utility, indication and goals of the long-term monitoring of the epilepsies with the Video-EEG

14.3 Video-EEG for Epilepsy Surgery

The Video-EEG goal is implementing video with EEG and vice versa, increasing the diagnostic yield beyond EEG alone, when events are recorded [9].

Detection of ictal events and/or interictal abnormalities is increased if standard EEGs are extended up to 60 min compared to routine 30-min studies [24].

Brief Video-EEG, lasting no more than 24 h [25] is a less costly and resource-demanding technique in comparison to long-term monitoring (LTM) with Video-EEG and can differentiate among epilepsy and other paroxysmal disorders in more than 50% of the patients [26, 27].

Polygraphy during Video-EEG should be always used, as a dynamic adapted short-term diagnostic test, tailored to a patient-specific clinical process. Therefore, additional physiologic parameters evaluated during Video-EEG monitoring include the electrocardiogram, blood pressure, heart rate, electrooculogram, electromyogram, and respiratory functions [9].

The presurgical work-up is often mandatory to record and analyze with me clinical and neurophysiological features of the seizures when clinical history, interictal EEG, and neuroimaging findings are not enough to reach a precise localizing diagnosis in focal epilepsies. Recording the seizures is necessary for long-term Video-EEG including also nocturnal sleep.

It is essential to carefully analyze and describe the seizure semiology throughout the ictal and the post-ictal phase, considering the chronological occurrence of the symptoms and signs. The clinical description has to be related to the reporting of the EEG modifications, and the electroclinical correlations permit to formulate a localizing hypothesis of the EZ and its eventual propagation to close or distant cortical areas. Finally, Video-EEG gives the opportunity to identify polymorphic or bilateral seizures, defining their EEG seizure onset and the corresponding anatomical localization. Consequent diagnostic exams (i.e. invasive recordings or direct epilepsy surgery) are evaluated on these assumptions.

14.3.1 Testing the Patient During Seizures

The potential for dangerous consequences exists during LTM because patients' seizures are induced [28]. Generalized seizure emergencies, falls, injury, and postictal psychosis, among others, are possible safety risks [28, 29]. Standardized protocols are recommended for guarantee basic patient safety [30, 31].

Trained staff and/or caregivers into the Epilepsy Monitoring Unit (EMU) should accurately supervise the patient, during all the Video-EEG monitoring, in order to notify, record, and examine the clinical events. Seizures can be detected by supervising staff, or by the patient himself, with the handy push button or verbal warning. Patients must be trained to warn speaking or using the push button, immediately after the onset of the subjective manifestation, if any. The supervising personnel should have the possibility to investigate seizure types and frequency. Of course, available standards applied to personnel and their role in caring for a variety of complex patients during LTM are resource dependent and significantly vary throughout the world [10].

During and after every ictal event, it is mandatory to:

- Take care of the patient, avoiding dangerous behavior
- Guide the patient in describing the seizures especially subjective manifestations and very discrete changes not evident on video recordings (e.g. pallor, flushing, sweating, piloerection, and sialorrhea)
- Observe the patient carefully, verbally describing everything that might be relevant, the neurological functions, testing the level of awareness, the language, the muscle tone, and eventual deficit of the visual field

Ictal and postictal testing is a cardinal element to approach an accurate diagnosis associated with the Video-EEG recording. Every EMU should have a protocol for testing patients systematically. All the tests must be adaptive to the dynamics of the individual seizures.

Once back to the baseline, the patient and the caregivers must be asked whether the event was typical and its severity [19, 32–34].

14.3.2 Management of Anti-Seizure Medications

Anti-seizure medications (ASMs) are often tapered to obtain the habitual seizures and try to reduce the duration of long-term Video-EEG monitoring [7, 35–40]. More than half of EMUs reduced or completely withdrawn one or more ASMs in 75–100% of patients, even if fast ASM reduction is considered less safe. At least 4–7 days of recording are needed to store a sufficient number of seizures, but sometimes a longer recording is necessary, varying on the basis of the indication for the EMU study [41–44]. Many factors can influence the strategy and the rapidity of ASM tapering/withdrawal, such as seizure frequency and type, number, kinetic, dosage and type of ASMs, past history of status epilepticus, and/or seizure clusters [45]. In patients without a history of status epilepticus or frequent daily seizures, tapering ASM by 30–50% daily should be considered aggressive: ASM taper may result in capturing non-habitual seizures, obscuring localizing information on ictal EEG, or producing seizure clustering and status epilepticus [10].

14.3.3 Safety

During Video-EEG monitoring, patients are vulnerable to potential injury, being observed for the whole recording time by competent personnel in only 80% of the EMU and having the pharmacological treatment quite always withdrawn. Ictal events are potentially dangerous leading to convulsions and seizure emergencies such as falls, injury, and postictal psychosis, among others, or death [31, 46–50]. It is therefore recommended to have and follow safety protocols [13]. Nevertheless, the safety protocols are lacking in more than 20% of the EMU, particularly concerning the management of generalized seizures, status epilepticus and clustering episodes, post-ictal psychosis, and related injuries.

Every EMU should have guidelines for patients in and out of bed, with adequate levels of nursing to provide quality care, after a formal educational program [51]. Protocol for the management of cardiac arrest and prompt access to intensive care unit should be required, as for avoiding Sudden Unexpected Death in Epilepsy (SUDEP) [52]. The prevalent complications are status epilepticus in 80% of patients, injuries in more than 70%, and postictal psychoses in 67% [30, 52].

Prevention consists of bed protection in 96% of EMUs, but antisuffocation pillows are employed in only 21%, and environmental protections in monitoring rooms and in bathrooms are present in less than 35% of centers [30].

14.3.4 Final Report

The Video-EEG reporting should consist of different principal parts [31, 53–55], starting with a brief statement about the clinical pathology requiring LTM (brief summary of the anamnestic data, a review of current ASMs and other comorbidities). The purpose of the LTM (diagnostic study, presurgical evaluation, and control after drug modification/introduction) has to be clearly stated.

The technical aspects of the recording (channel number, type and location of electrodes, and eventual polygraphy) should be documented. Activation procedures (Hyperventilation (HV), Intermittent Photic Stimulation (IPS), drug injection, suggestion, sleep deprivation, clap, and other stimuli) have to be fully described. The reduction of medications, especially those intended to increase the incidence of seizures, should be described.

Interictal findings as a description of the neurophysiological data (at rest, during sleep, and activation techniques) should be reported. A precise description about the interictal pathological slowing and spiking abnormalities, explaining the frequency of occurrence, the topographic distribution, and the eventual propagation, should be reported (Fig. 14.3).

Providing graphic display of EEG samples [56], including the most significant ictal and interictal patterns, enhances the reproducibility of EEG portions of the LTM report to facilitate patient management as for clinical research [56]. Video-EEG aims to record behaviors and clinical features so a great importance is due to the clinical description of ictal events, following the chronologic order of appearance of each symptom. Both semiology and ictal EEG reporting should consider a chronological order using standardized terminology (IFCN glossary for EEG; ILAE glossary for semiology [57]).

On the clinical point of view, a precise account is needed about the patient behavior, including the eventual activity preceding the episode, the features of the clinical onset (warned by the patient or signaled by the personnel), the chronological evolution of ictal symptoms and signs, the end of the episode, and the ictal and postictal behaviors. All the neurological functions are to be reported, including responsiveness, time–space orientation, language, memory, and motor function.

Once identified the first clinical signs or symptoms, preceding and/or contemporary EEG modifications are to be reported including a brief description of background activity and interictal discharges preceding the clinical event; then morphology, location, evolution, and eventual spread; and the post-ictal changes (Fig. 14.4).

The ictal EEG patterns may be difficult to interpret because of myogenic and movement artifacts, such as eye blinking or body movement. All information has to be included into the conclusions. An interpretation is required, stating the overall impressions, the clinical significance, and the electrographic and behavioral correlations, composing an interpretive synthesis. A conceiv-

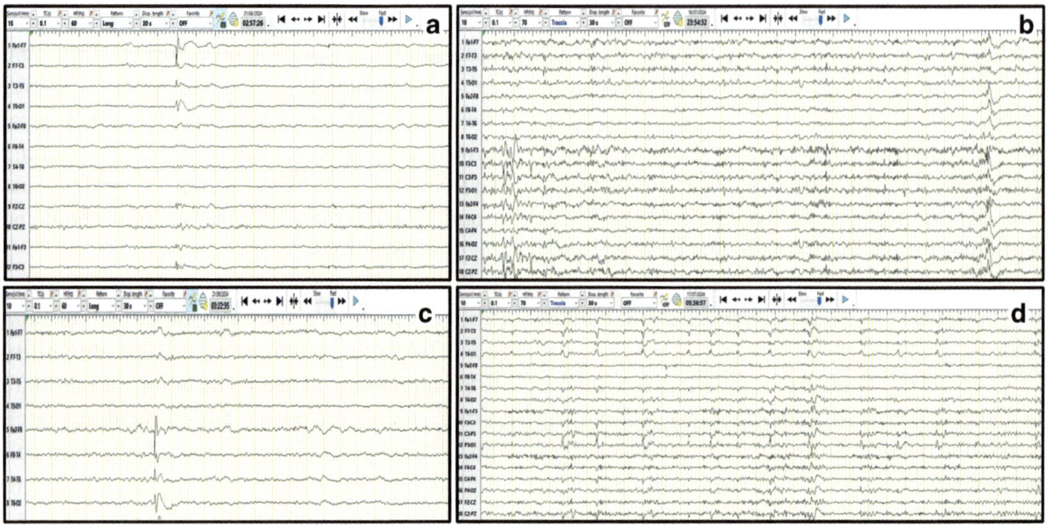

Fig. 14.3 Interictal abnormalities: (**a**) Spike and wave on left temporal areas, on the anteromedial temporal regions (longitudinal montage). (**b**) Left temporal theta slowing intermingled with frequent low-amplitude polyspikes mainly on mid-posterior temporal regions with diffusion to posterior suprasylvian regions (longitudinal montages). (**c**) Spike and wave on the right temporal areas, particularly on anterior temporal region (temporal leads only). (**d**) Repetitive spikes and waves on the left temporal regions, with a phase reversal on T3, and spread to the parieto-occipital homolateral regions

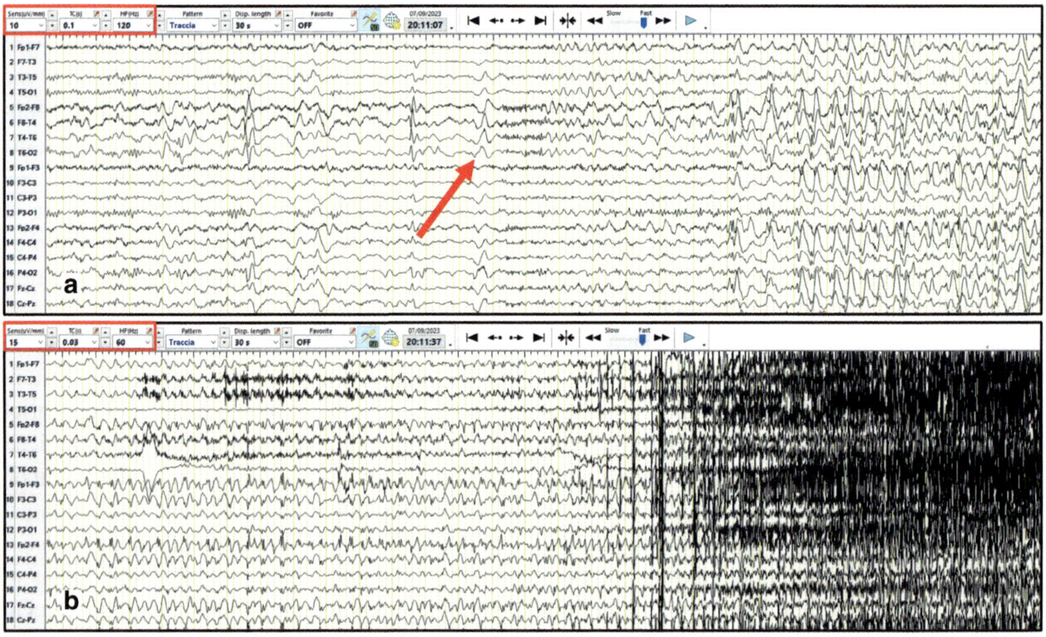

Fig. 14.4 A right temporal lobe seizure characterized by an initial high amplitude slow wave, followed by a fast activity (**a**, red arrow) with evolution in bilateral tonic–clonic seizure (**b**) (longitudinal montages). Note how the filters can be modified to better visualize the ictal discharge (red boxes)

able diagnosis has to be stated, associated with a proposed treatment.

In the presurgical Video-EEG monitoring, a hypothesis on the location of the Epileptogenic Zone should be detailed, including further surgical possibilities or exclusion.

14.4 Invasive EEG Recordings: Definition

In most complicated cases, invasive EEG recordings (iEEGs) are requested. All invasive EEG recordings share common recording features, being very close to and/or within the source, permitting a very high spatial resolution [58]. However, we have to keep in mind just a small portion of brain tissue can be sampled by each electrode, being the risk that of unnecessary recording due to misplacement of the electrodes or an insufficient number [59]. It is advisable to sample the selected targeted areas with a large number of electrodes, located on one or more clear hypotheses on the anatomo-functional network(s) involved at the seizure onset and during seizure propagation [60, 61]. The epileptologist, based on anatomo-electro-clinical correlations, derived from the presurgical work-up.

Within this background, the MRI-negative patients, and those with complex cortical developmental malformations (polymicrogyria, periventricular nodular heterotopia), lesions related to ischemic and/or hemorrhagic events, present the commonest indication for iEEG studies, even if, in cases in which the MRI reveals a discrete lesion consistent with focal cortical dysplasia (FCD), tuberous sclerosis, or benign developmental glioneuronal tumors, some epilepsy surgery centers prefer to explore the patient as the EZ could be more extensive [62–65]. It remains to be established the role of iEEG in other discrete lesional substrates such as acquired vascular lesions or Sturge–Weber syndrome and polymicrogyria [66, 67]; some studies promote extending the resection to the lesional cortex and/or beyond the anatomic lesion to achieve higher rates of seizure freedom, but others plain in favor of a not complete involvement of the malformative cortex in generating ictal discharges [68]. In some cases, such as where the MR lesion can be hemispheric but other noninvasive data suggest that seizures arise from only a restricted region, iEEG may allow focal successful resections. In deep-seated cortical malformation, as periventricular nodular heterotopia, Stereo-EEG remains the gold standard for exploring both the nodules and the overlying cortex and also for the therapeutic aspect (SEEG-guided THC [69]).

Several different methods are available to perform iEEG studies. The study can be achieved during surgery (electrocorticography [ECoG]) or extraoperatively following surgical implantation of electrodes; the latter, based on the type of electrodes, employ subdural electrodes or stereotactically implanted intracerebral electrodes, Stereo-EEG. While grids and strips of subdural electrodes provide a large coverage over only the surface of the cerebral cortex, they are often implanted in only one hemisphere, not reaching deeper brain structures (e.g. cingulate or insular cortex) and particularly the bottom of the sulci. By comparison, Stereo-EEG intracerebral electrodes can enable bilateral contemporary monitoring of superficial and deep cortical structures, but with limited cortical surface sampling [61].

14.4.1 The Epileptogenic Zone: Meaning and Definition

To define the EZ (the hypothetical minimum cortical area required to be resected or disconnected in order to stop the patient's seizures [70]), iEEG is considered as the gold standard. Invasive recordings should allow precisely defining and then removing different cortical areas. The classical Bancaud's definition underlined and described diverse cortical areas besides the EZ (lesional, irritative, and functional), not always exactly superimposable except few cases, and completely tailored on each patient. The lesional area delineated as the region in which slow waves were present and the background activity is absent; the irritative area is where spikes and waves are numerous and frequent. The high functional areas were identified by both functional examination (functional MRI (fMRI) and Magnetoencephalography (MEG)) and the results of the invasive electrical stimulations.

The epileptologist has to not only consider the presence of interictal paroxysms, the onset and propagation of the discharges during the spontaneous seizures, and the results of the electrical stimulations but also incorporate all the precedent regions as well as the nature and topography of any lesion [71, 72]. Furthermore, in order to be able to describe and define a pathological activity, the epileptologist should be sufficiently experienced to know all physiological activities with respect to their cortical localization.

The ictal clinical signs and symptoms may reflect either the cortical function or the dysfunction (in a positive or negative sense), or a simple propagation both cortical and/or subcortical, of the ictal discharges, or the involvement of a neuronal network. Consequently, defined on the basis of anatomo-electro-clinical correlations, the EZ should be delineated in different ways, as the cortical area(s) in which ictal discharges originate, or whose removal is sufficient to cure the patient, or coincident with the anatomical lesion, or, finally, the brain area responsible for the symptomatogenic zone.

Since the origins, the complexity of iEEG patterns was well recognized [73] and their use in synthesizing the EZ or the epileptogenic network is often based on empirical criteria and individual experience. The challenge in an individual case is to identify the minimum amount of tissue to remove and balance between a bigger resection for seizure control versus a smaller, less risky resection.

Abnormal interictal EEG data and seizure onsets, both on the subdural (SDE) and intracerebral electrodes, depending on the underlying pathology, can vary the spatial relationship between the lesion and the cortical surface and the correct placement of the electrodes [74–76], as ictal patterns may differ in term of morphology, shape, and rapidity on the different iEEG methods.

Different patterns are commonly observed during the initial phase including focal low-voltage fast activity in the gamma bands, spike/polyspikes or runs of spikes, flattening, and delta brush [73, 75–79]. Very high-frequency modifications or slow Direct Current Shift (DC) shifts can be observed, using a modification of the conventional filters. The presence of a localized low-voltage fast activity seems to be the most important positive prognostic factor [80].

Many variables can influence the morphologies of ictal discharges, reflecting a variety of EEG transition mechanisms, modulating the brain-involved structure, the underlying substrate, age, the type and the distance from the EZ of the recording electrodes, and the eventual withdrawal of antiepileptic medication (Fig. 14.5).

The "European" definition comprehends the notion the EZ represents the "site of the beginning and of the primary organization of the epileptic seizures" [81] emphasizing that great attention must be paid to those structures that are involved not only at seizure onset but also during the course of the seizure [82].

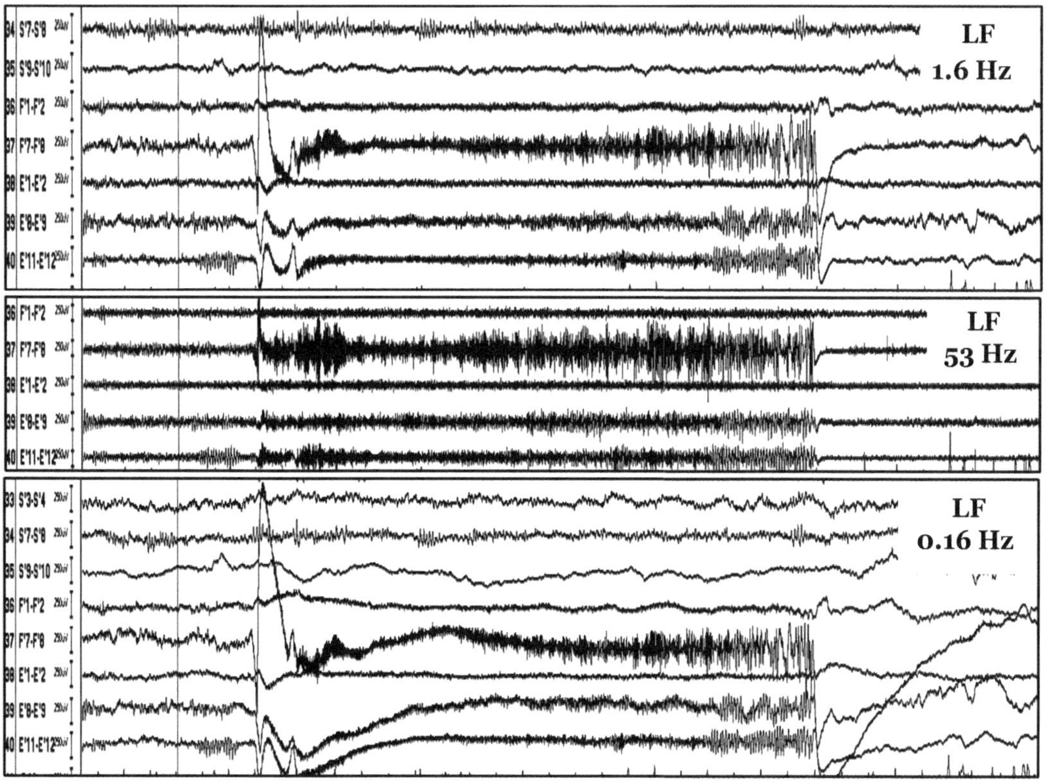

Fig. 14.5 Scheme of an electrode exploring, in its mesial portion the parietal cingulate gyrus (first red parenthesis), then the white matter (second parenthesis) and finally the cortex of the inferior parietal lobe. The white matter, in this case far from the EZ, was chosen as ground and reference

14.5 Stereo-EEG

14.5.1 Surgical Technique

The European Stereo-EEG methodology belonging to Talairach and Bancaud [83–86] employs only intracerebral electrodes. Based on anatomo-electro-clinical correlations, the planning of electrode insertion aims at verifying the hypothesis of localization of the EZ and studying not only the origin of ictal discharges but also their propagation and the organization of the whole epileptogenic network. As every patient is unique, each exploration is tailored and personalized.

Stereo-EEG electrodes (platinum and iridium) are flexible, semirigid, or rigid wires or catheters, generally with a diameter of less than 1 mm. Each electrode contains several contacts disposed of at regular intervals along the wire. Between 5 and 20 electrodes are implanted per patient under general anesthesia, during several hours, inserted through a burr hole, and placed (sometimes after an angiography is acquired and coregistered with 3D MRI) either with or without a stereotactic frame. The actual position of the electrodes is reconstructed using CT or MRI or directly visualized on MRI. A brief recording may be performed at the end of the surgical procedure to verify electrode integrity and the cortical targeting. Stereo-EEG recordings are performed during wakefulness and sleep, with the main aim to record spontaneous seizures. In children and adults, the duration of the recordings may vary from 1 to 28 days [87–89].

14.5.2 Stereo-EEG Recordings

Many of the clinical practices are common to Video-EEG recordings. The monitoring system should allow to record simultaneously at least 256 EEG channels in each patient, with a minimum sampling rate of 200–256 Hz; 512 Hz (or above) is needed to assess higher frequencies and research projects. The reference is chosen as far as possible from the EZ, possibly in a nonfunctional white matter (Fig. 14.6). Stereo-EEG channels are displayed in bipolar recordings, canceling the influence of the external reference [90].

Because the number of channels can easily exceed 100, being beyond the limits of adequate visual resolution, a synthetic montage should be chosen, including the most pathological contacts and each anatomical explored area. Adjunctive scalp electrodes can provide information on the regions unexplored by the intracerebral electrodes or to help sleep stage scoring. An additional polygraphy is assessed, depending on the clinical characteristics of the patient and seizure semiology [90].

A correct interpretation of Stereo-EEG recordings needs a 3D brain surface reconstruction permitting the placement of the electrodes in the multiplanar MRI [91]. Electrodes are removed either under sedation (in children) or without any anesthesia in the operating room (or at the bedside with little discomfort), and resective surgery is performed in a second step, after a complete analysis of all Stereo-EEG recordings and after checking the response to the Stereo-EEG guided thermocoagulations.

The radio frequency SEEG-guided thermocoagulation can be performed as a therapeutic option to "burn" limited well-defined cortical areas, previously identified by the Stereo-EEG recordings. The methodology offers a negligible possibility of rendering seizure-free patient, except those related to periventricular nodular heterotopia, even in bilateral cases [92]. In patients in whom conventional resective surgery is not feasible for a multifocality or the EZ is located into high functioning cortical areas [93], THC can be considered as an alternative therapeutic option, even if only with the goal of interrupting the pathological network.

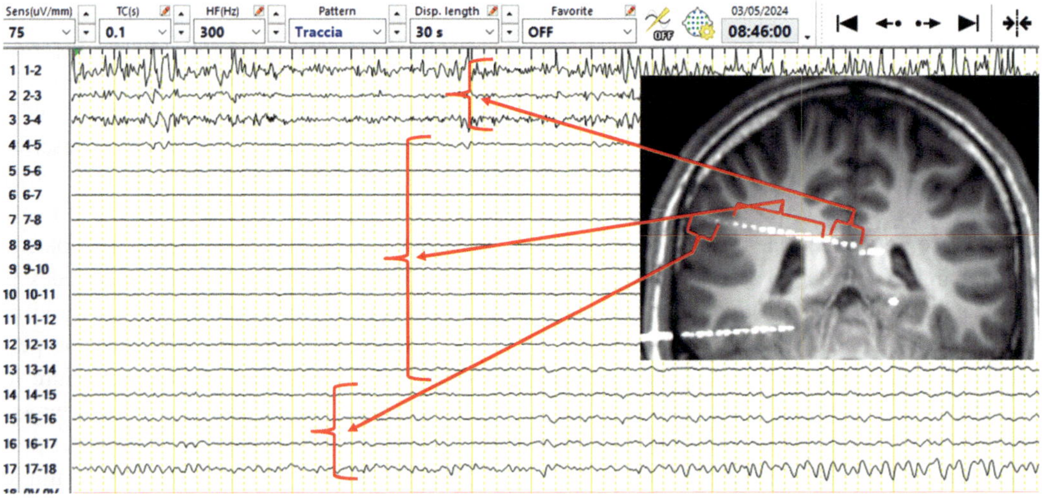

Fig. 14.6 Recording of a spontaneous seizure in Stereo-EEG. Examples of how filters can be modified to enhance the different parts of the ictal discharge. Keeping the HF open at 300 Hz, it is best to visualize baseline with the LF at 1.6 Hz. If, on the other hand, if we wish to enhance the LVFA, the LF should be set to 53 Hz. If we want to emphasize the initial slow deflection, the LF will have to be open to 0.16 Hz

14.5.3 Strong Points and Limits

Stereo-EEG can be accomplished without large craniotomies, providing a sampling of all cortical areas, simultaneously recording all cortical areas potentially involved in the onset, propagation, and clinical manifestations [70, 81, 94, 95].

The placement of a sufficient number of Stereo-EEG electrodes after a very careful analysis of all the noninvasive data permits to formulate one or more hypotheses concerning the site(s) of seizure onset and the pathways of preferential ictal spread. An adequate electrode placement must result from a suitable balance between the need to map cortical regions, the anatomical constraints, and the risks of parenchymatous damage. Stereo-EEG monitoring is feasible and safe, even in children less than 3 years of age [87, 89].

Stereo-EEG is particularly suited to record from all structures particularly deep and intrasulcal regions. A surgical procedure should be proposed only when it seems reasonable to lead to a satisfactory surgical outcome and is not indicated when a diffuse, bilateral, and multifocal epilepsy is suspected nor when there is not one or more defined localizing assumptions. Also, severe cognitive or behavioral disturbances that do not allow long-term monitoring and an age less than 2 years represent a relative contraindication to perform a Stereo-EEG study.

14.5.4 Risks

The morbidity using Stereo-EEG is predominantly related to hemorrhagic or infectious complications, varying from 0% to 7.5%. Only two studies reported specifically the risks of SEEG in children: the procedure also appears safe in this age group [87, 89]. In our population, 13 of 742 (1.8%) procedures were complicated, including four (0.5%) major events [96].

14.6 Subdural Electrodes and Combination of Subdural/Depth Electrodes

14.6.1 Surgical Technique

Subdural and depth electrodes (SDE) are generally made of platinum-iridium. The electrodes are shaped as rounds 4–5 mm in diameter and spaced 5–10 mm apart. They are secured in silastic strips (four to eight contacts) or square/rectangular grids (20–64 contacts); the option of special configuration may aid placement in particular regions [97].

The electrodes are implanted following a vast craniotomy, strips may be placed via burr holes. Following the removal of the bone and the opening of a dural flap, they are placed directly on the cortex or slid in the basal or interhemispheric aspect of the brain. A brief subdural electrode recording may be accomplished at the end of the implantation with the aim to check the electrode work or the abnormalities extend beyond the coverage permitting to modify their positioning. A dural flap is then wrapped around the cables; a dural graft may be required to guarantee adequate seal. Cables for each electrode are then tunneled under the scalp to minimize the risk of leaking and secured. Images of the cortex and the overlying electrodes help to define electrode placement but an MRI or a coregistered CT scan determine the actual location extraoperatively [98, 99]. Nowadays, Stereo-EEG has almost completely supplanted the use of subdural electrodes.

14.6.2 Recordings

Chronic iEEG recordings are generally performed both during wakefulness and sleep, the main aim being to record spontaneous seizures.

The recording software should allow to record simultaneously at least 128 channels, with a minimal sampling rate of 200–256 Hz; higher rates of 1024–2048 are required to permit signal analysis and research protocols. The reference is chosen among one of the contacts distant from the expected EZ. Recordings are displayed using bipolar montages but referential or special configurations focused on restricted areas of interest may be added. Additional scalp electrodes may be used to provide information on the regions unexplored by the intracranial electrodes or help for sleep stage scoring.

Spontaneous seizures are usually captured over 1–10 days; longer periods of up to a month may occasionally be accepted. As soon as one or more spontaneous seizures are acquired, medications are restored before the functional mapping with the electrical cortical stimulation [100].

14.6.3 Strong Points and Limits

Subdural electrodes permit sampling of large areas of the hemispheric surface; coverage over the convexity is generally easier than interhemispheric or basal or mesial cortices. Subdural electrodes are also largely suited especially for mapping functional regions, chiefly the motor and language cortex on the convexity. They can be used in all ages and are mostly well tolerated both in adults and children. Some trajectories (basal or interhemispheric) are difficult to manage, since the neurosurgeon tries to avoid veins and deflection of the electrodes [97]. The surgery eventually planned is inevitably performed at the end of the monitoring.

A combined depth/subdural electrode implantation can be suitable [100] in particular situation, as the necessity to cover large brain surface, or the need for precise mapping, or the evaluation of deep-seated anatomical lesion [101–104], even in the pediatric population [105], particularly in children with widespread or hemispheric cortical malformations.

14.6.4 Risks

General complications include wound infection, cerebrospinal fluid leak, intracranial bleeding, or symptomatic pneumocephalus [106–109], with a very rare risk of permanent neurological deficit or death. Neurologic infections were reported in 2.3%, bone or skin infections in 3%, intracranial hemorrhage in 4%, and augmented intracranial pressure in 2.4% [110]. Less than 3.5% of patients required additional surgical procedure(s) for the management of these adverse events. Risks are slightly higher in reoperated children [108] without a significant concern [111].

14.6.5 Electrical Stimulations and Cortical Mapping

In the presurgical evaluation of drug-resistant focal epilepsy, iEEG is often required to identify the EZ and the potential conflicts with neighboring eloquent regions. Cortical mapping consists of stimulations of two adjacent electrode contacts within the gray matter, exciting a cortical network or disrupting its functioning, and provoking a transient impairment. This data is predictive of a possible predictive prognosis of the functions that would be disturbed if the stimulated cortex were to be removed.

For cortical mapping by intracerebral electrodes (Stereo-EEG), two kinds of stimulations are usually employed in clinical practice: (1) low-frequency stimulations (LFS; 1 Hz, 15–60 s) usually provoke clinical signs in primary sensory cortex, and (2) high-frequency stimulations (HFS; 50 Hz) in brief trains (5–8 s), historically the main method for mapping the cortex beyond the motor or sensitive primary areas [112].

The principles of Electrical Stimulations (ES) are similar in the different employed methodologies: protocols are specifically used by individual centers. The cortical mapping is usually performed after almost a spontaneous seizure is acquired.

The duration of ES sessions may vary across several days. A constant current rectangular

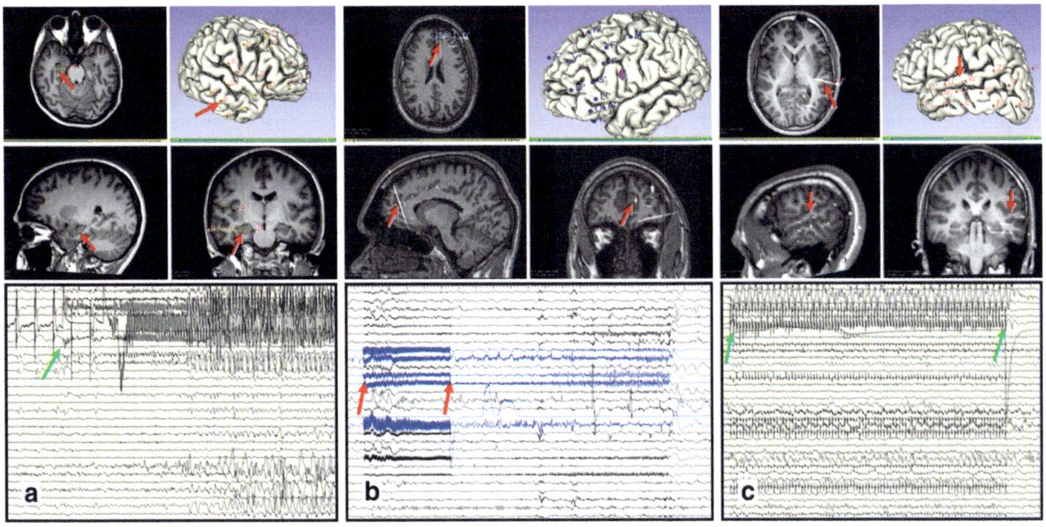

Fig. 14.7 Different type of stimulations: (**a**) 1-Hz low-frequency stimulation on the electrode B3–4 (anterior hippocampus, red arrows), 5-mA amperage, 20 seconds. During the last seconds of stimulation, we observe the onset of the seizure (green arrow). (**b**) 50-Hz high-frequency stimulation on X'9–10 (anterior mesial F1, dysplastic tissue), 0.6-mA amperage, 5 seconds. Seizure starts during stimulation (red arrows indicate onset and end of the stimulus). (c) 6-Hz intermediate-frequency stimulation on W'7–8 (Wernicke area, red arrows), 5-mA amperage, 15 seconds. Green arrows indicate onset and end of the stimulation, without post-discharges

biphasic pulse generator is the approved safe diagnostic stimulation of the human brain.

Largely speaking, with SDE ES is performed using 50 Hz pulses; infrequently 1 Hz may be used specially to differentiate motor behaviors. Pulse durations are maintained constant at 0.3 ms during trains for 0.5–10 s, starting with lower intensity, increasing in 0.5–1.0 mA until an electro-clinical modification occurs. The maximum intensity depends on the utilized stimulator.

With Stereo-EEG electrodes, ES can be performed either at low (LFES: 1–10 Hz; pulse width: 1–3 ms; 0.2–10 mA; duration: 30–40 s) (Fig. 14.7) or high frequency (HFES: 50–60 Hz; pulse width: 0.1–3 ms; 0.2–5 mA; duration: 3–10 s). LFES is favored when studying the motor cortex; HFES must be applied cautiously to avoid unpleasant sensations [77, 112]. Some cerebral structures as mesial temporal structures and FCD type II are more inclined to develop a seizure when stimulated. General limitations of ES are related to concerns of current spread and false and negative findings, requiring the cooperation of the patient. ES is secure; these parameters are shown to produce no structural damage [113].

Recent studies aim to demonstrate how a lower-frequency stimulation, using intermediate frequencies (IFS, i.e., between 6 and 15 Hz), is much better at analyzing the higher functions and especially the language). IFS permit longer duration of stimulations, enabling a more detailed assessment of the patient [114, 115].

14.7 Conclusion

The use of EEG in epilepsy diagnosis is essential and, in the presurgical stages, it becomes even more important to complete the presurgical work-up and establish the anatomo-electro-clinical correlations.

When necessary, the long-term Video-EEG monitoring should be considered and performed in expert and trained Epilepsy Centers. Epilepsy surgery involves a puzzling iter, and in most complex patients, invasive investigations are mandatory.

Nowadays invasive EEG recordings persist as the best practice in order to define the EZ and its related cortices. However, the assessment about the employed methodology is primarily determined by both the center experience and the economic possibilities, even if the Stereo-EEG methodology is spreading widely around the world.

Non invasive EEG recording strategy should be proposed in the absence of anatomo-electro-clinical data to guide the implantation, and an extremely accurate evaluation of all the noninvasive tests must be taken into account.

References

1. Berger H. Uber das Elektrenkephalogramm des Menschen. Arch Psychiat Nervenkr. 1929;87:527–70.
2. Adrian ED, Matthews BHC. The Berger rhythm, potential changes from the occipital lobe in man. Brain. 1934;57:345–59.
3. Gibbs FA, Gibbs EL, Lennox WG. Electro-encephalographic classification of epileptic patients and control subjects. Arch Neurol (Chicago). 1943;50:111–28.
4. Gibbs FA, Gibbs EL. Clinical and pharmacological correlates of fast activity in electroencephalography. J Neuropsychiatr. 1962;3(Suppl 1):S73–8.
5. Vannemreddy P, Stone JL, Slavin KV. Frederic Gibbs and his contributions to epilepsy surgery and electroencephalography. Neurosurgery. 2012;70(3):774–82.
6. Perry TR, Gumnit RJ, Gates JR, Leppik IE. Routine EEG vs. intensive monitoring in the evaluation of intractable epilepsy. Public Health Rep. 1983;98(4):384–9.
7. Engel J Jr, Burchfield J, Ebersole J, et al. Long term monitoring for epilepsy. Report of an IFCN committee. Electroencephal Clin Neurophysiol. 1993;87:437–58.
8. Burgess RC. Design and evolution of a system for long-term electroencephalographic and video monitoring of epilepsy patients. Methods. 2001;25:231–48.
9. Cascino GD. Clinical indications and diagnostic yield of videoelectroencephalographic monitoring in patients with seizures and spells. Mayo Clin Proc. 2002;77:1111–20.
10. Tatum WO, Mani J, Jin K, Halford JJ, Gloss D, Fahoum F, Maillard L, et al. Minimum standards for inpatient long-term video-electroencephalographic monitoring: a clinical practice guideline of the International League Against Epilepsy and International Federation of Clinical Neurophysiology. Epilepsia. 2022;63(2):290–315.
11. Alving J, Beniczky S. Diagnostic usefulness and duration of the inpatient long-term video-EEG monitoring: findings in patients extensively investigated before the monitoring. Seizure. 2009;18:470–3.
12. Asano E, Pawlak C, Shah A, Shah J, Luat AF, Ahn-Ewing J, et al. The diagnostic value of initial Video-EEG monitoring in children – review of 1000 cases. Epilepsy Res. 2005;66:129–35.
13. Velis D, Plouin P, Gotman J, da Silva FL, and the members of the ILAE DMC Subcommittee on Neurophysiology. Recommendations regarding the requirements and applications for long-term recordings in epilepsy. Epilepsia. 2007;48(2):379–84.
14. Beniczky SA, Fogarasi A, Neufeld M, Andersen NB, Wolf P, van Emde BW, et al. Seizure semiology inferred from clinical descriptions and from video recordings. How accurate are they? Epilepsy Behav. 2012;24:213–5.
15. Benbadis SR, Agrawal V, Tatum WO. How many patients with psychogenic nonepileptic seizures also have epilepsy? Neurology. 2001;57:915–7.
16. Reuber M. Psychogenic nonepileptic seizures: answers and questions. Epilepsy Behav. 2008;12: 622–35.
17. Devinsky O, Gazzola D, LaFrance WC. Differentiating between nonepileptic and epileptic seizures. Nat Rev Neurol. 2011;7:210–20.
18. Mostacci B, Bisulli F, Alvisi L, Licchetta L, Baruzzi A, Tinuper P. Ictal characteristics of psychogenic nonepileptic seizures: what we have learned from video/EEG recordings—a literature review. Epilepsy Behav. 2011;22:144–53.
19. Shih JJ, Spector AR, Taha FA, Mills RM. Clinically significant bradyarrhythmias in video-EEG patients. Seizure. 2011;20:225–30.
20. Popkirov S, Jungilligens J, Grönheit W, Wellmer J. Diagnosing psychogenic nonepileptic seizures: video-EEG monitoring, suggestive seizure induction and diagnostic certainty. Epilepsy Behav. 2017;73:54–8.
21. Appleton RE, Beirne M. Absence epilepsy in children: the role of EEG in monitoring response to treatment. Seizure. 1996;5:147–8.
22. Mohan KK, Markand ON, Salanova V. Diagnostic utility of Video-EEG monitoring in paroxysmal events. Acta Neurol Scand. 1996;47(suppl 1):26–30.
23. Ghougassian DF, d'Souza W, Cook MJ, O'rien TJ. Evaluating the utility of inpatient Video-EEG monitoring. Epilepsia. 2004;45(8):928–32.
24. Burkholder DB, Britton JW, Rajasekaran V, Fabris RR, Cherian PJ, Kelly-Williams KM, et al. Routine vs extended outpatient EEG for the detection of interictal epileptiform discharges. Neurology. 2016;86:1524–30.
25. Bubrick EJ, Yazdani S, Pavlova MK. Beyond standard polysomnography: advantages and indications for use of extended 10–20 EEG montage

during laboratory sleep study evaluations. Seizure. 2014;23:699–702.
26. Tinuper P, Grassi C, Bisulli F, Provini F, Plazzi G, Zoni E, et al. Split-screen synchronized display. A useful video-EEG technique for studying paroxysmal phenomena. Epileptic Disord. 2004;6:27–30.
27. Tallawy HN, Fukuyama H, Kader AA, Kamel NF, Badry R. Role of short-term out patient video electroencephalography in diagnosis of paroxysmal disorders. Epilepsy Res. 2010;88:179–82.
28. Kobulashvili T, Höfler J, Dobesberger J, Ernst F, Ryvlin P, Cross JH, et al. Current practices in long-term video-EEG monitoring services: a survey among partners of the E-PILEPSY pilot network of reference for refractory epilepsy and epilepsy surgery. Seizure. 2016;38:38–45.
29. Hamandi K, Beniczky S, Diehl B, Kandler RH, Pressler RM, Sen A, with ILAE British Chapter Workshop Attendees, et al. Current practice and recommendations in UK epilepsy monitoring units. Report of a national survey and workshop. Seizure. 2017;50:92–8.
30. Rubboli G, Beniczky S, Claus S, Canevini MP, Kahane P, Stefan H, et al. A European survey on current practices in epilepsy monitoring units and implications for patients' safety. Epilepsy Behav. 2015;44:179–84.
31. Shafer PO, Buelow J, Ficker DM, Pugh MJ, Kanner AM, Dean P, et al. Risk of adverse events on epilepsy monitoring units: a survey of epilepsy professionals. Epilepsy Behav. 2011;20:502–5.
32. Touloumes G, Morse E, Chen WC, Gober L, Dente J, Lilenbaum R, et al. Human bedside evaluation versus automatic responsiveness testing in epilepsy (ARTiE). Epilepsia. 2016;57(1):e28–32.
33. Beniczky S, Neufeld M, Diehl B, Dobesberger J, Trinka E, Mameniskiene R, et al. Testing patients during seizures: a European consensus procedure developed by a joint taskforce of the ILAE—Commission on European Affairs and the European Epilepsy Monitoring Unit Association. Epilepsia. 2016;57:1363–8.
34. Kinney MO, Kovac S, Diehl B. Structured testing during seizures: a practical guide for assessing and interpreting ictal and postictal signs during video EEG long term monitoring. Seizure. 2019;72:13–22.
35. Marciani MG, Gotman J, Andermann F, Olivier A. Patterns of seizure activation after withdrawal of antiepileptic medication. Neurology. 1985;35:1537.
36. So N, Gotman J. Changes in seizure activity following anticonvulsant drug withdrawal. Neurology. 1990;40:407.
37. Marks DA, Katz A, Scheyer R, Spencer SS. Clinical and electrographic effects of acute anticonvulsant withdrawal in epileptic patients. Neurology. 1991;41:508.
38. Haut SR, Swick C, Freeman K, Spencer S. Seizure clustering during epilepsy monitoring. Epilepsia. 2002;43:711–5.
39. Zhou D, Wang Y, Hopp P, Kerling F, Kirchner A, Pauli E, Stefan H. Influence on ictal seizure semiology of rapid withdrawal of carbamazepine and valproate in monotherapy. Epilepsia. 2002;43:386–93.
40. Thomas P, Valton L, Genton P. Absence and myoclonic status epilepticus precipitated by antiepileptic drugs in idiopathic generalized epilepsy. Brain. 2006;129:1281–92.
41. Zarkou S, Grade M, Hoerth MT, Noe KH, Sirven JI, Drazgowski JF. Indeterminate EMU admissions: does repeating the admission help? Epilepsy Behav. 2011;20:706–8.
42. Herman ST, Abend NS, Bleck TP, Chapman KE, Drislane FW, Emerson RG, Critical Care Continuous EEG Task Force of the American Clinical Neurophysiology Society, et al. Consensus statement on continuous EEG in critically ill adults and children, part II: personnel, technical specifications, and clinical practice. J Clin Neurophysiol. 2015;32:96–108.
43. Struck AF, Cole AJ, Cash SS, Westover MB. The number of seizures needed in the EMU. Epilepsia. 2015;56:1753–9.
44. Gazzola DM, Thawani S, Agbe-Davies O, Carlson C. Epilepsy monitoring unit length of stay. Epilepsy Behav. 2016;58:102–5.
45. Cox FME, Reus EEM, Visser GH. Timing of first event in inpatient long-term video-EEG monitoring for diagnostic purposes. Epilepsy Res. 2017;129:91–4.
46. Noe KH, Drazkowski JF. Safety of long-term video-electroencephalographic monitoring for evaluation of epilepsy. Mayo Clin Proc. 2009;84:495–500.
47. Rheims S, Ryvlin P. Patients' safety in the epilepsy monitoring unit. Curr Opin Neurol. 2014;27:213–8.
48. Spritzer SD, Riordan KC, Berry J, Corbett BM, Gerke JK, Hoerth MT, et al. Fall prevention and bathroom safety in the epilepsy monitoring unit. Epilepsy Behav. 2015;48:75–8.
49. Sauro KM, Wiebe N, Macrodimitris S, Wiebe S, Lukmanji S, Jetté N. Quality and safety in adult epilepsy monitoring units: a systematic review and metaanalysis. Epilepsia. 2016;57:1754–70.
50. van Griethuysena R, Hofstraa WA, van der Salma SMA, Bourez-Swarta MD, de Weerda AW. Safety and efficiency of medication withdrawal at home prior to long-term EEG video-monitoring. Seizure. 2018;56:9–13.
51. Stecker M, Stecker M. The effect of education on nurses' assessments in an epilepsy monitoring unit. Can J Neurosci Nurs. 2012;34:23–32.
52. Ryvlin P, Nashef L, Lhatoo SD, Bateman LM, Bird J, Bleasel A, et al. Incidence and mechanisms of cardiorespiratory arrests in epilepsy monitoring units (MORTEMUS): a retrospective study. Lancet Neurol. 2013;12:966–77.
53. Blume WT, Luders HO, Mizrahi E, Tassinari C, van Emde Boas W, Engel J Jr. Glossary of descriptive terminology for ictal semiology: report of the

ILAE Task Force on Classification and Terminology. Epilepsia. 2001;42:1212–8.
54. Flink R, Pedersen B, Guekht AB, Malmgren K, Michelucci R, Nevilleet B, (ILAE Commission Report), et al. Guidelines for the use of EEG methodology in the diagnosis of epilepsy. Acta Neurol Scand. 2002;106:1–7.
55. American Clinical Neurophysiology Society. Guideline twelve: guidelines for long-term monitoring for epilepsy. Am J Electroneurodiagnostic Technol. 2008;48:265–86.
56. Kaplan PW, Benbadis SR. How to write an EEG report: dos and don'ts. Neurology. 2013;80(1 Suppl 1):S43–6.
57. Kane N, Acharya J, Benickzy S, Caboclo L, Finnigan S, Kaplan PW, et al. A revised glossary of terms most commonly used by clinical electroencephalographers and updated proposal for the report format of the EEG findings. Revision 2017. Clin Neurophysiol Pract. 2017;2:170–85.
58. von Ellenrieder N, Beltrachini L, Muravchik CH. Electrode and brain modeling in stereo-EEG. Clin Neurophysiol. 2012;123(9):1745–54.
59. Gloor P. Contributions of electroencephalography and electrocorticography to the neurosurgical treatment of the epilepsies. Adv Neurol. 1975;8:59–105.
60. Chassoux F, Navarro V, Catenoix H, Valton L, Vignal JP. Planning and management of SEEG. Neurophysiol Clin. 2018;48(1):25–37.
61. Katz JS, Abel TJ. Stereoelectroencephalograp hy versus subdural electrodes for localization of the epileptogenic zone: what is the evidence? Neurotherapeutics. 2019;16(1):59–66.
62. Giulioni M, Rubboli G, Marucci G, Martinoni M, Volpi L, Michelucci R, et al. Seizure outcome of epilepsy surgery in focal epilepsies associated with temporomesial glioneuronal tumors: lesionectomy compared with tailored resection. J Neurosurg. 2019;111(6):1275–82.
63. Englot DJ, Berger MS, Barbaro NM, Chang EF. Factors associated with seizure freedom in the surgical resection of glioneuronal tumors. Epilepsia. 2012;53(1):51–7.
64. Thom M, Blumcke I, Aronica E. Long-term epilepsy-associated tumors. Brain Pathol. 2012;22(3):350–79.
65. Guo Z, Zhang C, Wang X, Liu C, Zhao B, Mo J, et al. Is intracranial electroencephalography mandatory for MRI-negative neocortical epilepsy surgery? J Neurosurg. 2022;138(6):1720–30.
66. Maillard LG, Tassi L, Bartolomei F, Catenoix H, Dubeau F, Szurhaj W, et al. Stereoelectroencephalography and surgical outcome in polymicrogyria-related epilepsy: a multicentric study. Ann Neurol. 2017;82(5):781–94.
67. Maillard L, Ramantani G. Epilepsy surgery for polymicrogyria: a challenge to be undertaken. Epileptic Disord. 2018;20(5):319–38.
68. Cossu M, Pelliccia V, Gozzo F, Casaceli G, Francione S, Nobili L, et al. Surgical treatment of polymicrogyria-related epilepsy. Epilepsia. 2016;57(12):2001–10.
69. Mirandola L, Mai RF, Francione S, Pelliccia V, Gozzo F, Sartori I, et al. Stereo-EEG: diagnostic and therapeutic tool for periventricular nodular heterotopia epilepsies. Epilepsia. 2017;58(11):1962–71.
70. Kahane P, Dubeau F. Intracerebral depth electrodes encephalography (stereoelectroencephalography). In: Ebersole JE, editor. Current practice of clinical encephalography. 4th ed; 2014.
71. Kahane P, Landré E, Minotti L, Francione S, Ryvlin P. The Bancaud and Talairach view on the epileptogenic zone: a working hypothesis. Epileptic Disord. 2006;8(Suppl 2):S16–26.
72. Najm IM, Tassi L, Sarnat HB, Holthausen H, Russo GL. Epilepsies associated with focal cortical dysplasias (FCDs). Acta Neuropathol. 2014;128(1):5–19.
73. Jayakar P. Chronic intracranial EEG monitoring in children: when, where and what? J Clin Neurophysiol. 1999;16(5):408–18.
74. Herberhold T, Pieper T, Kudernatsch M, Pascher B, Hartlieb T, Weber K, et al. Invasive pre-surgical epilepsy diagnostics in children: the advantage of depth electrodes combined with subdural grids in the evaluation of focal cortical lesions. Klinische Neurophysiologie. 2014;45(01) https://doi.org/10.1055/s-0031-1273985.
75. Perucca P, Dubeau F, Gotman J. Intracranial electroencephalographic seizure-onset patterns: effect of underlying pathology. Brain. 2014;137:183–96.
76. Di Giacomo R, Uribe-San-Martin R, Mai R, Francione S, Nobili L, Sartori I, et al. Stereo-EEG ictal/interictal patterns and underlying pathologies. Seizure. 2019;72:54–60.
77. Fisher RS, Webber WR, Lesser RP, Arroyo S, Uematsu S. High frequency EEG activity at the start of seizures. J Clin Neurophysiol. 1992;9:441–8.
78. Ikeda A, Terada K, Mikuni N, Burgess RC, Comair Y, Taki W, et al. Subdural recordings of ictal DC shifts in neocortical seizures in humans. Epilepsia. 1996;37(7):662–74.
79. Schiller Y, Cascino GD, Sharbrough FW. Chronic intracranial EEG monitoring for localizing the epileptogenic zone: an electroclinicalcorrelation. Epilepsia. 1998;39(12):1302–8.
80. Singh S, Sandy S, Wiebe S. Ictal onset on intracranial EEG: do we know it when we see it? State of the evidence. Epilepsia. 2015;56(10):1629–38.
81. Munari C, Bancaud J. The role of stereo-electro-encephalography (SEEG) in the evaluation of partial epileptic patients. In: Porter RJ, Morselli PL, editors. The epilepsies. London: Butterworths; 1987. p. 267–306.
82. Bartolomei F, Bettus G, Stam CJ, Guye M. Interictal network properties in mesial temporal lobe epilepsy: a graph theoretical study from intracerebral recordings. Clin Neurophysiol. 2013;124(12):2345–53.
83. Bancaud J, Talairach J, Bonis A, Schaub C, Szikla G, Morel P, et al. La stéréoencéphalographie dans l'épilepsie. In: Informations neuro-physio-

pathologiques apportées par l'investigation fonctionnelle stéréotaxique. Paris: Masson; 1965.
84. Bancaud J, Talairach J, Geier S, Scarabin JM. EEG et SEEG dans les tumeurs cérébrales et l'épilepsie. Paris: Edifor; 1973. p. 351.
85. Talairach J, Bancaud J. Stereotaxic approach to epilepsy. Methodology of anatomofunctional stereotaxic investigations. Prog Neurol Surg. 1973;5:297–354.
86. Talairach J, Bancaud J, Szikla G, Bonis A, Geier S. Approche nouvelle de la neurochirurgie de l'épilepsie. Méthodologie stéréotaxique et résultats thérapeutiques. Neurochirurgie. 1974;20(Suppl 1):1–240.
87. Cossu M, Schiariti M, Francione S, Fuschillo D, Gozzo F, Nobili L, et al. Stereoelectroencephalography in the presurgical evaluation of focal epilepsy in infancy and early childhood. J Neurosurg Pediatr. 2012;9:290–300.
88. Fohlen M, Jalin C, Soufflet C, Bulteau C, Dorfmuller G, Oliver V, et al. Electrophysiological investigations in childhood epilepsy surgery. Neurochirurgie. 2008;54:347–52.
89. Dorfmüller G, Ferrand-Sorbets S, Fohlen M, Bulteau C, Archambaud F, Delalande O, et al. Outcome of surgery in children with focal cortical dysplasia younger than 5 years explored by stereo-electroencephalography. Childs Nerv Syst. 2014;30(11):1875–83.
90. Isnard J, Taussig D, Bartolomei F, Bourdillon P, Catenoix H, Chassoux F, et al. French guidelines on stereoelectroencephalography (SEEG). Neurophysiol Clin. 2018;48(1):5–13.
91. Cardinale F, Cossu M, Castana L, Casaceli G, Schiariti MP, Miserocchi A, et al. Stereoelectroencephalography: surgical methodology, safety, and stereotactic application accuracy in 500 procedures. Neurosurgery. 2013;72(3):353–66.
92. Cossu M, Fuschillo D, Cardinale F, Castana L, Francione S, Nobili L, et al. Stereo-EEG-guided radio-frequency thermocoagulations of epileptogenic grey-matter nodular heterotopy. J Neurol Neurosurg Psychiatry. 2014;85:611–7.
93. Guénot M, Isnard J, Catenoix H, Maguiére F, Sindou M. SEEG-guided RF-thermocoagulation of epileptic foci: a therapeutic alternative for drug-resistant non-operable partial epilepsies. Adv Tech Stand Neurosurg. 2011;36:61–78.
94. Kahane P, Minotti L, Hoffmann D, Lachaux JP, Ryvlin P. Invasive EEG in the definition of the seizure onset zone: depth electrodes. In: Rosenow F, Lüders HO, editors. Handbook of clinical neurophysiology, Presurgical assessment of the epilepsies with clinical neurophysiology and functional imaging, vol. 3. Amsterdam: Elsevier BV; 2004. p. 109–33.
95. Kahane P, Francione S. Stereoencephalography. In: Lüders HO, editor. Textbook of epilepsy surgery. London: Informa Healthcare; 2008. p. 649–58.
96. Cardinale F, Rizzi M, Vignati E, Cossu M, Castana L, d'Orio P, et al. Stereoelectroencephalography: retrospective analysis of 742 procedures in a single centre. Brain. 2019;142(9):2688–704.
97. Bekelis K, Radwan TA, Desai A, Moses ZB, Thadani VM, Jobst BC, et al. Subdural interhemispheric grid electrodes for intracranial epilepsy monitoring: feasibility, safety, and utility: clinical article. J Neurosurg. 2012;117(6):1182–8.
98. Serra C, Huppertz HJ, Kockro RA, Grunwald T, Bozinov O, Krayenbühl N, et al. Rapid and accurate anatomical localization of implanted subdural electrodes in a virtual reality environment. J Neurol Surg A Cent Eur Neurosurg. 2013;74(3):175–82.
99. Taimouri V, Akhondi-Asl A, Tomas-Fernandez X, Peters JM, Prabhu SP, Poduri A, et al. Electrode localization for planning surgical resection of the epileptogenic zone in pediatric epilepsy. Int J Comput Assist Radiol Surg. 2014;9(1):91–105.
100. Bulacio JC, Jehi L, Wong C, Gonzalez-Martinez J, Kotagal P, Nair D, et al. Long-term seizure outcome after resective surgery in patients evaluated with intracranial electrodes. Epilepsia. 2012;53(10):1722–30.
101. Nespeca M, Wyllie E, Luders H, Rothner AD, Hahn J, Awad I, et al. EEG recording and functional localization studies with subdural electrodes in infants and young children. J Epilepsy. 1990;3:107–24.
102. Surbeck W, Bouthillier A, Weil AG, Crevier L, Carmant L, Lortie A, et al. The combination of subdural and depth electrodes for intracranial EEG investigation of suspected insular (perisylvian) epilepsy. Epilepsia. 2011;52(3):458–66.
103. Bouthillier A, Surbeck W, Weil AG, Tayah T, Nguyen DK. The hybrid operculo-insular electrode: a new electrode for intracranial investigation of perisylvian/insular refractory epilepsy. Neurosurgery. 2012;70(6):1574–80.
104. Wellmer J, von der Groeben F, Klarmann U, Weber C, Elger CE, Urbach H, et al. Risks and benefits of invasive epilepsy surgery workup with implanted subdural and depth electrodes. Epilepsia. 2012;53(8):1322–32.
105. Taussig D, Dorfmüller G, Fohlen M, Jalin C, Bulteau C, Ferrand-Sorbets S, et al. Invasive explorations in children younger than 3 years. Seizure. 2012;21(8):631–8.
106. Burneo JG, Steven DA, McLachlan RS, Parrent AG. Morbidity associated with the use of intracranial electrodes for epilepsy surgery. Can J Neurol Sci. 2006;33(2):223–7.
107. Johnston JM Jr, Mangano FT, Ojemann JG, Park TS, Trevathan E, Smyth MD. Complications of invasive subdural electrode monitoring at St. Louis Children's Hospital, 1994–2005. J Neurosurg. 2006;105(5 Suppl):343–7.
108. Musleh W, Yassari R, Hecox K, Kohrman M, Chico M, Frim D. Low incidence of subdural gridrelated complications in prolonged pediatric EEG monitoring. Pediatr Neurosurg. 2006;42(5):284–7.
109. Vale FL, Pollock G, Dionisio J, Benbadis SR, Tatum WO. Outcome and complications of chronically implanted subdural electrodes for the treatment of

medically resistant epilepsy. Clin Neurol Neurosurg. 2013;115(7):985–90.
110. Arya R, Mangano FT, Horn PS, Holland KD, Rose DF, Glauser TA. Adverse events related to extraoperative invasive EEG monitoring with subdural grid electrodes: a systematic review and meta-analysis. Epilepsia. 2013;54(5):828–39.
111. Vadera S, Jehi L, Gonzalez-Martinez J, Bingaman W. Safety and long term seizure free outcomes of subdural grid placement in patients with a history of prior craniotomy. Neurosurgery. 2013;73(3):395–400.
112. Munari C, Kahane P, Tassi L, Francione S, Hoffmann D, Lo Russo G, et al. Intracerebral low frequency electrical stimulation: a new tool for the definition of the "epileptogenic area"? Acta Neurochir Suppl (Wien). 1993;58:181–5.
113. Gordon B, Lesser RP, Rance NE, Hart J Jr, Webber R, Uematsu S, et al. Parameters for direct cortical electrical stimulation in the human: histopathologic confirmation. Electroencephalogr Clin Neurophysiol. 1990;75(5):371–7.
114. Giovannelli G, Pelliccia V, Abarrategui B, Scarpa P, Revay M, Tassi L. Intermediate stimulation frequencies for language mapping using Stereo-EEG. Clin Neurophysiol. 2022;144:91–7.
115. Abarrategui B, Pelliccia V, Giovannelli G, Nichelatti M, Valenzano S, Mikulan E, et al. New stimulation procedures for language mapping in stereo-EEG. Epilepsia. 2024;65(6):1720–9.

15 EEG in Emergencies

Markus Leitinger, Stefano Meletti, and Eugen Trinka

15.1 General Considerations

Neurological emergencies are typically characterized by an acute decline in neurological function due to impaired cerebral activity. Despite the wide range of potential causes, the impact on electrographic brain activity is often limited to a few distinct patterns. Consequently, EEG patterns observed in these emergencies are in general nonspecific. The extent to which an agent affects brain function depends on its speed of action, targeted brain areas, and the individual's ability to cope, which is influenced by age, lifestyle, comorbidities (e.g., hypertension, alcohol use, and smoking), and trigger factors such as sleep deprivation.

In addition to clinical assessment, electroencephalogram (EEG) is a crucial diagnostic tool for evaluating brain function. From a practical perspective, interpreting EEG patterns helps determine the need for anti-seizure medication (ASM) or other neuroprotective measures, for instance cooling or immunomodulators. Effective management involves (i) addressing the underlying cause and (ii) preventing further brain damage from hypersynchronous epileptiform EEG activities, including patterns on the ictal-interictal

M. Leitinger (✉)
Department of Neurology, Neurocritical Care, and Neurorehabilitation, Centre for Cognitive Neuroscience, Member of European Reference Network EpiCARE, Christian Doppler University Hospital, Paracelsus Medical University, Salzburg, Austria

Neuroscience Institute, Centre for Cognitive Neuroscience, Christian Doppler University Hospital, Paracelsus Medical University, Salzburg, Austria
e-mail: ma.leitinger@salk.at

S. Meletti
Department of Biomedical Metabolic Sciences and Neurosciences, University of Modena and Reggio Emilia, Modena, Italy

Neuroscience Department, Neurophysiology Unit and Epilepsy Centre, Azienda Ospedaliera-Universitaria di Modena, Modena, Italy
e-mail: stefano.meletti@unimore.it

E. Trinka
Department of Neurology, Neurocritical Care, and Neurorehabilitation, Centre for Cognitive Neuroscience, Member of European Reference Network EpiCARE, Christian Doppler University Hospital, Paracelsus Medical University, Salzburg, Austria

Neuroscience Institute, Centre for Cognitive Neuroscience, Christian Doppler University Hospital, Paracelsus Medical University, Salzburg, Austria

Institute of Public Health, Medical Decision Making and Health Technology Assessment, University for Health Sciences, Medical Informatics, and Technology, Hall in Tirol, Austria
e-mail: eugen@trinka.at

F. Brigo, O. Mecarelli (eds.), *EEG: The First 100 Years*,
https://doi.org/10.1007/978-3-031-86878-8_15

continuum (IIC), electrographic and electroclinical seizure burden, and status epilepticus (SE).

Detailed investigations are required to assess the harmful effects of epileptiform activity and the benefits of treatment. Excessive epileptiform activity can lead to brain atrophy, particularly in structures like the hippocampus [1]. MRI can reveal surrogate markers of cerebral tissue damage, such as changes in T2/fluid-attenuated inversion recovery (FLAIR), diffusion-weighted imaging (DWI), apparent diffusion coefficient (ADC), or arterial spin labelling (ASL) [2–4].

Intracerebral multimodal neuromonitoring (MMN) can detect metabolic consequences like acidosis or reduced brain tissue oxygen saturation but is an invasive procedure requiring significant expertise [5].

As scientific data is collected in many centres around the globe, standardizing terminology is crucial for comparing data [6]. The American Clinical Neurophysiology Society (ACNS) has incorporated validated Salzburg EEG criteria for non-convulsive status epilepticus (NCSE) and developed the 'ACNS Standardized Critical Care EEG Terminology' to unify EEG terminology in critical care [7–11].

15.2 EEG in Non-epileptic Emergencies

Some neurological emergencies are caused by conditions affecting the CNS's highest emotional and intellectual functions. For example, in functional or dissociative seizures (formerly known as psychogenic non-epileptic seizures) (FDS), the EEG may appear normal despite severe clinical manifestations, such as a dissociative state of consciousness mimicking coma or intense motor phenomena [12]. This normal or near-normal EEG represents one end of the spectrum of EEG changes seen in emergencies, where the EEG is minimally affected. This distinction is clinically crucial; approximately 10% of patients in the Established Status Epilepticus Treatment Trial (ESETT) had non-epileptic functional/dissociative status and were misdiagnosed as having convulsive status epilepticus (CSE) [13]. At a quaternary-level teaching hospital in Australia, 26.5% of patients admitted for seizures had functional/dissociative seizures [14].

Another common cause of transient loss of consciousness is syncope. Among patients admitted to emergency rooms for transient loss of consciousness, 51% had syncope, while only 6% had epileptic seizures [15]. The EEG in patients with syncope is typically normal after the event. During syncope, reduced cerebral perfusion leads to diffuse slowing of background rhythms, increased delta activity, and potentially flattening or suppression of the EEG after several seconds, which is fully reversible [16, 17].

15.3 Status Epilepticus

In contrast to non-epileptic emergencies, seizures and status epilepticus represent the opposite end of the EEG spectrum, where epileptic burden adds to the burden exerted by the aetiology. While seizures are discussed elsewhere in this book, this chapter focuses on status epilepticus (SE).

The diagnostic criteria for non-convulsive status epilepticus (NCSE) have evolved over time (Fig. 15.1).

Initially, Young et al. suggested criteria for diagnosing NCSE [18]. Subsequent terminology replaced the concepts of incremental onset and decremental offset with "evolution" [19]. Kaplan modified the frequency of epileptiform discharges (EDs) from 3.0 Hz to 2.5 Hz, introduced the clinical correlate time-locked to an EEEG pattern, and proposed criteria for patients with pre-existing epileptic encephalopathy [20]. During the fourth London-Innsbruck Colloquium on Status Epilepticus and Acute Seizures in Salzburg in 2013, an expert consensus developed the "Salzburg criteria" [7]. For patients without pre-existing epileptic encephalopathy (Fig. 15.2), the criteria for diagnosing NCSE are (1) epileptiform discharges (EDs) >2.5 Hz (Fig. 15.3), or if EDs ≤2.5 Hz or rhythmic delta/theta activity (>0.5 Hz) is present, then (2) typical spatiotemporal evolution (Fig. 15.4), (3) subtle clinical ictal phenomena during the mentioned EEG patterns (Fig. 15.5), or (4) EEG and clinical improvement with intravenous anti-seizure medication (IV ASM trial) (Fig. 15.6).

For patients with known epileptic encephalopathy, NCSE is diagnosed if (1) there is an increase in prominence or frequency of the features com-

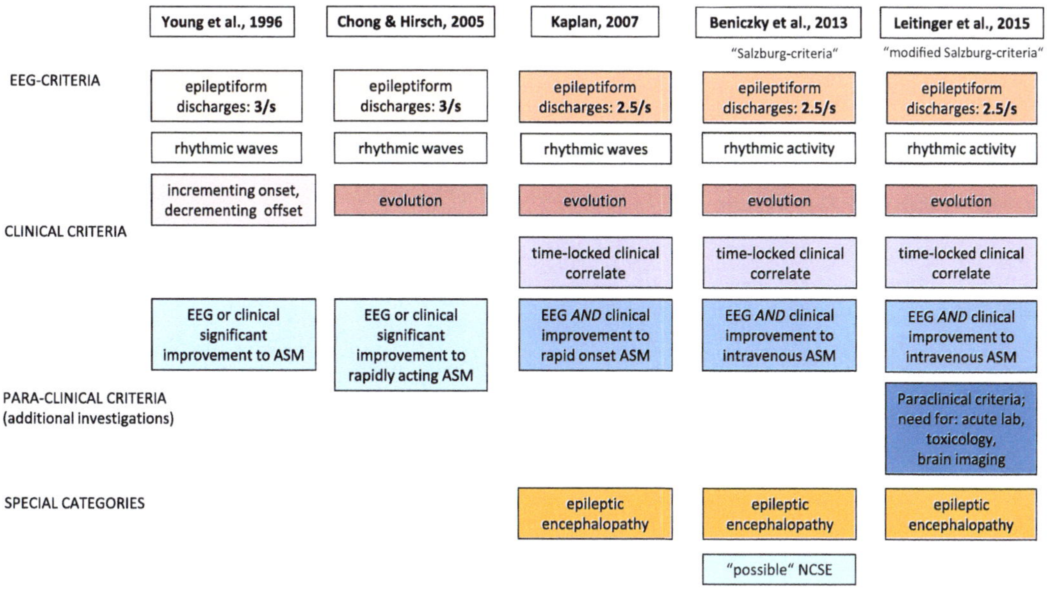

Fig. 15.1 The development of diagnostic criteria for non-convulsive status epilepticus (NCSE) over time

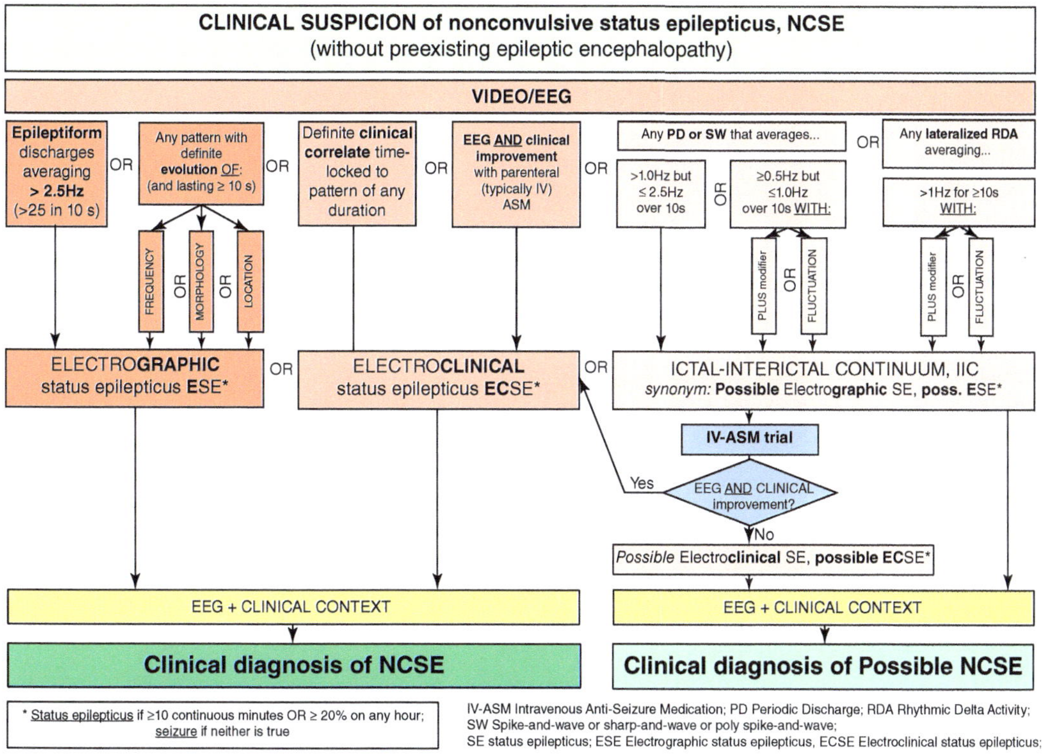

Fig. 15.2 Overview of criteria for non-convulsive status epilepticus (NCSE) including electrographic and electroclinical status epilepticus, and the ictal-interictal continuum (IIC) [7, 9, 11]. Modified with permission from [23]

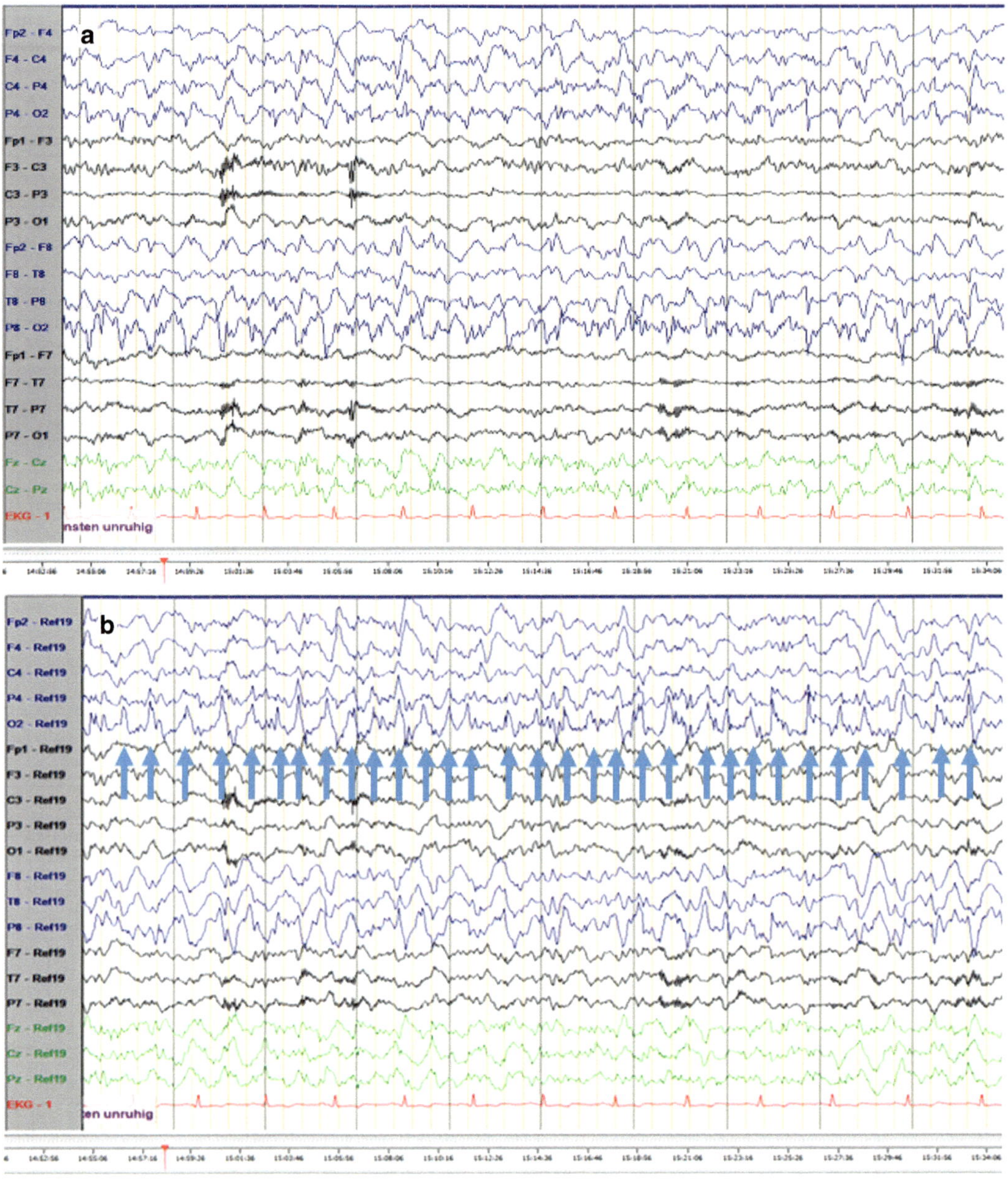

Fig. 15.3 Electrographic status epilepticus: criterion "epileptiform discharges (EDs) more than 25 per 10 seconds." The figure shows approximately 30 EDs per 10 s, i.e., ">2.5 Hz" over the right temporo-occipital region in a 65-year-old patient with bilateral tonic-clonic status for 10 min and confusion thereafter. Aetiology: meningoencephalitis, CSF 350 cells/μl, no germ identified. (**a**) Bipolar montage. (**b**) common average montage (blue arrows indicate EDs). (**a**) and (**b**): 30 mm/s, 7 μV/mm, 0.53 Hz/70 Hz

pared to baseline with observable changes in clinical state or (2) there is an EEG AND clinical response to IV ASM. If IV ASM leads to EEG improvement but not clinical response, "possible NCSE" is recommended, regardless of the presence of epileptic encephalopathy.

In a critical review at the fifth London-Innsbruck Colloquium, Leitinger et al. emphasized the need for clear definitions of terms such as "fluctuation" and "evolution" [8]. They suggested counting all epileptiform discharges (EDs) within 10 s to avoid inaccuracies in fre-

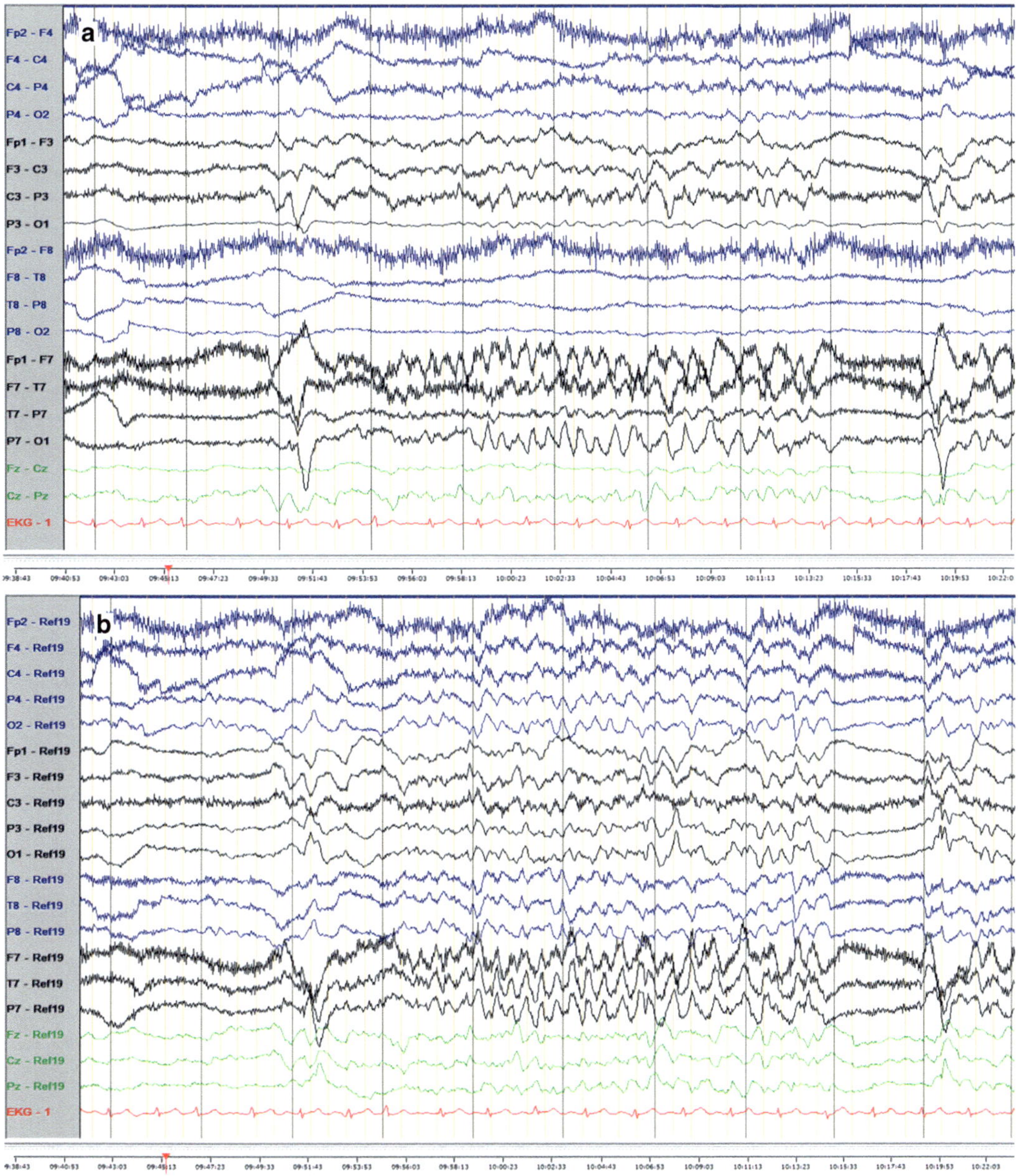

Fig. 15.4 Electrographic status epilepticus: criterion "evolution in frequency, location, or morphology." The figure shows evolution in frequency in an 85-year-old patient with bilateral tonic-clonic seizure activity of 5 min and subsequent coma. The figure shows highly repetitive patterns with evolution in frequency, over the left temporal region 8 Hz evolved to 6 Hz and later 3.5 Hz. Note that the individual evolution pattern is less than 10 s. However, consecutive evolution patterns are separated by only short gaps of 1–2 s. This results in a prevalence of about 70% exceeding the required minimum of 20%. (**a**) Bipolar montage. (**b**) common average montage. A and B: 30 mm/s, 7 μV/mm, 0.53 Hz/70 Hz

quency measurement. To reduce false positives, it was advised to identify situations where severe EEG changes are due to the underlying aetiology without a clear additional epileptic burden, such as in acute kidney injury or lithium intoxication, where a diagnostic trial with IV ASM may be preferable before aggressive status treatment. Comprehensive evaluation should include brain imaging, standard laboratory tests, toxicology, drug serum levels, and a

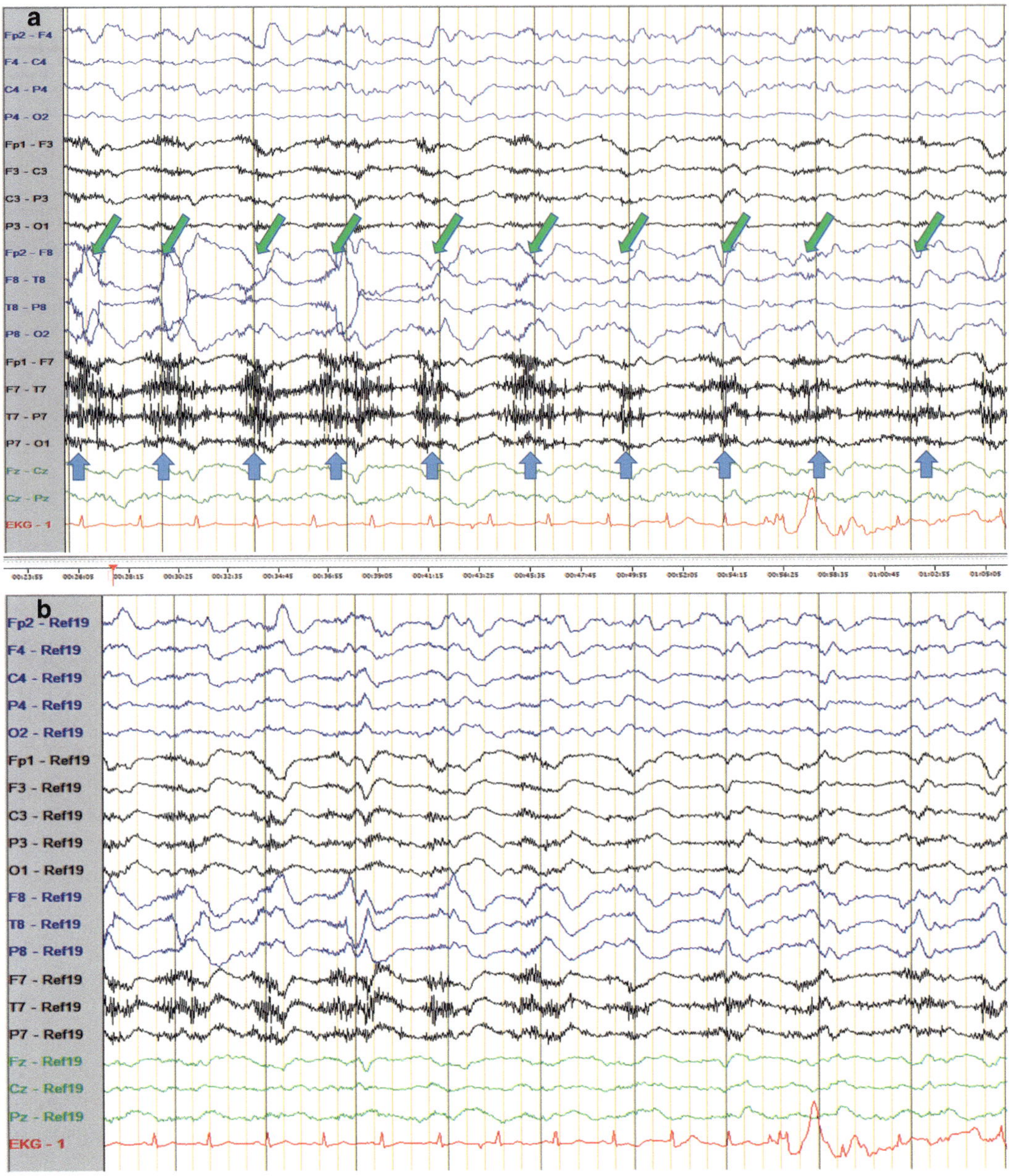

Fig. 15.5 Electroclinical status epilepticus: criterion "definite clinical correlate time locked to EEG pattern." The figure shows right temporal lateralized periodic discharges (LPDs, green oblique arrows) at a frequency of 1.0 Hz associated with clinical jerks in the left temporal muscle (vertical blue arrows) leading to regular jerks of the jaw in a 69-year-old patient suffering from diffuse cerebrovascular disease. Note the muscle artefacts in the left temporal region contralateral to the right temporal LPD recordings which provide a helpful clue for detecting and proving this criterion. (**a**) Bipolar montage. (**b**) common average montage. (**a**) and (**b**): 30 mm/s, 7 μV/mm, 0.53 Hz/70 Hz

detailed patient history [8] (Figs. 11, 12, and 14 in reference [21]).

The Salzburg criteria have been retrospectively validated in a multicentre study, showing high reliability and clinical applicability (sensitivity 97.7%, positive predictive value 84.0%, specificity 89.6%, negative predictive value 98.6%, and overall accuracy 92.5%) [9]. These criteria are applicable to patients with and without pre-existing epilepsy or epileptic encepha-

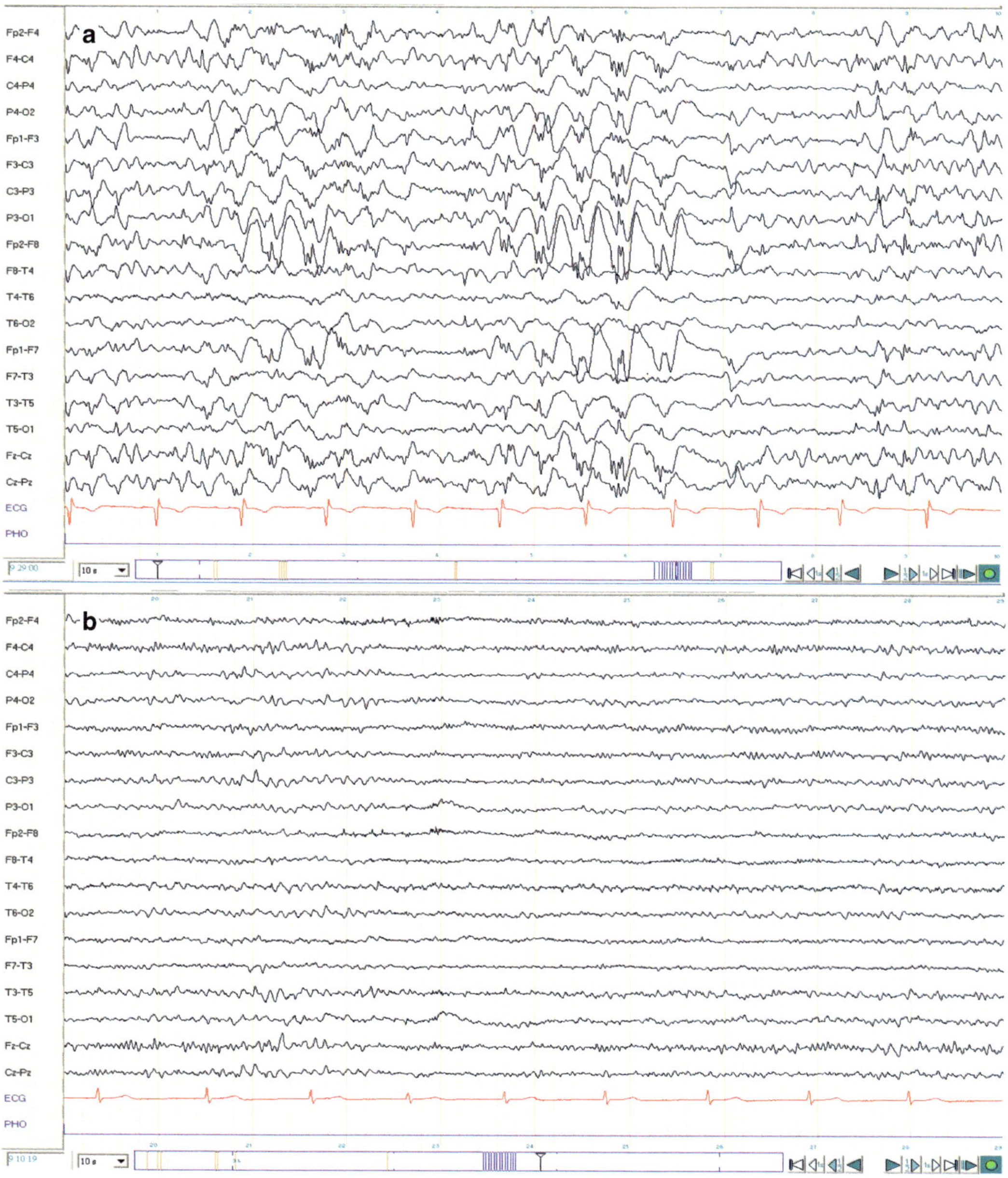

Fig. 15.6 Electroclinical status epilepticus: criterion "EEG *and* clinical improvement with parenteral (typically intravenous [IV]) anti-seizure medication (ASM)." (**a**) Generalized bifrontal spikes and polyspikes are present without fulfilling the diagnostic criteria for NCSE in a 60-year-old patient with traumatic brain injury 30 years ago presenting with confusion after a bilateral tonic-clonic status epilepticus for 10 min. (**b**) EEG normalized (EEG response) *and* confusion remitted (clinical response). (**a**) Bipolar montage. (**b**) common average montage. (**a**) and (**b**): 30 mm/s, 7 μV/mm, 0.53 Hz/70 Hz

lopathy, regardless of coma status or aetiology, and across different age groups.

Regarding the NCSE criterion "ED >2.5 Hz," evidence from multimodal neuromonitoring (MMN) in subarachnoid haemorrhage (SAH) patients shows that a drop in brain tissue oxygen partial pressure (PbtO2) occurs at 2.5 Hz [5]. Patterns with "ED >3.0 Hz" or "evolution in frequency, morphology, or location" were associated with cognitive impairment 3 months after

SAH [22]. The term "seizure burden" reflects the dose-dependent relationship between cognitive disturbance and the number of seizures. The American Clinical Neurophysiology Society (ACNS) integrated the Salzburg criteria into their Standardized Critical Care EEG Terminology in 2021 (Fig. 15.2) [11, 23]. According to ACNS criteria, a seizure must last at least 10 s, although shorter durations may be acceptable if clearly time locked to a clinical correlate. For NCSE, at least 10 continuous minutes or 20% of any 60-min recording period are required. The ACNS 2021 criteria also address the ictal-interictal continuum (IIC), covering the broader EEG spectrum spanning from benign patterns to seizures and status epilepticus.

15.4 The Ictal-Interictal Continuum

EEGs displaying periodic, rhythmic, or spike-and-wave, sharp-and-wave, or polyspike-and-wave patterns, are categorized under the ictal-interictal continuum (IIC) (Fig. 15.2) [11]. Technically, IIC patterns are first classified as generalized, lateralized, bilateral independent, unilateral independent, or multifocal (main term 1). In the next step, patterns are further described based on the presence of background activity between discharges: periodic (PD) patterns have intervening background, rhythmic (RDA) patterns show continuous discharges without background between the discharges, and spike-and-wave, sharp-and-wave, or polyspike-and-wave patterns (SW) represent specific discharge forms (main term 2).

Currently, research is focussed on determining which IIC patterns might benefit from prompt administration of anti-seizure medication (ASM). MRI with FLAIR, DWI, ADC, and ASL sequences is considered crucial for addressing this question. Interestingly, MRI, a structural and functional imaging method, complements and enhances the research of EEG, a functional method.

A key question is whether EEG changes directly reflect the underlying condition or are secondary epileptic phenomena triggered by the aetiology. The term "epileptic phenomenon" refers operationally to states where ASM may be beneficial, despite limited understanding of the underlying pathophysiology. In cases of IIC or other EEG patterns with high clinical suspicion of NCSE but not meeting NCSE criteria, a *diagnostic* intravenous ASM (IV ASM) should be administered to assess for "EEG AND clinical improvement," which would confirm NCSE, but a negative ASM trial in this occasion cannot exclude NCSE from a conceptional point of view (e.g. underdosed ASM trial or drug-resistant NCSE) [7, 8]. However, despite efforts to standardize, terms like "EEG improvement" or "clinical improvement," specific ASMs and respective doses for the "diagnostic IV ASM trial" remained undefined until recently. An expert panel at the eighth London-Innsbruck Colloquium on Status Epilepticus and Acute Seizures proposed specific criteria and recommendations [24]. Briefly, EEG response is defined as the cessation of the pattern for three times the longest spontaneous interval or at least one continuous minute. Clinical response is defined as a one-level enhancement on a 10-step NCSE-response scale (NRS) or a time-locked improvement of a focal deficit. Time criteria were also provided for when diagnostic IV ASM should result in the termination of the pattern or outline steps for further drug administration if this is not achieved.

Emergencies such as traumatic brain injury or subarachnoid haemorrhage are associated with high rates of non-convulsive seizures and NCSE, adding to the epileptic burden on brain tissue and necessitating continuous EEG (cEEG) monitoring [22]. cEEG is typically performed in intensive care units as part of multimodal neuromonitoring. A key aspect of cEEG is the use of fast Fourier analysis, which enables real-time detection of seizures and changes in frequency composition (digital spectral analysis) due to cerebral function variations, such as changes in the alpha-delta ratio in patients with subarachnoid haemorrhage (SAH) who are at risk for delayed cerebral ischaemia [25]. A detailed discussion of cEEG features is beyond the scope of this chapter.

15.5 Traumatic Brain Injury

In traumatic brain injury (TBI), EEG patterns vary based on injury severity and involvement of specific brain structures, such as the brainstem or white matter tracts. A study involving 165 EEGs from 140 patients with posttraumatic coma and secondary midbrain and bulbar brain syndrome observed that the disappearance of sleep or sleep-like activities, the appearance of alternating patterns, and loss of reactivity indicated a poor prognosis [26].

In animal models of mild traumatic brain injury (MTBI), EEG initially shows brief epileptiform activity, followed by suppressed activity lasting 1–2 min, then diffuses slowing, and typically returns to normal within an hour [27]. In humans, early changes are challenging to record. Many patients, particularly those injured in shipyard accidents, had normal EEGs recorded as early as 10–15 min post-injury [28]. However, patients who had EEGs recorded as soon as possible often showed diffuse slowing that resolved within an hour. Among boxers, reduced EEG amplitude and increased irregular theta activity were noted within 15–30 min after trauma, especially following a knockout [29]. A study of 344 adults found that 51% had initial EEG abnormalities 3 days post-injury, with 43% showing generalized slowing and 23% focal slowing [30]. In 40% of MTBI patients, diffuse slowing of the posterior dominant rhythm (PDR) was only observed retrospectively, with a noted increase from 9 Hz to 10 Hz in PDR over a year of follow-up. Focal EEG abnormalities typically resolved within 6 months. Amnesia exceeding 8 h was consistently associated with EEG abnormalities, while patients without amnesia generally had normal EEGs. No correlation was found between EEG abnormalities and symptoms like vertigo, dizziness, or nystagmus [31]. The correlation with MRI, CT, or SPECT abnormalities was poor [32].

15.6 Ischaemic Stroke

From a pathophysiological perspective, the EEG reflects cerebral function, metabolism, and cerebral blood flow [33]. Similar to syncope, reversible stages of cerebral ischaemia involve the loss of fast beta frequencies, followed by slowing of background theta activity, and in more severe cases, delta-range slowing before complete suppression of frequencies occurs [34]. One of the earliest studies on hemispheric acute ischaemic stroke (AIS) described widespread polymorphic delta activity as the mildest form of neuronal dysfunction [35]. More severe involvement is characterized by attenuation or loss of alpha and beta activity, and the loss of sleep spindles. In severe cases, all EEG frequencies may be suppressed. Mass effect following cerebral oedema often leads to contralateral frontal delta activity or intermittent bursts of delta activity. Large subcortical AIS can present with focal or generalized intermittent rhythmic delta [36]. Lacunar infarcts are typically associated with a normal EEG or subtle focal theta activity [37].

15.7 Hepatic Failure

The earliest reports on EEG changes in hepatic encephalopathy were published by Bickford and Butt in the mid-1950s [38]. Mild stages presented with theta and partly disorganized alpha activity, progressing to triphasic waves with usually absent alpha. In more severely affected patients, delta waves predominated with absent alpha. A 30-year-old woman in her fifth month of pregnancy who developed acute hepatic failure showed slow, irregular delta waves on EEG [39]. A 44-year-old man with acute fulminant hepatic failure of unknown aetiology displayed diffuse background slowing on day one, evolving to alpha coma, and then spindle coma over the next 2 days before the patient succumbed on day four [40].

EEG classifications for hepatic encephalopathy range from normal to coma, with diagnostic sensitivity varying from 43% to 100% [41]. Initial EEG changes include slowing of the posterior dominant alpha rhythm, followed by the gradual appearance of theta and high-amplitude irregular delta waves as the condition progresses. In a study of 12 patients with Grade II to IV encephalopathy, EEG revealed slow activity (six patients), low voltage (five patients), or a flat line (two patients) [42]. Three patients with EEG suppression due to severe hepatic encephalopathy between 24 and 72 h fully recovered following successful treatment of brain swelling [43].

A more recent classification outlines six stages: (1) normal or low-frequency alpha rhythm disturbed by random theta waves; (2) background theta activity with rare high-voltage delta waves, reduced or absent eye-opening response, and possible bursts of rhythmic delta and triphasic waves (Fig. 15.7); (3) severe EEG disorganization with asynchronous theta and delta waves, triphasic waves common; (4) high-voltage arrhythmic delta activity; (5) decreasing frequency and amplitude of arrhythmic delta; and

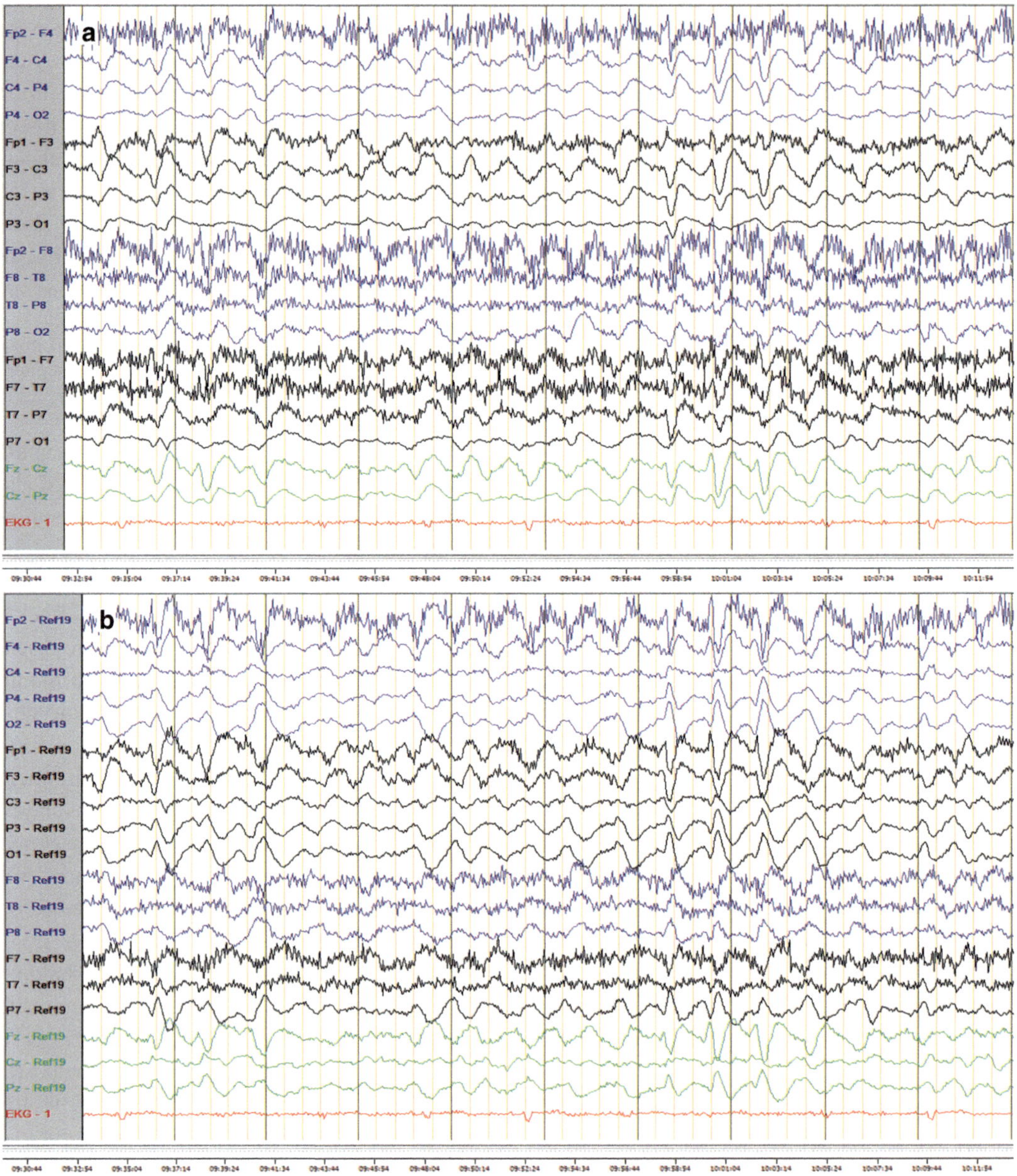

Fig. 15.7 The figure shows bifrontal triphasic waves in a 70-year-old patient with hepatic encephalopathy due to alcohol-mediated liver cirrhosis with serum ammonia level of 118 mmol/l (normal range 17–47). (**a**) Bipolar montage. (**b**) common average montage. (**a**) and (**b**): 30 mm/s, 7 μV/mm, 0.53 Hz/30 Hz

(6) a flat EEG [44]. In a 9-year cohort study, 12% of patients with triphasic waves had liver insufficiency [45].

15.8 Renal Failure

In eight patients with acute renal failure, EEG findings included irregular or slow background activity in six patients, paroxysmal bilaterally synchronous slow waves in five, and paroxysmal responses to eye opening in two [46]. In a cohort of 105 patients exhibiting triphasic waves, renal insufficiency was found in 50% [45]. The same cohort showed cerebral white matter lesions in 60% of patients and brain atrophy in 59%. Absent EEG background reactivity and respiratory failure were independently associated with mortality. In a case of acute kidney injury due to *Dioscorea quinqueloba* intoxication, the patient exhibited decreased consciousness and seizures, with EEG revealing diffuse continuous slowing in the theta and delta ranges with a frontal maximum [47]. Aphasic status resulting from acute on chronic renal failure due to neglect of peritoneal dialysis showed four to five ictal rhythmic discharges per hour over the left frontotemporal area [48]. Four patients with end-stage renal disease on peritoneal dialysis developed acute non-convulsive status epilepticus (NCSE) characterized by diffuse sharp waves and spikes; three of these patients had active systemic infections and were receiving antibiotic treatment [49]. EEG improvement was observed following the administration of anti-seizure medications (ASMs).

15.9 Disturbances of Electrolytes: Hyponatraemia

Serum sodium levels below 125 mmol/L are considered severe hyponatraemia [50]. Hyponatraemia leads to posterior slowing, which in more severe cases progresses to diffuse delta activity [51, 52]. The alpha rhythm may be interrupted by high-voltage rhythmic delta activity, and central high-voltage activity of 6–7 Hz may occur [53]. In even more severe cases, EEG may display triphasic waves and periodic lateralized discharges [54]. Recurrent absence status due to hyponatraemia has been reported in a patient with polydipsia [55].

15.10 Intoxications

15.10.1 Antibiotics

Non-convulsive status epilepticus (NCSE) has been reported in patients treated with higher-than-recommended doses of cefepime, with generalized spike-and-wave discharges at 2.0–2.5 Hz, which showed EEG and clinical improvement following intravenous (IV) anti-seizure medication (ASM) [56]. Discontinuation of cefepime led to rapid full recovery. In a series of 19 patients with renal impairment treated with cefepime, severe encephalopathy manifested as prolonged confusional states, with EEG revealing diffuse rhythmic nonreactive triphasic sharp waves at 1.5–2.0 Hz, possibly indicative of the ictal-interictal continuum (IIC) [57]. Discontinuation of cefepime led to resolution of electroclinical symptoms. NCSE was also reported with several IV cephalosporins (ceftriaxone, ceftazidime, and cefepime) in patients with renal failure [58]. Cefepime caused NCSE in the form of generalized periodic discharges (GPDs) at 2.0 Hz in a patient with normal renal function [59]. Other antibiotics, including levofloxacin (in one of three cases combined with piperacillin), ofloxacin (with piperacillin), ciprofloxacin, co-amoxiclav, and ceftazidime, have also been implicated in causing status epilepticus [60]. A systematic review of 37 publications involving 135 patients found that NCSE was present in 25%, myoclonic status epilepticus in 7%, triphasic waves in 40%, and focal sharp waves in 39% [61]. Approximately a quarter of patients experienced neurotoxicity despite appropriate dosing, while almost half were overdosed. The median time to onset of symptoms was 4 days after starting the medication, with resolution occurring a median of 2 days after discontinuation. White matter lesions were identified as a significant factor in cefepime-induced encephalopathy [62]. A patient treated with clarithromycin presented with generalized periodic discharges at 1.5–2.0 Hz, compatible with NCSE, and progressive recovery followed the cessation of the antibiotic [63]. The distinction between encephalopathy and NCSE has been debated; currently, such cases would be addressed as IIC, with NCSE diagnosed if both EEG and clinical improvement are observed following a diagnostic IV ASM trial [64]. (Fig. 15.8).

15.10.2 Ifosfamide

Non-convulsive status epilepticus was reported as a complication of ifosfamide chemotherapy, presenting as a periodic triphasic EEG pattern that resolved promptly with IV diazepam [65]. In two patients with ovarian carcinoma, EEG showed continuous, generalized, irregular slowing with admixed sharp activity and continuous generalized slow activity with admixed sharp

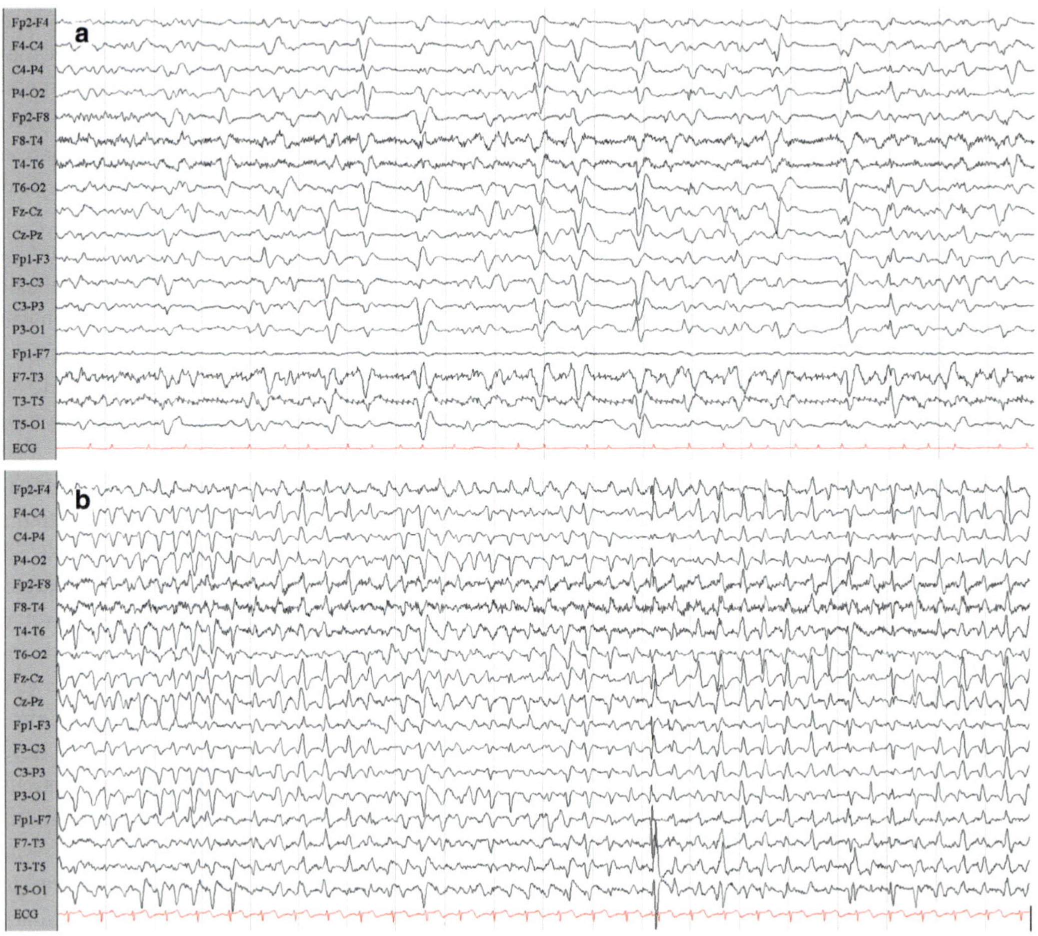

Fig. 15.8 Two examples of cefepime encephalopathy are shown. In (**a**) 75-year-old woman with fever, chronic renal failure being treated with cefepime for 5 days. From the morning the patient is stuporous. The EEG documents the presence of generalized periodic slow complexes at a frequency of less than 1 Hz. In (**b**) a man of 80 years with fever, diarrhoea, chronic renal failure that was treated with cefepime for 3 days. He has been in a coma-like state for 12 h. In this case the EEG pattern documents the presence of complexes with sharp morphology with a frequency approaching 2.5 Hz. In both cases diazepam infusion resolved the EEG pattern and lead to a clinical response (awakening) in B. Filters: hi-pass 0.3 Hz; low-pass: 30 Hz. Amplitude: 10 uV/mm. Each page shows 20 s

activity, respectively [66]. Both patterns resolved within minutes after IV diazepam administration, leading the authors to coin the term "diazepam-sensitive encephalopathy." This rapid response to diazepam was also observed in two patients with non-Hodgkin lymphoma [67]. Methylene blue effectively reversed ifosfamide-induced encephalopathy, which initially showed diffuse slowing on EEG [68]. However, methylene blue was only effective in one out of five children with grade II–IV neurological toxicity according to the National Cancer Institute (NCI) classification [69]. In a series of seven patients with NCSE due to ifosfamide, EEG patterns included generalized high-amplitude sharp wave complexes (1.0–2.0 Hz) in four patients, slow background and generalized triphasic sharp wave complexes in two, and diffuse theta background with posteriorly predominant triphasic complexes in one [70]. NCSE diagnosis was established based on EEG and clinical improvement with IV midazolam [71]. In a series of five patients, two experienced generalized tonic-clonic seizures lasting 1–4 min, with initial EEG showing isolated or intermittent epileptiform discharges [71]. In a series of 16 patients, EEG findings included generalized periodic discharges with or without triphasic morphology (n = 9), intermittent background attenuation (n = 5), background slowing with intermittent rhythmic delta (n = 4), bursts of bilateral synchronized delta activity (n = 2), and frontal intermittent delta activity (n = 1) [72]. NCSE was diagnosed in one patient with focal seizures.

15.10.3 Lithium

Lithium appears to have both cortical and subcortical effects [73]. Acute and chronic intoxication can occur even with serum levels within the "therapeutic range" [74]. EEG changes in lithium intoxication typically include diffuse slowing with rhythmic trains at 2–3 Hz [75]. In another patient, mild excess background slow activity was observed in the bilateral centroparietal regions on the third day, with triphasic waves emerging during recovery when serum lithium levels were reduced to a third of the original level on the eighth day [76]. Marked bilateral symmetrical slow-wave activity was present on the 14th day, persisting until the 23rd day after consciousness disturbances and choreoathetosis resolved [77]. Case reports have documented bifrontal delta activity up to 2 Hz [78], bifrontal generalized periodic discharges of slightly more than 1 Hz [79], and EEG findings suggestive of electrographic status epilepticus [80]. Lorazepam did not improve mental status or EEG abnormalities, but both improved upon discontinuation of lithium. Lithium intoxication was associated with a background 3–4 Hz delta rhythm and diffuses sharp discharges [81]. NCSE was diagnosed based on prompt EEG and clinical response following IV ASM. Re-exposure to lithium at the previous dosage resulted in a recurrence of symptoms after 1 week (Fig. 15.9).

15.10.4 Baclofen

Several case reports and series document both convulsive (bilateral tonic-clonic) and non-convulsive status epilepticus (NCSE) associated with baclofen treatment, typically due to accidental overdose [82–85]. The EEG findings in these cases often reveal continuous multifocal pseudo-periodic sharp waves and other patterns consistent with NCSE [86].

15.10.5 Snake Bites

EEG changes were observed in 26 patients with snake bites [87]. The predefined stages of EEG alterations included reduced alpha activity and increased theta/beta activity or sharp waves (grade 1), sharp waves or spikes and slow waves (grade 2), and diffuse delta activity (grade 3). Grade 1 was noted in 62% of cases, grade 2 in 31%, and grade 3 in 4%. These changes appeared within hours and persisted for several days.

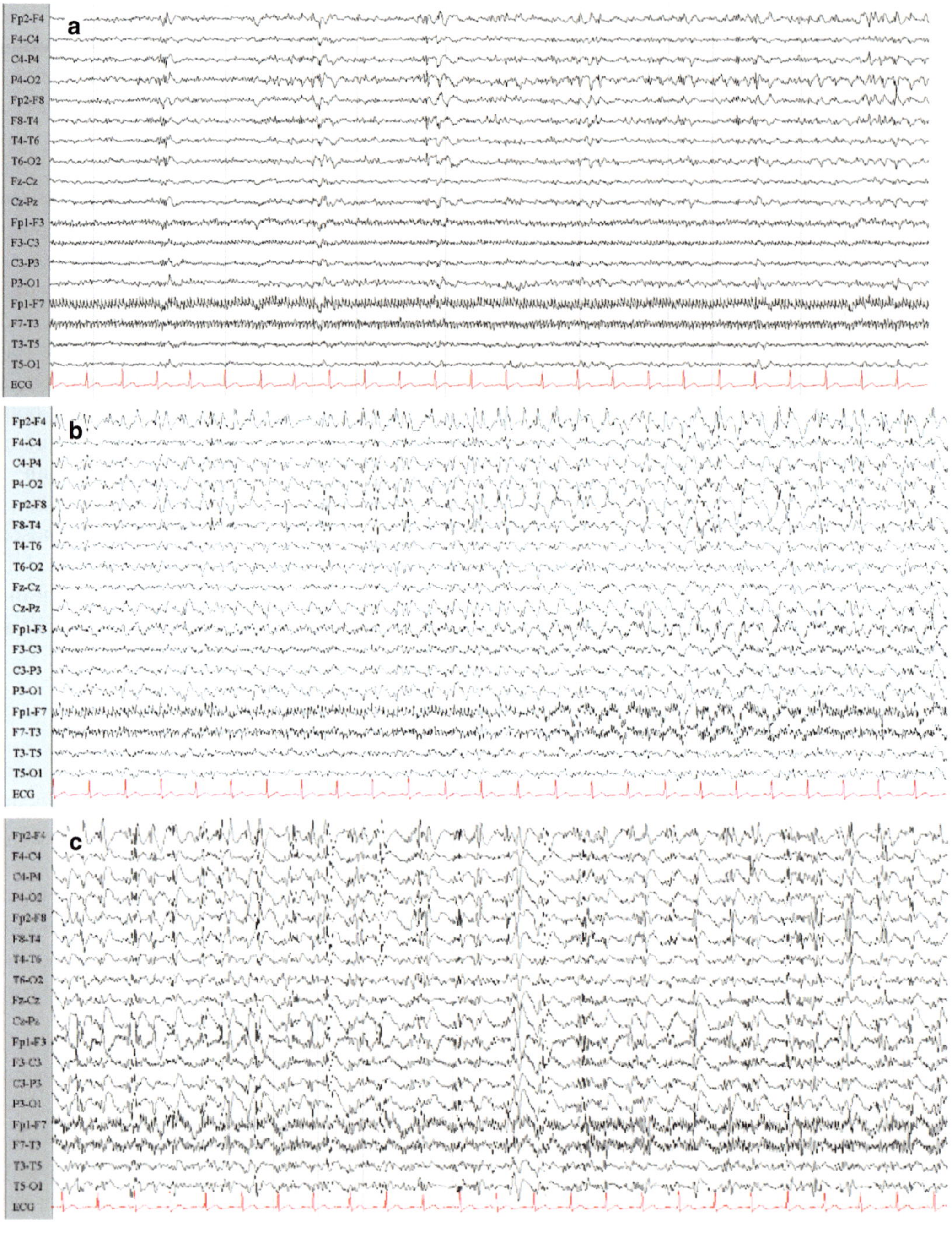

Fig. 15.9 Acute encephalopathy associated with lithium intoxication (> 3 mEq/l). A 35-year-old woman suffering from drug-resistant bipolar depression. In the last 2 days, progressive ideomotor slowing, mutism, and stuporous state. The EEG tracing shows a definite non-convulsive status epilepticus. It should be noted that the pattern shown in (**c**) characterized by generalized IEDs with frequency > 2.5 Hz is preceded at onset by focal activity (right frontotemporal) with evolution in frequency and location (**a** and **b**). The picture showed electrical and clinical resolution after administration of intravenous diazepam bolus. Filters: hi-pass 0.3 Hz; low-pass: 30 Hz. Amplitude: 10 uV/mm. Each page shows 20 s

Notably, one patient also experienced acute renal failure, and another had hyponatraemia. All snake species were identified.

15.10.6 Thyrotoxicosis

Among 3382 patients with hyperthyroidism, seven (0.2%) experienced seizures [88]. Four patients had primary generalized tonic-clonic seizures, two had focal impaired awareness seizures evolving into bilateral tonic-clonic seizures, and one experienced a focal seizure. Initial EEG findings varied: generalized slow activity was observed in four patients, two had normal EEGs, and one had diffuse generalized beta activity. Following treatment that achieved a euthyroid state, all EEGs normalized, and seizures ceased within 18–24 months. In a case report of another patient with thyrotoxicosis, no alternative aetiology was found after thorough investigation [89]. The EEG did not reveal an epileptogenic focus. A literature review identified only 13 published cases from 1956 to 1992 [89]. A patient with both focal and generalized seizures due to thyrotoxic encephalopathy presented with triphasic waves on EEG [90]. Seizures, encephalopathy, and EEG abnormalities resolved after successful treatment of hyperthyroidism. Triphasic waves associated with thyrotoxicosis were noted in several cases in studies on hyperthyroid patients [91, 92], sometimes accompanied by psychotic symptoms [93]. However, relapses of hyperthyroidism were associated with worsening EEG findings [94]. Withdrawal of anti-thyroid medication led to bilateral tonic-clonic seizures [95]. Juvenile myoclonic epilepsy was diagnosed in two patients with Graves' disease [96]. Although ASMs did not improve the epilepsy, seizures completely remitted after effective treatment of hyperthyroidism, even after discontinuation of ASMs.

A 68-year-old patient experienced de novo convulsive status epilepticus that initially presented with bilateral tonic-clonic seizures and later evolved into a non-convulsive confusional state [97]. The EEG showed a continuous irregular polyspike wave pattern. Status epilepticus and EEG abnormalities did not improve with ASMs but resolved after initiation of thyreostatic treatment. In a study of 26 hyperthyroid patients, 81% had EEG changes before treatment, and 68% still had abnormal EEGs 10 years after successful treatment [98]. Notably, seven out of 11 neuropsychological tests indicated significant intellectual impairment, suggesting irreversible brain dysfunction in this population [98].

15.11 Combination of Different Aetiologies

In clinical practice, multiple potential aetiologies often coexist, complicating the determination of the primary cause, particularly in cases with metabolic conditions (Fig. 15.8). Examples include concomitant acute renal and hepatic failure [99], cefepime use in patients with renal failure [100–102], ciprofloxacin administration in a patient with thyrotoxicosis [103], and intrathecal baclofen in a patient with acute renal failure secondary to vancomycin use [104]. Comorbidities also play a role, such as aztreonam use combined with acute kidney injury in a patient with preexisting dementia [105].

15.12 Limitations

Currently, EEG analysis is primarily conducted visually and qualitatively, particularly to identify patterns indicative of status epilepticus or the ictal-interictal continuum. Even with cEEG, condensed signals must be verified using standard EEG displays to avoid misinterpretation of artefacts. Clinicians often rely on the morphological appearance of EEG patterns to infer potential underlying pathomechanisms. However, there is a significant need for further evidence to support these sometimes arbitrary interpretations. This existing knowledge gap should be considered when discussing guidelines for non-convulsive status epilepticus (NCSE) and the ictal-interictal continuum (IIC).

15.13 Future Aspects

Continuous EEG is expected to become more precise and more frequently used in daily ICU practice, compared to the less than 1% utilization in recent years [106]. Experimental advancements include the use of endovascular catheters to record EEG from within blood vessels, either arterial or venous [107]. This innovation could be particularly beneficial in emergency situations, such as with super-refractory status epilepticus, allowing for targeted administration of antiseizure medications while minimizing side effects and reducing the need for mechanical ventilation and ICU stay.

15.14 Conclusion

It is reasonable to assume that any acute disruption in cerebral homoeostasis affects cerebral electrical activity, which is captured by EEG across approximately one-third of the brain's surface. The clinical utility of EEG is especially valuable in conditions where cerebral imaging and laboratory tests do not provide sufficient information for effective patient management. Therefore, the insights gained from EEG should lead to therapeutic actions in acute care settings. The critical question is whether brain damage or disturbances in cerebral function evoke an additional burden of specific hypersynchronous (epileptiform) activity, as this activity is treatable with ASMs. Additionally, non-epileptiform activity may also be clinically significant, although treatment in such cases would focus on addressing the underlying cause.

Future developments in EEG include broader use of cEEG in the ICU, with more refined patient selection to enhance diagnostic value. Ideally, the harmfulness of each EEG pattern will be better defined. Advanced algorithms and artificial intelligence may assist in real-time integration of all data from invasive and noninvasive monitoring at the bedside. Moreover, endovascular EEG recording could become a viable option for patients with super-refractory status epilepticus, helping to optimize ASM administration and minimize side effects.

References

1. Vespa PM, McArthur DL, Xu Y, Eliseo M, Etchepare M, Dinov I, Alger J, Glenn TP, Hovda D. Nonconvulsive seizures after traumatic brain injury are associated with hippocampal atrophy. Neurology. 2010;75(9):792–8. https://doi.org/10.1212/WNL.0b013e3181f07334. PMID: 20805525; PMCID: PMC2938965
2. Bonduelle T, Ollivier M, Trin K, Thomas B, Daubigney A, Michel V, De Montaudouin M, Marchal C, Aupy J. Association of Peri-ictal MRI abnormalities with mortality, Antiseizure medication refractoriness, and morbidity in status epilepticus. Neurology. 2023;100(9):e943–53. https://doi.org/10.1212/WNL.0000000000201599. Epub 2022 Nov 28. PMID: 36443013; PMCID: PMC9990431
3. Bosque Varela P, Machegger L, Oellerer A, Steinbacher J, McCoy M, Pfaff J, Trinka E, Kuchukhidze G. Imaging of status epilepticus: making the invisible visible. A prospective study on 206 patients. Epilepsy Behav. 2023;141:109130. https://doi.org/10.1016/j.yebeh.2023.109130. Epub 2023 Feb 18
4. Bosque Varela P, Tabaee Damavandi P, Machegger L, Prüwasser T, Zimmermann G, Oellerer A, Steinbacher J, McCoy M, Pfaff J, Trinka E, Kuchukhidze G. Magnetic resonance imaging fingerprints of status epilepticus: a case-control study. Epilepsia. 2024;65(6):1620–30. https://doi.org/10.1111/epi.17949. Epub 2024 Mar 20
5. Witsch J, Frey HP, Schmidt JM, Velazquez A, Falo CM, Reznik M, Roh D, Agarwal S, Park S, Connolly ES, Claassen J. Electroencephalographic periodic discharges and frequency-dependent brain tissue hypoxia in acute brain injury. JAMA Neurol. 2017;74(3):301–9. https://doi.org/10.1001/jamaneurol.2016.5325. PMID: 28097330; PMCID: PMC5548418
6. Nascimento FA, Jing J, Beniczky S, Benbadis SR, Gavvala JR, Yacubian EMT, Wiebe S, Rampp S, van Putten MJAM, Tripathi M, Cook MJ, Kaplan PW, Tatum WO, Trinka E, Cole AJ, Westover MB. One EEG, one read – a manifesto towards reducing interrater variability among experts. Clin Neurophysiol. 2022;133:68–70. https://doi.org/10.1016/j.clinph.2021.10.007. Epub 2021 Nov 5. PMID: 34814017; PMCID: PMC8926459
7. Beniczky S, Hirsch LJ, Kaplan PW, Pressler R, Bauer G, Aurlien H, Brøgger JC, Trinka E. Unified EEG terminology and criteria for nonconvulsive status epilepticus. Epilepsia. 2013;54(Suppl 6):28–9. https://doi.org/10.1111/epi.12270.

8. Leitinger M, Beniczky S, Rohracher A, Gardella E, Kalss G, Qerama E, Höfler J, Hess Lindberg-Larsen A, Kuchukhidze G, Dobesberger J, Langthaler PB, Trinka E. Salzburg consensus criteria for non-convulsive status epilepticus–approach to clinical application. Epilepsy Behav. 2015;49:158–63. https://doi.org/10.1016/j.yebeh.2015.05.007. Epub 2015 Jun 17
9. Leitinger M, Trinka E, Gardella E, Rohracher A, Kalss G, Qerama E, Höfler J, Hess A, Zimmermann G, Kuchukhidze G, Dobesberger J, Langthaler PB, Beniczky S. Diagnostic accuracy of the Salzburg EEG criteria for non-convulsive status epilepticus: a retrospective study. Lancet Neurol. 2016;15(10):1054–62. https://doi.org/10.1016/S1474-4422(16)30137-5. Epub 2016 Aug 8
10. Hirsch LJ, LaRoche SM, Gaspard N, Gerard E, Svoronos A, Herman ST, Mani R, Arif H, Jette N, Minazad Y, Kerrigan JF, Vespa P, Hantus S, Claassen J, Young GB, So E, Kaplan PW, Nuwer MR, Fountain NB, Drislane FW. American clinical neurophysiology society's standardized critical care EEG terminology: 2012 version. J Clin Neurophysiol. 2013;30(1):1–27. https://doi.org/10.1097/WNP.0b013e3182784729.
11. Hirsch LJ, Fong MWK, Leitinger M, LaRoche SM, Beniczky S, Abend NS, Lee JW, Wusthoff CJ, Hahn CD, Westover MB, Gerard EE, Herman ST, Haider HA, Osman G, Rodriguez-Ruiz A, Maciel CB, Gilmore EJ, Fernandez A, Rosenthal ES, Claassen J, Husain AM, Yoo JY, So EL, Kaplan PW, Nuwer MR, van Putten M, Sutter R, Drislane FW, Trinka E, Gaspard N. American clinical neurophysiology Society's standardized critical care EEG terminology: 2021 version. J Clin Neurophysiol. 2021;38(1):1–29. https://doi.org/10.1097/WNP.0000000000000806. PMID: 33475321; PMCID: PMC8135051
12. Hubsch C, Baumann C, Hingray C, Gospodaru N, Vignal JP, Vespignani H, Maillard L. Clinical classification of psychogenic non-epileptic seizures based on video-EEG analysis and automatic clustering. J Neurol Neurosurg Psychiatry. 2011;82(9):955–60. https://doi.org/10.1136/jnnp.2010.235424. Epub 2011 May 10
13. Kapur J, Elm J, Chamberlain JM, Barsan W, Cloyd J, Lowenstein D, Shinnar S, Conwit R, Meinzer C, Cock H, Fountain N, Connor JT, Silbergleit R, NETT and PECARN Investigators. Randomized trial of three anticonvulsant medications for status epilepticus. N Engl J Med. 2019;381(22):2103–13. https://doi.org/10.1056/NEJMoa1905795. PMID: 31774955; PMCID: PMC7098487
14. Lehn A, Watson E, Ryan EG, Jones M, Cheah V, Dionisio S. Psychogenic nonepileptic seizures treated as epileptic seizures in the emergency department. Epilepsia. 2021;62(10):2416–25. https://doi.org/10.1111/epi.17038. Epub 2021 Aug 15
15. Kapoor WN. Evaluation and outcome of patients with syncope. Medicine (Baltimore). 1990;69(3):160–75. https://doi.org/10.1097/00005792-199005000-00004.
16. Barolin GS, Scherzer E, Schnobert G. Die zerebrovaskulär bedingten Anfälle: unter besonderer Berücksichtigung der Anfälle im höheren Lebensalter. In: Aktuelle Probleme in der Psychiatrie, Neurologie, Neurochirurgie, Bd. 12. Bern: Verlag Hans Huber; 1975. p. 58–89.
17. Brenner RP. Electroencephalography in syncope. J Clin Neurophysiol. 1997;14(3):197–209. https://doi.org/10.1097/00004691-199705000-00004.
18. Young GB, Jordan KG, Doig GS. An assessment of nonconvulsive seizures in the intensive care unit using continuous EEG monitoring: an investigation of variables associated with mortality. Neurology. 1996;47(1):83–9. https://doi.org/10.1212/wnl.47.1.83.
19. Chong DJ, Hirsch LJ. Which EEG patterns warrant treatment in the critically ill? Reviewing the evidence for treatment of periodic epileptiform discharges and related patterns. J Clin Neurophysiol. 2005;22(2):79–91. https://doi.org/10.1097/01.wnp.0000158699.78529.af.
20. Kaplan PW. EEG criteria for nonconvulsive status epilepticus. Epilepsia. 2007;48(Suppl 8):39–41. https://doi.org/10.1111/j.1528-1167.2007.01345.x. Erratum in: Epilepsia 2007;48(12):2383
21. Trinka E, Leitinger M. Which EEG patterns in coma are nonconvulsive status epilepticus? Epilepsy Behav. 2015;49:203–22. https://doi.org/10.1016/j.yebeh.2015.05.005. Epub 2015 Jul 4
22. De Marchis GM, Pugin D, Meyers E, Velasquez A, Suwatcharangkoon S, Park S, Falo MC, Agarwal S, Mayer S, Schmidt JM, Connolly ES, Claassen J. Seizure burden in subarachnoid hemorrhage associated with functional and cognitive outcome. Neurology. 2016;86(3):253–60. https://doi.org/10.1212/WNL.0000000000002281. Epub 2015 Dec 23. PMID: 26701381; PMCID: PMC4733156
23. Trinka E, Leitinger M. Management of Status Epilepticus, refractory status epilepticus, and super-refractory status epilepticus. Continuum (Minneap Minn). 2022;28(2):559–602. https://doi.org/10.1212/CON.0000000000001103.
24. Leitinger M, Gaspard N, Hirsch LJ, Beniczky S, Kaplan PW, Husari K, Trinka E. Diagnosing nonconvulsive status epilepticus: defining electroencephalographic and clinical response to diagnostic intravenous antiseizure medication trials. Epilepsia. 2023;64(9):2351–60. https://doi.org/10.1111/epi.17694. Epub 2023 Jul 10
25. Guo Y, Fang S, Wang J, Wang C, Zhao J, Gai Y. Continuous EEG detection of DCI and seizures following aSAH: a systematic review. Br J Neurosurg. 2020;34(5):543–8. https://doi.org/10.1080/02688697.2019.1630547. Epub 2019 Jun 17
26. Rumpl E. Elektro-neurologische Korrelationen in den frühen Phasen des posttraumatischen Komas. I. Das EEG in den verschiedenen Phasen des akuten traumatischen sekundären Mittelhirn- und

Bulbärhirnsyndroms [Electro-neurological correlations in early stages of posttraumatic comatose states. I. The EEG at different stages of acute traumatic secondary midbrain and bulbar brain syndrome (author's transl)]. EEG EMG Z Elektroenzephalogr Elektromyogr Verwandte Geb. 1979;10(3):148–57. Germanica
27. Nuwer MR, Hovda DA, Schrader LM, Vespa PM. Routine and quantitative EEG in mild traumatic brain injury. Clin Neurophysiol. 2005;116(9):2001–25. https://doi.org/10.1016/j.clinph.2005.05.008.
28. Dow RS, Ulett G, Raaf J. Electroencephalographic studies immediately following head injury. Am J Psychiatry. 1944;101:174–83.
29. Larsson LE, Melin KA, Nordstrom-Ohrberg G, Silfverskold BP, Ohrberg K. Acute head injuries in boxers; clinical and electroencephalographic studies. Acta Psychiatr Neurol Scand Suppl. 1954;95:2–42.
30. Koufen H, Dichgans J. Häufigkeit und Ablauf von traumatischen EEG-Veränderungen und ihre klinischen Korrelationen: Systematische Verlaufsuntersuchungen bei 344 Erwachsenen [Frequency and course of posttraumatic EEG-abnormalities and their correlations with clinical symptoms: a systematic follow up study in 344 adults (author's transl)]. Fortschr Neurol Psychiatr Grenzgeb. 1978;46(4):165–77. Germanica
31. Scherzer E. Das flache EEG als bioelektrischer Ausdruck der Erwartungsspannung (psychogene Alphareduktion) [The low voltage EEG as bioelectric expression of expectancy tension (psychogenic alpha reduction)]. Psychiatr Neurol (Basel). 1966;152(3):207–12. Germanica, 152, 207.
32. Kant R, Smith-Seemiller L, Isaac G, Duffy J. Tc-HMPAO SPECT in persistent post-concussion syndrome after mild head injury: comparison with MRI/CT. Brain Inj. 1997;11(2):115–24. https://doi.org/10.1080/026990597123700.
33. Jordan KG. Emergency EEG and continuous EEG monitoring in acute ischemic stroke. J Clin Neurophysiol. 2004;21(5):341–52.
34. Gastaut H, Fischer-Williams M. Electroencephalographic study of syncope; its differentiation from epilepsy. Lancet. 1957;273(7004):1018–25. https://doi.org/10.1016/s0140-6736(57)92147-5.
35. Cohn R, Raines GN, et al. Cerebral vascular lesions; electroencephalographic and neuropathologic correlations. Arch Neurol Psychiatr. 1948;60(2):165–81. https://doi.org/10.1001/archneurpsyc.1948.02310020061005.
36. Macdonell RA, Donnan GA, Bladin PF, Berkovic SF, Wriedt CH. The electroencephalogram and acute ischemic stroke. Distinguishing cortical from lacunar infarction. Arch Neurol. 1988;45(5):520–4. https://doi.org/10.1001/archneur.1988.00520290048013.
37. Alberto P, Elisabetta F, Paola R, Uberto R, Alfredo B. The EEG in lacunar strokes. Stroke. 1984;15(3):579–80.
38. Bickford RG, Butt HR. Hepatic coma: the electroencephalographic pattern. J Clin Invest. 1955;34(6):790–9. https://doi.org/10.1172/JCI103134. PMID: 14381508; PMCID: PMC1072609
39. Chajek T, Berry EM, Friedman G, Abramsky O. Treatment of acute hepatic encephalopathy with L-dopa. Postgrad Med J. 1977;53(619):262–5. https://doi.org/10.1136/pgmj.53.619.262. PMID: 876934; PMCID: PMC2496557
40. Das A, Cherian A, Dash GK, Radhakrishnan A. Alpha coma evolving into spindle coma in a case of acute fulminant hepatic failure: what does it signify? Ann Indian Acad Neurol. 2010;13(1):69–71. https://doi.org/10.4103/0972-2327.61285. PMID: 20436753; PMCID: PMC2859594
41. Nabi E, Bajaj JS. Useful tests for hepatic encephalopathy in clinical practice. Curr Gastroenterol Rep. 2014;16(1):362. https://doi.org/10.1007/s11894-013-0362-0. PMID: 24357348; PMCID: PMC3918211
42. Shiotani S, Shimada M, Soejima Y, Yoshizumi T, Uemoto S, Kiuchi T, Tanaka K, Maehara Y. S100 beta protein: the preoperative new clinical indicator of brain damage in patients with fulminant hepatic failure. Transplant Proc. 2004;36(9):2713–6. https://doi.org/10.1016/j.transproceed.2004.09.030.
43. Hunter GR, Young GB. Recovery of awareness after hyperacute hepatic encephalopathy with "flat" EEG, severe brain edema and deep coma. Neurocrit Care. 2010;13(2):247–51. https://doi.org/10.1007/s12028-010-9391-9.
44. Amodio P, Montagnese S. Clinical neurophysiology of hepatic encephalopathy. J Clin Exp Hepatol. 2015;5(Suppl 1):S60–8. https://doi.org/10.1016/j.jceh.2014.06.007. Epub 2014 Aug 3. PMID: 26041960; PMCID: PMC4442865
45. Sutter R, Stevens RD, Kaplan PW. Significance of triphasic waves in patients with acute encephalopathy: a nine-year cohort study. Clin Neurophysiol. 2013;124(10):1952–8. https://doi.org/10.1016/j.clinph.2013.03.031. Epub 2013 May 15
46. Jacob JC, Gloor P, Elwan OH, Dossetor JB, Paters VR. Eelectroencephalographic changes in chronic renal failure. Neurology. 1965;15:419–29. https://doi.org/10.1212/wnl.15.5.419.
47. Kang KS, Heo ST. A case of life-threatening acute kidney injury with toxic encephalopathy caused by Dioscorea quinqueloba. Yonsei Med J. 2015;56(1):304–6. https://doi.org/10.3349/ymj.2015.56.1.304. PMID: 25510780; PMCID: PMC4276772
48. Kye MS, Lee JJ, Kim BK, Kwon O, Park JM, Kang K, Lee WW. Aphasic status epilepticus associated with uremia. J Epilepsy Res. 2017;7(2):115–7. https://doi.org/10.14581/jer.17018. PMID: 29344470; PMCID: PMC5767488
49. Chow KM, Wang AY, Hui AC, Wong TY, Szeto CC, Li PK. Nonconvulsive status epilepticus in peritoneal dialysis patients. Am J Kidney

Dis. 2001;38(2):400–5. https://doi.org/10.1053/ajkd.2001.26105.
50. Nardone R, Brigo F, Trinka E. Acute symptomatic seizures caused by electrolyte disturbances. J Clin Neurol. 2016;12(1):21–33. https://doi.org/10.3988/jcn.2016.12.1.21. PMID: 26754778; PMCID: PMC4712283
51. Zwang HJ, Cohn D. Electroencephalographic changes in acute water intoxication. Clin Electroencephalogr. 1981;12(1):35–40. https://doi.org/10.1177/155005948101200107.
52. Kaplan PW. The EEG in metabolic encephalopathy and coma. J Clin Neurophysiol. 2004;21(5):307–18.
53. Crawford JD, Dodge P. Complications of fluid therapy in neurologic disease: water intoxication and hypertonic dehydration. Pediatr Clin N Am. 1964;11:1029–52. https://doi.org/10.1016/s0031-3955(16)31636-4.
54. Itoh N, Matsui N, Matsui S. Periodic lateralized epileptiform discharges in EEG during recovery from hyponatremia: a case report. Clin Electroencephalogr. 1994;25(4):164–9. https://doi.org/10.1177/155005949402500411.
55. Primavera A, Fonti A, Giberti L, Cocito L. Recurrent absence status epilepticus and hyponatremia in a patient with polydipsia. Biol Psychiatry. 1995;38(3):189–91. https://doi.org/10.1016/0006-3223(95)00145-7.
56. Dixit S, Kurle P, Buyan-Dent L, Sheth RD. Status epilepticus associated with cefepime. Neurology. 2000;54(11):2153–5. https://doi.org/10.1212/wnl.54.11.2153.
57. Jallon P, Fankhauser L, Du Pasquier R, Coeytaux A, Picard F, Hefft S, Assal F. Severe but reversible encephalopathy associated with cefepime. Neurophysiol Clin. 2000;30(6):383–6. https://doi.org/10.1016/s0987-7053(00)00234-3.
58. Martínez-Rodríguez JE, Barriga FJ, Santamaria J, Iranzo A, Pareja JA, Revilla M, dela Rosa CR. Nonconvulsive status epilepticus associated with cephalosporins in patients with renal failure. Am J Med. 2001;111(2):115–9. https://doi.org/10.1016/s0002-9343(01)00767-7.
59. Maganti R, Jolin D, Rishi D, Biswas A. Nonconvulsive status epilepticus due to cefepime in a patient with normal renal function. Epilepsy Behav. 2006;8(1):312–4. https://doi.org/10.1016/j.yebeh.2005.09.010. Epub 2005 Nov 8
60. Misra UK, Kalita J, Chandra S, Nair PP. Association of antibiotics with status epilepticus. Neurol Sci. 2013;34(3):327–31. https://doi.org/10.1007/s10072-012-1001-5. Epub 2012 Mar 7
61. Payne LE, Gagnon DJ, Riker RR, Seder DB, Glisic EK, Morris JG, Fraser GL. Cefepime-induced neurotoxicity: a systematic review. Crit Care. 2017;21(1):276. https://doi.org/10.1186/s13054-017-1856-1. PMID: 29137682; PMCID: PMC5686900
62. Husari KS, Ritzl EK, Kaplan PW. Acute toxicity and Triphasic waves-the example of Cefepime. J Clin Neurophysiol. 2022;39(5):419–26. https://doi.org/10.1097/WNP.0000000000000791. Epub 2020 Oct 20
63. Bandettini di Poggio M, Anfosso S, Audenino D, Primavera A. Clarithromycin-induced neurotoxicity in adults. J Clin Neurosci. 2011;18(3):313–8. https://doi.org/10.1016/j.jocn.2010.08.014.
64. Tchapyjnikov D, Luedke MW. Cefepime-induced encephalopathy and nonconvulsive status epilepticus: dispelling an artificial dichotomy. Neurohospitalist. 2019;9(2):100–4. https://doi.org/10.1177/1941874418803225. Epub 2018 Oct 15. PMID: 30915188; PMCID: PMC6429673
65. Wengs WJ, Talwar D, Bernard J. Ifosfamide-induced nonconvulsive status epilepticus. Arch Neurol. 1993;50(10):1104–5. https://doi.org/10.1001/archneur.1993.00540100089024.
66. Simonian NA, Gilliam FG, Chiappa KH. Ifosfamide causes a diazepam-sensitive encephalopathy. Neurology. 1993;43(12):2700–2. https://doi.org/10.1212/wnl.43.12.2700.
67. Primavera A, Audenino D, Cocito L. Ifosfamide encephalopathy and nonconvulsive status epilepticus. Can J Neurol Sci. 2002;29(2):180–3.
68. Raj AB, Bertolone SJ, Jaffe N. Methylene blue reversal of ifosfamide-related encephalopathy. J Pediatr Hematol Oncol. 2004;26(2):116. https://doi.org/10.1097/00043426-200402000-00010.
69. Dufour C, Grill J, Sabouraud P, Behar C, Munzer M, Motte J, Oberlin O, Paci A, Hartmann O. Encéphalopathie induite par l'ifosfamide: 15 observations [Ifosfamide induced encephalopathy: 15 observations]. Arch Pediatr. 2006;13(2):140–5. https://doi.org/10.1016/j.arcped.2005.10.021. Epub 2005 Dec 20, 13, 140. French
70. Taupin D, Racela R, Friedman D. Ifosfamide chemotherapy and nonconvulsive status epilepticus: case report and review of the literature. Clin EEG Neurosci. 2014;45(3):222–5. https://doi.org/10.1177/1550059413500777.
71. Fountain NB, Waldman WA. Effects of benzodiazepines on triphasic waves: implications for nonconvulsive status epilepticus. J Clin Neurophysiol. 2001;18(4):345–52. https://doi.org/10.1097/00004691-200107000-00006.
72. Gusdon AM, Malani R, Chen X. Clinical and EEG characteristics of Ifosfamide-related encephalopathy. J Clin Neurophysiol. 2019;36(2):150–4. https://doi.org/10.1097/WNP.0000000000000539. PMID: 30694943; PMCID: PMC7856922
73. Mayfield D, Brown RG. The clinical laboratory and electroencephalographic effects of lithium. J Psychiatr Res. 1966;4(3):207–19. https://doi.org/10.1016/0022-3956(66)90008-2.
74. Sheean GL. Lithium neurotoxicity. Clin Exp Neurol. 1991;28:112–27.
75. Bartha L, Marksteiner J, Bauer G, Benke T. Persistent cognitive deficits associated with lithium intoxication: a neuropsychological case description.

Cortex. 2002;38(5):743–52. https://doi.org/10.1016/s0010-9452(08)70041-8.
76. Shibasaki Warabi Y, Idezuka J, Yamazaki M, Onishi Y. Triphasic waves detected during recovery from lithium intoxication. Intern Med. 2003;42(9):908–9. https://doi.org/10.2169/internalmedicine.42.908.
77. Wada K, Sasaki T, Yoshimura Y, Erabi H. Reversible choreoathetosis associated with lithium intoxication. Seishin Shinkeigaku Zasshi. 2003;105(9):1206–12. Japanese
78. Habermeyer B, Hess M, Kozomara-Hocke P, Mager R, Kawohl W. Schwere lithiumintoxikationen bei normalen serumspiegeln [Lithium intoxications at normal serum levels]. Psychiatr Prax. 2008;35(4):198–200. https://doi.org/10.1055/s-2007-986226. Germanica Epub 2007 Nov 6, 35, 198
79. Suda M, Kubota F, Aihara Y, Hiraoka T, Aoyama Y, Hattori S, Fukuda M, Mikuni M. A case of lithium intoxication with periodic sharp waves. Pharmacopsychiatry. 2009;42(3):122–3. https://doi.org/10.1055/s-0028-1112131. Epub 2009 May 18
80. Gansaeuer M, Alsaadi TM. Lithium intoxication mimicking clinical and electrographic features of status epilepticus: a case report and review of the literature. Clin Electroencephalogr. 2003;34(1):28–31. https://doi.org/10.1177/155005940303400110.
81. Bellesi M, Passamonti L, Silvestrini M, Bartolini M, Provinciali L. Non-convulsive status epilepticus during lithium treatment at therapeutic doses. Neurol Sci. 2006;26(6):444–6. https://doi.org/10.1007/s10072-006-0530-1.
82. Saltuari L, Marosi MJ, Kofler M, Bauer G. Status epilepticus complicating intrathecal baclofen overdose. Lancet. 1992;339(8789):373–4. https://doi.org/10.1016/0140-6736(92)91697-7.
83. Zak R, Solomon G, Petito F, Labar D. Baclofen-induced generalized nonconvulsive status epilepticus. Ann Neurol. 1994;36(1):113–4. https://doi.org/10.1002/ana.410360122.
84. Delhaas EM, Brouwers JR. Intrathecal baclofen overdose: report of 7 events in 5 patients and review of the literature. Int J Clin Pharmacol Ther Toxicol. 1991;29(7):274–80.
85. Berger B, Vienenkoetter B, Korporal M, Rocco A, Meinck HM, Steiner T. Accidental intoxication with 60 mg intrathecal baclofen: survived. Neurocrit Care. 2012;16(3):428–32. https://doi.org/10.1007/s12028-011-9669-6.
86. Boutte C, Vercueil L, Durand M, Vincent F, Alvarez JC. Apport de l' EEG dans le diagnostic d'une intoxication au baclofène [EEG contribution to the diagnosis of baclofen overdose]. Neurophysiol Clin. 2006;36(2):85–9. https://doi.org/10.1016/j.neucli.2006.03.004. French. Epub 2006 Apr 7, 36, 85
87. Ramachandran S, Ganaikabahu B, Pushparajan K, Wijesekera J. Electroencephalographic abnormalities in patients with snake bites. Am J Trop Med Hyg. 1995;52(1):25–8. https://doi.org/10.4269/ajtmh.1995.52.25.
88. Song TJ, Kim SJ, Kim GS, Choi YC, Kim WJ. The prevalence of thyrotoxicosis-related seizures. Thyroid. 2010;20(9):955–8. https://doi.org/10.1089/thy.2009.0276.
89. Lin CS, Yiin KT, Lin WH, Huang SY. Thyrotoxicosis accompanied with periodic seizure attacks a case report and review of literature. Zhonghua Yi Xue Za Zhi (Taipei). 1992;50(4):335–7.
90. Primavera A, Brusa G, Novello P. Thyrotoxic encephalopathy and recurrent seizures. Eur Neurol. 1990;30(4):186–8. https://doi.org/10.1159/000117342.
91. Van Zandycke M, Orban LC, Vander Eecken HV. Ondes lentes triphasiques dans deux cas d'encéphalopathie thyréotoxique [Occurrence of triphasic waves in two cases of thyrotoxic crisis (author's transl)]. Acta Neurol Belg. 1977;77(2):115–20. French
92. Scherokman BJ. Triphasic delta waves in a patient with acute hyperthyroidism. Arch Neurol. 1980;37(11):731. https://doi.org/10.1001/archneur.1980.00500600079018.
93. Hirano M, Endo M, Kubo T. Triphasic delta waves in a case of hyperthyroidism with psychotic symptoms. Clin Electroencephalogr. 1982;13(2):97–102. https://doi.org/10.1177/155005948201300205.
94. Leubuscher HJ, Herrmann F, Hambsch K, Langpeter D, Mueller P, Sorger D. EEG changes in untreated hyperthyroidism and under the conditions of thyreostatic treatment. Exp Clin Endocrinol. 1988;92(1):85–90. https://doi.org/10.1055/s-0029-1210785.
95. Korczyn AD, Bechar M. Convulsive fits in thyrotoxicosis. Epilepsia. 1976;17(1):33–4. https://doi.org/10.1111/j.1528-1157.1976.tb03378.x.
96. Su YH, Izumi T, Kitsu M, Fukuyama Y. Seizure threshold in juvenile myoclonic epilepsy with graves disease. Epilepsia. 1993;34(3):488–92. https://doi.org/10.1111/j.1528-1157.1993.tb02589.x.
97. Kahaly G, Tettenborn B, Mihaljevic V, Beyer J, Krämer G. Petit-mal-Status bei Hyperthyreose [Petit mal status in hyperthyroidism]. Dtsch Med Wochenschr. 1989;114(42):1607–11. https://doi.org/10.1055/s-2008-1066803. German. PMID: 2806091
98. Perrild H, Hansen JM, Arnung K, Olsen PZ, Danielsen U. Intellectual impairment after hyperthyroidism. Acta Endocrinol. 1986;112(2):185–91. https://doi.org/10.1530/acta.0.1120185.
99. Baerg RD, Kimberg DV. Centrilobular hepatic necrosis and acute renal failure in "solvent sniffers". Ann Intern Med. 1970;73(5):713–20. https://doi.org/10.7326/0003-4819-73-5-713.
100. Alpay H, Altun O, Biyikli NK. Cefepime-induced non-convulsive status epilepticus in a peritoneal dialysis patient. Pediatr Nephrol. 2004;19(4):445–7. https://doi.org/10.1007/s00467-003-1333-8. Epub 2004 Feb 21
101. Garces EO, Andrade de Anzambuja MF, da Silva D, Bragatti JA, Jacoby T, Saldanha Thomé F. Renal

failure is a risk factor for cefepime-induced encephalopathy. J Nephrol. 2008;21(4):526–34.
102. Nguyen DD, Lai S. Prolonged Cefepime-induced neurotoxicity in a patient with end-stage renal disease. Am J Case Rep. 2022;23:e934083. https://doi.org/10.12659/AJCR.934083. PMID: 35067669; PMCID: PMC8800466
103. Agbaht K, Bitik B, Piskinpasa S, Bayraktar M, Topeli A. Ciprofloxacin-associated seizures in a patient with underlying thyrotoxicosis: case report and literature review. Int J Clin Pharmacol Ther. 2009;47(5):303–10. https://doi.org/10.5414/cpp47303.
104. Gold J, Zhao K, Abraham M, Behmer Hansen R, Lad M, Mammis A. Encephalopathy of unknown origin in a baclofen patient: case report and review of the literature. World Neurosurg. 2020;136:136–9. https://doi.org/10.1016/j.wneu.2020.01.044. Epub 2020 Jan 16
105. Billnitzer A, Kaplan PW. Generalized periodic discharges with Triphasic morphology in the setting of Aztreonam neurotoxicity. Clin EEG Neurosci. 2021;52(1):66–8. https://doi.org/10.1177/1550059420917842. Epub 2020 May 13
106. Hill CE, Blank LJ, Thibault D, Davis KA, Dahodwala N, Litt B, Willis AW. Continuous EEG is associated with favorable hospitalization outcomes for critically ill patients. Neurology. 2019;92(1):e9–e18. https://doi.org/10.1212/WNL.0000000000006689. Epub 2018 Nov 30. PMID: 30504428; PMCID: PMC6336162
107. Fujimoto A, Matsumaru Y, Masuda Y, Sato K, Hatano K, Numoto S, Hotta R, Marushima A, Hosoo H, Araki K, Okanishi T, Ishikawa E. Endovascular electroencephalography (eEEG) can detect the laterality of epileptogenic foci as accurately as subdural electrodes. Heliyon. 2024;10(3):e25567. https://doi.org/10.1016/j.heliyon.2024.e25567. PMID: 38327423; PMCID: PMC10847992

Multimodal Neurophysiological Tests as a Prognostic Indicator for Coma

16

Antonello Grippo, Giovanni Salvestrini, and Maenia Scarpino

16.1 Introduction

Following major improvements in prehospital management, comatose patients with severe acquired brain injury (SABI) are increasingly being admitted to intensive care units (ICUs). One of the primary challenges encountered in ICUs is the evaluation of the long-term neurological prognosis of patients who have experienced coma after SABI, such as cardiac arrest (CA) and traumatic brain injury (TBI). The ability to *predict* the outcome of a comatose or poorly responsive patient represents a significant challenge for ICU teams. This is due to the necessity of providing families with accurate information and assisting in the selection of the most appropriate therapeutic option. To facilitate the decision-making process, clinicians require reliable criteria, particularly during the acute phase of coma. It is increasingly common for these patients not to be allowed to continue investigations or treatments with unreasonable stubbornness if they appear to be useless or disproportionate or to have no effect other than artificially maintaining life. The prognosis of neurological outcomes for patients who have not regained consciousness following sedation clearance is a central responsibility of neurologists, clinical neurophysiologists, and neurointensivists.

This task is often challenging and typically necessitates the utilisation of a range of tools, including neurological examinations, neurophysiological tests, neuroimaging, and biomarkers. It is recommended that results be evaluated collectively, as reliance on any single test may lead to erroneous conclusions regarding the likelihood of unfavourable outcomes.

A forecast of a poor outcome typically results in the withdrawal of life-sustaining therapy (WLST) and death. Finally, the availability of reliable unfavourable prognostic indices may make it possible to identify non-heart beating donors. Therefore, the prediction precision, typically reported as specificity, should be as precise as possible. However, in the patients still sedated, the reliability of a clinical examination is inherently compromised, thereby rendering instrumental tests the sole available option. In this context, electrophysiological recordings, i.e. electroencephalogram (EEG) and evoked potentials (EPs), are important, as they constitute an extension of the clinical examination and provide information that is supplementary to that yielded by neuroimaging.

Neurophysiological tests are also beneficial when clinical assessment is inadequate for evaluating the degree of consciousness of a patient or their capacity to interact with their surroundings.

A. Grippo (✉) · G. Salvestrini · M. Scarpino
Azienda Ospedaliero Universitaria Careggi, Florence, Italy
e-mail: antonello.grippo@unifi.it; giovanni.salvestrini@unifi.it; scarpinom@aou-careggi.toscana.it

F. Brigo, O. Mecarelli (eds.), *EEG: The First 100 Years*,
https://doi.org/10.1007/978-3-031-86878-8_16

Neurophysiological tests can provide evidence of the integrity of sensory tracts and demonstrate signs of consciousness, prompting caregivers and relatives to develop suitable communication strategies.

In the context of clinical neuroscience care, which may include neurologists, neurosurgeons, neuro-radiologists, or neurointensivists with expertise in neuroprognosis, there is an increasing request for neurophysiologists to perform electrophysiological recordings. In this context, neurophysiologists are confronted with several challenging questions, as outlined by Harvey [1]. These include determining which types of recordings are most pertinent in addressing clinicians' queries, identifying optimal methods for EEG or EP recordings in the ICU, establishing the appropriate timing for conducting these tests in various clinical scenarios, and interpreting the results in a meaningful manner. Additionally, it is crucial to understand the limitations of different tests regarding their prognostic value.

EEG and EPs offer a noninvasive approach for bedside monitoring of neuronal and synaptic activity within the brain. The interpretation of the signal is a challenging aspect, as it requires a specific expertise that is not always readily available. Additionally, there is a risk of artefacts, which can affect the reliability of the results. Furthermore, ensuring optimal timing of the recording is crucial, as is the translation of traditional EEG nomenclature into treatment implications. Finally, both EEG and EPs must be implemented in a multimodal prognostication model.

Neurophysiological tests, including somatosensory evoked potentials (SSEPs) and EEG, are the most employed techniques in the assessment of altered states of consciousness in critically ill patients. When used in conjunction, these tests can provide valuable insights into the diagnosis and prognosis of these patients. Guidelines exist which describe their use to assess the probability of patient recovery. Examples of such guidelines include those produced by the European Resuscitation Council (ERC) and the American Academy of Neurology [2] as well as recommendations produced by the French Society of Clinical Neurophysiology [3].

In the initial section of this chapter, we provide a comprehensive account of the most frequently utilised electrophysiological modalities, namely, EEG and SSEP. This is accompanied by methodological recommendations, an outline of the expected responses and their clinical significance, as well as an examination of the prognostic value and limitations of each test in terms of their performance and interpretation. In the second part, we put forward recommendations regarding the strategy and timing of neurophysiological evaluation in the context of different clinical scenarios.

16.2 Electroencephalogram

EEG is the most commonly available electrophysiological tool in most hospitals. Its ease of repetition at the patient's bedside, noninvasiveness, and ability to provide valuable and safe monitoring of brain activity make it an invaluable tool in healthcare. Nevertheless, the acquisition and interpretation of ICU EEGs necessitate significant human, technological, and financial resources. Furthermore, several significant logistical issues must be addressed to guarantee the accurate performance and interpretation of ICU EEGs. While there are several guidelines pertaining to EEG in ICU settings [4, 5], these provide limited guidance on the logistical challenges faced by many institutions.

In the ICU setting, where patients may require multiple devices for assistance or monitoring and are frequently transported for brain imaging, the development of CT and MRI-compatible electrodes, coupled with advances in quantitative analysis, makes it feasible to utilise EEG at the bedside.

It is of the utmost importance that nurses receive comprehensive training in the placement and surveillance of cEEG, to guarantee optimal signal quality. Conversely, the SSEP requires the availability of expert personnel, which has resulted in this technique being currently only employed in expert centres.

16.2.1 Recording

In intensive care units, disk and cup scalp electrodes are the electrodes of choice for EEG recording. In certain patient populations, the presence of craniotomies or intracranial pressure monitoring devices may necessitate the use of subdermal needle electrodes [6, 7]. In centres where continuous EEG is available, tests that require patient transport (e.g. neuroimaging) should ideally be completed before the patient is connected to the EEG. Specialised electrodes made of conductive plastics and nonferrous metals that are compatible with both CT and MR imaging are available [7, 8], but unfortunately, many centres may lack access to them. This necessitates patients being repeatedly disconnected and reconnected to the EEG to facilitate testing, which is a significant drain on resources, costly, and time-consuming for technologists and physicians. Even when specialised electrodes are used, there is a significant risk of these becoming detached from the patient during transport. It is of paramount importance to conduct daily inspections for the occurrence of skin breakdown and scalp infection. Electrode impedance and artefacts require lead maintenance and should be assessed regularly.

Once the patient's scalp has been adequately prepared, it is recommended that, where possible, 19 scalp electrodes, in accordance with the international 10–20 system, be used in conjunction with two additional electrocardiogram (ECG) electrodes, at least for the initial recording. The most common type of EEG used after SABI is the full-montage routine EEG, which typically lasts 20–30 min. This is a widely available procedure during office hours at most hospitals.

The use of full-montage, 21-channel for continuous EEG (c-EEG) is a resource-consuming and labour-intensive process. When cEEG is performed, a close collaboration between intensivists and neurologists or clinical electrophysiologists is essential. Consequently, strategies to simplify cEEG recording and interpretation have been explored. A reduction in the number of electrodes will result in a decrease in the burden of montage. In the event of data acquisition difficulties and/or for further recordings, the number of electrodes can be reduced to a minimum of nine scalp electrodes (including Cz) plus ECG recording. In the case of long-term c-EEG monitoring, it is recommended that the number of electrodes be kept to a minimum. The use of ten or six electrodes in place of a full montage did not result in any change to the accuracy of outcome prediction or the assessment of background patterns. Even if this approach may be less sensitive for seizure detection [9–11], the use of a limited-channel cEEG monitoring (four or six frontal electrodes) could detect the most common EEG patterns associated with poor and good outcomes [12, 13]. However, it should be noted that eye movement artefacts over bifrontal electrodes could induce false positive detection of periodic discharges [12].

It is of paramount importance that ICU staff are in place to guarantee the quality of prolonged registration. While attempting to diagnose seizures, ICU personnel may be tasked with administering medications for benzodiazepine or antiseizure trials. It is not uncommon for myoclonus or other patient movements that may have necessitated an EEG to result in the generation of profound muscle and movement artefacts. In such instances, the administration of neuromuscular blocking agents may be required (when deemed safe and appropriate) to suppress these artefacts and allow for the visualisation of the EEG signal underneath. It is the responsibility of ICU personnel to be familiar with the process of EEG annotation. This is because the annotation of relevant information, such as doses and titration of sedation/neuromuscular blocking agents, administration of ASM, and clinical signs that may not be evident in the video recording, is beneficial for the interpretation of the EEG. Bedside and central monitoring stations can facilitate the identification of changes that require urgent assessment by personnel who have received the requisite training. For this purpose, telemedicine can help [14].

16.2.2 Interpretation Consideration

It is widely acknowledged that EEG interpretation necessitates a greater degree of training than the assessment of alternative prognostic indicators, such as clinical examination and SSEPs. The impact of the wording of EEG reports on clinical decision-making regarding seizure or myoclonus management and neuroprognosis should be well considered. The interpretation of EEG results and their incorporation into clinical decision making should ideally be guided by individuals with content expertise. The potential for non-expert EEG interpretation to compromise its overall prognostic value is a significant concern.

To address these issues, several EEG patterns and grading systems have been put forth as potential predictors, particularly in comatose patients who have experienced CA or TBI. Most EEG grading systems that have been developed over the past few decades have employed a variety of features, as evidenced by the work of Hockaday [15], Hughes [16], Synek [17], Young [18], and Rae-Grant [19] (Table 16.1).

All the classification systems that were proposed considered background voltage to be a relevant prognostic feature. However, different terms were employed to describe this feature, including low output and low voltage. Additionally, there was a paucity of data regarding the provision of voltage values as cut-offs for the different EEG patterns. Consequently, the same EEG could be classified in a disparate manner according to the scale employed. Furthermore, there were inconsistencies across studies about the grading systems and definitions of malignant and benign EEG patterns. Consequently, in literature reviews or meta-analyses, it was not feasible to assess the prognostic value of EEG pictures due to the utilisation of disparate grading systems by different authors.

This is why, in the first guidelines pertaining to the prognosis of comatose survivors following CA [20], EEG was not included among the prognostic indicators, given that its prognostic accu-

Table 16.1 EEG features considered for the various classifications of EEG in comatose patients after Cardiac Arrest or Traumatic Brain Injury

Author	Hockaday et al. [15]	Hughes et al. [16] 1976	Synek [17] 1990	Young et al. [18]	Rae Grant et al. [19]	Hirsch et al. [21]
Coma etiology	**HIE**	**TBI-HIE**	**TBI-HIE**	**TBI-HIE**	**TBI**	
EEG feature						
Symmetry	No	No	No	No	Yes	Yes
Background activity (frequency)	Yes	Yes	Yes	Yes	Yes	Yes
AP gradient	Yes	No	No	No	Yes	Yes
Variability	Yes	No	No	No	Yes	Yes
Reactivity	Yes	Yes	Yes	Yes	Yes	Yes
Voltage	Yes	Yes	Low amplitude, low output, isoelectric	Suppression	Low voltage pattern, Electrocerebral inactivity	Normal Low Suppressed
Special patterns	No	No	Spindles, alpha/theta coma	Spindles, alpha/theta coma	Spindles, theta coma	Stage II sleep transients
Continuity	Yes	No	Yes	Burst-suppression	ELAE, burst suppression	Continuous, nearly continuous, discontinuous, burst-suppression

racy was deemed insufficient based on the available evidence. It would therefore be preferable to have standardised terminology that evaluates all important features to assess which individual features of the EEG are the most predictive. A classification scheme based on clear, consistent, and standardised definitions of patterns should be employed to facilitate comparison across different studies. The variability in classification and rating of EEG findings has been partially addressed by the introduction of standardised critical care EEG terminology by the ACNS, which has demonstrated high inter-rater agreement, particularly for periodic or rhythmic activity [21]. This terminology was updated in 2021 [22].

In delineating the parameters of background continuity with respect to the percentage of EEG where attenuation or suppression is observed, the ACNS furnished a comprehensive account of the descriptive characteristics of background activity. The more important features are continuity and voltage. Fig. 16.1 shows examples of EEG traces classified according to ACNS.

Concerning *voltage*, the ACNS defines normal background (> 20uV), low voltage (<20uV but >10 uV), and suppression as a background voltage<10 μV for>99% of the record [22].

Concerning *continuity*, an EEG is defined as *nearly continuous* if attenuation or suppression is present for less than 10% of the recording and as discontinuous if attenuation or suppression is present for more than 10% but less than 50% of the recording and burst suppression as a suppressed (<10 μV) pattern present for 50–99% of the record.

Burst Suppression (BS) consists of alternating periods of suppression and normal voltage activity, with 50–99% of the recording consisting of suppression [22]. In the last version of ACNS classification [22], BS patterns have been classified into two categories: those with synchronous bursts (which are often epileptiform and/or identical) (BSIB) and those with heterogenous bursts. Bursts are considered identical if the first 500 ms of the bursts are identical, irrespective of the subsequent amplitude or duration of the bursts [23] (Fig. 16.2).

EEG Reactivity (EEG-r) is defined as the occurrence of transient reproducible changes in the amplitude or frequency of the EEG background in response to external stimuli [22]. A variety of stimuli must be employed, including auditory stimuli (such as shouting or clapping close to the left and right ears) and painful stimulation of the face and limbs. The use of nipple pinching as a method of painful stimulation is no longer recommended in favour of nail bed pressure. The utilisation of standardised stimuli ensures optimal reliability. In the event of the occurrence of stereotypic movements (spinal reflex or posture) that have the potential to contaminate the EEG, verification of reactivity to nociceptive stimuli can be conducted following the administration of a muscle relaxant agent. The type of stimulus employed allows for the testing of multiple afferent pathways [24], so examination of EEG-r requires integrity of peripheral to central structures [25]. To enhance the sensitivity of EEG-r as a prognostic index, it is essential to evaluate the response to all the types of stimulation. One of the problems for prognostic use of EEG-r is the different protocols of stimulation that have been used [26, 27]. Admiraal et al. [26] put forth a novel definition of EEG-r, wherein any of five stimuli (sternal rub, clapping, calling the patient's name, nasal tickle, and eye opening) induces a change in EEG at least twice out of the three applications (Fig. 16.3). Muscle activities, eye blinks, and stimulus-induced rhythmic or periodic discharges (SIRPIDS) are usually not qualified as EEG-r [28, 29].

Rhythmic and Periodic Patterns (RPPs) are defined as repetitions of a waveform of relatively uniform morphology [22]. Depending on the interval between the waveforms, these can be classified as rhythmic (no interval) or periodic (intervals between waveforms).

Fig. 16.1 Examples of EEG patterns according to ACNS classification: benign patterns: (**a**) continuous, normal voltage; (**b**) nearly-continuous; (**c**) discontinuous background; (**d**) continuous low voltage background; (e) Generalized Periodic Discharges; (**f**) burst suppression; (**g**) suppressed backgound; (**h**) electrocerebral inactivity. Sweep time: 1 s/div; sensitivity: 7 μV/mm for all except for panel (**h**), where sensitivity is 2 μV/mm

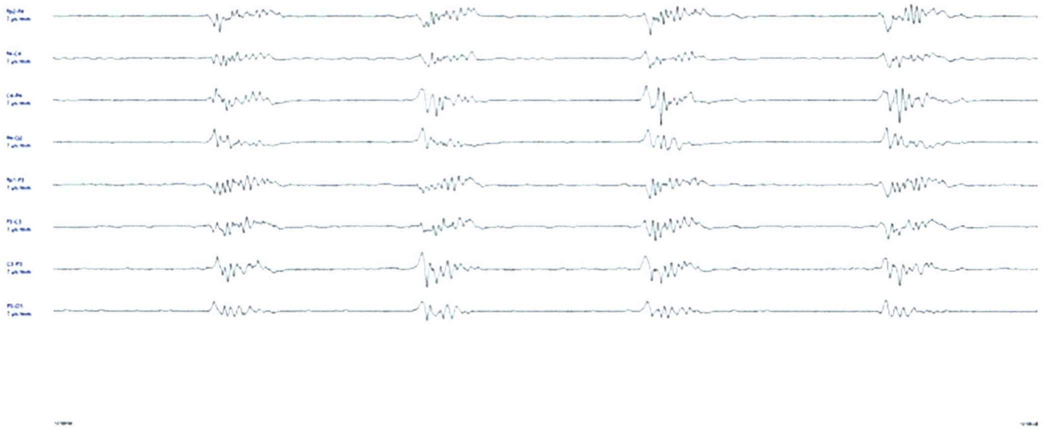

Fig. 16.2 Example of identical and synchronous bursts

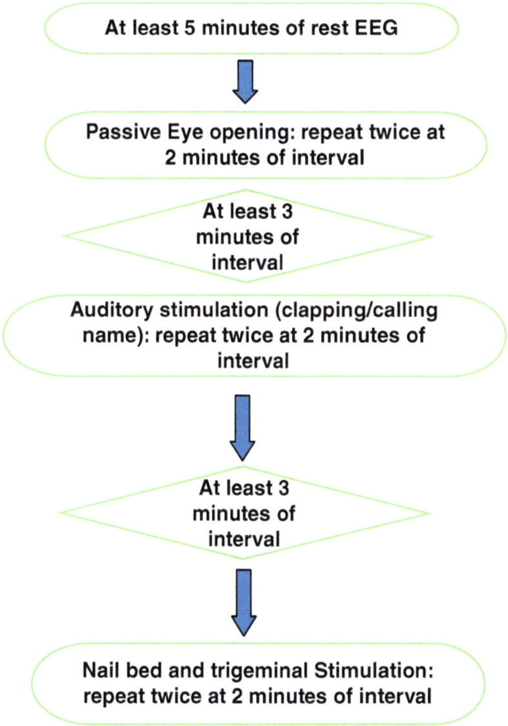

Fig. 16.3 Recommended stimulation protocol for assessing EEG reactivity in comatose patients

16.2.3 Potential Confounding Factors for EEG Evaluation" is the main heading for two paragraphs: Temperature and Sedation

Over the past decade, *therapeutic hypothermia (TH)* with a target temperature of 32–34 °C has become the standard of care in the management of patients following CA and has also been proposed TBI. The levels of hypothermia remain below the current recommended thresholds for TTM management, as outlined by Sandroni [30]. Furthermore, temperature management at 36 °C does not result in a notable alteration of the EEG [3]. Moderate therapeutic hypothermia (TH) at a temperature of 32 °C does not exert a considerable impact on the continuity or voltage of EEGs. Only at temperatures below 29 °C do we observe the progressive appearance of periodic complexes, burst suppression (at approximately 24 °C), and electrocerebral silence (at approximately 22 °C) [31, 32]. It is the metabolic effects of body temperature that are of greater consequence. A reduction in body temperature has been observed to result in a 7–22% decrease in the activity of hepatic enzymes for each degree below 37 °C. This can lead to a significant increase in anaesthetic levels and a prolongation of the effect of neuromuscular blockade. Furthermore, it can be challenging to ascertain whether residual sedation affects clinical and EEG reactivity, even when recordings are conducted 24 h after normothermia has been restored. Additionally, factors present in patients undergoing therapeutic hypothermia (TH) may influence the interpretation of EEG results. For instance, artefacts resulting from muscle shivering, mechanical ventilation, or electrical equipment commonly associated with TH may potentially impact the reliability of EEG interpretation. Nevertheless, it is evident that a considerable number of EEG patterns observed during moderate TH have comparable prognostic implications to those observed during normothermia [33].

16.2.4 Sedation

In the aftermath of an acute brain injury, sedation may be employed for several purposes, including synchronising the patient with artificial ventilation, controlling seizures, or treating intracranial hypertension. To achieve one or more of these objectives, the doses of sedative agents are adjusted, as are the required plasma concentrations. The effects of ongoing sedative infusions, the lingering effects of drugs or medications taken in excess, and the influence of profound metabolic disturbances can all potentially impact the EEG activity. In the context of excluding seizures, the presence of confounders may result in false negatives, with the potential for the emergence of electrographic or non-convulsive seizures as these confounders dissipate.

16.2.5 EEG Duration

The optimal duration of EEG recording remains a topic of contention, with differing opinions regarding the appropriate length of time for EEG monitoring, depending on the clinical indication. A minimum of 20 min of EEG recording is typically required for a standard examination. However, recent data suggest that continuous EEG (cEEG) monitoring may offer benefits [5]. Continuous EEG monitoring allows for the evaluation of the evolution of the EEG over time, which may facilitate the early detection of epileptiform features. cEEG monitoring is becoming more prevalent in the management of severe postanoxic encephalopathy, as it enables more accurate assessment of EEG evolution following return of spontaneous circulation (ROSC) and enhances the sensitivity for detecting potentially treatable seizures in hypoxic-ischaemic brain injury (HIBI) compared to a late routine EEG [34]. Nevertheless, there is no evidence to suggest that cEEG improves the outcome of HIBI [35].

Two large prospective studies have demonstrated that cEEG does not confer any advantage over intermittent EEG about neurological outcome or time to death. It is possible that this may not apply to the subgroup of patients with ‘true’ electrographic status epilepticus, who may respond to therapy and benefit from early detection. Nevertheless, international guidelines recommend the use of cEEG for seizure detection, treatment monitoring, and the prognostic assessment of HIBI when available [3, 5]. However, cEEG monitoring is resource intensive and not available in all centres [36, 37].

The generation of a substantial corpus of data from continuous EEG recordings necessitates the regular interpretation of this data by neurophysiologists to guarantee the timely delivery of treatments. The raw EEG signal can be subjected to computational analysis through the application of various mathematical techniques, resulting in the generation of quantitative EEG (qEEG). Several commercially available software packages can produce a variety of graphs, or trends, which display different EEG features (such as frequency, voltage/amplitude, rhythmicity, the presence of discharges, and so forth) on the Y and Z axes and time on the X axis. The most employed trends include colour and/or density spectral array (CSA, CDSA, DSA), rhythmicity spectrograms, relative asymmetry spectrograms, amplitude-integrated EEG, suppression ratios, alpha/delta ratio, and relative alpha variability. The use of qEEG allows for the screening of extended periods of EEG, the detection of gradual changes over time, and the identification of clinically relevant events that necessitate a review of the raw EEG data.

16.2.6 Resource Limitations and Bridging the Gap

The potential constraints on resources may impede the timely recording and interpretation of EEGs, which could subsequently delay the necessary escalation or de-escalation of care. A variety of technologies are currently available, including peel-and-stick electrodes, electrode caps and bands, abbreviated montages, and user-friendly EEG machines [38]. The evidence indicates that training bedside personnel to utilise these technologies to perform EEGs (as opposed to EEG technologists) results in a statistically significant reduction in the time delay to EEG, with a mean reduction of 86–667 min. There is no difference in the quality of the EEG. Most importantly, there are changes in the diagnostic work-up and treatment of patients, occurring 49% and 40–42% of the time respectively. In institutions where EEG technologists may not be available on a 24-h basis, these devices may facilitate the timely performance of EEGs by bedside ICU care providers. As an alternative, in institutions where access to formal EEG is limited, these devices may assist in identifying patients who require transfer to other centres with the capacity to facilitate EEG monitoring [38]. In most centres, continuous EEG tracings are only reviewed remotely by neurophysiologists a few times per day.

16.3 Somatosensory Evoked Potentials (SSEPs)

Evoked potentials are defined as neural responses that occur in a time-locked manner to the administration of a standardised stimulus. EPs differ from EEG signals in that they are stimulus-induced and represent the summed activities of large populations of neurons firing in synchrony. This requires the average of numerous stimulations. EP components are named according to their polarity (N for negative and P for positive) and their latency from the stimulation (in milliseconds). While several different types of evoked potential exist, the median nerve somatosensory evoked potential (SSEP) is the most frequently utilised in ICU to assess patients with sABI. This is because stimulation of the median nerve (a mixed sensory and motor nerve) provides the most robust responses.

16.3.1 Stimulation and Recording Parameters

It is recommended that the median nerve at the wrist be stimulated using a bipolar electrical stimulator, with the cathode positioned in proximity to the nerve. The site of stimulation may be modified according to the accessibility of the wrist. In cases where the wrist is inaccessible due to the presence of intravenous infusion sites, or chalk, stimulation can be performed at the elbow. The stimulus is set at the motor threshold value, capable of eliciting a distal twitch, or, if the patient is undergoing treatment with a muscle relaxant medication. It is recommended that at least two blocks be averaged to facilitate the reproducibility of SSEP responses [39].

In an ICU setting, *subdermal needle electrodes* are the optimal choice for recording SSEPs to improve the quality of the signal. To record the main components of median nerve SSEPs, it is recommended that an EP machine with at least four channels is used [40].

16.3.2 When Should SSEPs Be Recorded?

SSEPs can be recorded as early as 24 h after the onset of coma in the context of CA. Nevertheless, experts advise that an interval of 48 h should be observed, and anaesthetic drug treatments should be discontinued for a minimum of 6 h prior to the performance of SSEP recordings [2]. A small number of patients with postanoxic coma have been observed to exhibit a complete absence of N20 responses at an early stage, with a subsequent reappearance a few hours later. This phenomenon highlights the potential for false negative results in recordings conducted prematurely, particularly in the context of therapeutic hypothermia [41, 42]. In the context of TBI, the prognostic value of SSEPs is entirely distinct, as initially abolished N20 responses lack 100% specificity for the risk of non-awakening. In this context, N20 responses may reappear up to 10 days after the trauma. In cases where initially absent cortical SSEPs are observed in the absence of brainstem compromise, serial recordings are recommended to increase the confidence with which a poor prognosis can be confirmed.

A recent study evaluated the reliability of SSEPs recorded within 24 h of ICU admission to predict brain death in comatose patients with acquired brain injury of various aetiologies. The primary outcome of this study was that severe SSEP alteration (bilateral absence of SSEPs or unilateral absence associated with a pathological cortical response in the other hemisphere) in early recordings (within 72 h following the onset of coma) was able to predict an evolution towards brain death in 78% of comatose patients suffering from acute brain injury of different aetiologies [43].

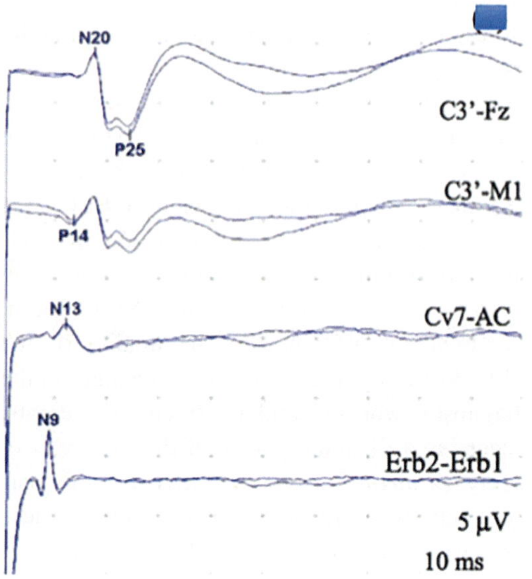

Fig. 16.4 Example of a normal Somatosensory Evoked Potential obtained after Median Nerve Stimulation. Typical Components are shown from the top (Cortical) to the bottom (peripheral). Abbreviations for peak labels are described in the text. C3' and Fz are located according 10-20 international System of EEG. M1 is extracephalic electrode located at the mastoid bone. CV7 is located over the most prominent spinous process at the base of the neck when the neck is flexed. AC is located on the skin surface of the supra-glottal region on the midline. ERB's electrodes are located within the angle formed by the posterior border of the clavicular head of the sternocleidomastoid muscle and the clavicle, 2± 3 cm above the clavicle

16.3.3 Response Analysis and Interpretation

Median SSEPs assess the functional state of the somatosensory pathways at different levels: the proximal peripheral portion of the median nerve, the dorsal (posterior) columns of the spinal cord, and the lemniscal somatosensory afferent pathways via the brainstem, the diencephalon, the thalamocortical tracts, and the primary sensory cortex (S1) [3, 44, 45]. The distinct responses are delineated by their polarity (denoted as N for negative and P for positive) and by their mean latency in normal subjects (see Fig. 16.4). The principal SSEP components are as follows:

The *N9* is a proximal peripheral response recorded at the supraclavicular Erb's point. It provides information on the conduction time concerning the median nerve and the brachial plexus.

The *N13* component corresponds to the segmental postsynaptic activity of the dorsal horn of the cervical spinal cord. The *P14* component arises from the cervicomedullary junction and the lower part of the brainstem and mainly reflects the activity of bulbo-thalamic neurons.

The *N18* potential has its origin in subcortical (thalamic) sources [46].

The *N20* represents the earliest neural activities in the parietal primary somatosensory cortex contralateral to the stimulation. The N20 component is typically followed by a positivity that exhibits variability in latency and is referred to by several names due to its inherent variability: P22, P24, P25, or P27. When a lengthy recording time window is employed, certain subsequent SSEP cortical components, such as N70, may be recorded; however, this does not offer significant clinical utility.

16.3.4 SEP Classification

SSEP appears to be more straightforward to interpret than the EEG. However, it is essential that those responsible for interpreting the data are adequately trained. Unfortunately, there is currently no consensus in the literature regarding the establishment of a standardised grading scale for SSEP abnormalities in ICU patients. Adherence to the following diagnostic criteria serves to minimise interobserver variability and false positives [47].

It is recommended that the upper normal limits be defined based on the mean plus 2.5 SD (or plus 3 SD) of latency (or interpeak latency) values obtained in a group of healthy controls in each laboratory. Therefore, SSEP latencies (or interpeak latencies) may be classified in a binary manner as either normal or delayed. Nevertheless, SSEP components may also be deemed absent (abolished) if their amplitude is <0.2 uV and indistinguishable from background noise.

SSEP waveforms can be grouped into several patterns based on their presence and amplitude in both hemispheres of the main and early cortical complex (N20/P25). SEPs are classified as normal (N), pathological (P), or absent (A). A normal SEP has a normal N20/P25 amplitude. A pathological SEP has an abnormally prolonged central conduction time (CCT) and/or a N20/P25 amplitude less than 1.2 mV or a left-right amplitude asymmetry greater than 50%. An absent SEP has no cortical responses with preserved cervical N13 [48]. Examples of these SEP patterns are reported in Fig. 16.5.

A reduction in SSEP amplitude may be a more accurate predictor of a more severe neurological outcome than higher SSEP amplitudes. However, there is currently no evidence to suggest that a relevant cut-off value exists.

The recording of SSEPs and their subsequent interpretation in the ICU can be influenced by several factors. The first one is the aetiology of coma. The absence of cortical SSEP responses in patients with peripheral neuropathy or traumatic spinal cord injury may be attributed to peripheral nerve lesions or spinal cord dysfunction. However, this should not be confused with the absence of cortical lesions. It is essential to collate any information pertaining to the patient's past medical or neurological history, as this may potentially lead to misinterpretation of the SSEP results. Furthermore, data obtained from brain imaging, such as magnetic resonance imaging (MRI) or computerised tomography (CT), should be considered to ascertain the presence of any lesions that may impair the ascending volley in the posterior column, the thalamocortical radiations, and the primary somatosensory cortex. This is particularly crucial in cases of TBI or suspected stroke.

The primary *non-neurological factors* that may influence SSEP recordings include pharmacological agents and body temperature. The most significant effects on SSEPs are caused by lipophilic agents, such as halogenated gases, propofol, and thiopental. These agents affect neuronal membranes, delay SSEP latencies, and prolong central conduction time [49] but at the infusion rates commonly used in ICUs have been observed to have a minimal impact on SSEP amplitude and

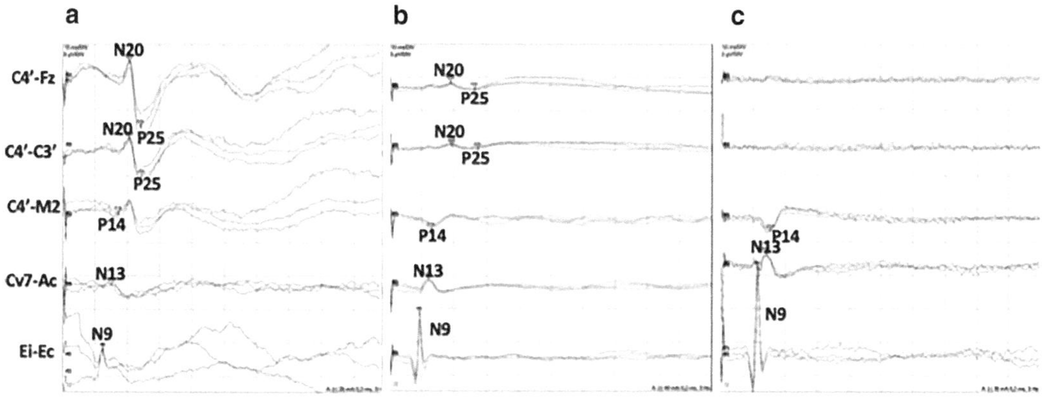

Fig. 16.5 Examples of cortical patterns of SEPs (**a**) Normal; (**b**) Pathological for amplitude (<1.2 μV); (**c**) Absent

latency. Moreover, agents that solely interact with brain receptors, such as opioids and benzodiazepines, exert no influence on subcortical conduction time, SSEP latencies, and amplitudes [49]. Muscle relaxant drugs don't affect cortical SSEPs and can be used to relax patients and prevent muscle contraction artefacts when needed [39].

Hypothermia has a depressant effect on SSEP cortical response amplitudes and prolongs conduction times [41, 49–52], but only severe hypothermia (20–25 °C) will abolish the N20 response [53]. Given the temperature values of therapeutic hypothermia (TH), SSEP accuracy for poor prognosis remains accurate within 24 h of CA [50]. To prevent any confounding effects, it is recommended that SSEPs should not be recorded if the body temperature is below 35 °C.

16.3.5 Traumatic Brain Injury

Two main factors partially limit the usefulness of neurophysiological data in TBI patients: the multiple mechanisms of injury and the use of high doses of anaesthetics to control intracranial pressure (ICP). The primary function of the EEG in the context of TBI is the identification of acute seizures that occur immediately following the injury and especially non-convulsive seizures or status epilepticus (NCSE). Furthermore, the EEG can be useful in determining the prognosis, although this is limited by the timing of the examination, given that all comatose patients are treated at an early stage with high doses of propofol or midazolam to control ICP, which can result in unreliable EEG findings. Consequently, the presence of malignant patterns (e.g. burst suppression) should be interpreted with caution, as they may be related to ongoing sedation rather than cerebral injury. However, given the effects of sedation on EEG activity, a comparison between the dosage of the sedatives and the expected EEG pattern should be considered. Approximately 1 week after the onset of post-traumatic coma, an EEG may be predictive of a favourable outcome in the presence of reactivity [54]. In a study by Logi et al. [55], the Synek classification was used in the post-acute phase to demonstrate that patients who exhibited a benign EEG pattern (dominant theta and alpha activity, delta activity with sleep spindles, and high-amplitude frontally predominant rhythmic delta activity in a background of alpha and theta activity) regained consciousness. The absence of EEG reactivity is an indicator of a poor prognosis in TBI patients when associated with severe SSEP alterations [48].

Regarding SSEPs, their principal advantage is that they are only minimally affected by sedation. The position of electrodes can be affected in patients who have had a craniectomy or bone injury. These factors may change how SSEP is acquired (e.g. by making N20 difficult to record). This can happen in up to 15% of cases, so it is important to check for it [56].

It has been observed that 90% of patients who demonstrate the preservation of the cervical N13 responses and the bilateral abolition of the cortical N20 responses ultimately evolve towards a vegetative state or evolve to brain death [43, 48, 57]. The patients with bilateral SSEP abolition ultimately regained consciousness yet exhibited a poor functional outcome in every instance [48]. The abolition of the N20 waveform may be attributed to focal lesions within the somatosensory pathway, such as axonal lesions [3]. Additionally, instances of late normalisation, occurring several weeks after the initial injury, have been documented [58].

Conversely, the presence of N20 responses is predictive of a favourable outcome (Glasgow Outcome Scale, GOS 4 and 5) or awakening in 86–93% of patients in postanoxic coma and 60 to 78% of patients in post-traumatic coma [57]. Nevertheless, the existing literature does not provide sufficient evidence to make a recommendation in this regard.

In the initial stages following a traumatic brain injury (TBI), patients at risk of developing intracranial hypertension are frequently sedated with the aim of controlling intracranial hypertension or optimising ventilation. Although electroencephalogram (EEG) changes may be influenced by ongoing sedation, the somatosensory evoked potential (SSEP) cortical response can still be observed and may be employed for the early detection of brain injury progression [59, 60].

Indeed, neurophysiology might assess more accurately the actual metabolic mismatch during critical moments of lesion extension. SSEP monitors with ad hoc protocols were adapted to the technical constraint of ICU, to overcome a rare use in clinical practice [61]. The continuous monitoring of evoked potential outside the operating room is currently confined to expert centres. Despite representing a promising technique for patients with rostrocaudal brain herniation (SSEPs), it should be noted that recording and interpretation require extensive training. Moreover, the introduction of such monitoring in routine neuro-ICU settings would require further investigation to demonstrate a significant improvement in outcome for selected cases.

16.3.6 Cardiac Arrest

In patients with out-of-hospital CA who achieve return of spontaneous circulation (ROSC), more than 80% will be comatose an hour later, with approximately half remaining comatose 72 h following ROSC (or 72 h from rewarming when hypothermia is used [62]. Despite improvements in post-cardiac arrest care, approximately two-thirds of patients who are comatose after ROSC remain dependent in their daily activities [63, 64]. In these patients, the prognostication of long-term neurological outcome is of paramount importance to inform clinicians and patients' relatives and ensure the provision of appropriate care [65].

Since the initial recommendations were put forth by Wyidicks, neurophysiological findings have been incorporated into the predictive factors. The utility of neurophysiological techniques, including EEG and SEP, has been the subject of growing investigation, with findings that align with those of numerous case series and can be incorporated into recent national [65] and intersocietal recommendations [2, 66].

16.3.7 EEG

The utilisation of disparate EEG classification systems has historically constrained the field of neurocritical care. However, the advent of the ACNS Critical Care EEG Terminology [22] has constituted a pivotal stride towards the standardisation of EEG assessments and research. In fact, the use of EEG to assist in neuroprognostication has grown considerably, with the EEG being the most frequently utilised tool for assessing the severity of HIBI, with a prevalence of 63% in Europe and 94.4% in the United States [67, 68]. Considering the data published in recent years, EEG has the potential to not only predict a poor prognosis but also to indicate a favourable outcome. Prediction of good outcomes may be important when making decisions about prolonging life support in patients whose outcome appears indeterminate.

16.3.7.1 Description of the EEG Predictors

Background Amplitude

The presence of an *isoelectric* EEG background (amplitude <2 uV), within 72 h of ROSC, is an accurate predictor of a poor neurological prognosis (CPC 3–5), with a high degree of precision (FPR of 0–1.4% and an upper limit of 95% confidence interval of <5%).

A *suppressed* (<10 uV) EEG is indicative of an unfavourable prognosis but depends on the time point after CA. Indeed, a suppressed pattern may be transient in the initial hours following CA, contingent upon timely synaptic recovery, even in patients with a favourable outcome. A small number of patients have been described in the literature with a suppressed pattern for between 24 and 72 h who subsequently recovered GOS 3–5 at 3 months after the arrest [69] or Cerebral Performance Category (CPC) 1–2 at 1 month [70]. Conversely, background suppression identified beyond 72 h from ROSC is indicative of a poor neurological prognosis with an FPR of 0% and high precision.

Background Continuity

A *continuous* EEG with a normal amplitude within 12–24 h after CA is indicative of a high probability of good neurological recovery achieving a sensitivity of 19% and a specificity of 98% [30, 71, 72]. At later time points, the presence of continuous background activity is a favourable sign, with the sensitivity for a good outcome ranging from 63% to 98% within 120 h [73–77]. This is because some EEG background activity appears in almost all patients after 24 h, even in those with severe encephalopathy and a poor outcome. Over time, synaptic activity in patients with severe postanoxic encephalopathy can recover to a certain extent, with measurable EEG rhythms, but without clinically relevant functional recovery of the brain [73, 75]. Furthermore, some false positive cases occur after 24 h due to the potential emergence of an α-theta coma pattern. This is characterised by frontally predominant, monotonous, and unreactive activity in the low α range and is strongly linked to a poor outcome.

The favourable predictive value of *nearly continuous* EEG in the initial 24-h period is comparable to that of continuous EEG [75].

The predictive value of *discontinuous* EEG is negligible. The presence of early background discontinuity during targeted temperature management does not preclude the possibility of a favourable outcome.

Burst Suppression (BS) has been associated with a variable extent of brain damage [78], but BSIB is associated with severe and widespread damage of the cerebral cortex and deeper brain structures such as the hippocampus and thalamus [78].

A single study [79] has demonstrated that the occurrence of BSIB recorded up to 72 h post-arrest is predictive of an unfavourable outcome, with an FPR of 0% (upper limit 95% CI < 5%) and that the presence of heterogenous bursts recorded beyond 24 h post-arrest is predictive of an unfavourable neurological outcome, with an FPR of 0–1.4% and high precision. Other studies that assessed BS without specifying whether synchronous or heterogenous have demonstrated that when present within and beyond 24 h from return of ROSC, they have an FPR of 2% (95% confidence interval (CI), 1–8%) and 1% (95% CI, 0–2%), respectively, for predicting a poor neurological outcome from hospital discharge to 6 months [80, 81].

Two considerations must be considered regarding the prognostic significance of BS. First, BS pattern may be artificially induced by sedation and hypothermia, particularly within the initial 72 h. It is therefore recommended to wait 72 h from ROSC, or rewarming, to identify these patterns, to minimise false positives while maintaining sensitivity. On the other hand, BS is often a transient finding [74] that relies on continuous EEG, which is not universally available [23, 82].

Background Reactivity

The value of EEG reactivity testing for the prediction of outcomes following cardiac arrest remains a topic of contention in the literature [77]. While reactivity testing may enhance the sensitivity for predicting favourable outcomes, it has a restricted capacity for predicting unfavourable outcomes [27, 83].

Reactive EEG is time-dependent. When it is evident within 12 h after CA, it has a high positive predictive value for a favourable outcome, with an FPR of 0 [74, 75]. However, only a small subgroup of patients with continuous or nearly continuous EEG shows a reactive EEG in this time frame, likely due to sedation. The presence of EEG reactivity 12–24 h from ROSC achieved a sensitivity of 75% and a specificity of 65% for the prediction of good long-term outcome [84]. If reactivity is present in later assessment (48–72 h after CA), predictive positive was about 55.6% [83]. Other authors have reported a sensitivity of up to 93% [85].

The absence of EEG-r was identified as a potential indicator of a poor outcome regardless of concomitant sedation [25, 86]. However, alternative findings have also been documented [74, 87]. Therefore, the FPR remains heterogeneous across studies (FPR = 0–50% CI95% (0–70.9)) limiting the use of this criteria as a robust marker of Poor prognosis [27, 83].

There are several factors that affect its prognostic accuracy. First, in the early post-CA period, there is a potential impact of sedation and/or TH. Another limiting factor is the moderate inter-rater agreement [27, 88, 89]. To overcome this limitation, it is recommended to use video-EEG association and standard stimulation to optimise accuracy. Some authors have proposed a quantitative assessment of EEG reactivity [84]. Finally, all studies exploring the value of EEG reactivity for prognostication have evaluated this feature as an 'isolated characteristic' without including additional analysis of the background EEG. Indeed, a reactive EEG may be associated with a continuous background pattern with physiological rhythms, whereas a non-reactive EEG may represent a low-voltage background pattern. Therefore, since the EEG pattern may already contain sufficient information for reliable prognostication, the additional predictive value of reactivity per se is unclear. For all the above, it is recommended that the *absence* of reactivity should not be used alone for poor prognosis, as it is a 'minor criterion' of severe HIBI [79].

Rhythmic and Periodic Patterns (RPPs) Although the definitions of seizures and status epilepticus vary between studies, reports have indicated that RPPs meeting the criteria for seizures or status epilepticus have been observed in up to one-third of comatose patients following CA. The mortality rate has been reported to be between 80% and 100% [90–92].

Of all possible superimposed discharges, RPPs superimposed on a suppressed background have been found to predict a poor neurologic prognosis when detected within 72 h post-arrest with an FPR of <5% and high precision. However, this finding has been inconsistent across studies. Other discharges, including periodic discharges on a non-suppressed background, rhythmic activity, or sporadic epileptiform discharges, yielded inconsistent results. It is unclear whether all RPPs represent a condition that can be treated with antiseizure drugs to improve outcome or merely an epiphenomenon of severe brain damage in which treatment would be futile.

The Treatment of Electroencephalographic Status Epilepticus after Cardiopulmonary Resuscitation (TELSTAR) trial demonstrated that intensive antiseizure treatment for a minimum of 48 h was not associated with an improved outcome in unselected patients with RPPs. Consequently, this treatment is likely to be futile in most comatose patients who have experienced cardiac arrest with RPPs [93]. Nevertheless, exploratory subgroup analyses indicated the potential for benefit in patients with RPPs meeting the criteria for 'true' electrographic status epilepticus. It has been suggested that background continuity, the timing of RPP appearance, and reactivity may serve as predictors of outcome in comatose patients following CA with RPPs [92].

Epileptiform Activity is typically observed after a period of 24 h [79, 87]. However, the presence of epileptiform discharges does not necessarily indicate a poor outcome. For instance, a poor outcome has been observed in TH-treated patients at any time point [23, 72, 94]. This is why many studies do not include epileptiform activity in the prognostic categories of EEG patterns in HIBI coma. The evidence that this pattern is less severe than other malignant patterns is demonstrated by the different percentages of the absence of cortical SSEPs in epileptiform (approximately 20%) and noncontinuous patterns (>65% for burst suppression, suppression, and isoelectric) [87]. Therefore, it is conceivable that the former reflects a different grade of encephalopathy severity [95–97].

Repetitive electrographic epileptiform discharges, including seizures or status epilepticus, occur in 20–30% of patients who survive CA [11] and have been associated with a poor outcome. The presence of discrete electrographic seizures is not a reliable predictor of outcome [86]. It is estimated that at least 10% of these patients may regain consciousness with the appropriate administration of treatment [98]. The identification of *electrographic status epilepticus* can also be subjective, and there is an insufficient body of evidence evaluating the prognostic value of status epilepticus strictly defined using standardised criteria [29, 91]. The EEG background from which status epilepticus evolves, or reverts to following treatment, may be a more important factor in predicting outcome [29, 92, 93]. Similarly, the EEG background may be of greater importance than the presence of superimposed generalised or lateralised periodic discharges, a frequent finding in the first 48–72 h following ROSC [86]. A continuous, reactive EEG background with peri-midline epileptiform discharges [95] after targeted temperature management, and with preserved brainstem reflexes [11, 76, 99] might identify patients with a better prognosis.

Conversely, absent bilateral cortical SSEPs identify patients in whom treatment seems futile [85].

In Fig. 16.6 are shown two patients with epileptiform patterns but a different cortical response of SSEP. A summary of the prognostic value of EEG patterns is given in Table 16.2.

The question of whether continuous cEEG should be preferred to standard EEG (30 min) for prognosis in postanoxic coma remains unresolved [33, 100]. The studies by Sivaraju [94] identified the times at which different EEG patterns acquire significant prognostic value and demonstrated that it is possible to predict both good and poor outcomes at an early stage. However, cEEG is not as widely available as standard EEG, and its cost is higher. The study by Crepeau et al. [33] did not provide evidence that cEEG is a more effective prognostic tool than standard EEG in clinical practice. It has been shown that standard EEG can be as informative as cEEG in predicting outcome, provided that the timing of the recording is considered. Several centres have adopted the practice of cEEG monitoring, starting shortly after ROSC, with the aim of identifying and managing seizures and aiding neuroprognosis. Given that the EEG typically evolves over time, reflecting changes in sedative concentrations [73], it is recommended that assessments be performed on a continuous or repeated basis over the first 48–72 h. As cEEG is not yet widely available, routine EEG, typically lasting less than an hour and repeated once a day for the first 72 h after cardiac arrest, may be an alternative for hospitals with limited resources.

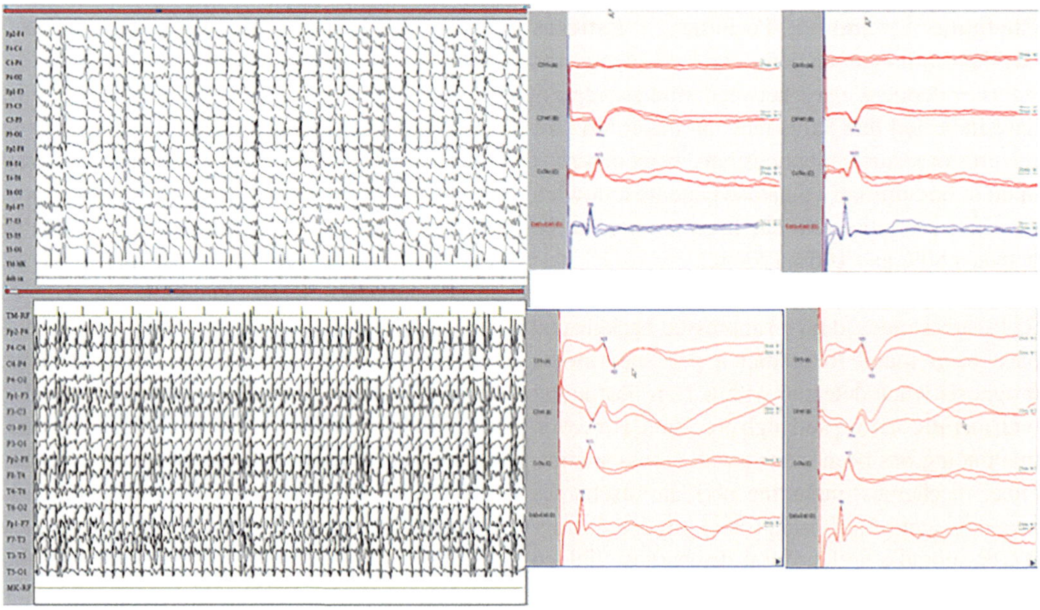

Fig. 16.6 The illustration depicts two patients in a comatose state following a cardiac arrest, both exhibiting an epileptiform EEG pattern. However, one patient displays no cortical SSEP, while the other demonstrates a normal cortical SSEP

Table 16.2 EEG Pattern predicting a poor (CPC 3–5) and a favorable (CPC 1–2) neurological prognosis

Pattern	Timing Postarrest	FPR	CI upper limit	Sensitivity	Outcome	Notes
Burst suppression Syncronous	6–96 h	0%	<5%		CPC 3–5	
Burst suppression non-Syncronous	>24 h	0–1.4%			CPC 3–5	Consider sedation
Isoelectric background	Within 72 h	0–1.4%	<5%		CPC 3–5	
Suppressed background	>72 h	0%		50%	CPC 3–5	No consistency of data <72 h
Rhythmic & Periodic Patterns	0–120 h	0–33%			CPC 3–5	Multiple patterns considered
Generalized periodic discharges superimposed on suppressed background	>24 h	0–3%	34%		CPC 3–5	Inconsistent data across studies
Electrographic seizures/ status epilepticus		0–17%	26%		CPC 3–5	Good prognosis possible if no other markers of poor prognosis present
Reactivity absent		0–50%	70.9%		CPC 3–5	Sedation
Continuous/nearly continuous backgrounds	<24 h	2%	20%	56–90%	CPC 1–2	
Reactivity present	12 h 48–72 h	0–11% 55%		40–91%	CPC 1–2	Fair interrater reliability

16.3.8 Somatosensory Evoked Potentials

The presence of the N20 wave is indicative of the integrity of both the cortex and the thalamus. Conversely, the absence of this wave bilaterally is likely to reflect the severity of HIBI, as evidenced by the fact that all patients presenting with a bilateral absence of the N20 wave also exhibited cortical and thalamic injuries according to the pathophysiology of anoxic-ischaemic brain damage [78]. The absence of cortical SSEPs is a robust predictor of poor outcome with a specificity of nearly 100% with a grade B level of and is thus included among the six prognostic markers of ERC/ESICM algorithm of anoxic-ischaemic encephalopathy [2] and for the assessment of the initiation of withdrawal of life-sustaining treatment (WLST) [101].

It has been reported that neurological recovery has occurred despite a bilaterally absent N20. An analysis of 35 studies involving 3816 patients suggested that specificity may be as low as 93% when WLST triggered by SSEP results is considered [102]. A nonsystematic review of patients who survived despite the absence of cortical SSEPs showed that the specificity approached 100% when technical and other issues affecting the results were considered [47, 103]. The voltage of the evoked potential signal is considerably lower than that of EEG or muscular activity. Therefore, the use of neuromuscular blocking drugs and meticulous avoidance of artefacts are advised when SSEPs are recorded for prognostication [2].

In addition, interobserver variability seems less important for SSEPs than for EEG except in case of high noise level [104, 105].

In contrast to its specificity, the sensitivity of the SSEP in predicting poor outcome is relatively low, rarely exceeding 30%. Recent literature has focused on the absolute amplitude values of the N20 component, encompassing both favourable and unfavourable outcomes.

Histopathological evidence has demonstrated that the N20 SSEP amplitude (N20Amp) is inversely correlated with the severity of HIBI [106]. Shao and colleagues [107] recently demonstrated that the N20-baseline amplitudes were significantly correlated with the degree of thalamic damage, as reflected in the resting-state functional MRI (rs-fMRI) BOLD signal. It is noteworthy that post-mortem histopathology of patients who had experienced postanoxic coma revealed that the absence of SSEP responses was invariably accompanied by thalamic damage [78].

All these findings provided a rationale for the utilisation of N20Amp in predicting outcomes following CA. Combining a low N20Amp with the conventional dichotomous classification (i.e. presence or absence of the N20 wave) yielded an increase of sensitivity for prediction of poor outcome of up to 70% with a high specificity (93–100%) [75, 92, 108–113].

However, the studies employed disparate N20Amp cut-off values (typically within the range of 0.5–1 uV) to dichotomise their findings. Furthermore, the N20Amp was assessed at various time points between 12 and 72 h post-ROSC, and there is evidence that the N20Amp exhibits changes over time in the initial days following arrest [109, 114].

Conversely, patients with a high N20–P25 amplitude (>2.30 to 4 μV) do not have severe brain damage [108, 111] and have favourable outcome with a high specificity (85–96%) but a moderate sensitivity (30–61%) [108–112, 115, 116]. The underlying pathophysiological mechanism of this phenomenon may be elucidated by resting-state functional MRI (rs-fMRI) Bold Signal Findings [107]. Although the intracerebral conduction pathways for SSEP remain intact, abnormal activity in other localised brain regions associated with conscious perception and processing may also result in an unfavourable neurological prognosis.

Standardisation of the method (i.e. SSEP N20-baseline or N20–P25 peak-to-peak measure) has been suggested considering amplitude as a prognostic marker. A recent study highlighted that N20–P25 presented a higher prognostic value (AUC = 0.85) compared to N20-baseline (AUC = 0.70) [110].

It should be noted that a bilateral absence of the N20 wave is unlikely to result from *sedation* [117]; however, the N20 amplitude is reduced by up to 40% under midazolam anaesthesia [118]. At present, no definitive cut-off value can be recommended for the prediction of poor outcome in HIBI using N20Amp.

Thus, N20 amplitude could be assessed as a continuum rather than a categorical variable, the underlying concept being that the amplitude of SSEP components could be inversely related to the severity of neurological injury. However, further validation is required before the implementation of guidelines.

Other researchers have evaluated the potential of intermediate latency components (n70) for favourable long-term functional outcomes.

Nevertheless, the widespread availability of SSEP is suboptimal due to a lack of recording software and trained personnel. As a result of the unavailability of SSEP, some authors investigated the correlation between benign (continuous normal voltage) EEG and the presence of cortical SSEP [119]. They found that all patients with a benign EEG exhibited N20 potentials, suggesting that SSEP may be omitted in these patients to conserve resources. As an alternative, they propose that the recording of SSEPs should be restricted to those cases with malignant or highly malignant EEGs, given that these patterns are associated with both the presence and absence of N20 potentials.

We believe that it is more useful to have both methods in view of both multimodal prognosis and to be able to verify inconsistencies of PES findings in relation to the EEG picture [111].

An element of uncertainty is inherent in neuroprognostication, even with predictors considered ‘reliable’. Therefore, it is critical that a thorough assessment be performed with every patient and that the entirety of the clinical picture be considered. This uncertainty is related in part to the inability of studies to fully account for the self-fulfilling prophecy while evaluating the reliability of a predictor [102]. In addition, the possibility of technical, operator, or interpretation error while evaluating a predictor is always present [47] and the ability to account for confounders, such as sedation, is imperfect. Moreover, discordance in prognostication mostly reflects selective brain damage. Clinical examination after cardiac arrest focuses on the detection of consciousness, myoclonic status, and assessment of brain stem reflexes. Enduring absence of brain stem reflexes is strongly associated with poor outcome. Brain stem reflexes reflect brainstem function, whereas EEG records cortical activity. Therefore, brain stem reflexes can be intact despite a severely disturbed EEG [120].

The utilisation of a combination of assessment modalities, including clinical examination, imaging, and electrophysiological studies, serves to reduce the potential for error arising from reliance on a single modality. Information derived from these modalities should be largely consistent. For instance, the reliability of an absent cortical SEP should be called into question in a comatose cardiac arrest survivor with a withdrawal motor response, reactive EEG, and brain imaging at 72 h. When tests from different areas are combined, the results may be concordant or discordant. The presence of concordant tests enhances specificity and mitigates the likelihood of an erroneous pessimistic prediction. However, these tests may also be perceived as superfluous [121].

The European Resuscitation Council (ERC) and the European Society of Intensive Care Medicine (ESICM) have published guidelines for post-resuscitation care which recommend the use of a multimodal approach to predicting the outcome of patients with postanoxic coma [2].

To ensure the accurate prognostic use of the EEG, it is essential to consider the natural evolution and underlying pathophysiology of EEG changes following CA and return to spontaneous circulation. The electrophysiological changes following CA have been the subject of study in animal models, demonstrating a clear evolution of changes over time. These include an initial period of complete EEG suppression, a subsequent burst suppression pattern, and a later transition into continuous cortical activity. The duration of EEG suppression and the latency of emergence of a continuous background have been demonstrated to correlate with neurological outcome and neuropathological injury [122]. Changes in humans may follow a similar pattern.

In contrast to SSEPs, EEG is a time-dependent indicator, given that the prognostic value of a single pattern is subject to change over time, at least until 72 h following CA. This factor has a significant impact on the optimal timing of EEG recording and the interpretation of distinct EEG patterns, which in turn assists with prognostication in post-CA patients. In each temporal frame, the application of standardised definitions of EEG patterns allows for the identification of at least one EEG pattern that is specific to good and poor outcomes. From a practical standpoint, three temporal frames can be distinguished: 12, 24, and 72 h.

During the early phase following CA (*12 h*), continuous and near-continuous EEGs indicate a favourable prognosis (FPR = 0) [33, 87, 123, 124]. The value of these EEG patterns for a good prognosis at a very early stage following CA is strengthened when the EEG is used in combination with SSEPs. In fact, both EEG (continuous and quasi-continuous) and SSEP (bilaterally absent) patterns have an FPR = 0, but with opposite prognostic significance (indeed each of these findings is mutually exclusive of the other). The EEG and SSEP recordings in the same patient increase the number of correctly classified cases (65.2%) in comparison with the number of patients classified by each test individually (EEG predicted a good prognosis in 30.5%; SSEPs predicted a poor prognosis in 34.7%). Moreover, continuous and near-continuous EEGs are always associated with the presence of cortical SSEPs, whereas bilaterally absent cortical SSEPs are always associated with malignant EEG patterns. This consistency of findings increases the reliability of prediction in a single patient. In literature, there have been reports, albeit rarely, of patients with bilaterally absent cortical SSEPs who have recovered consciousness [125], but no indication of the EEG pattern in these patients was reported. The discovery of a continuous or nearly continuous electroencephalogram (EEG) within 12 h following cardiac arrest (CA) in patients with bilaterally absent cortical somatosensory evoked potentials (SEPs) necessitates a cautious approach to interpretation. A repeat SEP recording is recommended to prevent an erroneous pessimistic prognosis. From a pathophysiological standpoint, the lack of correlation between SSEPs and EEG during the early stages following CA may be attributed to selective resistance to hypoxia/ischaemia of their generator systems. Somatosensory neurons possess both 'receiving' and 'sending' properties, and thus the preservation of N20 reflects an intact 'receiving mode'. In contrast, the EEG generation primarily reflects the activity of multiple cortical synaptic functions, with a greater involvement of 'sending mode' [126].

In all cases, the presence of continuous and nearly continuous EEG within the first 12 h was associated with the occurrence of cortical SSEPs, indicating a relatively low degree of severity of HIE. Concurrently, the absence of cortical SSEPs bilaterally was consistently associated with malignant EEG patterns, which serve as an indicator of severe HIE. Furthermore, the presence of malignant EEG in conjunction with preserved cortical SSEPs suggests an intermediate degree of HIE, which may account for the favourable outcome observed in some patients.

At approximately *24 h* the EEG shows less prognostic utility. Continuous quasi-continuous patterns begin to lose the specificity of 100% for good prognosis. Suppression/burst suppression patterns show a specificity of 100% for poor prediction at 24 h; however, since these EEG patterns occur in a small percentage (4%) of patients with preserved cortical SSEPs [127], the EEG can slightly increase the prediction of a poor outcome as compared with SSEPs. Moreover, at this time point, the suppressed pattern has not yet reached 100% specificity for a poor outcome.

At *72 h*, both malignant EEG patterns and bilaterally absent cortical SSEPs indicate a poor prognosis, with an FPR = 0.

This finding suggests two potential scenarios. On the one hand, the presence of both unfavourable EEG and SSEP patterns serves to confirm and reinforce their prognostic value. Conversely, the co-occurrence of malignant EEG and SSEPs in some patients allows for the prediction of an unfavourable prognosis in a greater number of patients. Continuous or near-continuous patterns in the later time frames are typically indicative of favourable outcomes. However, it is important to note that some patients exhibiting continuous and near-continuous EEGs in these time frames may still have a poor outcome. These discrepancies in prediction can be attributed to the emergence of alpha-theta coma patterns. The aetiology of ATC is believed to involve alternative generators and subcortical pathways, including the caudate nucleus, amygdala, hypothalamus, brainstem, and basal forebrain [128]. It is conceivable that the organisation of these subcortical circuits, which are responsible for the appearance of a de novo alpha rhythm, does not occur immediately following CA but takes more time to establish, so that a continuous pattern invariably predicts a good outcome only if observed within the first 12 h following CA. To avoid a false optimistic prediction, the SSEP pattern should be considered; if cortical responses are absent in both hemispheres, then the prognosis will be poor. If cortical SSEP responses are present, it would be useful to repeat the EEG after a few days. In the case of a true ATC pattern, an evolution towards the malignant patterns (suppressed or low voltage) is to be expected.

16.4 Conclusions

In patients with an unreliable neurological examination because of ongoing sedation or consequent brain injuries, electrophysiological evaluation can be a reliable surrogate of neurological function. It can be continuous (EEG) or discontinuous (EEG or SSEP). All electrophysiological raw data signals should be reviewed to ensure its quality and look for abnormal activities. Then, quantitative data can be computed and displayed over time to evaluate slower changes and tailor therapeutic strategies. qEEG is useful to guide sedation and look for seizures or new cortical injuries. To achieve this, anaesthesiologist and intensivist should be trained in electrophysiological analyses. As the interpretation of EEG and evoked potential signals can be challenging when they are not normal, intensivists should work in close collaboration with neurophysiologist, especially when outcome predictions are discussed as no *WLST* should be decided on an isolated result.

In the context of predicting poor neurological outcomes, it is desirable to have a test with high specificity, that is, a very low rate of falsely pessimistic predictions that could potentially lead to inappropriate WLST. In an ideal scenario, an index test would be 100% specific, with an FPR of 0. However, this is challenging to achieve in practice. Conversely, for a favourable prognosis, a satisfactory level of sensitivity against a non-absolute specificity is sufficient.

References

1. Harvey D, Butler J, Groves J, Manara A, Menon D, Thomas E, et al. Management of perceived devastating brain injury after hospital admission: a consensus statement from stakeholder professional organizations. Br J Anaesth. 2018;120(1):138–45.
2. Nolan JP, Sandroni C, Böttiger BW, Cariou A, Cronberg T, Friberg H, et al. European Resuscitation Council and European Society of Intensive Care Medicine guidelines 2021: post-resuscitation care. Intensive Care Med. 2021;47(4):369–421.
3. André-Obadia N, Zyss J, Gavaret M, Lefaucheur JP, Azabou E, Boulogne S, et al. Recommendations for the use of electroencephalography and evoked potentials in comatose patients. Neurophysiol Clin. 2018;48(3):143–69.
4. Herman ST, Abend NS, Bleck TP, Chapman KE, Drislane FW, Emerson RG, et al. Consensus statement on continuous EEG in critically ill adults and children, Part I: indications. J Clin Neurophysiol. 2015;32(2):87–95.
5. Claassen J, Taccone FS, Horn P, Holtkamp M, Stocchetti N, Oddo M. Recommendations on the use of EEG monitoring in critically ill patients: consensus statement from the neurointensive care section of the ESICM. Intensive Care Med. 2013;39(8):1337–51.
6. Sinha SR, Sullivan L, Sabau D, San-Juan D, Dombrowski KE, Halford JJ, et al. American Clinical Neurophysiology Society Guideline 1: minimum technical requirements for performing clinical electroencephalography. J Clin Neurophysiol. 2016;33(4):303–7.
7. Mirsattari SM, Lee DH, Jones D, Bihari F, Ives JR. MRI compatible EEG electrode system for routine use in the epilepsy monitoring unit and intensive care unit. Clin Neurophysiol. 2004;115(9):2175–80.
8. Abend NS, Dlugos DJ, Zhu X, Schwartz ES. Utility of CT-compatible EEG electrodes in critically ill children. Pediatr Radiol. 2015;45(5):714–8.
9. Tjepkema-Cloostermans MC, Hofmeijer J, Hom HW, Bosch FH, Van Putten MJAM. Predicting outcome in postanoxic coma: are ten EEG electrodes enough? J Clin Neurophysiol. 2017;34(3): 207–12.
10. Backman S, Cronberg T, Rosén I, Westhall E. Reduced EEG montage has a high accuracy in the post cardiac arrest setting. Clin Neurophysiol. 2020;131(9):2216–23.
11. Backman S, Westhall E, Dragancea I, Friberg H, Rundgren M, Ullén S, et al. Electroencephalographic characteristics of status epilepticus after cardiac arrest. Clin Neurophysiol. 2017;128(4):681–8.
12. Pati S, McClain L, Moura L, Fan Y, Westover MB. Accuracy of limited-montage electroencephalography in monitoring Postanoxic comatose patients. Clin EEG Neurosci. 2017;48(6):422–7.
13. Touchard C, Cartailler J, Vellieux G, de Montmollin E, Jaquet P, Wanono R, et al. Simplified frontal EEG in adults under veno-arterial extracorporeal membrane oxygenation. Ann Intensive Care. 2021;11(1):76.
14. Stipa G, Gabbrielli F, Rabbito C, Di Lazzaro V, Amantini A, Grippo A, et al. The Italian technical/administrative recommendations for telemedicine in clinical neurophysiology. Neurol Sci. 2021;42(5):1923–31.
15. Hockaday JM, Potts F, Epstein E, Bonazzi A, Schwab RS. Electroencephalographic changes in acute cerebral anoxia from cardiac or respiratory arrest. Electroencephalogr Clin Neurophysiol. 1965;18(6):575–86.
16. Hughes JR, Boshes B, Leestma J. Electro-clinical and pathologic correlations in comatose patients. Clin Electroencephalogr. 1976;7(1):13–30.
17. Synek VM. Prognostically important EEG coma patterns in diffuse anoxic and traumatic encephalopathies in adults. J Clin Neurophysiol. 1988;5(2):161–74.
18. Young GB, McLachlan RS, Kreeft JH, Demelo JD. An electroencephalographic classification for coma. Can J Neurol Sci J Can Sci Neurol. 1997;24(4):320–5.
19. Rae-Grant AD, Barbour PJ, Reed J. Development of a novel EEG rating scale for head injury using dichotomous variables. Electroencephalogr Clin Neurophysiol. 1991;79(5):349–57.
20. Wijdicks EFM, Hijdra A, Young GB, Bassetti CL, Wiebe S. Practice parameter: prediction of outcome in comatose survivors after cardiopulmonary resuscitation (an evidence-based review) [RETIRED]: report of the quality standards Subcommittee of the American Academy of Neurology. Neurology. 2006;67(2):203–10.
21. Hirsch LJ, LaRoche SM, Gaspard N, Gerard E, Svoronos A, Herman ST, et al. American Clinical Neurophysiology Society's standardized critical care EEG terminology: 2012 version. J Clin Neurophysiol. 2013;30(1):1.

22. Hirsch LJ, Fong MWK, Leitinger M, LaRoche SM, Beniczky S, Abend NS, et al. American clinical neurophysiology Society's standardized critical care EEG terminology: 2021 version. J Clin Neurophysiol. 2021;38(1):1.
23. Hofmeijer J, Tjepkema-Cloostermans MC, Van Putten MJAM. Burst-suppression with identical bursts: a distinct EEG pattern with poor outcome in postanoxic coma. Clin Neurophysiol. 2014;125(5):947–54.
24. Azabou E, Navarro V, Kubis N, Gavaret M, Heming N, Cariou A, et al. Value and mechanisms of EEG reactivity in the prognosis of patients with impaired consciousness: a systematic review. Crit Care. 2018;22(1):184.
25. Benghanem S, Mazeraud A, Azabou E, Chhor V, Shinotsuka CR, Claassen J, et al. Brainstem dysfunction in critically ill patients. Crit Care. 2020;24(1):5.
26. Admiraal MM, Van Rootselaar AF, Horn J. International consensus on EEG reactivity testing after cardiac arrest: towards standardization. Resuscitation. 2018;131:36–41.
27. Admiraal MM, Van Rootselaar A, Hofmeijer J, Hoedemaekers CWE, Van Kaam CR, Keijzer HM, et al. Electroencephalographic reactivity as predictor of neurological outcome in postanoxic coma: a multicenter prospective cohort study. Ann Neurol. 2019;86(1):17–27.
28. Alvarez V, Sierra-Marcos A, Oddo M, Rossetti AO. Yield of intermittent versus continuous EEG in comatose survivors of cardiac arrest treated with hypothermia. Crit Care. 2013;17(5):R190.
29. Caporro M, Rossetti AO, Seiler A, Kustermann T, Nguepnjo Nguissi NA, Pfeiffer C, et al. Electromyographic reactivity measured with scalp-EEG contributes to prognostication after cardiac arrest. Resuscitation. 2019;138:146–52.
30. Sandroni C, D'Arrigo S, Cacciola S, Hoedemaekers CWE, Westhall E, Kamps MJA, et al. Prediction of good neurological outcome in comatose survivors of cardiac arrest: a systematic review. Intensive Care Med. 2022;48(4):389–413.
31. Stecker MM, Cheung AT, Pochettino A, Kent GP, Patterson T, Weiss SJ, et al. Deep hypothermic circulatory arrest: I. Effects of cooling on electroencephalogram and evoked potentials. Ann Thorac Surg. 2001;71(1):14–21.
32. Crepeau AZ, Rabinstein AA, Fugate JE, Mandrekar J, Wijdicks EF, White RD, et al. Continuous EEG in therapeutic hypothermia after cardiac arrest: prognostic and clinical value. Neurology. 2013;80(4):339–44.
33. Crepeau AZ, Fugate JE, Mandrekar J, White RD, Wijdicks EF, Rabinstein AA, et al. Value analysis of continuous EEG in patients during therapeutic hypothermia after cardiac arrest. Resuscitation. 2014;85(6):785–9.
34. Elmer J, Coppler PJ, Solanki P, Westover MB, Struck AF, Baldwin ME, et al. Sensitivity of continuous electroencephalography to detect ictal activity after cardiac arrest. JAMA Netw Open. 2020;3(4):e203751.
35. Urbano V, Alvarez V, Schindler K, Rüegg S, Ben-Hamouda N, Novy J, et al. Continuous versus routine EEG in patients after cardiac arrest: analysis of a randomized controlled trial (CERTA). Resuscitation. 2022;176:68–73.
36. Gavvala J, Abend N, LaRoche S, Hahn C, Herman ST, Claassen J, et al. Continuous EEG monitoring: a survey of neurophysiologists and neurointensivists. Epilepsia. 2014;55(11):1864–71.
37. Hilkman DMW, Van Mook WNKA, Mess WH, Van Kranen-Mastenbroek VHJM. The use of continuous EEG monitoring in intensive care units in The Netherlands: a national survey. Neurocrit Care. 2018;29(2):195–202.
38. Kromm J, Fiest KM, Alkhachroum A, Josephson C, Kramer A, Jette N. Structure and outcomes of educational programs for training non-electroencephalographers in performing and screening adult EEG: a systematic review. Neurocrit Care. 2021;35(3):894–912.
39. Koenig MA, Kaplan PW. Clinical applications for EPs in the ICU. J Clin Neurophysiol. 2015;32(6):472–80.
40. Struck AF, Tabaeizadeh M, Schmitt SE, Ruiz AR, Swisher CB, Subramaniam T, et al. Assessment of the validity of the 2HELPS2B score for inpatient seizure risk prediction. JAMA Neurol. 2020;77(4):500.
41. Leithner C, Ploner CJ, Hasper D, Storm C. Does hypothermia influence the predictive value of bilateral absent N20 after cardiac arrest? Neurology. 2010;74(12):965–9.
42. Taccone FS, Cronberg T, Friberg H, Greer D, Horn J, Oddo M, et al. How to assess prognosis after cardiac arrest and therapeutic hypothermia. Crit Care. 2014;18(1):202.
43. Scarpino M, Lanzo G, Carrai R, Lolli F, Migliaccio ML, Spalletti M, et al. Predictive patterns of sensory evoked potentials in comatose brain injured patients evolving to brain death. Neurophysiol Clin Neurophysiol. 2017;47(1):19–29.
44. Kimura J, Ishida T, Suzuki S, Kudo Y, Matsuoka H, Yamada T. Far-field recording of the junctional potential generated by median nerve volleys at the wrist. Neurology. 1986;36(11):1451.
45. Yamada T. The anatomic and physiologic bases of median nerve somatosensory evoked potentials. Neurol Clin. 1988;6(4):705–33.
46. Desmedt JE, Cheron G. Non-cephalic reference recording of early somatosensory potentials to finger stimulation in adult or aging normal: differentiation of widespread N18 and contralateral N20 from the prerolandic p22 and N30 components. Electroencephalogr Clin Neurophysiol. 1981;52(6):553–70.
47. Rothstein TL. SSEP retains its value as predictor of poor outcome following cardiac arrest in the era of therapeutic hypothermia. Crit Care. 2019;23(1):327.

48. Guérit JM. Neurophysiological testing in neurocritical care. Curr Opin Crit Care. 2010;16(2):98–104.
49. Bouwes A, Binnekade JM, Zandstra DF, Koelman JHTM, Van Schaik IN, Hijdra A, et al. Somatosensory evoked potentials during mild hypothermia after cardiopulmonary resuscitation. Neurology. 2009;73(18):1457–61.
50. Lang M, Welte M, Syben R, Hansen D. Effects of Hypothermia on Median Nerve Somatosensory Evoked Potentials During Spontaneous Circulation: J Neurosurg Anesthesiol. 2002;14(2):141–5.
51. Zanatta P, Bosco E, Comin A, Mazzarolo AP, Di Pasquale P, Forti A, et al. Effect of Mild Hypothermic Cardiopulmonary Bypass on the Amplitude of Somatosensory-evoked Potentials. J Neurosurg Anesthesiol. 2014;26(2):161–6.
52. Markand ON, Warren C, Mallik GS, King RD, Brown JW, Mahomed Y. Effects of hypothermia on short latency somatosensory evoked potentials in humans. Electroencephalogr Clin Neurophysiol Potentials Sect. 1990;77(6):416–24.
53. Fischer C, Mutschler V. Traumatic brain injuries in adults: from coma to wakefulness. Neurophysiological data. Ann Readaptation Med Phys Rev Sci Soc Francaise Reeducation Fonct Readaptation Med Phys. 2002;45(8):448–55.
54. Luaute J, Cotton F, Lemaire J-J, et al. Let live or let die after traumatic coma. Neurol Clin Pract 2012; 2:24–32.
55. Logi F, Pasqualetti P, Tomaiuolo F. Predict recovery of consciousness in post-acute severe brain injury: The role of EEG reactivity. Brain Inj. 2011;25(10):972–9.
56. Cruccu G, Aminoff MJ, Curio G, Guerit JM, Kakigi R, Mauguiere F, et al. Recommendations for the clinical use of somatosensory-evoked potentials. Clin Neurophysiol. 2008;119(8):1705–19.
57. Logi F, Fischer C, Murri L, Mauguière F. The prognostic value of evoked responses from primary somatosensory and auditory cortex in comatose patients. Clin Neurophysiol. 2003;114(9):1615–27.
58. Claassen J, Hansen HC. Early recovery after closed traumatic head injury: somatosensory evoked potentials and clinical findings. Crit Care Med. 2001;29(3):494–502.
59. Amantini A, Fossi S, Grippo A, Innocenti P, Amadori A, Bucciardini L, et al. Continuous EEG-SEP monitoring in severe brain injury. Neurophysiol Clin Neurophysiol. 2009;39(2):85–93.
60. Amantini A, Amadori A, Fossi S. Evoked potentials in the ICU. Eur J Anaesthesiol. 2008;25:196–202.
61. Baars JH, von Klitzing JP. Easily applicable SEP-monitoring of the N20 wave in the intensive care unit. Neurophysiol Clin Neurophysiol. 2017;47(1):31–4.
62. Dragancea I, Horn J, Kuiper M, Friberg H, Ullén S, Wetterslev J, et al. Neurological prognostication after cardiac arrest and targeted temperature management 33°C versus 36°C: results from a randomised controlled clinical trial. Resuscitation. 2015;93:164–70.
63. Moulaert VRM, Van Heugten CM, Gorgels TPM, Wade DT, Verbunt JA. Long-term outcome after survival of a cardiac arrest: a prospective longitudinal cohort study. Neurorehabil Neural Repair. 2017;31(6):530–9.
64. Yan S, Gan Y, Jiang N, Wang R, Chen Y, Luo Z, et al. The global survival rate among adult out-of-hospital cardiac arrest patients who received cardiopulmonary resuscitation: a systematic review and meta-analysis. Crit Care. 2020;24(1):61.
65. Taccone FS, Baar I, De Deyne C, Druwe P, Legros B, Meyfroidt G, et al. Neuroprognostication after adult cardiac arrest treated with targeted temperature management: task force for Belgian recommendations. Acta Neurol Belg. 2017;117(1):3–15.
66. Rajajee V, Muehlschlegel S, Wartenberg KE, Alexander SA, Busl KM, Chou SHY, et al. Guidelines for neuroprognostication in comatose adult survivors of cardiac arrest. Neurocrit Care. 2023;38(3):533–63.
67. Friberg H, Cronberg T, Dünser MW, Duranteau J, Horn J, Oddo M. Survey on current practices for neurological prognostication after cardiac arrest. Resuscitation. 2015;90:158–62.
68. Maciel CB, Barden MM, Youn TS, Dhakar MB, Greer DM. Neuroprognostication practices in postcardiac arrest patients: An international survey of critical care providers. Crit Care Med. 2020;48(2):e107–14.
69. Jiang M, Su Y, Liu G, Huang H, Tian F. EEG pattern predicts awakening of comatose patients after cardiopulmonary resuscitation. Resuscitation. 2020;151:33–8.
70. Benarous L, Gavaret M, Soda Diop M, Tobarias J, De Ghaisne De Bourmont S, Allez C, et al. Sources of interrater variability and prognostic value of standardized EEG features in post-anoxic coma after resuscitated cardiac arrest. Clin Neurophysiol Pract. 2019;4:20–6.
71. Sugiyama K, Kashiura M, Akashi A, Tanabe T, Hamabe Y. Prognostic value of the recovery time of continuous normal voltage in amplitude-integrated electroencephalography in out-of-hospital cardiac arrest patients treated with therapeutic hypothermia: a retrospective study. J Intensive Care. 2016;4(1):25.
72. Amorim E, Rittenberger JC, Baldwin ME, Callaway CW, Popescu A. Malignant EEG patterns in cardiac arrest patients treated with targeted temperature management who survive to hospital discharge. Resuscitation. 2015;90:127–32.
73. Ruijter BJ, Van Putten MJAM, Van Den Bergh WM, Tromp SC, Hofmeijer J. Propofol does not affect the reliability of early EEG for outcome prediction of comatose patients after cardiac arrest. Clin Neurophysiol. 2019;130(8):1263–70.
74. Spalletti M, Carrai R, Scarpino M, Cossu C, Ammannati A, Ciapetti M, et al. Single electroencephalographic patterns as specific and time-dependent indicators of good and poor out-

come after cardiac arrest. Clin Neurophysiol. 2016;127(7):2610–7.
75. Scarpino M, Carrai R, Lolli F, Lanzo G, Spalletti M, Valzania F, et al. Neurophysiology for predicting good and poor neurological outcome at 12 and 72 h after cardiac arrest: the ProNeCA multicentre prospective study. Resuscitation. 2020;147:95–103.
76. Rossetti AO, Tovar Quiroga DF, Juan E, Novy J, White RD, Ben-Hamouda N, et al. Electroencephalography predicts poor and good outcomes after cardiac arrest: a two-center study. Crit Care Med. 2017;45(7):e674.
77. Johnsen B, Jeppesen J, Duez CHV. Common patterns of EEG reactivity in post-anoxic coma identified by quantitative analyses. Clin Neurophysiol. 2022;142:143–53.
78. Van Putten MJAM, Jansen C, Tjepkema-Cloostermans MC, Beernink TMJ, Koot R, Bosch F, et al. Postmortem histopathology of electroencephalography and evoked potentials in postanoxic coma. Resuscitation. 2019;134:26–32.
79. Sandroni C, D'Arrigo S, Cacciola S, Hoedemaekers CWE, Kamps MJA, Oddo M, et al. Prediction of poor neurological outcome in comatose survivors of cardiac arrest: a systematic review. Intensive Care Med. 2020;46(10):1803–51.
80. Coppler PJ, Kusztos AE, Andreae M, Butcher BW, Doshi A, Baldwin ME, et al. Awakening from post anoxic coma with burst suppression with identical bursts. Resusc Plus. 2021;7:100151.
81. Peluso L, Legros B, Caroyer S, Taccone FS, Gaspard N. Falsely predictive EEG and clinical signs after post-anoxic brain injury under sevoflurane anesthesia. Clin Neurophysiol. 2021;132(5):1080–2.
82. Muhlhofer W, Szaflarski JP. Prognostic value of EEG in patients after cardiac arrest—an updated review. Curr Neurol Neurosci Rep. 2018;18(4):16.
83. Admiraal MM, Horn J, Hofmeijer J, Hoedemaekers CWE, Van Kaam CR, Keijzer HM, et al. EEG reactivity testing for prediction of good outcome in patients after cardiac arrest. Neurology. 2020;95(6):e653–61.
84. Duez CHV, Ebbesen MQ, Benedek K, Fabricius M, Atkins MD, Beniczky S, et al. Large inter-rater variability on EEG-reactivity is improved by a novel quantitative method. Clin Neurophysiol. 2018;129(4):724–30.
85. Rossetti AO, Rabinstein AA, Oddo M. Neurological prognostication of outcome in patients in coma after cardiac arrest. Lancet Neurol. 2016;15(6):597–609.
86. Tsetsou S, Novy J, Oddo M, Rossetti AO. EEG reactivity to pain in comatose patients: importance of the stimulus type. Resuscitation. 2015;97:34–7.
87. Grippo A, Carrai R, Scarpino M, Spalletti M, Lanzo G, Cossu C, et al. Neurophysiological prediction of neurological good and poor outcome in post-anoxic coma. Acta Neurol Scand. 2017;135(6):641–8.
88. Westhall E, Rosén I, Rossetti AO, van Rootselaar AF, Wesenberg Kjaer T, Friberg H, et al. Interrater variability of EEG interpretation in comatose cardiac arrest patients. Clin Neurophysiol. 2015;126(12):2397–404.
89. Hermans MC, Westover MB, Van Putten MJAM, Hirsch LJ, Gaspard N. Quantification of EEG reactivity in comatose patients. Clin Neurophysiol. 2016;127(1):571–80.
90. Ruijter BJ, Van Putten MJAM, Hofmeijer J. Generalized epileptiform discharges in postanoxic encephalopathy: quantitative characterization in relation to outcome. Epilepsia. 2015;56(11):1845–54.
91. Rossetti AO, Oddo M, Liaudet L, Kaplan PW. Predictors of awakening from postanoxic status epilepticus after therapeutic hypothermia. Neurology. 2009;72(8):744–9.
92. Barbella G, Lee JW, Alvarez V, Novy J, Oddo M, Beers L, et al. Prediction of regaining consciousness despite an early epileptiform EEG after cardiac arrest. Neurology. 2020;94(16):e1675.
93. Ruijter BJ, Keijzer HM, Tjepkema-Cloostermans MC, Blans MJ, Beishuizen A, Tromp SC, et al. Treating rhythmic and periodic EEG patterns in comatose survivors of cardiac arrest. N Engl J Med. 2022;386(8):724–34.
94. Sivaraju A, Gilmore EJ, Wira CR, Stevens A, Rampal N, Moeller JJ, et al. Prognostication of post-cardiac arrest coma: early clinical and electroencephalographic predictors of outcome. Intensive Care Med. 2015;41(7):1264–72.
95. Elmer J, Rittenberger JC, Faro J, Molyneaux BJ, Popescu A, Callaway CW, et al. Clinically distinct electroencephalographic phenotypes of early myoclonus after cardiac arrest. Ann Neurol. 2016;80(2):175–84.
96. Aicua Rapun I, Novy J, Solari D, Oddo M, Rossetti AO. Early Lance–Adams syndrome after cardiac arrest: prevalence, time to return to awareness, and outcome in a large cohort. Resuscitation. 2017;115:169–72.
97. Forgacs PB, Devinsky O, Schiff ND. Independent functional outcomes after prolonged coma following cardiac arrest: a mechanistic hypothesis. Ann Neurol. 2020;87(4):618–32.
98. Beretta S, Coppo A, Bianchi E, Zanchi C, Carone D, Stabile A, et al. Neurologic outcome of postanoxic refractory status epilepticus after aggressive treatment. Neurology. 2018;91(23):e2153.
99. Westhall E, Rosén I, Rundgren M, Bro-Jeppesen J, Kjaergaard J, Hassager C, et al. Time to epileptiform activity and EEG background recovery are independent predictors after cardiac arrest. Clin Neurophysiol. 2018;129(8):1660–8.
100. Hofmeijer J, Beernink TMJ, Bosch FH, Beishuizen A, Tjepkema-Cloostermans MC, van Putten MJAM. Early EEG contributes to multimodal outcome prediction of postanoxic coma. Neurology. 2015;85(2):137–43.
101. Soar J, Maconochie I, Wyckoff MH, Olasveengen TM, Singletary EM, Greif R, et al. International Consensus on Cardiopulmonary Resuscitation

and Emergency Cardiovascular Care Science with Treatment Recommendations: Summary From the Basic Life Support; Advanced Life Support; Pediatric Life Support; Neonatal Life Support; Education, Implementation, and Teams; and First Aid Task Forces. Circulation. 2019;140(24):e826–80.
102. Amorim E, Ghassemi MM, Lee JW, Greer DM, Kaplan PW, Cole AJ, et al. Estimating the false positive rate of absent somatosensory evoked potentials in cardiac arrest prognostication. Crit Care Med. 2018;46(12):e1213–21.
103. Bouwes A, Binnekade JM, Kuiper MA, Bosch FH, Zandstra DF, Toornvliet AC, et al. Prognosis of coma after therapeutic hypothermia: a prospective cohort study. Ann Neurol. 2012;71(2):206–12.
104. Pfeifer R, Weitzel S, Günther A, Berrouschot J, Fischer M, Isenmann S, et al. Investigation of the inter-observer variability effect on the prognostic value of somatosensory evoked potentials of the median nerve (SSEP) in cardiac arrest survivors using an SSEP classification. Resuscitation. 2013;84(10):1375–81.
105. Celani MG, Carrai R, Cantisani TA, Scarpino M, Ercolani MV, Lolli F, et al. Is there inter-observer variation in the interpretation of SSEPs in comatose cardiac arrest survivors? Further considerations following the Italian multicenter ProNeCa study. Resuscitation. 2020;155:207–10.
106. Endisch C, Westhall E, Kenda M, Streitberger KJ, Kirkegaard H, Stenzel W, et al. Hypoxic-ischemic encephalopathy evaluated by brain autopsy and Neuroprognostication after cardiac arrest. JAMA Neurol. 2020;77(11):1430.
107. Shao R, Wang T, Hang C, An L, Wang X, Zhang L, et al. Alteration in early resting-state functional MRI activity in comatose survivors of cardiac arrest: a prospective cohort study. Crit Care. 2024;28(1):260.
108. Endisch C, Storm C, Ploner CJ, Leithner C. Amplitudes of SSEP and outcome in cardiac arrest survivors: a prospective cohort study. Neurology. 2015;85(20):1752–60.
109. Glimmerveen AB, Keijzer HM, Ruijter BJ, Tjepkema-Cloostermans MC, Van Putten MJAM, Hofmeijer J. Relevance of somatosensory evoked potential amplitude after cardiac arrest. Front Neurol. 2020;11:335.
110. Benghanem S, Nguyen LS, Gavaret M, Mira JP, Pène F, Charpentier J, et al. SSEP N20 and P25 amplitudes predict poor and good neurologic outcomes after cardiac arrest. Ann Intensive Care. 2022;12(1):25.
111. Scarpino M, Lolli F, Lanzo G, Carrai R, Spalletti M, Valzania F, et al. SSEP amplitude accurately predicts both good and poor neurological outcome early after cardiac arrest; a post-hoc analysis of the ProNeCA multicentre study. Resuscitation. 2021;163:162–71.
112. Oh SH, Park KN, Choi SP, Oh JS, Kim HJ, Youn CS, et al. Beyond dichotomy: patterns and amplitudes of SSEPs and neurological outcomes after cardiac arrest. Crit Care. 2019;23(1):224.
113. van Soest TM, van Rootselaar AF, Admiraal MM, Potters WV, Koelman JHMT, Horn J. SSEP amplitudes add information for prognostication in postanoxic coma. Resuscitation. 2021;163:172–5.
114. Scarpino M, Lolli F, Lanzo G, Carrai R, Spalletti M, Valzania F, et al. Do changes in SSEP amplitude over time predict the outcome of comatose survivors of cardiac arrest? Resuscitation. 2022;181:133–9.
115. Bongiovanni F, Romagnosi F, Barbella G, Di Rocco A, Rossetti AO, Taccone FS, et al. Standardized EEG analysis to reduce the uncertainty of outcome prognostication after cardiac arrest. Intensive Care Med. 2020;46(5):963–72.
116. Carrai R, Scarpino M, Lolli F, Spalletti M, Lanzo G, Peris A, et al. Early-SEPs' amplitude reduction is reliable for poor-outcome prediction after cardiac arrest? Acta Neurol Scand. 2019;139(2):158–65.
117. Fossi S, Amantini A, Grippo A, Innocenti P, Amadori A, Bucciardini L, et al. Continuous EEG–SEP monitoring of severely brain injured patients in NICU: methods and feasibility. Neurophysiol Clin Neurophysiol. 2006;36(4):195–205.
118. Sloan TB, Fugina ML, Toleikis JR. Effect of midazolam on median nerve somatosensory evoked potentials. Br J Anaesth. 1990;64(5):590–3.
119. Fredland A, Backman S, Westhall E. Stratifying comatose postanoxic patients for somatosensory evoked potentials using routine EEG. Resuscitation. 2019;143:17–21.
120. Peluso L, Oddo M, Minini A, Citerio G, Horn J, Bernardini ED, et al. Neurological pupil index and its association with other prognostic tools after cardiac arrest: a post hoc analysis. Resuscitation. 2022;179:259–66.
121. Hoedemaekers C, van der Hoeven J. Concordance in multimodal prognostication after cardiac arrest: improving accuracy or comparing apples to oranges? Resuscitation. 2022;179:114–5.
122. Koenig MA, Kaplan PW, Thakor NV. Clinical neurophysiologic monitoring and brain injury from cardiac arrest. Neurol Clin. 2006;24(1):89–106.
123. Tjepkema-Cloostermans MC, Hofmeijer J, Trof RJ, Blans MJ, Beishuizen A, Van Putten MJAM. Electroencephalogram predicts outcome in patients with postanoxic coma during mild therapeutic hypothermia. Crit Care Med. 2015;43(1):159–67.
124. Carrai R, Grippo A, Scarpino M, Spalletti M, Cossu C, Lanzo G, et al. Time-dependent and independent neurophysiological indicators of prognosis in post-anoxic coma subjects treated by therapeutic hypothermia. Minerva Anestesiol. 2016;82(9):940–9.
125. Pfeiffer G, Pfeifer R, Isenmann S. Cerebral hypoxia, missing cortical somatosensory evoked potentials and recovery of consciousness. BMC Neurol. 2014;14(1):82.

126. Van Putten MJAM. The N20 in post-anoxic coma: are you listening? Clin Neurophysiol. 2012;123(7):1460–4.
127. Scarpino M, Lanzo G, Lolli F, Carrai R, Moretti M, Spalletti M, et al. Neurophysiological and neuroradiological multimodal approach for early poor outcome prediction after cardiac arrest. Resuscitation. 2018;129:114–20.
128. Fossi S, Amantini A, Grippo A, Cossu C, Boni N, Pinto F. Anoxic-ischemic alpha coma:prognostic significance of the incomplete variant. Neurol Sci. 2004;24(6):397–400.

Electroencephalogram (EEG) and Sleep Disorders

17

Paola Proserpio, Ramona Cordani, and Lino Nobili

17.1 The Macrostructure and Microstructure of Sleep

Sleep is a fundamental biological process crucial for overall health and well-being, which can be broadly categorized into non-rapid eye movement (NREM) and rapid eye movement (REM) sleep. A comprehensive understanding of its structure, encompassing both macro- and microstructure, is essential for addressing sleep disorders and optimizing sleep quality.

The macrostructure of sleep refers to the organization of sleep stages over a typical sleep period, which cycles approximately every 90 minutes and includes alternating phases of NREM and REM sleep, each characterized by distinct physiological and behavioral patterns that follow a predictable sequence. The typical sleep architecture in a healthy adult involves four to six cycles of NREM and REM sleep per night, with a gradual reduction in N3 and an increase in REM duration as the night progresses (Fig. 17.1).

The microstructure of sleep refers to transient and phasic brief events occurring within sleep stages, often revealed through high-resolution electroencephalogram (EEG) analysis. These micro-events provide insights into the dynamic processes underlying sleep regulation and stability.

NREM sleep is subdivided into three stages: N1, N2, and N3, each with specific features (Fig. 17.2).

Stage N1 marks the transition from wakefulness to sleep, characterized by a decrease in alpha waves (8–12 Hz) and the appearance of theta waves (4–7 Hz) on the EEG. During this stage, electrooculography (EOG) may show slow eye movements, and electromyography (EMG) shows a reduction in muscle tone [1] (Fig. 17.2). This stage involves the thalamus and cortex transitioning to sleep and is crucial for reducing responsiveness to external stimuli. It represents a light sleep stage with low arousal thresholds.

Stage N2 constitutes approximately 50% of total sleep time and is defined by the presence of K-complexes and sleep spindles, superimposed on a background of theta and delta waves. During Stage N2, EOG shows minimal or no eye move-

Paola Proserpio and Ramona Cordani contributed equally with all other contributors.

P. Proserpio
IRCCS San Raffaele Scientific Institute, Department of Clinical Neurosciences, Neurology – Sleep Disorders Center, Milan, Italy
e-mail: Proserpio.paola@hsr.it

R. Cordani · L. Nobili (✉)
IRCCS Istituto G. Gaslini, Child Neuropsychiatry Unit, Genoa, Italy

Department of Neurosciences, Rehabilitation, Ophthalmology, Genetics and Maternal and Child Health (DINOGMI), University of Genova, Genoa, Italy
e-mail: Ramonacordani@gaslini.org; Lino.nobili@unige.it

F. Brigo, O. Mecarelli (eds.), *EEG: The First 100 Years*,
https://doi.org/10.1007/978-3-031-86878-8_17

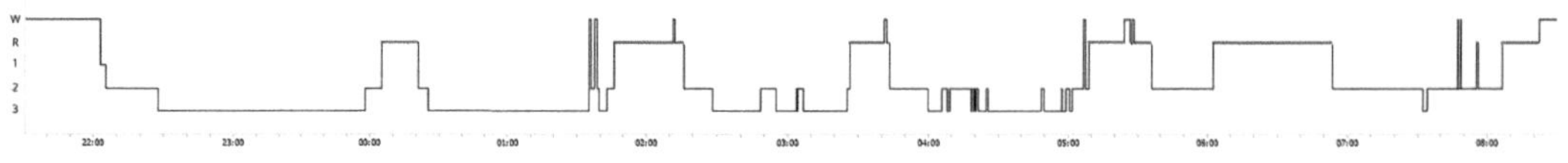

Fig. 17.1 Hypnogram, graphical representation of the stages of sleep over a night, illustrating transitions between wakefulness (W), NREM (1, N1; 2, N2; 3, N3), and REM (R) sleep stages

ment, and EMG continues to show decreased muscle activity (Fig. 17.2).

K-complexes (KCs) are high-amplitude biphasic waves, occurring spontaneously or in response to external stimuli [1]. Considered as *type I* slow waves, they play a protective role in sleep by inducing widespread cortical down-states in response to external stimuli, facilitating sensory gating, and maintaining sleep integrity [2]. *Type I* slow waves may reflect a fast and efficient neuronal synchronization process. The generation of these slow waves may depend on a "bottom-up" subcortico-cortical synchronization mediated by arousal-promoting structures. Input from diffusely projecting subcortical sources should trigger a near-simultaneous depolarization in widespread regions of the cortex, followed by a massive hyperpolarization (down-state) producing large negative waves in the EEG signal. Conversely, as sleep deepens, smaller and shallower *type II* slow waves become predominant. Characterized by reduced amplitude, slope, and localized cortical distribution, these waves likely result from a less efficient cortico-cortical "horizontal" synchronization process. *Type II* slow waves exhibit homeostatic regulation, with decreasing amplitude and slope as sleep pressure diminishes throughout the night. They are linked to spindle activity, promoting sleep stability and cortical disengagement from environmental stimuli [2]. Sleep spindles are bursts of oscillatory brain activity (12–16 Hz) lasting approximately 0.5–2 seconds [1]. They are hallmark of NREM sleep, primarily seen during N2 sleep. Generated by the thalamic reticular nucleus, sleep spindles are critical EEG phenomena that play a crucial role in synaptic plasticity and memory consolidation by activating calcium-dependent pathways essential for long-term potentiation. They synchronize with hippocampal ripples and cortical slow oscillations, facilitating memory transfer across brain regions [3]. Spindles regulate arousal and maintain disconnection from the external environment through a *thalamic gate* which is essential for sustaining deep sleep. Their topographical distribution varies between and within NREM sleep cycles, indicating a dynamic role in sleep maintenance and cognitive functions [4].

Stage N3, also referred to as slow-wave sleep (SWS), is marked by high-amplitude delta waves (0.5–4 Hz) in the EEG (Fig. 17.2) [1]. As the deepest stage of NREM sleep, it is distinguished by the highest arousal thresholds and is essential for restorative processes, including tissue repair and immune function, and involves widespread cortical and subcortical networks. Slow waves play a crucial role in sleep homeostasis, the process that regulates sleep intensity and duration based on prior wakefulness. This is reflected in slow-wave activity (SWA), which increases following periods of wakefulness and decreases after sleep, indicating the brain's recovery from wakefulness-induced neural activity. Neuronal synchrony varies with sleep pressure. During early sleep, when sleep pressure is high, neuronal firing is highly synchronized, resulting in large, steep slow waves. As sleep pressure decreases, neuronal firing becomes less synchronized, and slow waves become smaller and less steep, showing how sleep restores neural function. Extended wakefulness leads to higher neuronal firing rates in all states. The systematic increase in neuronal firing during wakefulness and decrease during sleep indicate that sleep is crucial for restoring brain cell function. The correlation between neuronal firing patterns and SWA supports the idea that sleep counterbalances accumulated neural activity, highlighting a cellular mechanism for sleep homeostasis [5]. Delta power during N3 represents the strength and number of slow-wave oscillations and is a marker of sleep depth and quality, indicating the recuperative aspects of

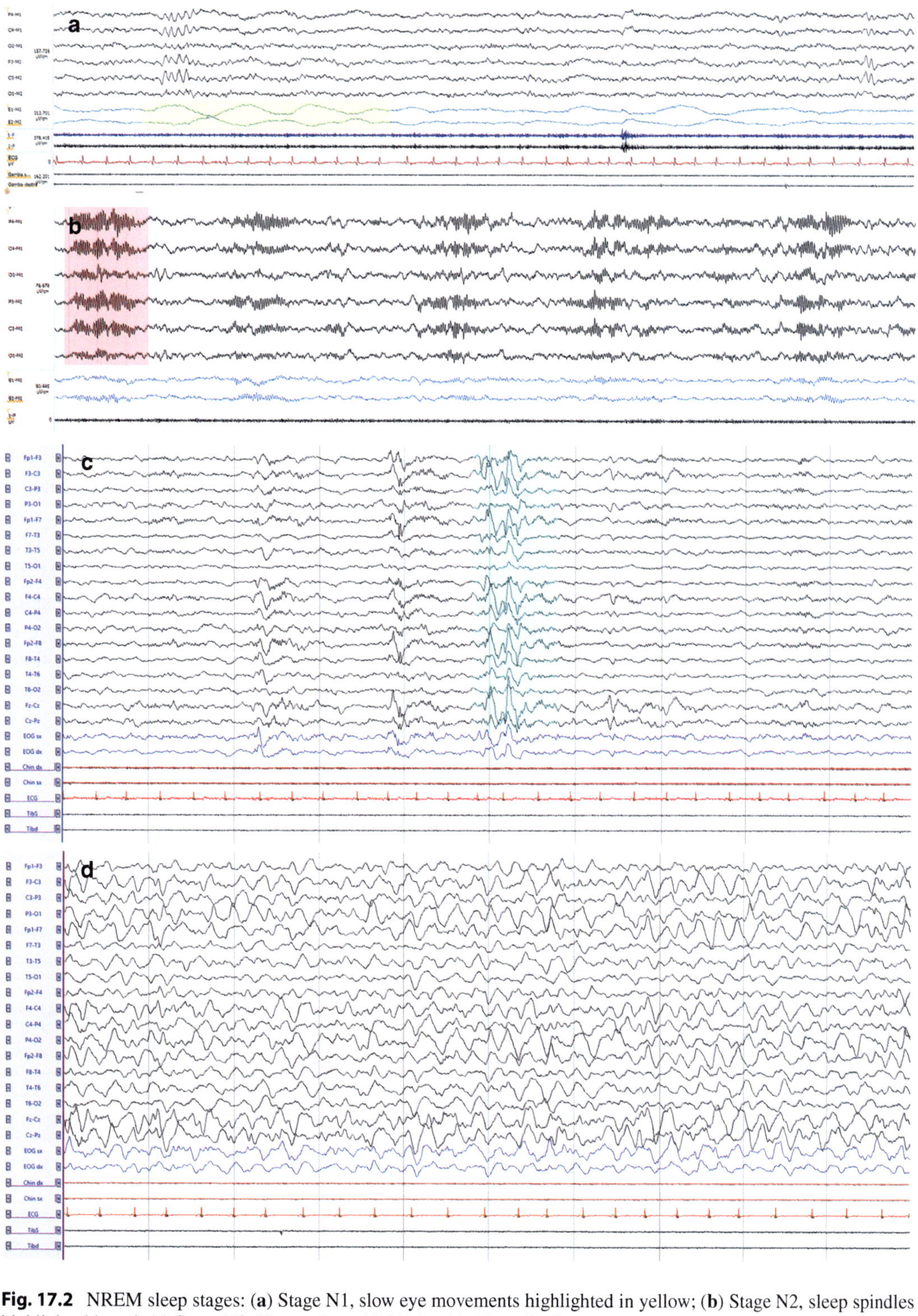

Fig. 17.2 NREM sleep stages: (**a**) Stage N1, slow eye movements highlighted in yellow; (**b**) Stage N2, sleep spindles highlighted in red; (**c**) Stage N2, K-complexes highlighted in blue; (**d**) Stage N3

sleep. The distribution of delta power can also vary between different brain regions, providing insights into localized brain activity during deep sleep. Studies have shown that delta power tends to be higher in frontal brain regions, which are involved in executive functions and cognitive processes. This localized increase in delta power may reflect the brain's need to recuperate and strengthen neural connections critical for cognitive performance.

REM sleep, constituting 20–25% of adult human sleep, displays a desynchronized, low-amplitude, mixed-frequency EEG pattern akin to wakefulness. The EOG shows rapid eye movements, while EMG demonstrates muscle atonia, except for diaphragmatic and ocular muscles (Fig. 17.3). This phase in adults typically begins about 90 minutes after sleep onset and recurs every 90–120 minutes, with each REM period lengthening as the night progresses (Fig. 17.1).

REM sleep is associated with vivid dreaming, emotional processing, and memory consolidation, primarily involving a brain network that includes the brainstem (particularly the pons) and the limbic system (including the hippocampus and amygdala) [5]. REM sleep is characterized by the alternating phasic and tonic microstates, each with unique neurophysiological features. Tonic REM is characterized by periods without rapid eye movements, exhibiting low muscle tone, and is associated with the maintenance of sleep and facilitation of brain plasticity. Phasic REM is distinguished by bursts of rapid eye movements, myoclonic twitches, sawtooth waves, and irregularities in respiratory and cardiac activity, linked to ponto-geniculo-occipital (PGO) waves.

Phasic REM involves widespread activation of the brain, including the pons and occipital cortex. These microstates differ in terms of environmental responsiveness and cortical activity, with unique spontaneous and evoked patterns of cortical activity in phasic compared to tonic REM [6]. This distinction also influences information processing, underscoring the complex nature of REM sleep and the importance of distinguishing

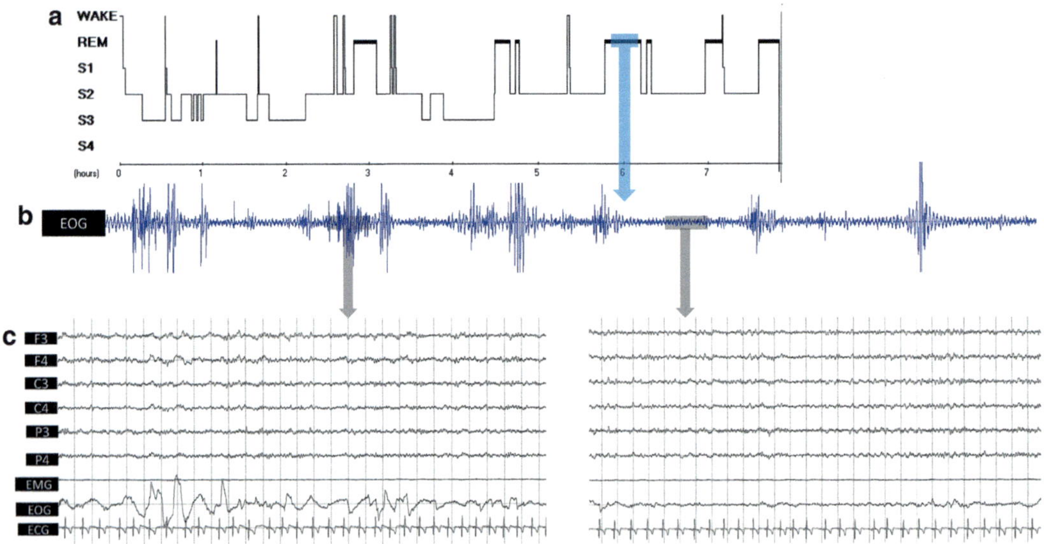

Fig. 17.3 Phasic and tonic REM microstates. (**a**) The hypnogram of a healthy human subject. REM sleep cycles reappear four to five times a night and last approximately 20 min. (**b**) A 15-min long REM sleep segment of EOG activity. Periods with rapid eye movements (phasic REM) as reflected by bursts of increased EOG amplitudes, and periods without eye movements (tonic REM) as reflected by reduced EOG amplitudes alternate within the REM segment. (**c**) A 30-s long period of phasic (on the left) and tonic (on the right) REM sleep. The time series represent frontal (F3, F4), central (C3, C4), and parietal (P3, P4) EEG and electromyographic (EMG), electrooculographic, (EOG), and electrocardiographic (ECG) activity. (Reproduced from Simor et al., Sleep Med Rev. 2020 [6])

its phasic and tonic components to better understand its mechanisms and functions in both healthy and pathological conditions.

17.1.1 Arousals and the Cyclic Alternating Pattern (CAP)

Arousals are defined by the American Academy of Sleep Medicine as an abrupt shift in EEG frequency, which may include theta, alpha, and /or frequencies greater than 16 Hz, which is at least 3 seconds in duration [1]. A certain number of spontaneous arousals appear to be an intrinsic component of physiological sleep. However, their frequent occurrences can impair sleep quality as it happens in various sleep disorders, such as obstructive sleep apnea and periodic limb movement disorder. Besides arousals, other transient EEG changes have been observed alongside vegetative modifications, such as alterations in heart rate and blood pressure. These EEG changes can exhibit various patterns, including K-complexes and delta bursts. The cyclic alternating pattern is the term used to characterize these transient EEG patterns that arise from background activity with a quasi-periodic mode of occurrence. The cyclic alternating pattern (CAP) is indeed a marker of sleep instability reflecting fluctuations in sleep depth during NREM sleep. CAP is composed by cyclic sequences of cerebral activation (phase A) followed by periods of deactivation (phase B) which separate two successive A phases with an interval < 1 min. (Figs. 17.4 and 17.5) [7].

A phases can be further subdivided into three types: A1, A2, and A3. A1 phases are characterized by predominant slow waves, minimal changes in muscle tone, and subtle autonomic activity. These phases help maintain sleep continuity with minimal disruption; the number of A1 phase tends to decrease during the successive sleep cycle within the night, following the dynamics of delta activity decay, thus suggesting their relationship with the homeostatic sleep pressure [8]. A2 phases are characterized by a mixture of slow waves and low-voltage fast activities, moderate increases in muscle tone, and autonomic activation. In A3 phases, fast activities and desynchronization are predominant together with a clear-cut increases in muscle tone and autonomic activation. A2 and A3 phases reflect a state of progressive heightened vigilance and potential readiness to respond to external stimuli (Fig. 17.5).

CAP sequences can be quantitatively assessed to evaluate sleep quality and the presence of sleep disturbances. A higher CAP rate is often associated with fragmented sleep and various sleep disorders, such as insomnia, sleep apnea, and periodic limb movement disorder. The analysis of CAP can provide insights into the underlying mechanisms of these disturbances and inform therapeutic interventions.

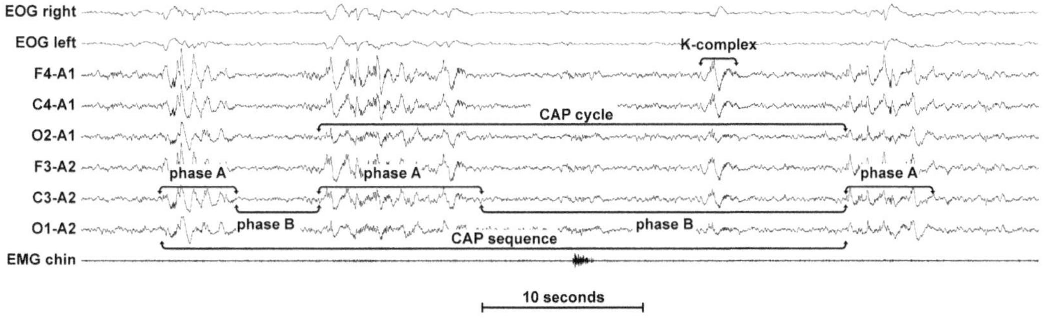

Fig. 17.4 Illustration of a CAP cycle: a sequence of two alternating stereotyped EEG patterns, each lasting more than 2 and less than 60 s, called phase A and phase B. At least two full CAP cycles in succession are needed to define a CAP sequence; thus, the minimum content of a sequence is A + B + A + B + A. Note that single K-complexes do not identify a CAP phase A. (Reproduced from Parrino et al., Sleep Med Rev. 2012 [36])

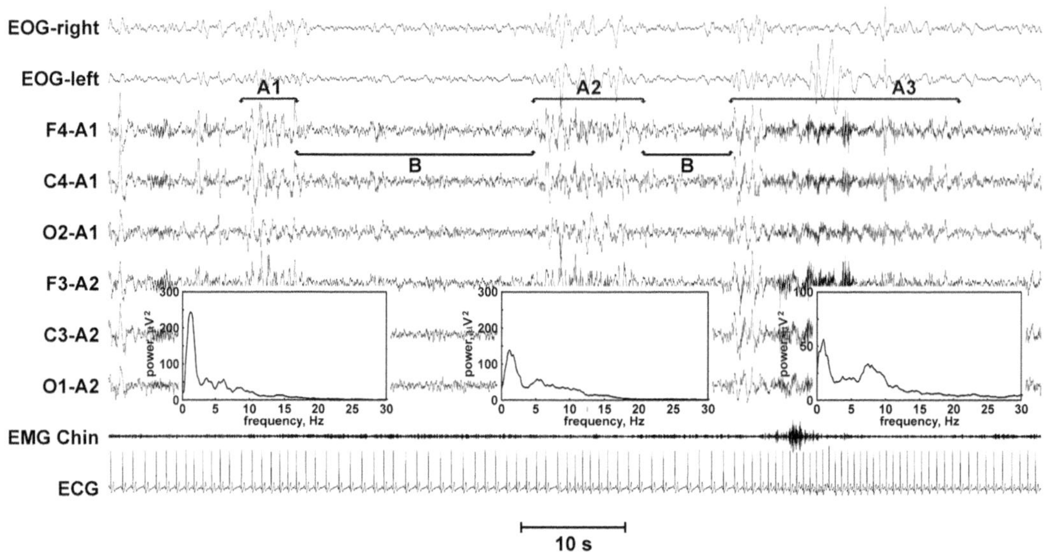

Fig. 17.5 CAP sequence composed by three A phases. Below each A phase, the corresponding power spectrum is shown, computed from the F4-A1 lead. CAP A1 subtype is dominated by the slow component in the delta frequency range; the CAP A2 subtype shows a clearly smaller delta peak with an emergency of a peak in the frequencies above 5 Hz; CAP A3 subtype shows a further reduction of the delta peak and a clear peak in the alpha frequency range. (Reproduced from Parrino et al., Sleep Med Rev. 2012 [36])

17.1.2 Clinical Implications

Understanding the macrostructure and microstructure of sleep is essential for comprehending the functional significance of various sleep stages and their role in health and disease. Alterations in the macrostructure, such as reduced N3 or REM sleep, are associated with conditions like insomnia, depression, and neurodegenerative disorders. Microstructural abnormalities, including altered sleep spindle density or increased arousal frequency and CAP rate, can indicate specific sleep pathologies and inform treatment strategies. The exploration of the macro- and microstructure of sleep provides a comprehensive understanding of its complex and dynamic nature. By elucidating the patterns and mechanisms of sleep, researchers and clinicians can develop more effective strategies to promote healthy sleep and address sleep disorders, ultimately enhancing overall health and quality of life.

17.2 Global and Local Aspects of Sleep EEG: A Comprehensive Analysis

In the last two decades, the traditional view of sleep as a global phenomenon orchestrated by specialized central neuronal networks has been challenged by evidence demonstrating that vigilance states are highly dynamic, regionally complex, and locally regulated. Although the concept of local sleep was proposed and experimentally shown many years ago, it has been more recently introduced in humans over the past 20 years [9, 10]. For a long time, sleep and wakefulness were considered mutually exclusive behavioral states. This view is still evident in the current classification of sleep stages, where each 30-second epoch of PSG recordings is labelled as either sleep or wakefulness. However, recent data show that different brain regions can display temporal or spatial discrepancies in their activities, making it difficult to assign a global vigilance state to such a mixture of dynamics.

At the beginning of this century, the first studies in humans proposing the concept of local sleep were primarily conducted using scalp EEG. These studies demonstrated that NREM sleep is characterized by large regional frequency-specific EEG differences, with a predominance of delta activity (1–4.5 Hz) in frontal regions [11–13]. Additionally, other studies showed that sleep EEG delta power can be locally modulated based on previous waking activity [14–16] highlighting that sleep is "use-dependent" also at a local level. The most evident examples of a local regulation of sleep come from some migrating birds and marine mammals, where one cerebral hemisphere displays low-amplitude high-frequency EEG activity (typical of wakefulness), while the other simultaneously exhibits high-amplitude slow waves (typical of sleep). The discovery of this unihemispheric sleep has various functional and evolutionary implications. In birds, it allows them to continue flying while keeping one eye open to monitor the environment, and in dolphins, it enables them to continue breathing while they swim [17].

Over the last 20 years, new and increasingly sophisticated techniques such as magnetoencephalography, high-density EEG, fMRI-EEG, and stereo-EEG have been developed to study sleep with higher spatial and/or temporal resolution compared to standard scalp EEG. These tools have deepened our understanding of the local aspects of sleep and provided insights into sleep physiology and pathophysiology.

In this chapter, we will summarize the main evidence supporting the local regulation of sleep derived from studies using intracranial EEG recording techniques. The recent spread of stereo-EEG investigations, mainly for the presurgical evaluation of drug-resistant epilepsy patients, has allowed researchers to overcome the inherent spatial limitations of scalp EEG, providing detailed information on the activity of various cortical and subcortical structures during both wakefulness and sleep.

17.2.1 Falling Asleep Locally

The transition from wakefulness to NREM sleep is characterized by a shift from low-amplitude high-frequency to high-amplitude low-frequency EEG activity, known as "EEG synchronization," leading to a deactivated state. However, through scalp EEG analysis, it is impossible to pinpoint a precise moment that characterizes the shift from wakefulness to sleep. Moreover, scalp EEG studies have demonstrated that the wake-to-sleep transition is regulated at the regional level. Specifically, quantitative analyses of sleep scalp EEG recordings during sleep onset show that the more anterior areas show an earlier synchronization with a predominance of slow frequencies below 4 Hz and spindle oscillations (i.e., they fall asleep first). Conversely, frequencies >8 Hz are prevalent in the occipital regions [18].

More recently, stereo-EEG investigations in humans have shown that the transition to sleep varies significantly in time across different subcortical and cortical structures. Subcortical regions such as the hippocampus and thalamus fall asleep minutes before cortical regions. For the first time, in 2010, Magnin et al., using stereo-EEG recordings, showed that during physiological sleep onset, the thalamus deactivates much earlier than other cortical areas [19]. Different cortical regions maintained an activated pattern for several minutes after thalamic deactivation, demonstrating marked topographical differences in cortical deactivation during sleep onset within and between patients.

A more recent stereo-EEG study involving simultaneous intracerebral recordings from the hippocampus and various neocortical areas in nine drug-resistant epilepsy patients found that sleep spindles occurred in the hippocampus several minutes before sleep onset. Hippocampal spindle detections consistently preceded neocortical events, with increasing delays along the cortical antero-posterior axis (Fig. 17.6) [20]. In one patient, SEEG activity was recorded from a bipolar contact in the anterior thalamic nucleus.

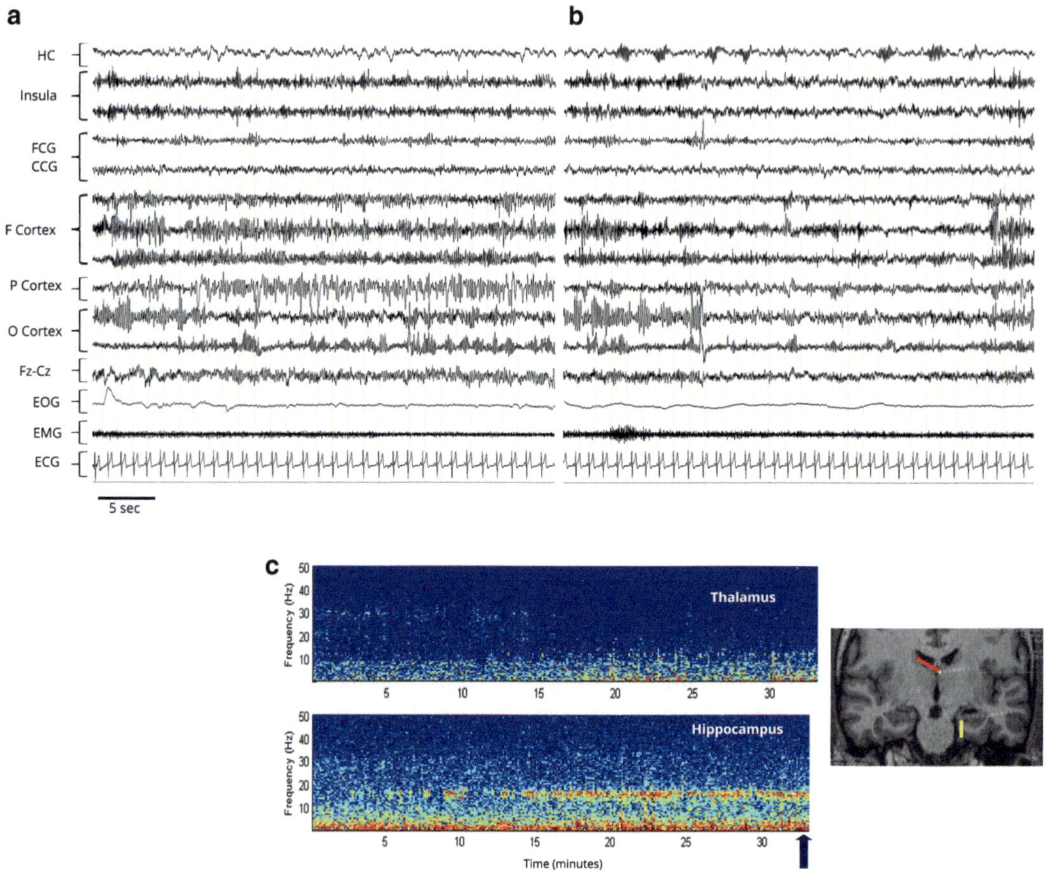

Fig. 17.6 (**a**, **b**) Stereo-EEG signals recorded from various brain regions. The regions include the hippocampus (HC), insula, frontal cingulate gyrus (FCG), central cingulate gyrus (CCG), frontal cortex (F cortex), parietal cortex (P cortex), occipital cortex (O cortex), scalp EEG (Fz–Cz), electrooculogram (EOG), electromyogram (EMG), and electrocardiogram (ECG). (**a**) Baseline recordings showing typical wake EEG activity across all channels. (**b**) Notable change in hippocampal activity characterized by the appearance of periodic spindles indicative of sleep-like activity, while all other channels continue to display wake EEG activity. (**c**) Time–frequency distribution of EEG patterns recorded from the thalamus and hippocampus. The data illustrates an increase in delta power in the thalamus minutes before sleep onset (indicated by the blue arrow). Correspondingly, there is an increase in spindle frequency activity (14–16 Hz) in the hippocampus. On the right, brain MRI indicating the positions of the electrodes. The red arrow points to the electrode in the thalamus, and the yellow arrow points to the electrode in the hippocampus. This figure demonstrates the distinct changes in hippocampal activity associated with a "local" sleep onset, contrasted with the wake EEG patterns observed in other brain regions. The time–frequency analysis further emphasizes the preparatory changes in the thalamus and hippocampus preceding the global sleep onset

During wakefulness, the thalamus exhibited a fast rhythm, peaking around 25 Hz. Some minutes before sleep onset, periodic slow oscillations in the low-delta range (0.5–2 Hz) appeared in the thalamic EEG recording, coinciding with the appearance of hippocampal spindle activity (Fig. 17.6) [21]. This observation was further confirmed by a seed-target analysis showing that hippocampal spindles were invariably associated with thalamic slow oscillations. These findings suggest significant interplay between subcortical and archicortical structures during the wake-to-sleep transition, highlighting that brain structures such as the thalamus and hippocampus tend to show sleep signs before the neocortex [20].

17.2.2 Local Activation During NREM Sleep

Stereo-EEG investigations have demonstrated that NREM sleep can be characterized by the coexistence of wake-like and sleep-like activity in different cortical areas. Nobili et al. analyzed intracerebral sleep EEG activity of the motor cortex, dorsolateral prefrontal cortex, and scalp EEG in five drug-resistant patients [22]. During NREM sleep, they found abrupt increases in high-frequency EEG activity (alpha and/or beta rhythm) over the motor cortex without evident motor behaviors (Fig. 17.7). The motor cortex exhibited higher rates of local activations during NREM sleep compared to the prefrontal cortex and scalp. These local activations increased across sleep cycles and within each cycle toward the end of each NREM period. Generally, the local activations lasted around 10 seconds but could extend up to 120 seconds, often occurring without EMG activation, either in isolated or cyclical patterns (Fig. 17.8). Cyclical patterns resembled the scalp EEG pattern of the cyclic alternating pattern. This fast activity in the motor cortex may reflect noradrenergic system activation, representing a form of arousal rarely captured by scalp EEG recordings and classical sleep

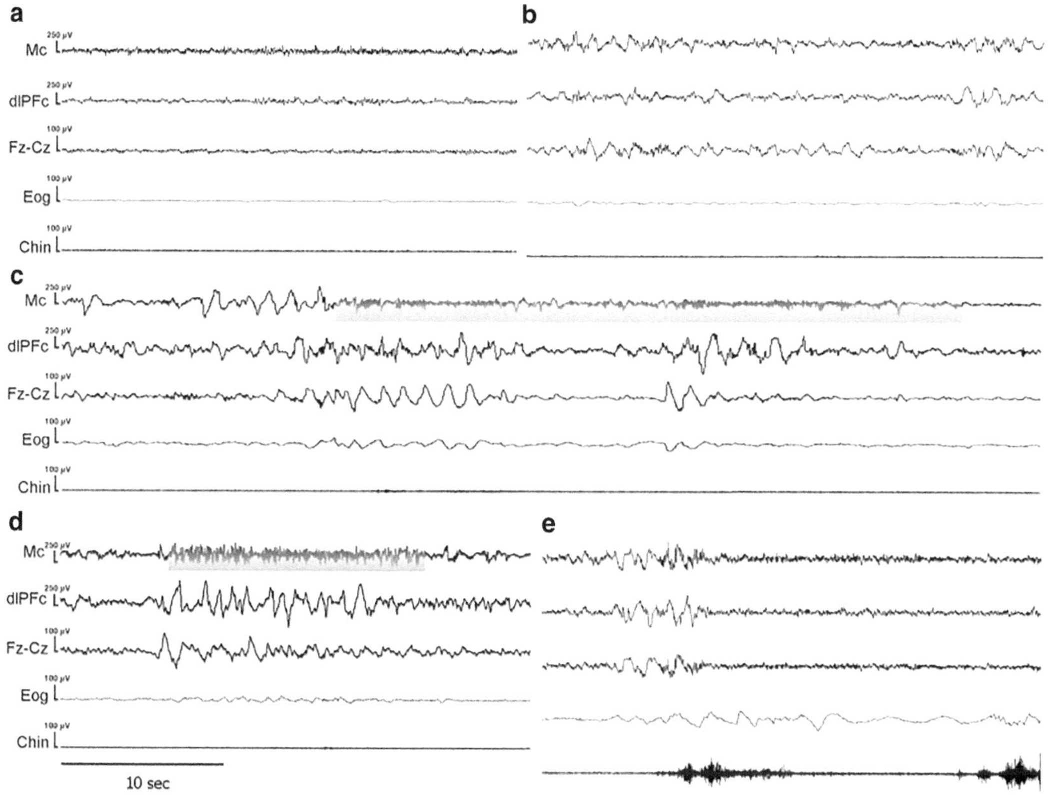

Fig. 17.7 Sample patterns of intracerebral EEG. The first three traces show EEG recordings from the motor cortex (Mc), dorsolateral prefrontal cortex (dlPFc), and scalp (Fz–Cz) and are followed by the electrooculogram (EOG) and electromyogram (EMG, chin). (**a**) Quiet wakefulness (eyes closed): fast rhythms prevail in all the three EEG tracings. (**b**) NREM sleep: slow waves prevail in all the three EEG tracings. (**c**) A local activation (LA, gray shadowed area), characterized by fast EEG activity, appears in the motor cortex derivation and continue for tens of seconds, while sleep EEG with slow waves prevails in the other tracings. (**d**) When a short LA appears in the motor cortex, a burst of slow waves characterizes the other two EEG derivations. (**e**) A diffuse short burst of delta waves followed by a complete awakening with fast rhythms in all the three EEG recordings. (Reproduced from Nobili et al., Neuroimage 2011 [22])

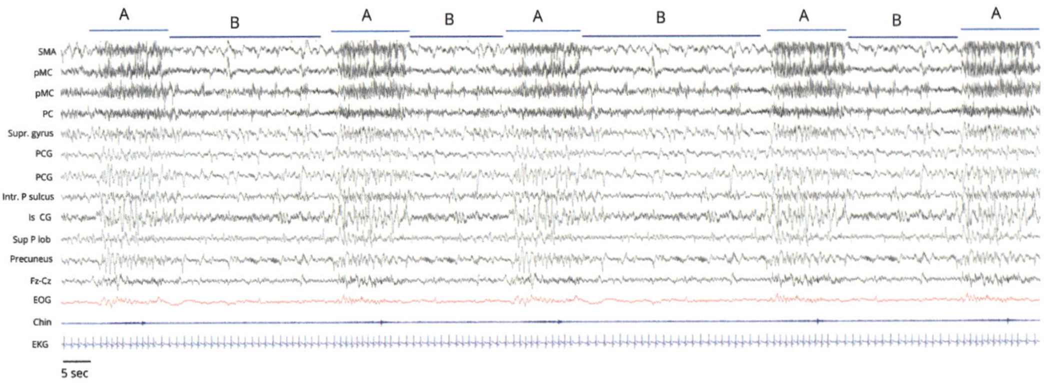

Fig. 17.8 Stereo EEG recordings from various cortical areas during NREM sleep. The recordings illustrate oscillatory arousal fluctuations characterized by alternating phases A and B of cyclic alternating pattern. During phase A, certain brain regions exhibit fast EEG activity indicative of wake-like activity, while other regions show slow waves characteristic of sleep-like activity. This coexistence of wake-like and sleep-like activity within the same phase highlights the complexity and regional specificity of brain dynamics during NREM sleep. SMA (supplementary motor area), pMC (primary motor cortices), and PC (parietal cortex) show fast EEG activity (wake-like activity) during A phases. Supr. Gyrus (supramarginal gyrus), Intr. P Sulcus (intraparietal sulcus), Is CG (isthmus of the cingulate gyrus), Sup P Lob (superior parietal lobule), and precuneus and scalp EEG electrodes (Fz–Cz) display slow waves during A phases A. EOG (electrooculogram). Chin (chin electromyogram). EKG (electrocardiogram). The figure underscores the phenomenon of oscillatory activity in the brain during NREM sleep, where wake-like and sleep-like activities may coexist and vary across different cortical regions

scoring. This local coexistence of sleep-like and wake-like activity could balance the need to continue sleeping with the importance of waking up to external stimuli.

Recent studies analyzed intracerebral dynamics leading to different arousal subtypes on scalp EEG. Sleep arousals are transient periods of increased vigilance during sleep without awareness or recollection, classically considered an all-or-none phenomenon representing an intermediate state between sleep and wakefulness. Arousals can exhibit a range of scalp EEG activity patterns depending on factors such as the ongoing sleep stage, the stimulus triggering the arousal, homeostatic factors, and the brain area involved. In 2015, Peter-Derex et al. analyzed SEEG activity in eight drug-resistant patients from the thalamus and primary, associative, or high-order cortical areas during spontaneous or nociceptive-induced arousals in NREM and REM sleep [23]. They found stereotyped thalamic activity during all arousals, corresponding to a state between wakefulness and sleep. Cortical activity patterns during arousals varied based on sleep stages, cortical areas, arousal modality ("spontaneous" vs. nociceptive-induced), and homeostasis. This suggests stereotyped thalamic arousals are associated with diverse cortical activation patterns depending on multiple regulation factors. The spectral composition of EEG signals during arousals differed from wakefulness, indicating arousals are not simply brief awakenings but represent a distinct brain state.

A recent study using combined polysomnography and SEEG investigated intracranial correlates of scalp EEG arousals during NREM sleep [24]. Authors differentiated arousals into slow-wave (SW) and non-SW arousals based on their co-occurrence with a scalp EEG SW event. Analyses in up to 26 brain regions from 26 drug-resistant patients revealed increased intracranial theta, alpha, sigma, and beta activities in up to 25 regions and decreased gamma and high-frequency activities in up to 18 regions across the 5 brain lobes for both arousal types. During SW arousals, intracranial delta power widely increased across five lobes, while it decreased during non-SW arousals. This suggests non-SW arousals correspond to higher brain activation levels than SW arousals, with decreased high-frequency activities possibly explaining variation in awareness and recollection during arousals.

17.2.3 NREM to REM Sleep Transition

As stated above, various studies have analyzed cortical and subcortical correlations to wake–sleep transitions; however, only a few works are available regarding NREM to REM sleep transitions. Results from scalp EEG studies suggest that the transition between NREM and REM sleep is characterized by a gradual increase in mechanisms determining the occurrence of REM sleep periods several minutes before REM onset and a decrease in the activation of NREM sleep networks only a few seconds before REM sleep onset. The gradual appearance of REM sleep PSG features (sawtooth waves, muscle atonia, and bursts of REMs) during the NREM to REM sleep transition is influenced by different factors, such as the instability of NREM sleep during the ascending slope of NREM episodes, the frequency of arousals, the progressive decrease in delta and sigma activity, and a progressive increase in alpha and high beta activities. Using SEEG data from the hippocampus and cortex, a recent study demonstrated that the hippocampus enters REM sleep before the neocortex and that epochs of REM sleep can occur within the hippocampus independently of the cortex. In particular, the authors found that the cortex is in REM sleep simultaneously with the hippocampus about 60% of the time [25]. Another study applied PSG combined with SEEG recordings in 29 epileptic patients to explore the dynamics of NREM to REM transitions [26]. The average transition time from all intracerebral channels to the first visually marked REM sleep epoch was 8 s ± 1 min 58 s, with great heterogeneity between brain areas. Moreover, this study demonstrated that REM sleep onset shows a reproducible cortical anatomical pattern. Indeed, the occipital cortex exhibits the first cortical manifestations of REM sleep, on average 2 minutes before the scalp EEG detects the transition and close to the first sawtooth wave marker. Conversely, regions with late transitions were the inferior frontal and orbital gyri.

17.2.4 Local Activations During REM Sleep

REM sleep represents a state of high cerebral and low physical activation. Indeed, dreams are frequently reported by subjects awakened from REM sleep, but this sleep stage is characterized by atonia of postural muscles. There are few studies directly analyzing the activity of the human motor cortex during REM sleep. However, animal models and clinical observations in patients with REM sleep behavior disorder (RBD) suggest a possible activation of the motor cortex during REM sleep.

As previously described, REM sleep can be subdivided into two different substates, characterized by different sensory processing, awakening thresholds, and cortical activity: phasic REM sleep (REM sleep with bursts of rapid eye movements, REMs) and tonic REM sleep (REM sleep without REMs) (Fig. 17.3) [6]. Interestingly, RBD-associated behaviors are reported to occur more frequently during phasic REM sleep than during tonic REM sleep, suggesting a higher level of motor cortex activation during phasic REM sleep.

In 2016, a study using SEEG data from seven epileptic patients evaluated the activity of the motor cortex and the dorsolateral prefrontal cortex during physiological REM sleep and wakefulness [27]. During tonic REM sleep, the motor cortex showed alpha-like oscillatory activity (mu-rhythm), which disappeared during phasic REM sleep, characterized by a desynchronized pattern (Fig. 17.9). The log-transformed power spectra of the relevant motor cortex EEG signals confirmed the visual analysis, finding a decrease of power in a large frequency band up to 25 Hz, with a slight increase of power above 25 Hz during phasic REM sleep. Post hoc comparisons showed that only the motor cortex presented higher mean frequency spectral values during phasic REM sleep compared to tonic REM sleep; this behavior was not evident in the prefrontal cortex. To evaluate possible similarities between the EEG activity of the motor cortex during REM

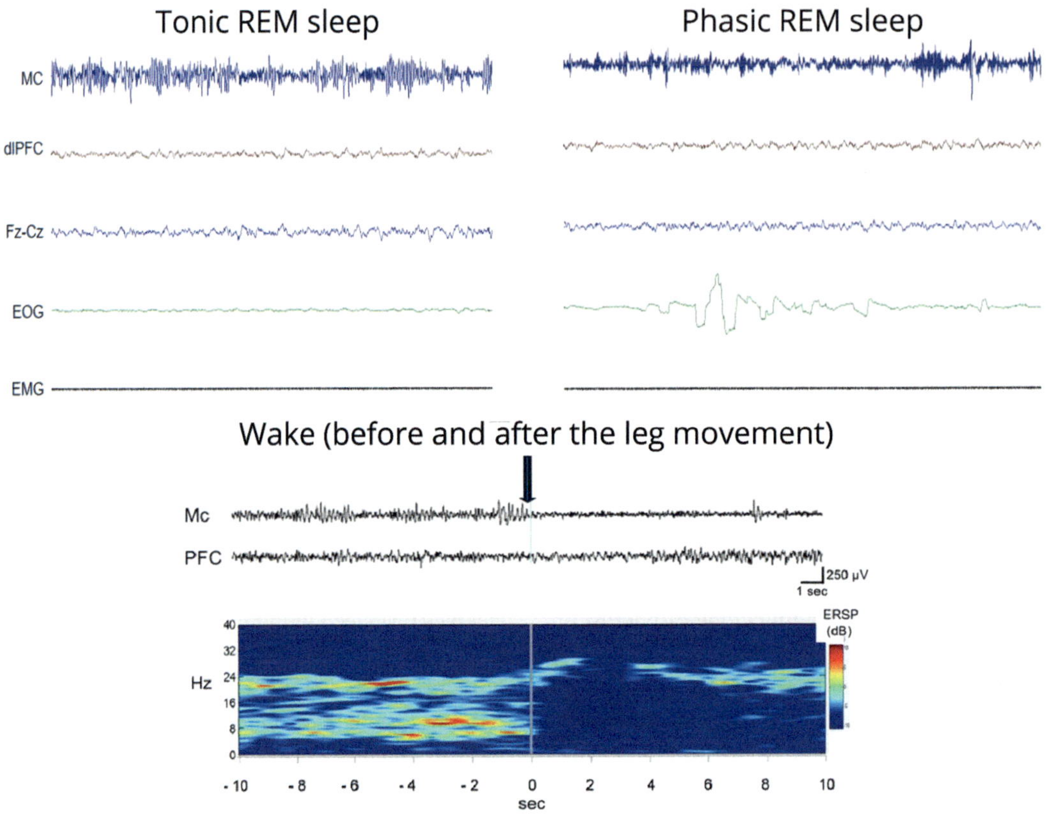

Fig. 17.9 Top. Example of tonic and phasic rapid eye movement (REM) 30-second epochs. Notice the EEG desynchronization, characterized by the disappearance of the mu-rhythm (alpha-like oscillatory activity) during phasic REM sleep. MC, motor cortex; dlPFC, dorsolateral prefrontal cortex; Fz–Cz, scalp EEG; EOG, electrooculogram, EMG, electromyogram. Bottom: Electroencephalographic (EEG) activity in the motor cortex (Mc) and in the dorsolateral prefrontal cortex (PFC) during rest and voluntary limb movements. EEG trace derived from a bipolar derivation in the Mc and PFC during a 20-second epoch around the onset of a voluntary leg movement (marked by the arrow and the gray vertical bar). Notice the EEG desynchronization, characterized by the disappearance of the mu-rhythm (alpha-like oscillatory activity) after movement in the Mc. Below: a time–frequency distribution of the amplitude of the EEG signal recorded from the Mc derivation and averaged among four 20-second epochs centered around leg movements. (Modified from De Carli et al., 2016 [27])

sleep and wakefulness, electrophysiological data were also acquired during voluntary leg movement in wakefulness. A clear mu-rhythm (6–12 Hz) characterized the activity of the motor cortex during the premovement epoch and disappeared at movement onset, followed by a predominant shift toward higher-frequency beta activity. Post hoc comparisons showed that only the motor cortex presented significant changes between conditions. In summary, this study demonstrated that similar EEG frequency values and changes were observed in the motor cortex during wakefulness (before and after movement) and during phasic and tonic REM sleep. Specifically, the motor cortex showed a similar increase in mean EEG frequency during phasic REM sleep and active wakefulness compared to tonic REM sleep and resting state, respectively. Therefore, it can be hypothesized that during phasic REM sleep and active wakefulness, the motor cortex exhibits a similar pattern of activation.

Other SEEG studies investigated the activity of other human brain structures during REM sleep substates, such as the limbic system and the thalamus. In particular, the activity of the amygdala was analyzed during spontaneous sleep in four epileptic patients with depth electrodes implanted in the temporal lobes [28]. This study found a transient activation of the amygdala time-locked to the onset of REMs during REM sleep, but not during waking eye movements. This finding suggests the central role of the amygdala in further activating the limbic–paralimbic network during REM sleep, highlighting the participation of this brain structure in the emotional content of dreams.

A recent study examined the activity of the anterior nuclei of the thalamus (ANTs) and their functional connectivity with scalp EEG recordings during REM microstates and wakefulness [29]. The study demonstrated that the heterogeneity of phasic and tonic REM sleep is not limited to cortical activity but is also manifested in subcortical regions. Indeed, enhanced high-alpha and beta frequency power emerged in the ANT during tonic compared with phasic REM, suggesting tonic REM as an intermediate state between phasic REM and wakefulness. Moreover, the study found increased thalamocortical synchronization in phasic compared with tonic REM sleep, especially in the slow and fast frequency ranges.

17.3 The Evolution of Sleep EEG: From Newborn to Old Age

The EEG provides critical insights into the development and maturation of the human brain throughout the lifespan. From newborns to the elderly, sleep EEG patterns undergo significant changes, reflecting the brain's evolving structure and function.

17.3.1 Evolution from Premature to Full-Term Neonate

The progression of sleep development in premature and neonatal infants follows a well-defined and reproducible ontogenetic trajectory. In premature infants, EEG recordings are typically characterized by discontinuous activity, with periods of quiescence interspersed with bursts of mixed-frequency activity. These patterns are essential indicators of brain maturation and can vary significantly based on gestational age. For instance, at around 24–28 weeks of gestation, EEG activity shows long periods of inactivity punctuated by brief bursts, a pattern known as *trace discontinu* [30] (Fig. 17.10). As gestational age progresses, the periods of inactivity shorten and the bursts become more frequent and com-

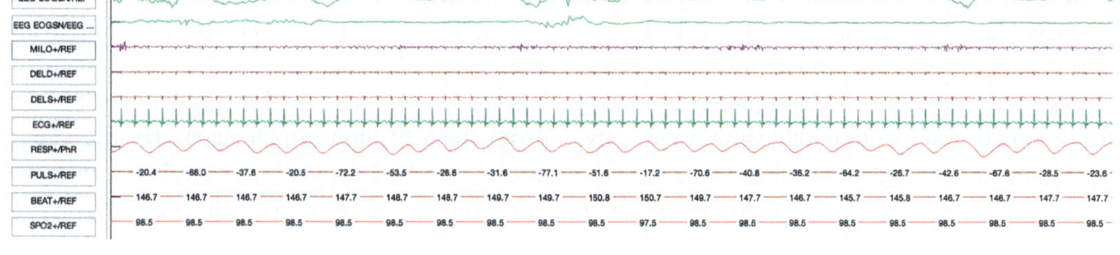

Fig. 17.10 *"Trace discontinu"* characterized by periods of inactivity interrupted by brief bursts

plex, indicating advancing cortical development. Early EEG in preterm infants also features delta brushes, which are high-amplitude delta waves with superimposed fast activity, primarily observed in the central and occipital regions.

At full-term, the EEG exhibits more continuous and organized patterns compared to the premature stage. Various EEG patterns identified in neonates during sleep include *tracé alternant* (TA), high-voltage slow (HVS) activity, mixed (M) activity, and low-voltage irregular (LVI) activity (Fig. 17.10). Frequent transitions between these sleep states reflect the rapid developmental changes occurring in the neonatal brain. Neonates exhibit many epochs of sleep that are not easily categorized into specific sleep states due to discordant features among physiological markers of sleep/wake. These periods, defined as transitional, indeterminate, or intermediate sleep, are markers of developmental maturation and decrease as the infant matures [31].

The neonatal EEG is characterized by three primary behavioral states: wakefulness, active sleep (akin to REM sleep in older children and adults), and quiet sleep (precursor to non-REM sleep) (Fig. 17.11). Wakefulness is characterized by open eyes, sometimes following visual targets, elevated chin EMG tone, rapid and irregular heart and respiratory rates, spontaneous movements, crying, and a continuous, irregular, mixed-frequency EEG with frequent movement artifacts. Active sleep is characterized by low-voltage mixed-frequency EEG activity, frequent limb or facial movements, rapid eye movements, and irregular respiration and heart rates (Fig. 17.11). Quiet sleep, on the other hand, is marked by a *trace alternant* pattern—alternating periods of high-voltage slow waves and low-voltage mixed-frequency activity—reduced movements, rare or absent eye movements, and regular respiration and heart rates (Fig. 17.11). This pattern gradually transitions to more continuous slow-wave activity by the end of the neonatal period [32]. The quality and organization of sleep during the neonatal stage may impact later cognitive and behavioral outcomes. Poor-quality sleep in both premature and term infants can result in long-term cognitive deficits, underscoring the importance of monitoring and promoting healthy sleep patterns from an early age [31].

17.3.2 Infancy (1–12 Months)

The first year of life is marked by rapid changes in EEG patterns, reflecting significant brain maturation. By 3 months, the EEG during wakefulness begins to show more rhythmic theta waves, and by 6 months, these rhythms increase in frequency, gradually giving way to early alpha rhythms [33]. Sleep spindles, which are bursts of rapid oscillatory brain activity, begin to appear around 2 months and become more pronounced and synchronous by 6 months. Similarly, other transient EEG activities, such as K-complexes and delta bursts, allow to define slow-wave sleep (SWS) at 4–4.5 months post-term.

Sleep in infants evolves with a gradual decrease in the proportion of REM sleep from around 50% at birth to approximately 30% by the end of the first year. The significant amount of REM sleep in early infancy underscores its crucial role in fostering optimal brain development, cognition, and behavior by providing sensory experiences in the absence of external stimuli, essential for developing sensory and motor pathways. This reduction is accompanied by an increase in NREM sleep stages, marked by the appearance of spindles, K-complexes, and more organized slow-wave activity [33].

17.3.3 Developmental Changes in Childhood

As children grow, notable changes occur in their sleep EEG patterns. Between the ages of one and three, the EEG maturation is evident, with the posterior dominant rhythm (PDR) transitioning from theta to alpha frequencies. By the age of three, the PDR typically ranges between 6 and 9 Hz. During drowsiness, hypnagogic hypersynchrony, characterized by high-voltage rhythmic theta activity, is commonly observed, gradually diminishing with age [32].

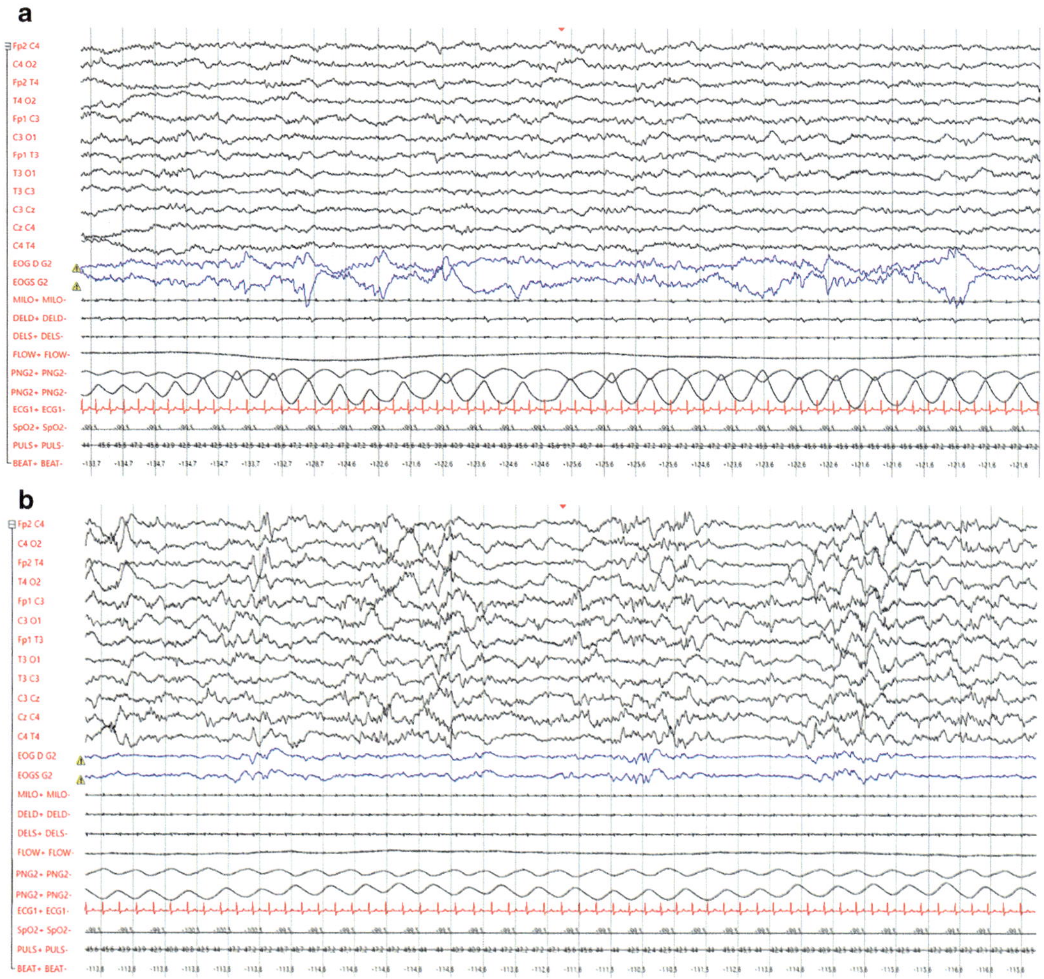

Fig. 17.11 Neonatal EEG patterns: (**a**) Active sleep marked by low-voltage mixed-frequency EEG activity, rapid eye movements, and irregular respiration and heart rates. (**b**) Quiet sleep characterized by a "trace alternant" pattern, with alternating periods of high-voltage slow waves and low-voltage mixed-frequency activity, rare eye movements, and regular respiration and heart rates

Delta waves dominate the EEG during slow-wave sleep, peaking in amplitude and frequency during early childhood and reflecting the high synaptic density and brain plasticity at this stage. In preschool children, the EEG shows further maturation of the PDR, which stabilizes within the alpha range (8–12 Hz). Sleep EEG patterns become more organized, with clear differentiation of sleep stages. Stage N2 is characterized by prominent sleep spindles and K-complexes, while stage N3 is marked by high-amplitude delta activity [32].

During school age, the EEG continues to mature, with the PDR increasing to around 10–11 Hz by the age of ten. The amplitude of the PDR is generally higher over the nondominant hemisphere. Sleep architecture becomes more stable, with the appearance of adultlike sleep patterns, including well-defined REM and NREM stages [32].

The EEG shows a gradual reduction in delta power, indicating a decrease in the intensity of slow-wave activity. This decline continues into adolescence, correlating with synaptic pruning and the maturation of neural networks. The structure of sleep becomes more stable, with well-defined sleep cycles and a decrease in total sleep duration as the child's need for sleep diminishes with age.

Adolescent EEG is characterized by a further increase in PDR frequency, stabilizing around 10 Hz. The amplitude of the PDR decreases compared to childhood, and there is a reduction in posterior slow waves. Sleep patterns also mature, with a more pronounced differentiation between REM and NREM stages and a reduction in the overall duration of REM sleep [32].

The changes in EEG patterns from infancy through adolescence reflect underlying neurophysiological maturation, including synaptic pruning, myelination, and cortical reorganization. These processes contribute to the development of cognitive functions, sensory processing, and motor coordination.

17.3.4 Adult Sleep EEG

In adulthood, the sleep EEG is characterized by the presence of distinct sleep stages: NREM stages N1, N2, and N3 and REM sleep. N1 is marked by a transition from alpha waves (8–12 Hz) to low-amplitude mixed-frequency activity. N2 shows sleep spindles (12–16 Hz) and K-complexes, superimposed on a background of theta waves (4–7 Hz). N3, or SWS, is dominated by high-amplitude delta waves (0.5–4 Hz), crucial for restorative processes such as tissue repair and memory consolidation. REM sleep in adults is characterized by low-amplitude, mixed-frequency EEG activity similar to wakefulness, with rapid eye movements and muscle atonia. The proportion of REM sleep remains relatively stable throughout adulthood, constituting about 20–25% of total sleep time. However, the overall amount of slow-wave sleep begins to decline in the late twenties, reflecting ongoing changes in brain structure and function [34].

17.3.5 EEG Changes in the Elderly

In older adults, sleep EEG patterns exhibit several notable changes that reflect the aging process and its impact on brain function [34, 35]. One of the most significant alterations is the decline in slow-wave sleep, with a corresponding decrease in delta power. This decline in delta activity is associated with reduced synaptic density and lower brain plasticity in the aging brain. Sleep spindles and K-complexes also show alterations, with spindles becoming less frequent and of lower amplitude. This reduction in spindle activity has been associated with declines in cognitive functions, particularly those related to memory consolidation. K-complexes also show diminished amplitude and frequency with age, further contributing to the fragmentation of sleep architecture.

The overall architecture of sleep in the elderly is characterized by increased sleep fragmentation, with more frequent awakenings and transitions between sleep stages. REM sleep percentage may decrease slightly, and the latency to REM sleep onset can be prolonged. These changes contribute to the overall lighter and more disrupted sleep observed in older adults, which can impact cognitive function and overall health.

The evolution of sleep EEG patterns from the neonatal period to old age reflects the intricate processes of brain maturation and aging. Understanding these changes is essential for diagnosing and managing sleep disorders across the lifespan. Research continues to uncover the mechanisms behind these EEG alterations, providing deeper insights into the relationship between sleep and brain health.

17.3.6 Modification of the Cyclic Alternating Pattern (CAP) Throughout Different Ages

As reported in the previous paragraphs, the cyclic alternating pattern (CAP) is an essential feature of NREM sleep, reflecting a marker of sleep stability during sleep [36]. The characteristics and prevalence of CAP change throughout the lifespan, influenced by neurodevelopmental and neurodegenerative processes. CAP rate, calculated as the percentage ratio of total CAP time to non-REM sleep time, evolves complexly during development. It is very low in the first months, gradually increases to peak during adolescence, decreases in adulthood, and increases again in the elderly. The first prototype of CAP appears at 46–55 weeks of conceptional age, with rates increasing in 3–4 month-old infants as sleep patterns mature. The high CAP rate in children seems to be associated with the high synaptic density and neuroplasticity characteristic of developing brains. In general, the changes in CAP rate parallel brain plasticity changes during biological growth, the onset of adolescence, the processes of maturation, and senescence, which in turn also manifest as modifications in sleep stability.

The percentage of A1 subtypes follows a bell-shaped curve, peaking in school and peripubertal children and decreasing in adults and the elderly. In contrast, A2 and A3 subtypes increase linearly with age. The high A1 to A2/A3 ratio in school-age children indicates optimal sleep quality due to length, continuity, and restorative features. Increased A2 in preschoolers may indicate higher sleep instability.

In school-aged children, increased CAP rate during SWS and a higher percentage of A1 subtypes suggest that more oscillations are needed to maintain restorative sleep.

Also, the number of CAP cycles varies with development, peaking in peripubertal children and decreasing in adults and the elderly. Mean CAP cycle duration is stable, with phase B dominating. The time structure of CAP, evaluated by the intervals between consecutive A phases, shows periodicity. A1 intervals peak at about 25 seconds, with no clear peaks for A2 and A3 phases. This periodicity is stable across ages, indicating consistent CAP components throughout development [36].

17.4 EEG in Sleep Disorders

17.4.1 Insomnia

Insomnia is the most common sleep disorder and is among the most prevalent of all mental health disorders. This condition is characterized by difficulty in initiating or maintaining sleep, or experiencing non-restorative sleep, which causes clinically significant distress or impairment in social, occupational, or other important areas of daily functioning. Polysomnography is not recommended for the routine evaluation of insomnia, as the diagnosis primarily relies on self-reported symptoms. According to the recent European Insomnia Guideline, the diagnosis of chronic insomnia disorder continues to be based solely on clinical interviews, encompassing sleep and medical history [37]. The use of sleep questionnaires and diaries (and physical examinations and additional measures where indicated) should support the diagnosis. Actigraphy is not recommended for the routine evaluation of insomnia but may be useful for differential diagnosis purposes. The recent European guidelines state that polysomnography is not necessary or sufficient for the diagnosis of insomnia per se. However, it could be indicated when (a) the diagnosis is uncertain, (b) a sleep-related breathing disorder or periodic limb movement disorder is suspected, or (c) the patient is refractory to treatment.

A meta-analysis of PSG studies showed that insomnia patients, compared with controls without sleep complaints, regularly display a significant reduction in total sleep time (by 24 minutes), a significantly longer sleep onset duration (by about 6 minutes), and more frequent nocturnal awakenings (about 6 more awakenings). Moreover, considering the sleep macrostructure, PSG detected 20 minutes less slow-wave sleep and 11 minutes less REM sleep in patients with insomnia compared to good sleepers, corresponding to an increase of about 22 minutes of wake after sleep onset (WASO) [38]. These differences, compared to good sleepers, however, are considerably smaller than would be expected from subjective measures of sleep, such as diaries or questionnaires. Furthermore, these PSG abnormalities are not specific to insomnia but can be found in several other sleep disorders and conditions.

Besides conventional PSG measures, more integrative and detailed information in patients with insomnia can be derived through sophisticated examination techniques such as sleep EEG spectral analysis or the detection of micro-arousals and cyclic alternating patterns (CAP). A recent meta-analysis of studies on EEG spectral features in insomnia revealed that during wakefulness, insomnia patients exhibited significant increases in absolute and overall theta power, relative beta power, and absolute gamma power compared to good sleepers [39]. During NREM sleep, they showed significant alterations in EEG power across all frequency bands except gamma. In these patients, NREM sleep was associated with decreases in relative and overall delta power and increases in absolute and overall beta power, as well as relative power of theta, alpha, sigma, and beta bands. Finally, during REM sleep, insomnia patients exhibited significant increases in absolute alpha power, absolute sigma power, and overall beta power. The finding of increased beta activity during both wakefulness and sleep in insomnia patients supports the concept of continuous hyperarousal underlying this disorder and confirms that insomnia represents a 24-hour pathology.

The cyclic alternating pattern is a useful microstructural measurement in insomnia, significantly correlated with subjective sleep quality. A high CAP rate indicates disturbed and poor-quality sleep. Insomnia patients typically show increased CAP time and rate, which can help quantify sleep quality and the effectiveness of hypnotic treatments. CAP analysis is particularly suitable for patients with "misperception insomnia," which shows a mismatch between subjective reports (poor sleep quality, nocturnal awakenings) and objective PSG parameters. Indeed, these patients show significantly higher amounts of CAP rate in N1 and N2 stages, but not in slow-wave sleep [40]. Moreover, misperceptors report fewer but longer subjective awakenings in contrast to objective findings. This mismatch seems to be related to the high amounts of CAP between successive awakenings, which are merged together by the patient into a single experience. Therefore, if sleep between two successive awakenings is superficial, unstable (characterized by a high CAP rate), and fragmented (increased arousal index), the time separating the two events may be perceived as continuous wakefulness by patients with misperception insomnia [40].

17.4.2 NREM Parasomnia

NREM sleep parasomnias represent a group of sleep disorders characterized by episodes of abnormal sleep-related complex movements, associated with various degrees of autonomic nervous system activation, inappropriate or limited responsiveness to the external environment, and partial to complete amnesia. These episodes usually occur during the first third of the major sleep period and are considered to result from an incomplete arousal from NREM sleep (Fig. 17.12). According to the International Classification of Sleep Disorders, Third Edition (ICSD-3), NREM sleep parasomnias encompass disorders of arousal (DOAs), which include three main clinical presentations: confusional arousals, sleep terrors, and sleepwalking (also known as somnambulism). Sleep-related abnormal sexual behavior, also known as sexsomnia, is a subtype of confusional arousals and sleepwalking. Within the group of NREM parasomnias, sleep-related eating disorder (SRED) is also included [41].

A clear clinical history is sufficient to diagnose NREM parasomnia in the majority of cases, but in some instances, only video-polysomnography (vPSG) recording can clarify the nature of the disorder. Although vPSG is not necessary for the diagnosis according to ICSD-3, it is recommended in the following situations: (a) when the clinical history is not completely suggestive of NREM parasomnia, (b) in the presence of injurious or extremely disruptive behaviors,

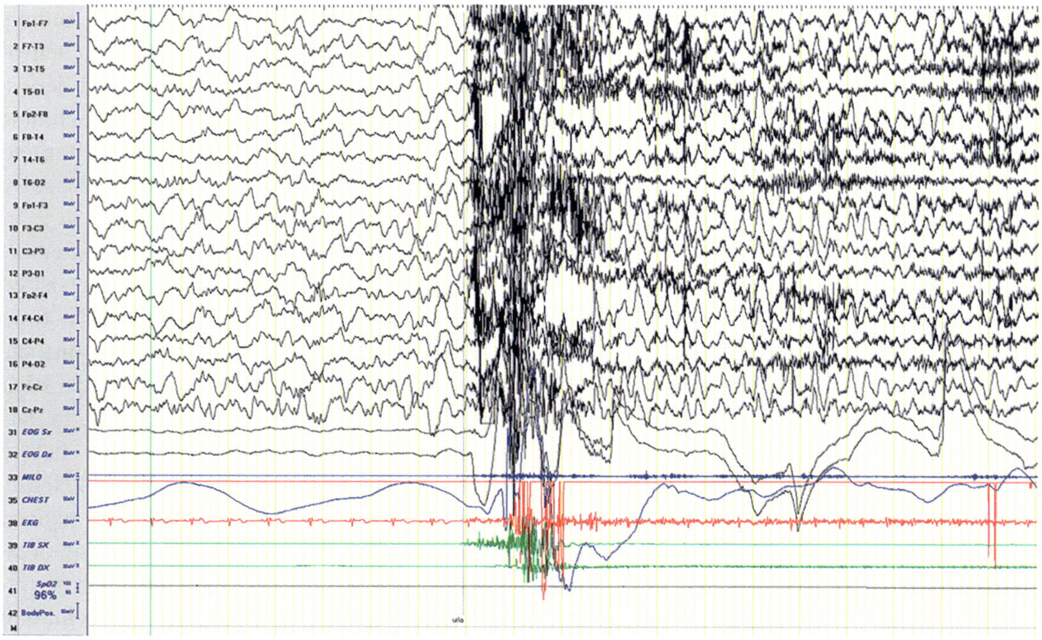

Fig. 17.12 NREM sleep and onset of confusional arousal in a child. This figure represents a 20-second segment of NREM sleep, capturing the onset of a confusional arousal in a child. The recordings highlight the persistence of rhythmic slow waves with superimposed muscular activity. This combination is indicative of the transitional state between sleep and wakefulness, often seen during confusional arousals. From 1–18 EEG channels, bipolar derivations; EOG, electrooculogram (dx, right; sx, left); Milo, electromyogram; Chest, thoracic respiratory band; EKG, electrocardiogram; TIB (tibial muscles, left, sx; and right, dx). SpO_2, oxygen saturation

(c) when an associated sleep disorder (e.g., sleep apnea or periodic limb movement) is suspected, or (d) when the parasomnia is associated with medical, psychiatric, or neurological symptoms or findings [41].

There are no consistently robust features in terms of overall sleep architecture and normal cycling among sleep stages that can be highly suggestive of patients with NREM parasomnia. However, subjects with NREM parasomnias exhibit NREM sleep instability. Polysomnographic studies in both children and adults have shown increased fragmentation of SWS, characterized by a higher number of awakenings and micro-arousals during SWS, even on nights without episodes [42, 43]. Additionally, an increase in cyclic alternating pattern rates has been observed even on nights when parasomnia episodes do not occur [44, 45].

Studies with high-density EEG and source modelling techniques in patients with DOAs compared to controls found a decrease in normalized EEG power during N2 and N3 sleep, which is prominent in the slow-wave activity range (1–4 Hz) and is localized mainly to the cingulate and motor regions in patients suffering from NREM parasomnia [46].

REM sleep has rarely been investigated in DOA, but a polysomnographic study showed excessive twitching during REM sleep in adults with DOA (but no tonic increase in muscle activity), which may result either from higher dreaming activity in REM sleep or from a more generalized NREM/REM motor dyscontrol [47].

The scalp EEG during a motor episode is often contaminated by muscle activity, but when visible, it may show hypersynchronous delta activity (HSD) (Fig. 17.12). HSD is defined as continuous high-voltage (>150 microV) delta waves occurring during slow-wave sleep or immediately prior to an episode and has been investigated for a long time as a possible diagnostic sign of NREM parasomnia. Notably, the spatial aspect of the EEG can differ on the frontal channels (containing diffuse slow waves) and the posterior channels (where EEG activity is faster). In some cases, a combination of delta and alpha activity (alpha–delta aspect) is observed on the same channel. HSD was investigated as a possible diagnostic marker in NREM parasomnia. However, more recent studies indicate that this EEG pattern does not appear to be a sensitive or specific diagnostic sign for NREM parasomnia. Indeed, HSD can also be detected in patients with other sleep disorders, such as sleep apnea or periodic leg movements. Moreover, HSD was absent in many sleepwalkers before episodes of complex behaviors.

The intracerebral electrophysiological aspects during NREM parasomnia episodes have been recorded by chance in epileptic (and more rarely non-epileptic) patients using stereo-EEG. In particular, in three case reports, Terzaghi et al. described the coexistence of sleep-like patterns (increased or persistent bursts of sleep-like delta waves) over frontoparietal associative networks and spindles in the hippocampus and wake-like patterns (low-voltage fast activity) over the motor cortex and limbic structures, such as the cingulate cortex (frontal and central cingulate gyrus), insular cortex, temporopolar cortices, and amygdala (Fig. 17.13) [48, 49]. Moreover, during a DOA episode, EEG activity within the ventromedial portion of the thalamus showed a slight decrease in delta power and a clear-cut emergence of beta activity, similar in frequency and amplitude to the thalamic activity recorded during wakefulness.

Based on this dissociated pattern, authors hypothesized that the frequent emotional activation and amnesia for parasomnias could derive, respectively, from the activation of the limbic system disengaged from the prefrontal cortex and the parallel deactivation of the frontal associative and hippocampal cortices [50].

More recently, Flamand et al. analyzed 19 episodes of DOA recorded in 5 drug-resistant epileptic patients suffering incidentally from arousal disorders with the aim of studying the precise temporospatial dynamics of EEG changes in different cortical areas [51]. In particular, during the last few seconds before behavior onset, delta activity inconsistently increased, predominantly in frontal regions (especially in the supplementary motor area, the dorsolateral premotor, and the anterior cingulate cortices), opercular cortex, and temporal regions. This aspect was interpreted as the general mode of reaction to external or internal stimuli during N3 sleep, whether in normal or parasomniac subjects. In contrast, during the behavioral phase of confusional arousals, a complex, widespread increase in activity in all frequency bands (albeit predominantly at very low frequencies) occurred in the frontal lobes, cingulate cortex, insular cortex, and precuneus. A few seconds later, very low (<1 Hz) frequencies ceased, while HSD arose in the frontoparietal cortices (marked in the opercular cortex), contrasting with high (alpha and beta) frequency activities in sensorimotor, orbitofrontal,

Fig. 17.13 Local dissociation of sleep and arousal captured by stereo-electroencephalogram (stereo-EEG). (**a**) Example of local activation in the motor, cingulate (frontal and central cingulate gyrus), insular, temporopolar, and amygdalar cortices in contrast to the presence of slow waves in the frontal and parietal dorsolateral cortices as well as persistent spindles in the hippocampal cortex. (**b**) Arousal local dissociation captured by stereo-EEG in relation to clinical manifestations: a 30-s episode during which, with his eyes open, the patient sat up in bed, uttering unintelligible words and failing to answer his mother's questions. (**c**) Undisturbed sleep taken during the same time of the night (for comparison of EEG traces between local arousal event and normal sleep). Each EEG trace is a bipolar derivation from contiguous contacts. Six derivations record from the frontal dorsolateral associative cortex (H10–H11, H15–H16, F12–F13, and F13–F14, middle frontal gyrus; G13–G14 and F1–F2, superior frontal gyrus); three from the hippocampus (B2–B3; B3–B4; B4–B5: note the persistence of spindle activity); one from the temporal pole (T3–T4); one from the insular cortex (R1–R2); two from the amygdala (A1–A2; A2–A3); seven from the motor and motor control cortex (M1–M2, supplementary sensory motor area; M6–M7, precentral gyrus; S12– S13, primary motor area; G1–G2, anterior cingulate gyrus; H1–H2 and H2–H3, frontal cingulate gyrus; S1–S2, central cingulate gyrus). EOG, electrooculogram; Chin, chin electromyography; EKG, electrocardiogram. (Reproduced from Terzaghi et al., 2012 [49])

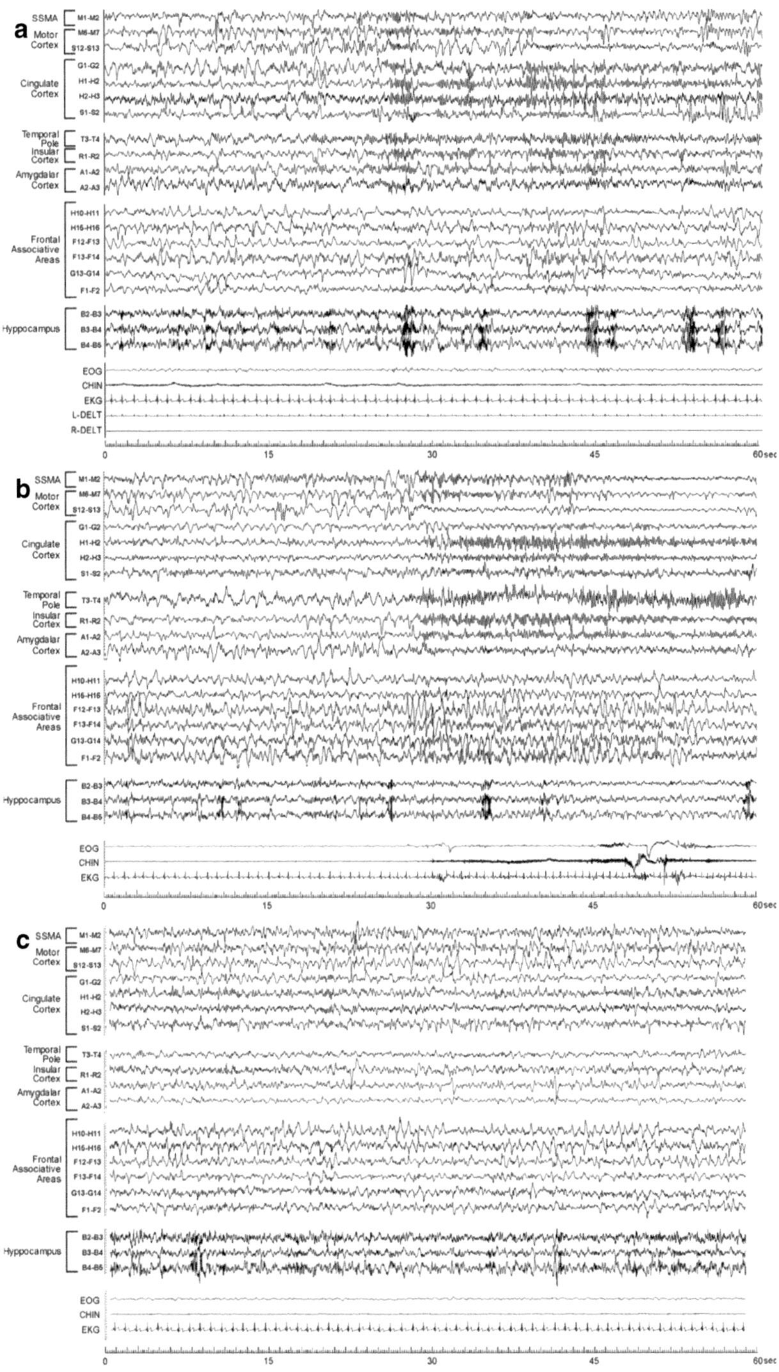
a
SSMA
Motor Cortex
M1-M2
M6-M7
S12-S13
Cingulate Cortex
G1-G2
H1-H2
H2-H3
S1-S2
Temporal Pole
T3-T4
Insular Cortex
R1-R2
Amygdalar Cortex
A1-A2
A2-A3
Frontal Associative Areas
H10-H11
H15-H16
F12-F13
F13-F14
G13-G14
F1-F2
Hyppocampus
B2-B3
B3-B4
B4-B5
EOG
CHIN
EKG
L-DELT
R-DELT
0
15
30
45
60sec
b
c

and temporal lateral (sometimes including the hippocampus) cortices. Authors hypothesized that the altered state of consciousness experienced by parasomniac patients might result from a disturbance of the activity (HSD) in a broad frontoparietal cortical network, especially the inferior frontal gyrus, which is involved in consciousness.

17.4.3 Sleep-Related Hypermotor Epilepsy (SHE)

Sleep-related hypermotor epilepsy (SHE), previously known as nocturnal frontal lobe epilepsy (NFLE), is a focal epilepsy characterized by asymmetric tonic/dystonic posturing and/or complex hyperkinetic seizures occurring mostly during sleep [52]. SHE represents a heterogeneous epileptic syndrome encompassing familial, sporadic, idiopathic, cryptogenic, and symptomatic cases. It is generally considered a benign form of epilepsy due to the relative ease of controlling seizures with antiseizure medications. However, severe and drug-resistant forms have been described.

Classically, SHE is associated with frequent complex motor seizures occurring during sleep. In a single night, patients may exhibit different sleep-related motor events of increasing complexity, ranging from short minor motor events (MME) and paroxysmal arousals (PAs) to longer-lasting major attacks.

SHE has been extensively studied as the archetype syndrome of hypermotor manifestations during sleep, initially with scalp and sphenoidal EEG, and more comprehensively with SEEG investigations. Unfortunately, the interictal and ictal scalp EEG features of SHE patients are often uninformative, especially in cases of seizures originating from deep brain regions, such as the mesial frontal cortex or insular cortex. However, large discrepancies in interictal epileptiform discharges (IEDs) can be appreciated during simultaneous recordings of scalp EEG and SEEG in patients with SHE (Fig. 17.14) [53]. Specifically, considering different sleep stages, IEDs from a focal cortical dysplasia recorded with SEEG, unlike scalp EEG findings, may be more numerous during wakefulness and REM sleep than during NREM sleep. During NREM sleep, individual IEDs become more widespread and of greater amplitude, increasing their visibility on scalp EEG (Fig. 17.15).

In symptomatic drug-resistant cases of SHE, the most frequent histopathological finding is type 2 focal cortical dysplasia (FCD) or Taylor-type. SEEG has enabled the characterization of the peculiar electrophysiological features of this FCD during different sleep/wake stages. During wakefulness, type II FCD produces characteristic rhythmic and subcontinuous spike- and polyspike-and-wave discharges that alternate with occasional short bursts of fast discharges (*brushes*), interrupted by electrical flattening. In NREM sleep, fewer spike-and-wave discharges

Fig. 17.14 Each EEG segment represents a 30-second epoch. (A) Wakefulness: the seizure onset zone (SOZ) exhibits the typical wake interictal activity of type II focal cortical dysplasia (FCD) consisting of subcontinuous rhythmic or sub-rhythmic spike- and polyspike-and-waves. Of note, this activity is not transferred to the scalp EEG recording over the SOZ. (**b**) NREM sleep: the SOZ produces the typical pseudo-rhythmic low-voltage fast discharges ("brushes") that are intermingled with spindle-like activity. Stereo-EEG contacts outside the lesional area and scalp EEG show periodic K-complexes, bursts of slow waves, and sleep spindles. (**c**) Paroxysmal arousal (PA): example of the observed "dissociated" pattern of coexistent sleep-like slow waves and fast wake-like activity on stereo-EEG. Following arousal, the lateral portion of the frontal lobe (sfg, mfg1, mfg2) produces high-voltage sleep-like slow waves, which are also detectable in the centro-parietal regions on scalp EEG. Meanwhile, the motor cortex (Mc), the somatosensory cortex (SSc), and the SOZ shift from delta rhythms to wake-like beta activity. (**d**) Seizure: a hypermotor seizure occurs in NREM sleep approximately 1 minute after the end of the PA. The SOZ initiates a low-voltage fast rhythmic epileptiform discharge (asterisk) that progressively evolves, reaching other contacts outside the SOZ at the end of the epoch (mfg1–2, mesial frontal gyrus contacts; sfg, superior frontal gyrus; pcl, paracentral lobule; Mc, motor cortex; SSc, somatosensory cortex; pcg, posterior cingulate gyrus; PC1–4, precuneus contacts; EKG, electrocardiogram; EOG, electrooculogram; EMG, electromyogram (chin). (Reproduced from Gibbs et al., 2015 [53])

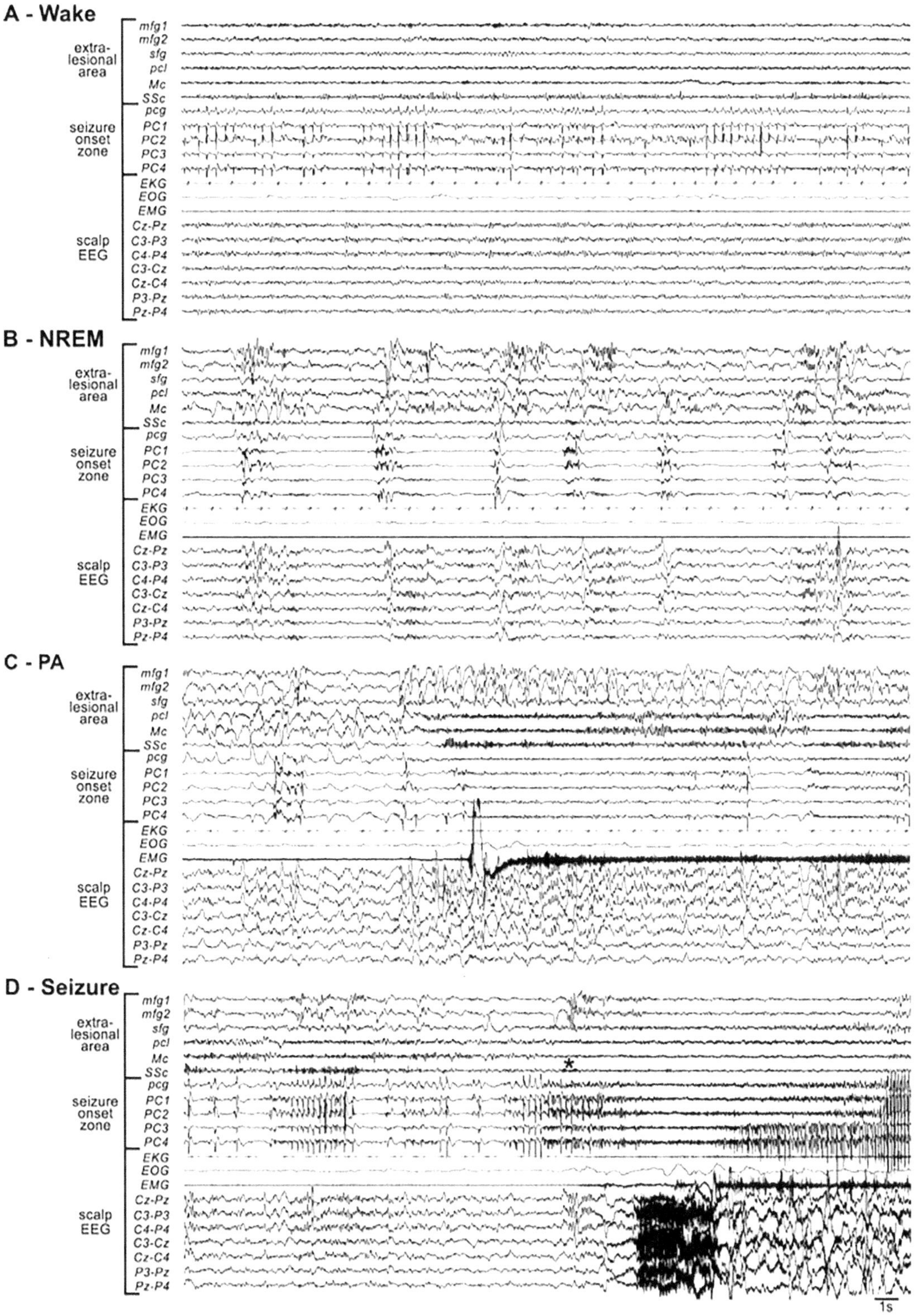
A - Wake
B - NREM
C - PA
D - Seizure
extra-lesional area
seizure onset zone
scalp EEG
mfg1
mfg2
sfg
pcl
Mc
SSc
pcg
PC1
PC2
PC3
PC4
EKG
EOG
EMG
Cz-Pz
C3-P3
C4-P4
C3-Cz
Cz-C4
P3-Pz
Pz-P4
*
1s

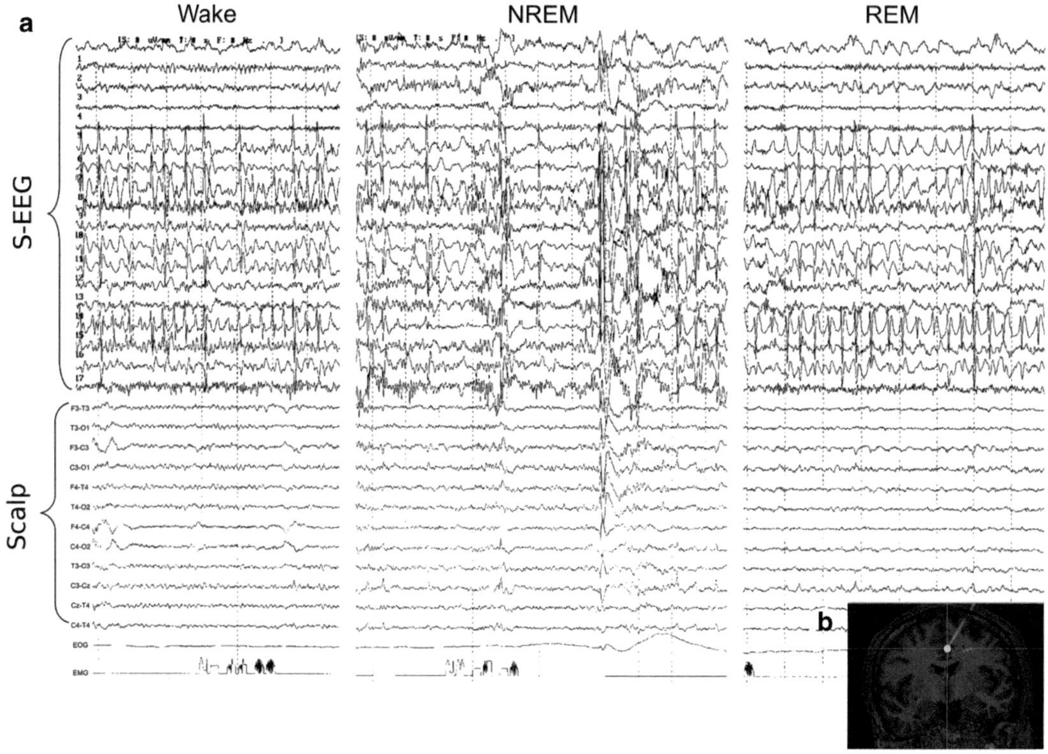

Fig. 17.15 (**a**) Simultaneous stereo-EEG (upper panel) and scalp EEG (lower panel) during wakefulness (wake), NREM sleep stage 2, and REM sleep. The epileptogenic zone was located medially in the left central cingulate gyrus (**b**, crosshair and dot). In all phases of the sleep–wake cycle, large discrepancies in interictal epileptiform discharges (IEDs) frequency between SEEG and scalp EEG are observed. The IEDs recorded with SEEG, contrary to scalp EEG findings, are more numerous during wakefulness and REM sleep than during NREM sleep. However, during NREM sleep, individual IEDs become more widespread and of greater amplitude increasing their visibility on scalp EEG. EOG, electrooculogram; EMG, electromyogram (chin). (Reproduced from Gibbs et al., 2016 [54])

are seen, replaced by considerably more frequent short bursts of low-voltage fast discharges inside the dysplasia that tend to recur pseudo-periodically, often spreading over the surrounding non-lesional areas and sometimes developing into a seizure (Fig. 17.16) [54].

The frequency of seizures in SHE patients can be very high, ranging from 1 to 20 attacks per night, and minor motor events or paroxysmal arousals may be even more frequent. SHE patients often complain of nocturnal sleep discontinuity and consequently excessive sleepiness. Furthermore, sleep deprivation can facilitate seizure appearance, perpetuating a cycle between seizures and triggering factors. Scalp EEG studies confirmed that SHE patients with excessive daytime sleepiness showed increased sleep instability, quantified by the analysis of CAP rate. Moreover, seizures occurred mostly during unstable sleep, suggesting a probable connection between sleep-related motor manifestations, sleep disruption, and daytime symptoms. SEEG studies confirmed this finding, suggesting that interictal epileptic discharges may act as an internal trigger inducing an increase in arousal level, which in turn favors the occurrence of motor events, both epileptic or not.

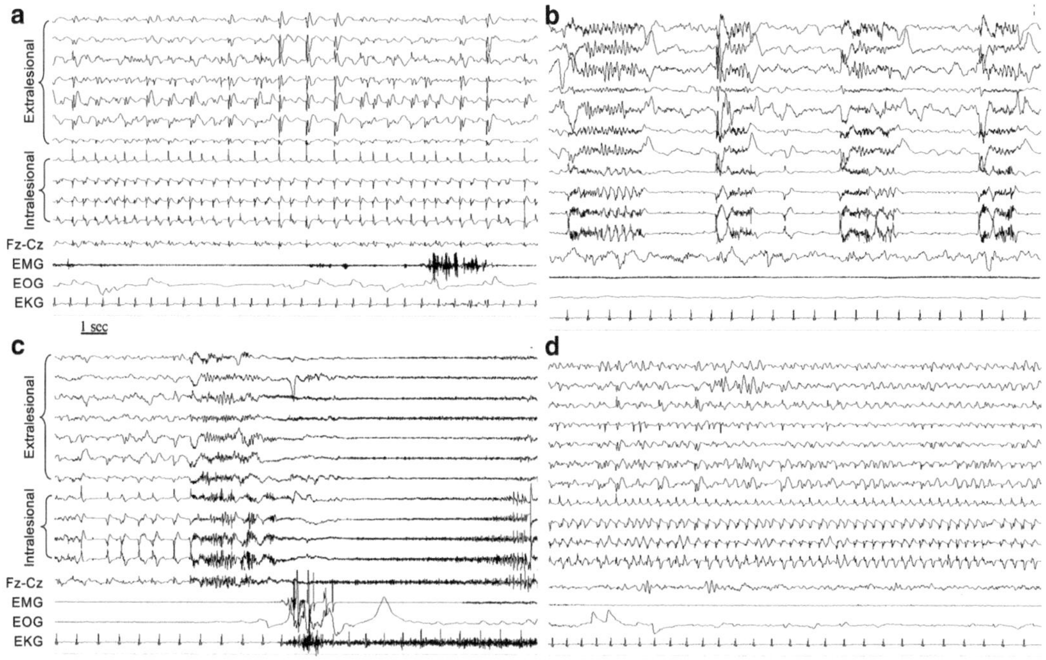

Fig. 17.16 SEEG recordings of a patient with type IIb focal cortical dysplasia (FCD) during wakefulness and sleep. The four EEG panels (**a**–**d**) are divided in extra- and intralesional bipolar channels. The intralesional contacts are within the type IIb FCD. (**a**) Pathognomonic activity of FCD during wakefulness with rhythmic and subcontinuous spike- and polyspike-and-wave discharges and occasional repetitive "brushes" inside the lesion. (**b**) During NREM sleep, FCD activity is composed of pseudo-periodic bursts of low-voltage fast discharges that spread to non-lesional areas. (**c**) The onset of a seizure during stage II NREM sleep. (**d**) Return of a rhythmic pattern of spike- or polyspike-and-wave during REM sleep, similar to wakefulness. Fz–Cz, scalp activity on the anterior vertex; EMG, electromyography (chin); EOG, electrooculogram; EKG, electrocardiogram. (Reproduced from Gibbs et al., 2016 [54])

17.4.4 REM Sleep Behavior Disorder (RBD)

REM sleep behavior disorder (RBD) represents the most frequent REM parasomnia. According to ICSD-3, this sleep disorder manifests as vivid, often frightening dreams accompanied by simple or complex motor behaviors during REM sleep. It is characterized by a loss of physiological muscle atonia. Patients with RBD seem to "enact their dreams" with their behaviors, probably mirroring the dream content.

Isolated RBD (iRBD) is considered an early manifestation of a neurodegenerative disease, especially synucleinopathies. Conversely, RBD can be found in a large proportion of both children and adults with narcolepsy, representing a form of REM sleep motor behavior dyscontrol.

Video-polysomnography (vPSG) is mandatory for the diagnosis of RBD. Diagnostic criteria include an excess of muscle tone during REM (REM without atonia or RSWA) on PSG, sleep-related injurious or potentially injurious disruptive behaviors by history and/or abnormal REM sleep behaviors during PSG, absence of epileptiform activity during REM sleep (unless RBD can be clearly distinguished from any concurrent REM sleep-related seizure disorder), and no better alternative explanation for the sleep disturbance (ICSD-3) (Fig. 17.17) [41].

RSWA (REM sleep without atonia) represents the neurophysiological and polysomnographic main feature of RBD. It consists of persistent muscle tone during REM sleep, resulting in either sustained (tonic) excessive activity during REM sleep in the chin EMG or intermittent (phasic)

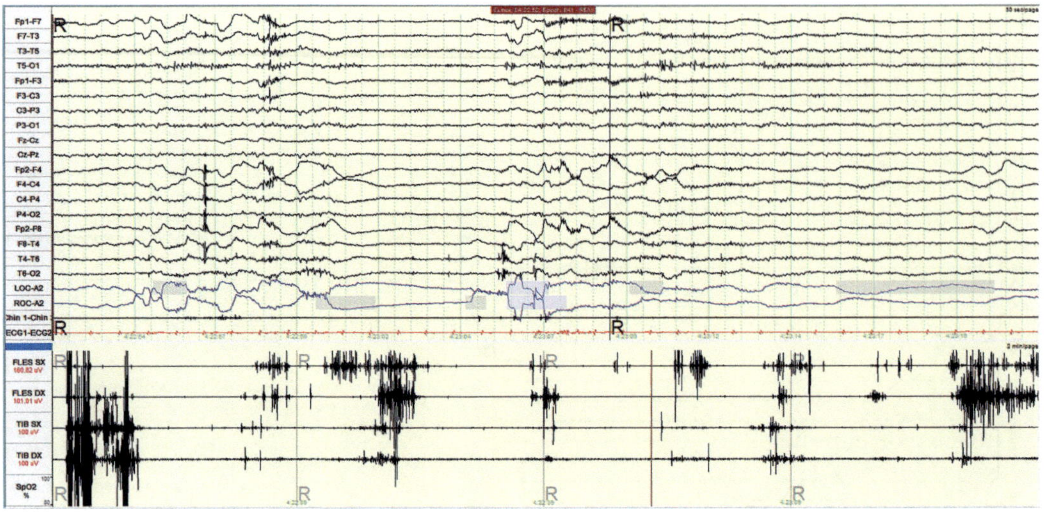

Fig. 17.17 A 2-minute polysomnogram (PSG) recording during REM sleep, capturing an episode of REM sleep behavior disorder (RBD). The figure highlights the presence of strong muscular activation across different muscle groups, which is a hallmark of RBD. Left side EEG bipolar EEG derivations: Fp1-F7, Fp1-F3, F3-C3, C3-P3, P3-O1; F7-T3, T3-T5, T5-O1; right side EEG bipolar derivations: Fp2-F4, F4-C4, C4-P4, P4-O2; F8-T4, T4-T6, T6-O2. Midline EEG derivations: Fz-Cz, Cz-Pz. LOC-A2, left ocular movement; ROC-A2, right ocular movement; Chin 1-Chin, chin EMG channel; ECG1-ECG2, electrocardiogram; FLES SX, flexor leg muscles, left; FLES DX, flexor leg muscles, right; TIB SX, tibial muscles, left; TIB DX, tibial muscles, right. SpO_2, blood oxygen saturation

excessive activity during REM sleep in the chin or limb EMG, or both. According to the latest American Academy of Sleep Medicine (AASM) scoring manual, a 30-second epoch of REM sleep is defined as "tonic" if at least 50% of its duration contains a chin EMG with an amplitude greater than the minimum amplitude observed during NREM sleep. Conversely, the "phasic" components of RSWA are defined as excessive transient muscle activity bursts on the chin and limb EMG channels lasting 0.1–5.0 seconds with an amplitude at least four times as high as the background EMG activity [1]. Moreover, a 30-second REM sleep epoch, subdivided into 3-second mini-epochs, is considered to have excessive phasic EMG activity when at least 50% of the 3-second mini-epochs contain bursts of transient EMG activity.

Although the presence of RSWA is mandatory for the diagnosis of RBD, both the AASM scoring manual and the ICSD-3 do not provide established cut-off values for RSWA. Therefore, different visual scoring methods to detect and quantify RSWA have been developed and validated. The most widely used visual methods are the "Montreal" method, the SINBAR method, and the Mayo Clinic method [55–57]. However, there is no consensus on which method is more efficient or which muscle group is better suited to identify or quantify RSWA and distinguish RBD from healthy subjects.

Patients with iRBD show some EEG features similar to those observed in neurodegenerative disorders, such as Parkinson's disease and dementia with Lewy bodies. In particular, they frequently present diffuse EEG slowing during wakefulness, more evident over the occipital scalp areas. Interestingly, higher absolute delta and theta power in all cortical regions can identify iRBD patients with a higher risk of short-term progression into alpha-synucleinopathies.

A study by Sunwoo et al. investigated EEG power spectral density and functional connectivity during phasic and tonic REM sleep in patients with iRBD and controls and found pathological motor cortex activation during phasic REM sleep that may explain the generation of dream-enacted behavior in iRBD patients [58].

Finally, a PSG study with high-density EEG in patients with RBD and controls conducted power spectra analyses in REM and NREM sleep, comparing tonic versus phasic REM sleep and the first versus last NREM cycle [59]. Patients with iRBD did not display a decrease in beta power during phasic versus tonic REM sleep, as observed in controls. The absence of suppression of beta rhythms during phasic REM sleep in RBD patients should increase cortical arousal, possibly favoring the occurrence of behavioral episodes. Moreover, RBD patients displayed a reduced decline in slow-wave amplitude from early to late NREM sleep, as well as reduced overnight changes in the distribution of slow-wave amplitude. This finding may suggest a reduced capacity for synaptic plasticity in patients with iRBD, thus favoring progression toward overt neurodegeneration.

17.4.5 Restless Leg Syndrome and Periodic Leg Movements

Restless legs syndrome (RLS), also called Willis–Ekbom disease (WED/RLS), is defined by clinical symptoms involving an urge to move the legs, which worsens with rest, is relieved by movement, and is most prominent in the evening and at night. The diagnosis is based on a clinical evaluation and cannot be made by questionnaire alone; a careful clinical evaluation is required to exclude mimics.

Patients with RLS frequently report disrupted nocturnal sleep. In particular, they complain of problems falling asleep and frequent awakenings due to sensory symptoms. These sleep problems occur in moderate-to-severe RLS patients but not in the majority of mild RLS patients. Sleep laboratory studies comparing moderate-to-severe RLS patients with control subjects indicate increased wake after sleep onset, decreased sleep time, decreased stage 2 sleep, and increased stage 1 sleep, with little change in REM or slow-wave sleep. Daytime fatigue is commonly manifested, but most RLS patients do not report the level of sleepiness expected for their degree of sleep loss. Thus, RLS appears to involve some increased arousal process counteracting the sleep-loss effects.

RLS is usually associated, once the patient eventually falls asleep, with periodic limb movements (PLMs). Periodic leg movements (PLMs) are characterized by extension of the big toe in combination with partial flexion of the ankle, knee, and hip. Although these movements predominantly occur in the legs, the upper extremities may also be involved. According to the AASM scoring criteria, PLMs consist of a series of four or more consecutive movements, each lasting 0.5–10 seconds with an interval of 5–90 seconds between movements and an amplitude greater than 8 mV above the resting baseline EMG signal. The PLM index is defined as the number of PLMs divided by the observation time in hours. In general, a PLM index is considered abnormal if there are more than 5 per hour in children and more than 15 per hour in adults. Approximately 5% of adults demonstrate incidental, asymptomatic PLMs on PSG, and in these cases, they are considered benign epiphenomena.

The consequences of PLMs on sleep are still a matter of debate. They are often associated with EEG arousals or with changes in heart rate, breathing rate, or blood pressure. The EEG arousal may precede, coincide with, or follow the movement. These observations suggest that limb jerks are part of a dynamic, physiologically oscillating process of sleep, with a 20–40-second periodicity involving different systems, including cortical, vegetative, and behavioral functions (Fig. 17.18).

17.4.6 Obstructive Sleep Apnea

Obstructive sleep apnea (OSA) is the most common sleep-related breathing disorder, characterized by intermittent airway collapse during sleep, resulting in obstructive apneas, hypopneas, and/or respiratory effort-related arousals. Risk factors for OSA include obesity, older age, male gender, and craniofacial/upper

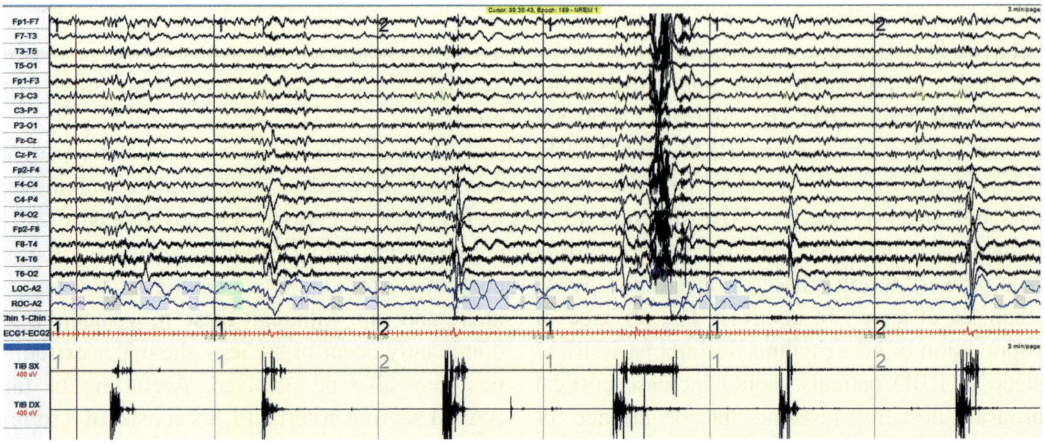

Fig. 17.18 A 3-minute polysomnogram (PSG) recording during NREM sleep, capturing episodes of periodic leg movements (PLMs) associated with periodic arousals. The figure highlights the characteristic leg movements and corresponding cortical arousals seen in patients with PLM disorder. Left side EEG bipolar EEG derivations: Fp1-F7, Fp1-F3, F3-C3, C3-P3, P3-O1; F7-T3, T3-T5, T5-O1; right side EEG bipolar derivations: Fp2-F4, F4-C4, C4-P4, P4-O2; F8-T4, T4-T6, T6-O2. Midline EEG derivations: Fz–Cz, Cz-Pz. LOC-A2, left ocular movement; ROC-A2, right ocular movement; Chin 1-Chin, chin EMG channel; ECG1-ECG2, electrocardiogram; TIB SX, tibial muscles, left; TIB DX, tibial muscles, right

airway abnormalities. The determination of apneas, hypopneas, and respiratory effort-related arousals is primarily conducted visually. According to the recent criteria from the American Academy of Sleep Medicine Manual for the Scoring of Sleep and Associated Events, apneas are defined as a drop in oronasal breathing amplitude by at least 90% for at least 10 seconds. Hypopneas must last at least 10 seconds, with a signal amplitude reduction of 30% or more, accompanied by either a drop in oxygen saturation of at least 3% or an associated arousal in the EEG (Fig. 17.19). More subtle respiratory events, known as RERAs, are defined by obstructive upper airway airflow reductions (not meeting the criteria for apnea or hypopnea) associated with progressive negative esophageal pressure lasting at least 10 seconds, culminating in an arousal. Sleep apnea severity is assessed using the respiratory disturbance index (RDI), which counts the number of apneas, hypopneas, and RERAs occurring per hour of sleep. An RDI of less than 5/hour is considered normal, while mild OSA ranges from 5 to 14.9/hour, moderate OSA from 15 to 29.9/hour, and severe OSA is defined as an AHI of 30/hour or more. The gold standard for diagnosing OSA is overnight polysomnography (PSG), which involves simultaneous recordings of multiple physiological signals, including EEG, electrooculogram (EOG), electromyogram (EMG), oronasal airflow, thoracic and abdominal effort, and oxyhemoglobin saturation. However, PSG is expensive and technically demanding. As a result, portable monitoring, which lacks neurological signals (EEG, EOG, EMG) and cannot identify arousals and RERAs, is becoming a cost-effective alternative.

Patients with OSA often show a reduction in deep sleep (N3) and REM sleep, with an increase in NREM stage 2. They also exhibit increased time spent awake after sleep onset and shortened sleep latency. Slow-wave activity (SWA) during NREM sleep is generally reduced compared to healthy individuals but increases with CPAP treatment. Studies have shown negative associations between SWA and apnea index in NREM sleep and positive associations between SWA and indices of hypoxemia in REM sleep. Findings also indicate slower dissipation of SWA across the night in OSA patients, suggesting disrupted sleep homeostasis due to sleep fragmentation. Regarding hallmark graphoelements of NREM sleep, spindle density and fre-

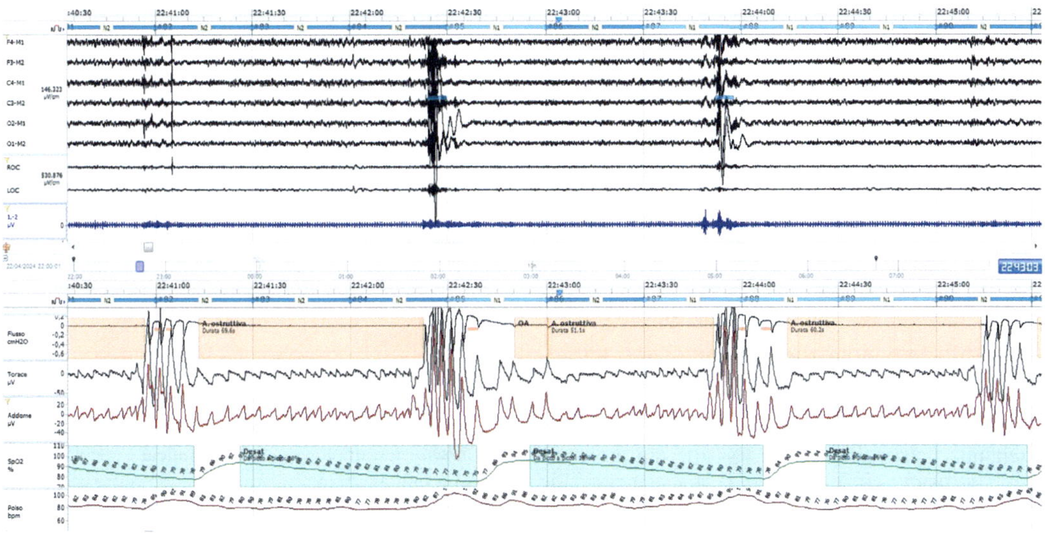

Fig. 17.19 Five-minute polysomnogram (PSG) recording during NREM sleep, capturing episodes of periodic obstructive sleep apnea (OSA) correlated with EEG arousals. F4-M1, F3-M2, C4-M1, C3-M2, O2-M1, O1-M2: EEG channels. ROC (right ocular movement); LOC (left ocular movement). EMG: electromyogram. Flusso (flow): measures airflow through the nose and mouth; Torace (thorax): measures chest movements. Addome (abdomen): measures abdominal movements. SpO_2: blood oxygen saturation

quency are lower in middle-aged patients with mild-to-moderate OSA compared to controls. Severe OSA patients (AHI > 30) also have lower spindle densities. Higher AHI and AI are associated with reduced spindle density. OSA patients tend to have shorter K-complex durations and smaller negative and total waveform amplitudes, potentially due to blunted cortical responses and synaptic strength resulting from structural brain injury caused by OSA. Compared to healthy subjects, OSA patients exhibit abnormal sleep microstructure characterized by an increased cortical arousal index and enhanced amounts of CAP, with longer and more desynchronized EEG patterns.

The best treatment for moderate-to-severe OSA is continuous positive airway pressure (CPAP) therapy. In OSA patients effectively treated with CPAP, therapy induces an immediate increase in deep sleep, more consolidated REM sleep, a reduction in stages N1 and N2, fewer arousals, and a lower CAP rate. However, normal sleep architecture is established only after sustained CPAP treatment.

17.4.7 Narcolepsy

According with ICSD3, hypersomnia of central origin encompasses Narcolepsy type 1, Narcolepsy type 2, Idiopathic Hypersomnia, and Kleine–Levin syndrome. Hypersomnia can also be due to medical disorders, medication, and psychiatric diseases or be secondary to insufficient sleep. Narcolepsy type 1 (NT1) is a chronic neurological disorder characterized by the classic narcoleptic tetrad, including excessive daytime sleepiness, cataplexy, sleep paralysis, and hypnagogic/hypnopompic hallucinations. Cataplexy consists in a generalized or focal sudden, short-lasting loss of muscle tone during wake provoked by intense emotion, particularly positive, without loss of consciousness. Daytime sleepiness is characteristically incoercible and sudden (sleep attacks) and can appear in situations of passive activity but sometimes may occur in more active circumstances. Usually, the daytime sleep lasts less than 20 min and it is deeply restoring. Disturbed nighttime sleep occurs frequently and is particularly disturbing for patients. It is associ-

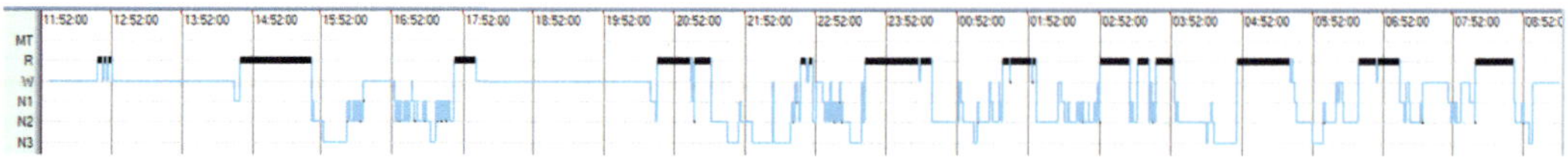

Fig. 17.20 Hypnogram of a patient affected by narcolepsy, illustrating the distribution of sleep stages during both the day and the night. The hypnogram highlights the occurrence of sleep-onset REM (rapid eye movement) periods, which are a hallmark of narcolepsy. These REM periods are seen both during daytime naps and the nocturnal sleep episode

ated with decreased quality of life and is considered a fifth symptom after the narcoleptic tetrad.

The cause of NT1 is a massive loss of orexin-producing neurons localized in the lateral hypothalamus, probably due to an autoimmune process. PSG is the gold standard for objectively assessing sleep quantity and sleep quality. Therefore, PSG measurement of nighttime sleep is mandatory before the execution of a multiple sleep latency test (MSLT) the following day. PSG allows for screening for other sleep disturbances and potential comorbidity diagnosis but also may reveal clinically relevant findings that are helpful for diagnosis of narcolepsy. Indeed, one of the hallmarks of narcolepsy is an early appearance, within 15 min after the sleep onset, of REM sleep (sleep-onset REM periods—SOREMPs). For diagnostic purposes, SOREMPs are explored both during the PSG but also during the MSLT, which consists of five nap opportunities. This test is performed the day after the video-PSG and allows to measure sleep latency (i.e., the propensity to fall asleep/an objective measurement of sleepiness). Narcoleptic subjects show a mean sleep latency below 8 min (characteristically <3 min) and the presence of at least two SOREMPs (for the diagnosis, a single SOREMP is sufficient if one SOREMP occurs during the PSG) (Fig. 17.20). The minimal duration of nocturnal PSG is 7 h.

Sleep macrostructure in narcolepsy is not dramatically altered (with a possible presence of a SOREMP), but an increase in stage N1 is reported. Shorter REM latency, sleep fragmentation with repetitive awakening, middle-of-the-night insomnia, or a REM without atonia with RDB can be found. As a translation of impaired arousal regulation, a significant decrease in CAP rate is reported.

References

1. American Academy of Sleep Medicine. The AASM manual for the scoring of sleep and associated events: rules, terminology and technical specifications (Version 3). 2023.
2. Bernardi G, Siclari F, Handjaras G, Riedner BA, Tononi G. Local and widespread slow waves in stable NREM sleep: evidence for distinct regulation mechanisms. Front Hum Neurosci. 2018;12:248.
3. Fernandez LMJ, Lüthi A. Sleep spindles: mechanisms and functions. Physiol Rev. 2020;100(2):805–68.
4. Andrillon T, Nir Y, Staba RJ, Ferrarelli F, Cirelli C, Tononi G, et al. Sleep spindles in humans: insights from intracranial EEG and unit recordings. J Neurosci. 2011;31(49):17821–34.
5. Vyazovskiy VV, Olcese U, Lazimy YM, Faraguna U, Esser SK, Williams JC, et al. Cortical firing and sleep homeostasis. Neuron. 2009;63(6):865–78.
6. Simor P, van der Wijk G, Nobili L, Peigneux P. The microstructure of REM sleep: why phasic and tonic? Sleep Med Rev. 2020;52:101305.
7. Terzano MG, Parrino L, Sherieri A, Chervin R, Chokroverty S, Guilleminault C, et al. Atlas, rules, and recording techniques for the scoring of cyclic alternating pattern (CAP) in human sleep. Sleep Med. 2001;2(6):537–53.
8. Terzano MG, Parrino L, Smerieri A, Carli F, Nobili L, Donadio S, et al. CAP and arousals are involved in the homeostatic and ultradian sleep processes. J Sleep Res. 2005;14(4):359–68.
9. Pigarev IN, Nothdurft HC, Kastner S. Evidence for asynchronous development of sleep in cortical areas. Neuroreport. 1997;8(11):2557–60.
10. Krueger JM, Huang YH, Rector DM, Buysse DJ. Sleep: a synchrony of cell activity-driven small network states. Eur J Neurosci. 2013;38(2):2199–209.
11. Ferrara M. Regional differences of the human sleep electroencephalogram in response to selective slow-wave sleep deprivation. Cereb Cortex. 2002;12(7):737–48.
12. Finelli LA, Borbély AA, Achermann P. Functional topography of the human nonREM sleep electroencephalogram. Eur J Neurosci. 2001;13(12):2282–90.
13. Werth E, Dijk DJ, Achermann P, Borbely AA. Dynamics of the sleep EEG after an early eve-

ning nap: experimental data and simulations. Am J Phys Regul Integr Comp Phys. 1996;271(3):R501–10.
14. Cantero JL, Atienza M, Salas RM, Dominguez-Marin E. Effects of prolonged waking-auditory stimulation on electroencephalogram synchronization and cortical coherence during subsequent slow-wave sleep. J Neurosci. 2002;22(11):4702–8.
15. Huber R, Ghilardi MF, Massimini M, Tononi G. Local sleep and learning. Nature. 2004;430(6995):78–81.
16. Huber R, Ghilardi MF, Massimini M, Ferrarelli F, Riedner BA, Peterson MJ, et al. Arm immobilization causes cortical plastic changes and locally decreases sleep slow wave activity. Nat Neurosci. 2006;9(9):1169–76.
17. Rattenborg NC, Amlaner CJ, Lima SL. Behavioral, neurophysiological and evolutionary perspectives on unihemispheric sleep. Neurosci Biobehav Rev. 2000;24(8):817–42.
18. Marzano C, Moroni F, Gorgoni M, Nobili L, Ferrara M, De Gennaro L. How we fall asleep: regional and temporal differences in electroencephalographic synchronization at sleep onset. Sleep Med. 2013;14(11):1112–22.
19. Magnin M, Rey M, Bastuji H, Guillemant P, Mauguière F, Garcia-Larrea L. Thalamic deactivation at sleep onset precedes that of the cerebral cortex in humans. Proc Natl Acad Sci USA. 2010;107(8):3829–33.
20. Sarasso S, Pigorini A, Proserpio P, Gibbs SA, Massimini M, Nobili L. Fluid boundaries between wake and sleep: experimental evidence from Stereo-EEG recordings. Arch Ital Biol. 2014;152(2–3):169–77.
21. Nobili L, De Gennaro L, Proserpio P, Moroni F, Sarasso S, Pigorini A, et al. Local aspects of sleep: observations from intracerebral recordings in humans. Prog Brain Res. 2012;199:219–32.
22. Nobili L, Ferrara M, Moroni F, De Gennaro L, Russo GL, Campus C, et al. Dissociated wake-like and sleep-like electro-cortical activity during sleep. NeuroImage. 2011;58(2):612–9.
23. Peter-Derex L, Magnin M, Bastuji H. Heterogeneity of arousals in human sleep: a stereo-electroencephalographic study. NeuroImage. 2015;123:229–44.
24. Wang YL, Avigdor T, Hannan S, Abdallah C, Dubeau F, Peter-Derex L, et al. Intracerebral dynamics of sleep arousals: a combined scalp-intracranial EEG study. J Neurosci. 2024;44(16):e0617232024.
25. Jang RS, Ciliberti D, Mankin EA, Poe GR. Recurrent Hippocampo-neocortical sleep-state divergence in humans. Proc Natl Acad Sci USA. 2022;119(44):e2123427119.
26. Peter-Derex L, Von Ellenrieder N, Van Rosmalen F, Hall J, Dubeau F, Gotman J, et al. Regional variability in intracerebral properties of NREM to REM sleep transitions in humans. Proc Natl Acad Sci USA. 2023;120(26):e2300387120.
27. De Carli F, Proserpio P, Morrone E, Sartori I, Ferrara M, Gibbs SA, et al. Activation of the motor cortex during phasic rapid eye movement sleep. Ann Neurol. 2016;79(2):326–30.
28. Corsi-Cabrera M, Velasco F, Del Río-Portilla Y, Armony JL, Trejo-Martínez D, Guevara MA, et al. Human amygdala activation during rapid eye movements of rapid eye movement sleep: an intracranial study. J Sleep Res. 2016;25(5):576–82.
29. Simor P, Szalárdy O, Gombos F, Ujma PP, Jordán Z, Halász L, et al. REM sleep microstates in the human anterior thalamus. J Neurosci. 2021;41(26):5677–86.
30. Bourel-Ponchel E, Gueden S, Hasaerts D, Héberlé C, Malfilâtre G, Mony L, et al. Normal EEG during the neonatal period: maturational aspects from premature to full-term newborns. Neurophysiol Clin. 2021;51(1):61–88.
31. Grigg-Damberger MM. The visual scoring of sleep in infants 0 to 2 months of age. J Clin Sleep Med. 2016;12(3):429–45.
32. Eisermann M, Kaminska A, Moutard ML, Soufflet C, Plouin P. Normal EEG in childhood: from neonates to adolescents. Clin Neurophysiol. 2013;43(1):35–65.
33. Scher MS. Ontogeny of EEG-sleep from neonatal through infancy periods. Sleep Med. 2008;9(6):615–36.
34. Ohayon MM, Carskadon MA, Guilleminault C, Vitiello MV. Meta-analysis of quantitative sleep parameters from childhood to old age in healthy individuals: developing normative sleep values across the human lifespan. Sleep. 2004;27(7):1255–73.
35. Crowley K. Sleep and sleep disorders in older adults. Neuropsychol Rev. 2011;21(1):41–53.
36. Parrino L, Ferri R, Bruni O, Terzano MG. Cyclic alternating pattern (CAP): the marker of sleep instability. Sleep Med Rev. 2012;16(1):27–45.
37. Riemann D, Baglioni C, Bassetti C, Bjorvatn B, Dolenc Groselj L, Ellis JG, et al. European guideline for the diagnosis and treatment of insomnia. J Sleep Res. 2017;26(6):675–700.
38. Baglioni C, Regen W, Teghen A, Spiegelhalder K, Feige B, Nissen C, et al. Sleep changes in the disorder of insomnia: a meta-analysis of polysomnographic studies. Sleep Med Rev. 2014;18(3):195–213.
39. Zhao W, Van Someren EJW, Li C, Chen X, Gui W, Tian Y, et al. EEG spectral analysis in insomnia disorder: a systematic review and meta-analysis. Sleep Med Rev. 2021;59:101457.
40. Parrino L, Milioli G, De Paolis F, Grassi A, Terzano MG. Paradoxical insomnia: the role of CAP and arousals in sleep misperception. Sleep Med. 2009;10(10):1139–45.
41. Sateia MJ. International classification of sleep disorders-third edition. Chest. 2014;146(5):1387–94.
42. Lopez R, Shen Y, Chenini S, Rassu AL, Evangelista E, Barateau L, et al. Diagnostic criteria for disorders of arousal: a video-polysomnographic assessment. Ann Neurol. 2018;83(2):341–51.
43. Lopez R, Laganière C, Chenini S, Rassu AL, Evangelista E, Barateau L, et al. Video-polysomnographic assessment for the diagnosis

of disorders of arousal in children. Neurology. 2021;96(1):e121–30.
44. Zucconi M, Oldani A, Ferini-Strambi L, Smirne S. Arousal fluctuations in non-rapid eye movement parasomnias: the role of cyclic alternating pattern as a measure of sleep instability. J Clin Neurophysiol. 1995;12(2):147–54.
45. Guilleminault C, Kirisoglu C, da Rosa AC, Lopes C, Chan A. Sleepwalking, a disorder of NREM sleep instability. Sleep Med. 2006;7(2):163–70.
46. Castelnovo A, Riedner BA, Smith RF, Tononi G, Boly M, Benca RM. Scalp and source power topography in sleepwalking and sleep terrors: a high-density EEG study. Sleep. 2016;39(10):1815–25.
47. Haridi M, Weyn Banningh S, Clé M, Leu-Semenescu S, Vidailhet M, Arnulf I. Is there a common motor dysregulation in sleepwalking and REM sleep behaviour disorder? J Sleep Res. 2017;26(5):614–22.
48. Terzaghi M, Sartori I, Tassi L, Didato G, Rustioni V, LoRusso G, et al. Evidence of dissociated arousal states during NREM parasomnia from an intracerebral neurophysiological study. Sleep. 2009;32(3):409–12.
49. Terzaghi M, Sartori I, Tassi L, Rustioni V, Proserpio P, Lorusso G, et al. Dissociated local arousal states underlying essential clinical features of non-rapid eye movement arousal parasomnia: an intracerebral stereo-electroencephalographic study. J Sleep Res. 2012;21(5):502–6.
50. Castelnovo A, Lopez R, Proserpio P, Nobili L, Dauvilliers Y. NREM sleep parasomnias as disorders of sleep-state dissociation. Nat Rev Neurol. 2018;14(8):470–81.
51. Flamand M, Boudet S, Lopes R, Vignal JP, Reyns N, Charley-Monaca C, et al. Confusional arousals during non-rapid eye movement sleep: evidence from intracerebral recordings. Sleep. 2018;41(10)
52. Tinuper P, Bisulli F, Cross JH, Hesdorffer D, Kahane P, Nobili L, et al. Definition and diagnostic criteria of sleep-related hypermotor epilepsy. Neurology. 2016;86(19):1834–42.
53. Gibbs SA, Figorilli M, Casaceli G, Proserpio P, Nobili L. Sleep related Hypermotor seizures with a right parietal onset. J Clin Sleep Med. 2015;11(08):953–5.
54. Gibbs SA, Proserpio P, Terzaghi M, Pigorini A, Sarasso S, Lo Russo G, et al. Sleep-related epileptic behaviors and non-REM-related parasomnias: insights from stereo-EEG. Sleep Med Rev. 2016;25:4–20.
55. Lapierre O, Montplaisir J. Polysomnographic features of REM sleep behavior disorder: development of a scoring method. Neurology. 1992;42(7):1371.
56. McCarter SJ, St. Louis EK, Duwell EJ, Timm PC, Sandness DJ, Boeve BF, et al. Diagnostic thresholds for quantitative REM sleep phasic burst duration, phasic and tonic muscle activity, and REM Atonia index in REM sleep behavior disorder with and without comorbid obstructive sleep apnea. Sleep. 2014;37(10):1649–62.
57. Iranzo A, Frauscher B, Santos H, Gschliesser V, Ratti L, Falkenstetter T, et al. Usefulness of the SINBAR electromyographic montage to detect the motor and vocal manifestations occurring in REM sleep behavior disorder. Sleep Med. 2011;12(3):284–8.
58. Sunwoo JS, Cha KS, Byun JI, Kim TJ, Jun JS, Lim JA, et al. Abnormal activation of motor cortical network during phasic REM sleep in idiopathic REM sleep behavior disorder. Sleep. 2019;42(2)
59. Valomon A, Riedner BA, Jones SG, Nakamura KP, Tononi G, Plante DT, et al. A high-density electroencephalography study reveals abnormal sleep homeostasis in patients with rapid eye movement sleep behavior disorder. Sci Rep. 2021;11(1):4758.

18 Polygraphic Recordings of Epileptic Motor Phenomena

Guido Rubboli, Elena Gardella,
Gaetano Cantalupo, and Carlo Alberto Tassinari

18.1 Introduction

The term "polygraphy" refers to the simultaneous recording of multiple physiologic measures. Polygraphic investigations can thus provide a correlation between a behavioral manifestation and a variable set of physiologic parameters. Ideally, a polygraphic study, particularly in epileptology, should be tailored to the specific clinical manifestation that is intended to be investigated in order to select the relevant and appropriate physiological parameters that can characterize the manifestation itself. The clinical yields and the variables that can be monitored by using polygraphic studies have been illustrated in several reviews and monographies [1–4].

G. Rubboli (✉)
Danish Epilepsy Center, Dianalund, Denmark

Institute of Clinical Medicine, University of Copenhagen, Copenhagen, Denmark
e-mail: guru@filadelfia.dk

E. Gardella
Danish Epilepsy Center, Dianalund, Denmark

University of Southern Denmark, Odense, Denmark
e-mail: elga@filadelfia.dk

G. Cantalupo
Child Neuropsychiatry Unit, Department of Children and Maternal Health, AOUI Verona, Verona, Italy
e-mail: gaetano.cantalupo@univr.it

C. A. Tassinari
University of Bologna, Bologna, Italy

In the study of epilepsy, polygraphic recordings can be essential to identify and describe the occurrence and characteristics of changes occurring in various physiologic functions in concomitance with epileptic abnormalities as documented by electroencephalogram (EEG) and clarify their temporal correlation with the EEG events. Thus, polygraphic features can be extremely relevant to describe the phenomenology of epileptic seizures, defining the semiological clinical features of different epileptic conditions and contributing to elucidate their physiopathogenetic mechanisms. In this respect, polygraphic studies might be crucial to establish and verify diagnostic hypotheses.

18.2 The Contribution of Polygraphic Studies in the Investigation of Epileptic Motor Phenomena

Motor phenomena frequently manifest as the most visible clinical features of an epileptic seizure. However, analyzing solely clinical data can make it challenging to define and interpret patterns of muscle activation accurately. Observers may overlook mild contractions or sporadic muscle twitches. Polygraphic techniques that record surface electromyography (EMG) activity prove invaluable for identifying and characterizing even subtle and seemingly subclinical muscular

F. Brigo, O. Mecarelli (eds.), *EEG: The First 100 Years*,
https://doi.org/10.1007/978-3-031-86878-8_18

manifestations, as well as correlating these phenomena with EEG activities.

18.2.1 Analysis of Motor Phenomena

Recording surface EMG can elucidate the nature of muscle contractions, allowing for distinctions between different types of motor phenomena. Various muscular phenomena can arise individually or in multiple combinations, contributing to the clinical manifestations of distinct seizure types.

18.2.1.1 Myoclonus

Positive myoclonus is characterized by a sudden, shock-like muscular contraction that affects one or more segments of the body. It appears in the EMG as a brief enhancement of muscle potentials, synchronized in both agonist and antagonist muscles, which may or may not have an associated EEG correlate (Fig. 18.1, upper panel). The opposite phenomenon, known as "negative" myoclonus, is defined as a transient interruption of a tonic muscle contraction and can sometimes be clinically indistinguishable from positive myoclonus. When associated with a time-locked epileptiform EEG event, it is classified as "epileptic negative myoclonus" [5–9] (Fig. 18.2). Epileptic negative myoclonus, characterized by the sudden cessation of tonic muscular activity for less than 500 milliseconds without preceding myoclonia, has been suggested as a seizure type [10, 11].

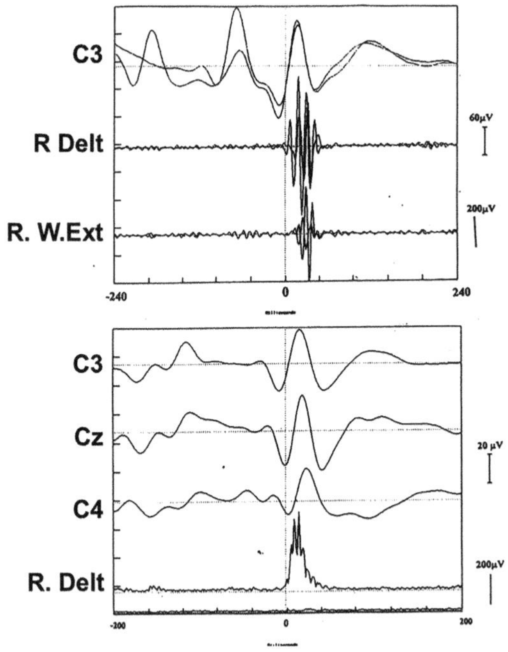

Fig. 18.1 Upper panel: superposition of two polygraphic (EEG-EMG) traces in a patient with Unverricht–Lundborg disease. Myoclonic potentials in the right deltoid (R. Delt) and right wrist extensor (R.W.Ext); in the contralateral central region (C3), a positive transient precedes the onset of the myoclonia. Lower panel: back-averaging of the EEG signal triggered by the onset of the myoclonic potential in the right deltoid. The rectified EMG signal of the myoclonic jerk is associated with a clear positive potential in the left central (C3) region and in the vertex (Cz), consistent with a cortical myoclonus

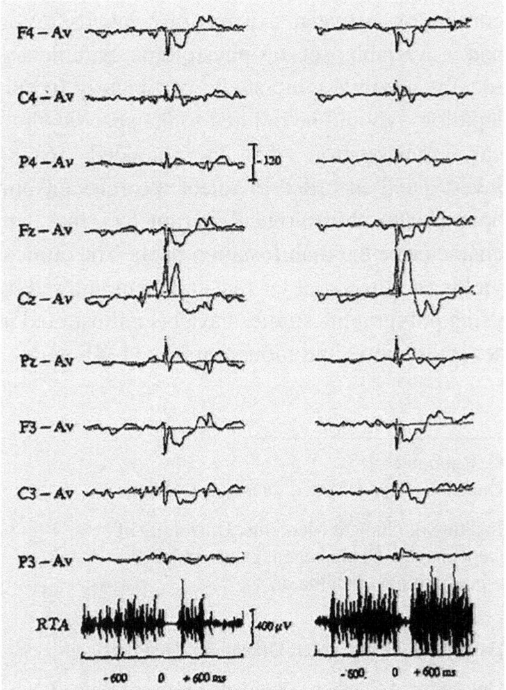

Fig. 18.2 Epileptic negative myoclonus in the right tibialis anterior (RTA): the EMG silent period in the RTA is time-locked to the spike of the spike-wave complex in Cz. The onset of the EMG silent period is not preceded by a positive myoclonia

18.2.1.2 Spasm

A spasm appears as a massive yet gradual contraction that peaks and then progressively decreases, typically involving the axial and proximal muscle groups. Polygraphic recordings employing multiple EMG leads have illustrated that epileptic spasms can involve complex muscle activation patterns, incorporating cranial as well as limb and axial muscles (Fig. 18.3). Clinically, this complex pattern of contraction pattern may appear similar across various spasms, despite differing sequences of muscle recruitment [12, 13]. On the EEG, spasms may not be associated with overt changes or may display flattening, diffuse low-voltage fast activity, or a slow wave [14–16].

18.2.1.3 Tonic Contraction

Tonic contractions consist of slow, sustained muscle contractions that endure over time and affect several muscle groups. These contractions are commonly associated with fast recruiting EEG activity.

18.2.1.4 Clonic Contraction

Clonic contractions are identified by a series of jerks and appear on EMG as hypersynchronous muscle potentials. They can vary in amplitude, symmetry, frequency, and topography, often correlating with contralateral EEG spikes [10, 11, 14].

18.2.1.5 Atonic Phenomena

Atonic phenomena are defined as a sudden global or localized loss of muscle tone, characterized on polygraphy by an abrupt flattening of EMG activity. Various types of EEG epileptiform discharges can be associated with them, including generalized spike-and-wave activity, diffuse fast rhythmic spikes, bilateral synchronous fast waves intermixed with slow waves, or even no EEG alterations [10, 14, 17].

18.2.2 Analysis of Motor Patterns

Polygraphy is essential for determining the characteristics of a motor manifestation during a seizure concerning (a) its relationship with concomitant EEG activity, (b) the muscular groups involved, (c) the timing and progression of motor pattern activation throughout the seizure, and (d) the presence of a consistent motor pattern across various seizure types.

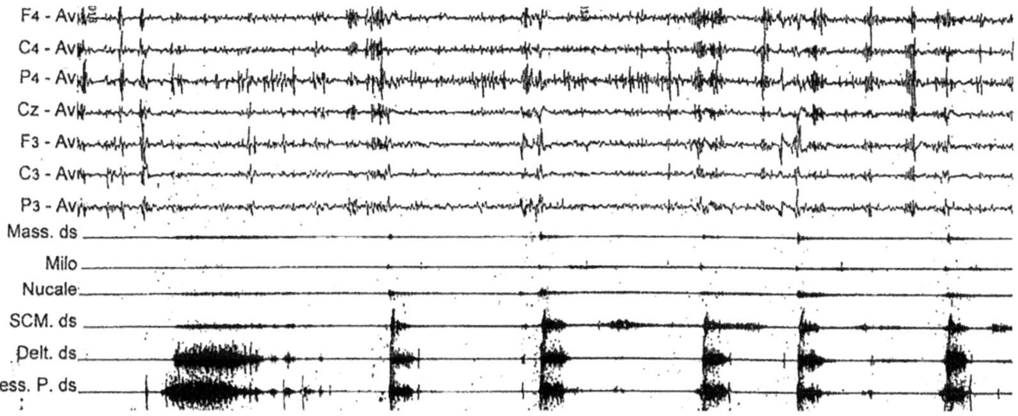

Fig. 18.3 Polygraphic recording of a cluster of spasms following a brief tonic seizure in an 8-month-old child with severe epileptic encephalopathy associated with cortical dysplasia. The multichannel electromyogram recording shows the involvement of cranial and limb muscles with a quite reproducible pattern. A slow complex associated with the spasms is detectable in the electroencephalogram. Delt., deltoid; ds, right; Fless. P., wrist flexor; Mass., masseter; Milo, mylohyoideus; Nucale, splenius capitis; SCM., sternocleidomastoideus

18.2.2.1 Generalized Tonic–Clonic Seizures

Polygraphic recordings enable the characterization of the complex features of generalized tonic–clonic seizures, which consist of a relatively stereotyped sequence of motor phenomena [18–20]. This sequence includes an initial tonic phase of continuous muscle contraction lasting between 10 and 20 s, involving all skeletal muscles. This phase features a characteristic body posture (a flexion followed by a prolonged extension phase), then a diffuse vibratory contraction appears, followed by clonic manifestations characterized by brief, maximal flexor contractions affecting the entire body with intervals between the jerks progressively increasing. Notably, polygraphic recordings demonstrate that shortly after the final clonic jerk, a new tonic phase usually reemerges, as intense as the initial one but with a different topographic distribution, primarily affecting facial and masticatory muscles [14, 19, 20]. The EEG correlates associated with a tonic–clonic seizure typically begin with an initial desynchronization, which may sometimes be preceded by generalized bursts of polyspike-and-wave activity, followed by a recruiting rhythm. These patterns are intermixed with slow waves that exhibit a decreasing frequency and an increasing amplitude, which at a certain point correspond to the cessation of the tonic massive contraction and the appearance of the clonic manifestations. Then, a flattening of the EEG activity follows, representing the phase known as "cortical extinction," which can last for several seconds [19–21].

18.2.2.2 Tonic Seizures

In the case of tonic seizures, the main clinical feature is a muscular tonic contraction lasting between 5 and 20 s, which is accompanied by a loss of consciousness and affects various muscle groups with different intensities and topographies, such as the head and trunk, the limb girdles, and, to a lesser degree, the legs, or the entire body [14, 22–24]. The accompanying polygraphic findings are represented by a variety of EEG patterns, including (a) a flattening of the tracing, (b) a fast activity at approximately 20 Hz that shows a progressive increase in amplitude, and (c) a recruiting rhythmic discharge occurring at around 10 Hz, which may sometimes present with high amplitude from the onset [3, 19, 20]. The electromyography (EMG) leads reveal an interference pattern across all muscles involved, corresponding to the tonic contraction (Fig. 18.4). Various patterns of muscular activation can be noted depending on the duration of the seizure: if the seizure is short, the contraction reaches its peak at the onset and then diminishes; conversely, if it lasts longer, there is a gradual increase in muscular activity that reflects the intensifying contraction. In global tonic seizures, there is a noticeable axial predominance of muscular activation. Asymmetries in the tonic contraction or the emergence of myoclonia at the conclusion of the tonic phase may also be observed. Tonic seizures are frequently linked with autonomic changes, which include alterations in heart rate and respiration, mydriasis, vasomotor phenomena, increased intravesicular pressure, and positive electrodermogram responses [3, 22, 23].

18.2.2.3 Myoclonic Absences

Myoclonic absences are characterized by a generalized 3-c/s spike-and-wave discharge that can last for several seconds and is associated with myoclonic jerks. These jerks are particularly noticeable in the upper limbs and are polygraphically characterized by myoclonic potentials that, a few seconds after the discharge begins, are superimposed on a progressively increasing tonic muscular contraction, primarily affecting the shoulders and deltoids, resulting in the abduction and elevation of the upper limbs [19, 20, 25, 26] (Fig. 18.5a). With the addition of extra leads, twitching can also be observed in the facial muscles, including the mentalis and orbicularis oris muscles, time-locked to the spike-and-wave complex. Oscilloscopic analysis of the correlation between EEG and EMG has shown that the positive transient of the 3-c/s spike-and-wave complex [27] corresponds to the occurrence of myoclonic jerks (Fig. 18.5b). The latency between the myoclonic jerk and the spike on

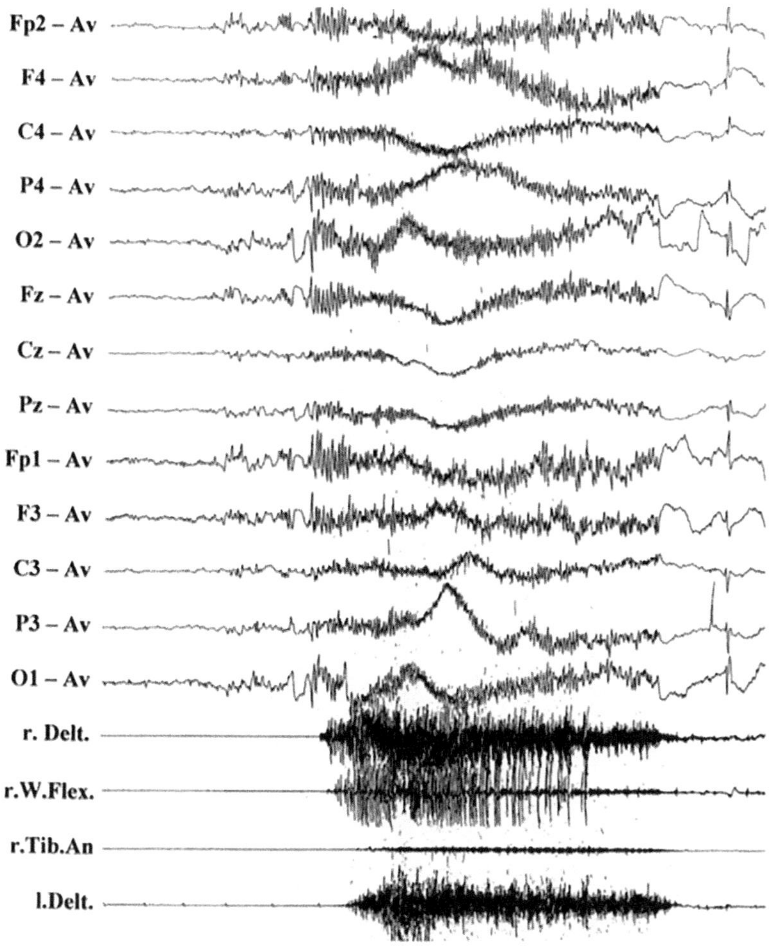

Fig. 18.4 Tonic seizure. The bilateral tonic contraction, more prominent in both deltoids, is associated with a fast rhythmic polyspike activity. Delt., deltoid; l, left; r., right; Tib. An., tibialis anterior; W. Flex., wrist flexor

the EEG is measured to be between 15 and 40 milliseconds for the upper limbs and 50–70 milliseconds for the lower limbs. Following the myoclonic bursts, there is a brief silent period lasting between 60 and 120 milliseconds that interrupts the tonic contraction [25, 26].

18.2.2.4 Juvenile Myoclonic Epilepsy

In juvenile myoclonic epilepsy, polygraphic recordings can be particularly valuable for confirming the diagnosis, as this condition is marked by brief myoclonic jerks—especially in the upper limbs—that are associated with polyspike-and-wave complexes occurring at a frequency of 3–3.5 c/s [28, 29]. Computerized polygraphic recordings have revealed that the myoclonic potentials in juvenile myoclonic epilepsy are linked to a cortical positive potential that is encompassed within the polyspike discharge of the polyspike-and-wave complex [30].

18.2.2.5 Epilepsia Partialis Continua

In epilepsia partialis continua, polygraphic studies allow the analysis of the rhythm, intensity, rate, and distribution of muscle jerks [31]. Myoclonic EMG potentials may be observed associated with slow focal abnormalities on the EEG, contralateral focal epileptiform discharges, or even the absence of any evident paroxysmal activity.

18.2.2.6 Epileptic Negative Myoclonus

In epileptic negative myoclonus (ENM), the disruption of muscular activity occurs synchronously in both agonist and antagonist muscles. When this form of myoclonus is focal, affecting

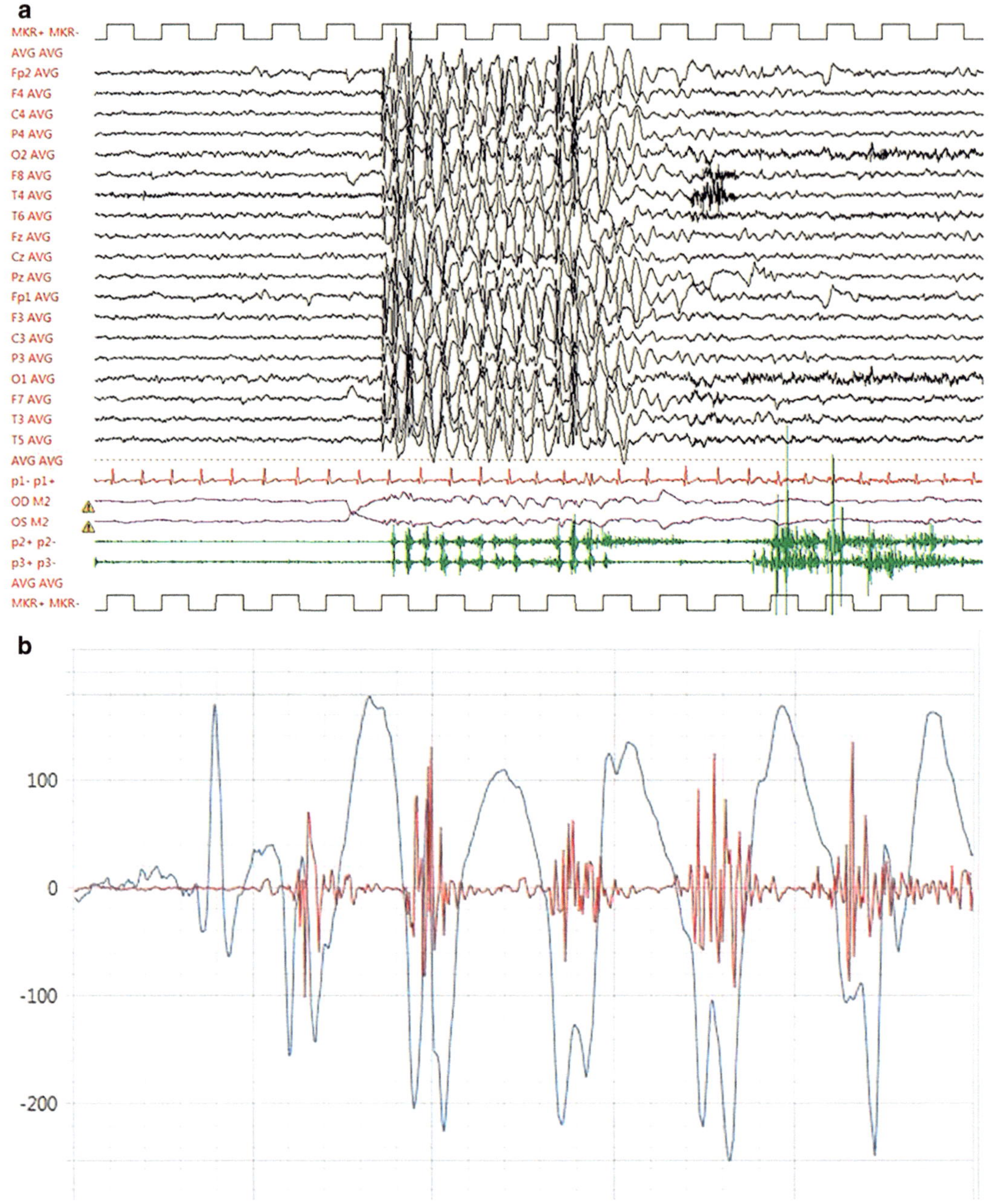

Fig. 18.5 Polygraphic recording of a myoclonic absence seizure. (**a**) A 3-c/s generalized spike-and-wave discharge is accompanied by rhythmic myoclonic jerks in both deltoids (p2+p2 and p3+p3), occurring at the same frequency of the spike-and-wave complexes and progressively associated with a tonic contraction. (**b**) Superposition of the EEG channel (F3) and the EMG channel (p2+p2) showing the strict temporal relationships with each spike-wave complex with a myoclonic burst

one or both limbs on the same side of the body, it is typically associated with a contralateral EEG spike located in the centroparietal region. Epileptic negative myoclonus can be observed across a broad range of epileptic disorders, including idiopathic, cryptogenic, and symptomatic types [8]. In some patients with epilepsy, ENM presents with a clinical picture that resembles epilepsia partialis continua that is frequent, subcontinuous jerks that become apparent when the patient maintains a specific posture, or it may involve a tonic contraction. Polygraphic recordings are crucial in clarifying the precise nature of this motor disorder, which depends on subcontinuous ENMs [3, 6, 8, 9, 32–34].

18.2.2.7 Progressive Myoclonic Epilepsies

Progressive myoclonic epilepsies (PME) exhibit polygraphic features that include the constant presence of action myoclonus, which is characterized by high-amplitude EMG potentials of short duration (ranging from 20 to 30 milliseconds) that occur synchronously in both agonist and antagonist muscles. These bursts are followed by a period of EMG silence lasting between 40 and 120 milliseconds, though it may occasionally extend to 300 milliseconds. The relationship between myoclonic bursts and silent periods with EEG spike-and-wave complexes and polyspike-and-wave complexes is inconsistent [35–37]. In patients with Unverricht–Lundborg disease, myoclonic seizures are characterized by generalized myoclonia, which is predominantly proximal in the upper limbs and exhibits a varying rhythm, often associated with generalized, symmetric polyspikes or polyspike-and-wave patterns. Lafora disease is identified polygraphically by the presence of abundant, asymmetric, asynchronous, and subtle myoclonia at rest, affecting all muscular groups, inconstantly related to an EEG correlate [38, 39] (Fig. 18.6). In several forms of PME, intermittent photic stimulation can trigger rapid polyspikes and polyspike-and-wave complexes that are associated with massive myoclonic jerks [37–41].

18.2.3 Evolution of Motor Manifestations in Relation to the Sleep–Wake Cycle

Epileptic manifestations are often entrained to the sleep–wake cycle across various epileptic disorders [42, 43]. As a result, the motor phenomena associated with epilepsy can exhibit changes dur-

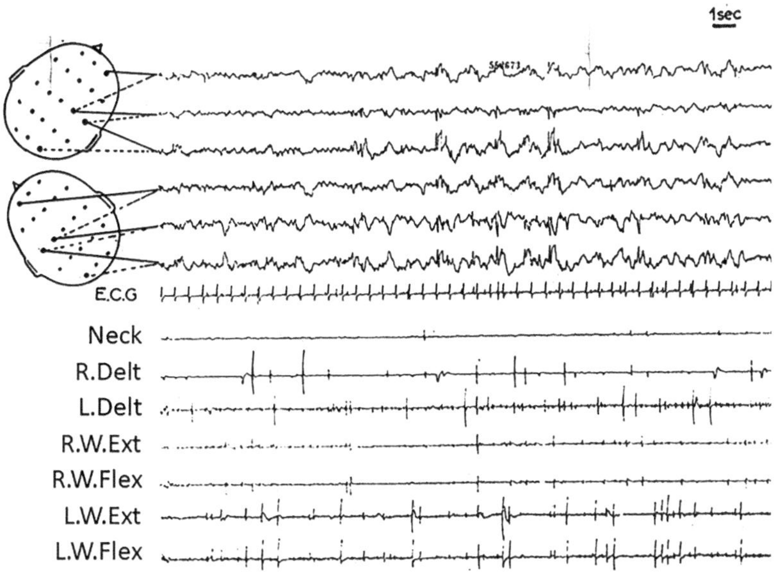

Fig. 18.6 Polygraphic recording in a patient with Lafora disease showing abundant, erratic myoclonic jerks, inconstantly related to EEG spikes. Neck, splenius capitis; Delt, deltoid; W.Ext, wrist extensor; W.Flex, wrist flexor; R, right; L, left

ing the transition from wakefulness to sleep. For instance, photosensitivity and light-induced myoclonus tend to diminish during non-REM sleep [44–46], while conflicting findings—likely due to variations in medication and stimulation methods—suggest either a decrease or an increase in photosensitivity-related myoclonus during REM sleep [44, 47, 48].

18.2.3.1 Lennox–Gastaut Syndrome

In individuals with Lennox–Gastaut syndrome [24, 49], nocturnal tonic seizures can occur so frequently during non-REM sleep that they may lead to a state of tonic status epilepticus. This latter condition is characterized polygraphically by a gradual reduction of tonic motor manifestations, accompanied by deterioration of vegetative parameters and impaired consciousness between seizures. In severe cases, this can progress to a coma-like state, marked by autonomic dysfunction and mild motor activities, such as hand lifting and slow eye movements. Thorough polygraphic monitoring is essential for diagnosing and managing this condition. Additionally, patients with Lennox–Gastaut syndrome may exhibit minor nocturnal seizures, which can involve eye-opening episodes, head nodding, and fluctuations in cardiac and respiratory rates. Often, these seizures are only identified through polygraphic recordings, which may show generalized polyspikes, rapid recruiting rhythms, or slow spike-and-wave patterns [22, 49]. Alterations in sleep architecture have also been documented in these patients [50].

18.2.3.2 Epilepsia Partialis Continua

Polygraphic investigations during sleep in epilepsia partialis continua have shown persistent occurrence of focal myoclonic jerks, which vary only in intensity and frequency throughout all stages of sleep [51].

18.2.3.3 Unverricht–Lundborg Disease

In Unverricht-Lundborg disease, sleep studies indicate (a) the absence of activation of generalized paroxysmal discharges and (b) the emergence of focal multiple fast spikes that appear in repetitive bursts, predominantly localized over the midline and centroparietal regions, and occurring more frequently during REM sleep, especially when eye movements are prevalent. These fast spikes may be time-locked to myoclonic jerks, particularly in muscles that exhibit pronounced action myoclonus during wakefulness [35, 37].

18.2.3.4 Lafora Disease

In Lafora disease, the organization of sleep is severely disrupted, with the various stages being barely distinguishable; paroxysmal activity does not appear to increase during sleep. There is a presence of diffuse multiple fast spikes with variable amplitude and topography, which can intermingle with fast activity, while posterior spikes persist during slow-wave sleep and may intensify during REM sleep [37, 39].

18.2.3.5 Myoclonic Encephalopathies

Myoclonic encephalopathies can be effectively studied through sleep polygraphy to identify myoclonus during sleep, particularly when other abnormal movements that occur during wakefulness are absent [52]. In patients with myoclonic epilepsy with ragged red fibers (MERRF), sleep polygraphic studies have revealed the presence of vertex and central spikes, the absence of physiological EEG elements characteristic of stage II sleep, and the persistence of generalized paroxysmal activity similar to that seen during wakefulness [53]. Sleep polygraphic examinations are also pertinent for investigating aminoacidopathies, where multifocal paroxysmal abnormalities and suppression burst patterns that are not evident during wakefulness can be revealed during sleep.

18.2.4 Computer-Assisted Analysis of Polygraphic Signals

The advent of computerized methods has introduced innovative tools for analyzing polygraphic signals. A primary area of application in epilepsy involves the simultaneous collection and processing of EEG and EMG signals. This integration facilitates the identification of cortical correlates time-locked to muscular phenomena that standard polygraphic techniques cannot detect, enabling a more accurate definition and quantification of the timing relationship between cortical and muscular events.

18.2.4.1 Jerk-Locked Averaging

A useful technique is jerk-locked averaging, a computerized EEG averaging method that was first utilized on polygraphic data by Shibasaki and Kuroiwa in 1975 [54]. This averaging technique is based on a trigger signal which is the EMG pulse represented by a myoclonic potential, and it allows the extraction of cortical potential associated with muscular events from the background EEG activity (Fig. 18.1, lower panel). It enables the characterization of various types of myoclonus based on the presence or absence of a cortical correlate, as well as the latency between the cortical correlate and the myoclonic jerk, as reported by Shibasaki and Neshige in 1987 [55] and Shibasaki et al. in 1978 [56].

18.2.4.2 Silent-Period Locked Averaging

A related technique, known as silent-period locked averaging, was introduced by Ugawa et al. in 1989 [57]. Unlike jerk-locked averaging, this method is triggered by the onset of an EMG silent period intermixed in a tonic contraction. Silent-period locked averaging allows for the investigation of different forms of negative myoclonus, including asterixis, and their association with cortical events.

18.2.4.3 Spike Averaging

Spike activity related to various muscular events can be analyzed using computerized EEG system that enables the visualization of the topographic distribution of EMG-related cortical potentials across the scalp. These methodologies have been employed to explore paroxysmal activities associated with ENM. The discovery of a frontal spike component suggests that inhibitory frontal areas play a role in generating negative motor phenomena [32], as confirmed by subsequent intracerebral electrical stimulation studies [33]. Spike-averaging techniques have revealed that spikes associated with ENM myoclonus differ from those not associated with ENM [32].

18.2.4.4 Computerized Analysis of EEG-EMG Correlations

EEG-EMG spectral analysis and measurements of coherence and phase functions can be performed using bivariate and time-varying autoregressive (AR) models. In PME, these analytical techniques have illustrated coherent cortical and EMG beta oscillations that are critically involved in myoclonus generation, in addition to revealing uneven spectral profiles that characterize the various forms of PME [58].

18.2.4.5 Quantitative Analysis of Electromyography

Quantitative analysis of EMG signals can be achieved through the rectification and integration of EMG waveforms. These techniques facilitate the calculation of the area between the EMG tracing and the baseline, with this area varying according to changes in amplitude, frequency, and duration of the potentials, thereby providing a quantification of EMG activity [4].

18.2.4.6 Multichannel Electromyography

Multichannel computerized EMG recordings, using only a limited number of EEG channels, can analyze the sequential activation patterns of different muscle groups involved in sudden motor events, such as massive myoclonic jerks, or in the complex motor manifestations associated with focal seizures affecting motor areas [59]. By analyzing the temporal sequence of muscle contraction onset in various cranial and limb muscles, researchers have proposed a differentiation between cortical and reticular myoclonus [34, 60–62].

18.2.4.7 Combination Recordings

The study of the modulation of the motor system during ictal and interictal paroxysmal discharges can be conducted by integrating polygraphic recordings with techniques that evaluate cortical excitability (such as transcranial magnetic stimulation) and spinal excitability (including H-reflex, stretch reflex, and F-wave) [4, 63]. During the EMG silent period of ENM, the inability of transcranial magnetic stimulation to elicit a motor response, coupled with the preservation of the F-wave, indicates a cortical origin for this motor disorder [8]. In patients experiencing severe partial epilepsies and falling seizures, H-reflex and motor-evoked potentials remain unchanged during slow spike-and-wave complexes but tend to increase during subclinical fast polyspike discharges [4]. A similar facilitation is observed at the onset of tonic seizures when the EMG is silent; during the resulting tonic contraction, H-reflex and motor-evoked potentials are significantly enhanced, decreasing remarkably only in the post-ictal phase. This differs with what has been observed during 3-c/s spike-and-wave discharges of typical absences or evoked by intermittent photic stimulation during which motor-evoked potentials or the stretch reflex are not altered.

18.3 Conclusions

Polygraphic recordings can be an invaluable tool for examining the changes occurring in a variable set of physiological parameters during epileptic phenomena and for supplying important data for both diagnostic and therapeutic purposes. The utility of polygraphy is particularly pronounced in delineating the various motor manifestations associated with different epileptic seizures and syndromes. Indeed, subtle motor phenomena—such as the diffuse myoclonia seen in Lafora disease, which often are barely noticeable during a clinical examination—can only be detected through polygraphic methods. Polygraphic techniques allow for the analysis of complex motor patterns that involve multiple muscle groups, and they enable the investigation of a stereotypical sequence of muscular activation. Consequently, it becomes feasible to describe not only "what moves first" but also "how it moves." In ENM, polygraphy is crucial for demonstrating the impact on a muscular contraction of paroxysmal activities, revealing that an apparent "interictal" epileptic abnormality may, in fact, be "ictal," depending on the patient's condition (whether relaxed or maintaining a muscular tone), highlighting the essential role of simultaneous EEG/EMG recording. The rapid advancements in technology will undoubtedly continue to yield new tools for collecting and processing various physiological parameters, thereby enhancing our capacity to explore the relationships among distinct anatomical and functional systems during an epileptic event.

References

1. Gastaut H, Broughton R. A clinical and polygraphic study of episodic phenomena during sleep. In: Wortis J, editor. Recent advances in biological psychiatry, vol. 7. New York: Plenum Press; 1965. p. 197–221.
2. Kamp A, Pfurtscheller G, Lopes da Silva F. Polygraphy. In: Niedermeyer E, Lopes Da Silva F, editors. Electroencephalography: basic principles, clinical applications and related fields. Baltimore: Williams & Wilkins; 1993. p. 757–63.
3. Tassinari CA, Cantalupo G, Rubboli G. Polygraphic recordings. In: Koutroumanidis M, Panaiyotopoulos C, editors. An Atlas of EEG in Epilepsy, vol. 2. Berlin: Springer Verlag; 2010. p. 723–40.
4. Tassinari CA, Gardella E, Cantalupo G, Rubboli G. Polygraphic recordings. In: Engel J, Moshe S, editors. Epilepsy. A comprehensive textbook. Philadelfphia: Walters Kluwer; 2023.
5. Tassinari CA, Regis H, Gastaut H. A particular form of muscular inhibition in epilepsy: the related epileptic silent period (RESP). Proc Aust Assoc Neurol. 1968;5:595–602.
6. Tassinari CA. New perspectives in epileptology. In: Trends in modern epileptology. proceedings of the international public seminar on epileptology. Tokyo: Japanese Epilepsy Association; 1981. p. 42–59.
7. Guerrini R, Dravet C, Genton P, et al. Epileptic negative myoclonus. Neurology. 1993;43:1078–83.
8. Tassinari CA, Rubboli G, Parmeggiani L, et al. Epileptic negative myoclonus. In: Fahn S, Hallett M, Marsden CD, editors. Negative motor phenomena. Advances in neurology, vol. 67. New York: Raven Press; 1995. p. 181–97.
9. Rubboli G, Tassinari CA. Negative myoclonus. An overview of its clinical features, pathophysiological mechanisms, and management. Neurophysiol Clin. 2006;36:337–43.
10. Blume WT, Luders HO, Mizrahi E, et al. Glossary of descriptive terminology for ictal semiology: report of the ILAE task force on classification and terminology. Epilepsia. 2001;42:1212–8.

11. Beniczky S, Tatum WO, Blumenfeld H, et al. Seizure semiology: ILAE glossary of terms and their significance. Epileptic Disord. 2022;24:447–95.
12. Bisulli F, Volpi L, Meletti S, et al. Ictal pattern of EEG and muscular activation in symptomatic infantile spasms: a videopolygraphic and computer analysis. Epilepsia. 2002;43:1559–63.
13. Vigevano F, Fusco L, Pachatz C. Neurophysiology of spasms. Brain Dev. 2001;23:467–72.
14. Gastaut H, Tassinari CA. Epilepsies. In: Remond A, editor. Handbook of electroencephalography and clinical neurophysiology, vol. 13A. Amsterdam: Elsevier; 1975.
15. Fusco L, Vigevano F. Ictal clinical electroencephalographic findings of spasms in West syndrome. Epilepsia. 1993;34:671–8.
16. Gobbi G, Bruno L, Pini A, et al. Periodic spasms: an unclassified type of epileptic seizures in childhood. Dev Med Child Neurol. 1987;29:766–75.
17. Dragoumi P, Chivers F, Brady M, et al. Epilepsy with myoclonic-atonic seizures (Doose syndrome): When video-EEG polygraphy holds the key to syndrome diagnosis. Epilepsy Behav Case Rep. 2015 Oct;28(5):31–3. https://doi.org/10.1016/j.ebcr.2015.10.001.
18. Gastaut H, Broughton R. Epileptic seizures: clinical and electrographic features. Springfield: Thomas; 1972.
19. Koutroumanidis M, Arzimanoglou A, Caraballo R, et al. The role of EEG in the diagnosis and classification of the epilepsy syndromes: a tool for clinical practice by the ILAE Neurophysiology Task Force (Part 1). Epileptic Disord. 2017a;19:233–98.
20. Koutroumanidis M, Arzimanoglou A, Caraballo R, et al. The role of EEG in the diagnosis and classification of the epilepsy syndromes: a tool for clinical practice by the ILAE Neurophysiology Task Force (Part 2). Epileptic Disord. 2017b;19:385–437.
21. Tatum WO, Rubboli G, Kaplan PW, et al. Clinical utility of EEG in diagnosing and monitoring epilepsy in adults. Clin Neurophysiol. 2018;129:1056–82.
22. Chatrian GE, Lettich E, Wilkus RJ, et al. Polygraphic and clinical observations on tonic–autonomic seizures. In: Broughton R, editor. Henri Gastaut and the Marseilles School's contribution to the neurosciences. Amsterdam: Elsevier; 1982. p. 101–24.
23. Gastaut H, Roger J, Ouachi S, et al. An electroclinical study of generalized epileptic seizures of tonic expression. Epilepsia. 1963;4:15–44.
24. Gastaut H, Roger J, Soulayrol R, et al. Childhood epileptic encephalopathy with diffuse slow spike-waves (otherwise known as "petit mal variant") or Lennox syndrome. Epilepsia. 1966;7:139–79.
25. Tassinari CA, Bureau M, Thomas P. Epilepsy with myoclonic absences. In: Roger J, Bureau M, Dravet C, et al., editors. Epileptic syndromes in infancy, childhood and adolescence. 2nd ed. London: John Libbey; 1992. p. 151–60.
26. Tassinari CA, Lyagoubi S, Santos V, et al. Etude des decharges de pointes ondes chez l'homme, II—Les aspects cliniques et electroencephalographiques des absences myocloniques. Rev Neurol. 1969;121:379–83.
27. Weir B. The morphology of the spike-wave complex. Electroencephalogr Clin Neurophysiol. 1965;19:284–90.
28. Janz D. The natural history of primary generalized epilepsies with sporadic myoclonias of the "impulsive petit mal" type. In: Lugaresi E, Pazzaglia P, Tassinari CA, editors. Evolution and prognosis of epilepsies. Bologna: Aulo Gaggi; 1973. p. 55–61.
29. Simonsen N, Mollgaard V, Lund V. A controlled clinical and electroencephalographic study of myoclonic epilepsy (impulsive petit-mal): preliminary report. In: Janz D, editor. Epileptology. Stuttgart: Thieme; 1976. p. 41–8.
30. Panzica F, Rubboli G, Franceschetti S, et al. Cortical myoclonus in Janz syndrome. Clin Neurophysiol. 2001;112:1803–9.
31. Thomas JE, Thomas JR, Klass DW. Epilepsia partialis continua. A review of 32 cases. Arch Neurol. 1977;34:266–75.
32. Rubboli G, Parmeggiani L, Tassinari CA. Frontal inhibitory spike component associated with epileptic negative myoclonus. Electroencephalogr Clin Neurophysiol. 1995;95:201–5.
33. Rubboli G, Mai R, Meletti S, et al. Negative myoclonus induced by cortical electrical stimulation in epileptic patients. Brain. 2006;129:65–81.
34. Apartis E, Vercueil L. To jerk or not to jerk: a clinical pathophysiology of myoclonus. Rev Neurol. 2016;172:465–76.
35. Tassinari CA, Coccagna G, Mantovani M, et al. Polygraphic study of dyssynergia cerebellaris myoclonica (Ramsay-Hunt syndrome) and of intention myoclonus (Lance-Adams syndrome) during sleep. Eur Neurol. 1973;9:105–20.
36. Tassinari CA, Bureau-Paillas M, Dalla Bernardina B, et al. Generalized epilepsies and seizures during sleep. A polygraphic study. In: Van Praag HM, Meinardi H, editors. Brain and Sleep. Amsterdam: De Erven Bohn; 1974. p. 154–66.
37. Avanzini G, Shibasaki H, Rubboli G, et al. Neurophysiology of myoclonus and progressive myoclonus epilepsies. Epileptic Disord. 2016;18(Suppl 2):11–27.
38. Roger J, Gastaut H, Boudouresques J, et al. Epilepsie myoclonique progressive avec corps de Lafora. Etude clinique et polygraphique. Controle anatomique ultrastructural. Rev Neurol. 1967;116:197–212.
39. Tassinari CA, Bureau-Paillas M, Dalla Bernardina B, et al. La maladie de Lafora. Rev EEG Neurophysiol. 1978;8:107–22.
40. Rubboli G, Franceschetti S, Berkovic SF. Clinical and neurophysiologic features of progressive myoclonus epilepsy without renal failure caused by SCARB2 mutations. Epilepsia. 2011;52:2356–66.
41. Rubboli G, Veggiotti P, Pini A. Spinal muscular atrophy associated with progressive myoclonic epilepsy:

A rare condition caused by mutations in ASAH1. Epilepsia. 2015;56:692–8.
42. Nobili L, de Weerd A, Rubboli G, et al. Standard procedures for the diagnostic pathway of sleep-related epilepsies and comorbid sleep disorders: A European Academy of Neurology, European Sleep Research Society and International League against Epilepsy-Europe consensus review. J Sleep Res. 2020 Dec;29:e13184. https://doi.org/10.1111/jsr.13184.
43. Wu T, Avidan AY, Engel J Jr. Sleep and epilepsy, clinical spectrum and updated review. Sleep Med Clin. 2021;16:389–408.
44. Corletto F, Gentilomo A, Rosadini G, et al. Comportamento durante il sonno della stimolazione luminosa intermittente. Riv Neurol. 1967;37:106–7.
45. Meyer-Ewert KH, Broughton R. Photomyoclonic response of epileptic and non-epileptic subjects during wakefulness, sleep and arousal. Electroencephalogr Clin Neurophysiol. 1967;23:142–51.
46. Rodin EA, Daly DD, Bickford R. Effects of photic stimulation during sleep: a study of normal subjects and epileptic patients. Neurology. 1955;5:149–59.
47. Scollo-Lavizzari G, Scollo-Lavizzari GR. Sleep, sleep deprivation, photosensitivity and epilepsy. Eur Neurol. 1974;11:1–21.
48. Yamamoto J, Furuya E, Wakamatsu H, et al. Modification of photosensitivity in epileptics during sleep. Electroencephalogr Clin Neurophysiol. 1971;31:509–13.
49. Arzimanoglou A, French J, Blume WT, et al. Lennox-Gastaut syndrome: a consensus approach on diagnosis, assessment, management, and trial methodology. Lancet Neurol. 2009;8:82–93.
50. Amir N, Shalev RS, Steinberg A. Sleep patterns in the Lennox-Gastaut syndrome. Neurology. 1986;6:1224–6.
51. Lugaresi E, Tassinari CA, Coccagna G, et al. Evoluzione nel sonno di un caso di epilessia parziale continua di Kojewnikoff. Riv Neurol. 1966;36:606–15.
52. Dalla Bernardina B, Dulac O, Bureau M, et al. Encephalopathie myoclonique precoce avec épilepsie. Rev EEG Neurophysiol. 1982;12:8–14.
53. Roger J, Pellissier JF, Dravet C, et al. Degenerescence spino-cerebelleuse-atrophie optique-epilepsie-myoclonies-myopathie mitochondriale. Rev Neurol. 1982;138:187–200.
54. Shibasaki H, Kuroiwa Y. Electroencephalographic correlates of myoclonus. Electroencephalogr Clin Neurophysiol. 1975;39:455–63.
55. Shibasaki H, Neshige R. Photic cortical reflex myoclonus. Ann Neurol. 1987;22:252–7.
56. Shibasaki H, Yamashita Y, Kuroiwa Y. Electroencephalographic studies of myoclonus. Myoclonus-related cortical spikes and high amplitude somatosensory evoked potentials. Brain. 1978;101:447–60.
57. Ugawa Y, Shimpo T, Mannen T. Physiological analysis of asterixis: silent period locked averaging. J Neurol Neurosurg Psychiatry. 1989;52:89–93.
58. Panzica F, Varotto G, Canafoglia L, et al. EEG-EMG coherence estimated using time-varying autoregressive models in movement-activated myoclonus in patients with progressive myoclonic epilepsies. Annu Int Conf IEEE Eng Med Biol Soc. 2010;2010:1642–5.
59. Meletti S, Rubboli G, Testoni S, et al. Early ictal speech and motor inhibition in fronto-mesial epileptic seizures: a polygraphic study in one patient. Clin Neurophysiol. 2003;114:56–62.
60. Hallett M, Chadwick D, Adam J, et al. Reticular reflex myoclonus: a physiological type of human post hypoxic myoclonus. J Neurol Neurosurg Psychiatry. 1977;40:253–64.
61. Hallett M, Chadwick D, Marsden CD. Cortical reflex myoclonus. Neurology. 1979;29:1107–25.
62. Rothwell JC, Obeso JA, Marsden CD. Electrophysiology of somatosensory reflex myoclonus. In: Fahn S, Marsden CD, Van Woert MH, editors. Myoclonus. Advances in neurology, vol. 43. New York: Raven Press; 1986. p. 385–98.
63. Tassinari CA, Rubboli G, Shibasaki H. Neurophysiology of positive and negative myoclonus. Electroencephalogr Clin Neurophysiol. 1998;107:181–95.

19 Electroencephalography (EEG) Combined with Neuroimaging and Neurostimulation

Anna Elisabetta Vaudano, Stefan Rampp, and Louis Lemieux

19.1 Introduction

Neuroimaging refers to a variety of noninvasive techniques and technologies that offer insights into both the structure and function of neural mechanisms. Key methods include electroencephalography (EEG), magnetoencephalography (MEG), functional magnetic resonance imaging (fMRI), positron emission tomography (PET), and structural MRI (MRI). Each technique has distinct advantages and limitations, particularly in terms of spatial and temporal resolution, safety, and accessibility [1]. For instance, while EEG provides high temporal resolution, it has lower spatial resolution; conversely, fMRI delivers high spatial resolution but lower temporal resolution. Additionally, complementary techniques such as transcranial magnetic stimulation (TMS) and other neuromodulation methods are frequently employed alongside neuroimaging to aid in source localization and stimulation.

Although each modality contributes unique perspectives on brain structures and activity, recent technological advancements have facilitated the combination of multiple techniques—termed multimodal neuroimaging—to mitigate the limitations of individual methods and develop a more holistic understanding of brain function. Notably, data fusion plays a crucial role in this area by integrating various imaging modalities to utilize cross-modal information [2]. Data fusion involves the simultaneous analysis of two or more brain imaging modalities, enhancing the interpretation of data. Importantly, capturing multiple datasets at the same time allows for complementary imaging information to be collected under consistent physiological and psychological conditions.

EEG, known for its excellent temporal resolution and portability, is often combined with other imaging techniques. For example, the integration of EEG (and MEG) with structural MRI improves the localization of neuronal sources (source imaging approaches). Meanwhile, pairing EEG with fMRI (EEG-fMRI) or metabolic imaging like PET (EEG-PET) deepens our understanding of brain functions in both healthy individuals and those with pathological conditions. Furthermore, combining EEG with TMS (EEG-TMS) or transcranial electrical stimulation (EEG-TES) opens

A. E. Vaudano (✉)
Neurophysiology Unit and Epilepsy Center, AOU Modena, Department of Biomedical, Metabolic and Neural Science, University of Modena and Reggio Emilia, Modena, Italy
e-mail: annaelisabetta.vaudano@unimore.it

S. Rampp
Department of Neurosurgery and Department of Neuroradiology, University Hospital Erlangen, Erlangen, Germany
e-mail: stefan.rampp@uk-erlangen.de

L. Lemieux
UCL Queen Square Institute of Neurology, University College London, London, UK
e-mail: louis.lemieux@ucl.ac.uk

F. Brigo, O. Mecarelli (eds.), *EEG: The First 100 Years*,
https://doi.org/10.1007/978-3-031-86878-8_19

up potential therapeutic avenues and monitoring strategies for brain disorders.

This chapter aims to provide a thorough overview of how EEG can be integrated with other neuroimaging techniques, focusing on both research and clinical applications. We will explore the simultaneous acquisition of EEG with other imaging modalities, addressing the technical challenges involved and highlighting their clinical relevance. Special attention will be given to the significant role these techniques play in epilepsy research and treatment while also considering their wider applications in clinical practice.

19.2 Source Imaging

19.2.1 Rationale for Source Imaging

Conventionally, EEG is interpreted visually by identifying specific graphoelements, patterns, and rhythms. The locations of their underlying neural sources are inferred by considering the topographical distribution of amplitudes over the recorded electrodes. This is achieved by applying montages and interpreting the different appearances, locations of maxima, and phase reversals. A classic example is the localization of the irritative zone in persons with focal epilepsy. Here, the channels showing the maximum negativity of interictal epileptic discharges (IED) in referential montages are thought to be closest to the generating neural populations. In bipolar montages, the same electrodes show a characteristic phase reversal.

With digital EEG enabling easy access to remontaging, this strategy is quick to implement in practice. However, there are many situations in which this approach yields inaccurate or even erroneous results. The maximum negativity and respectively the phase reversal always occurs in the electrodes located on the scalp along the orientation of the main EEG generators, i.e., the cortical pyramidal cells [3]. Pyramidal cells themselves are oriented perpendicular to the cortical surface. Correspondingly, the maximum negativity and phase reversal is located directly above the generators only if the pyramidal cells directly point to the scalp, i.e., they have a "radial" orientation. Purely radial sources, however, are limited to sulcal depths and gyral crowns. In sulcal walls, pyramidal cells have a "tangential" orientation, parallel to the scalp surface. These do not contribute in any relevant manner to the EEG recorded from electrodes above them. Instead, the maximum negativity/phase reversal is projected to more distant areas, sometimes several centimeters away or even over the contralateral hemisphere. In practice, generators of EEG signals may involve sulcal and gyral aspects. The corresponding orientation is then not solely radial or tangential but oblique with different contributions of both.

Localization of neural sources is further complicated by the influence of the head's anatomy. Scalp EEG does not directly reflect neuronal activity. Instead, it represents the correlate of neural activity that is conducted through the different tissues. Scalp, skull, CSF, etc. also have different electric resistances or conductivities for electric currents. With approx. 0.0042 S/m, for example, the skull is significantly less conductive than the scalp with 0.33 S/m or CSF with 1.79 S/m. Furthermore, there is considerable variability between individuals, age ranges, and studies [4]. These differences distort the EEG signal according to their anatomical distribution. Therefore, these influences are especially relevant in patients with large lesions, resection volumes and skull holes from previous surgery, etc.

Measuring the tiny magnetic fields that occur in conjunction with neuronal activity using magnetoencephalography (MEG) circumvents this issue. The conductivity of biological tissues for magnetic fields is nearly identical and similar to the conductivity of air. Correspondingly, MEG measurements are contactless and provide an almost undistorted view on the activity. However, the technique is only sensitive to tangential components, which results in amplitude distributions which are similarly complex as those of tangential EEG sources. Visual interpretation of MEG therefore suffers from similar issues as the EEG.

For accurate EEG/MEG interpretation regarding the location of sources, e.g., of epileptic foci or functional cortex, both source orientation and volume conduction must be considered. Accordingly, visual interpretation based solely on the raw signal in different montages is limited.

Source imaging is a technique to overcome these limitations, combining EEG (or MEG) recordings with structural neuroimaging to provide more accurate localizations. It projects the recorded signals from "electrode/sensor space" to the intracranial "source space" transforming neurophysiology to neuroimaging.

19.2.2 How Does Source Imaging Work?

Source imaging constructs a model of signal generation and volume conduction to estimate the location and characteristics of neural sources underlying measured activity. Such a model consists of three essential components in addition to the EEG/MEG data itself as well as information about the location of the electrodes: source model, source space, and volume conduction model.

The source model provides assumptions about the neural generators themselves, i.e., whether a single or few sources dominate the activity or whether a multifocal or more diffuse distribution is more likely. For source imaging of focal epileptic activity, for example, dipole models are often used. These assume that the generating neural population is focal, circumscribed, and highly synchronized and can thus be described sufficiently by only a location, an orientation, and activation (or "dipole moment" summarizing the latter two). In contrast, dipole models would be inadequate to model activity during cognitive processes, e.g., language comprehension or production. In these cases, source models assuming activation of distributed populations are better suited.

The source space defines the volume where the likely neural generators are expected to be located. Reminiscent of other neuroimaging techniques, such as fMRI or PET, a 3D voxel grid is often used, limited to intracranial space. Alternatively, locations of the modelled sources can be constrained to the cortical surface, excluding white matter and often also basal ganglia and the brainstem if these are not of interest for the given clinical or research question. Further approaches may limit the source space to only smaller regions, e.g., to save on computation time.

Finally, the volume conduction model provides information about the anatomy and the distribution of the different biological tissues. Source imaging uses this data to consider distortions due to different electric conductivities. A spectrum of approaches offers models with different degrees of detail. Single sphere models, for example, model the head as a sphere with a single "bulk" conductivity. Correspondingly, results have only limited accuracy. Multiple sphere models provide different compartments with respective conductivity values for, e.g., the skin, skull, CSF, and brain. More complex approaches replace spheres with shells adjusted to the individual patient's MRI ("boundary element models," BEM). Finally, the most detailed volume conduction models approximate anatomy with small elements, e.g., tetrahedra, based on MRI ("finite element" and "finite difference" models, FEM, FDM). They are also able to consider conduction anisotropy along and perpendicular to white matter fiber tracts. Detailed volume conduction models are especially relevant for EEG-based source imaging in patients with skull defects, resection volumes, or cysts, i.e., large alterations of the anatomy, whereas for MEG, simpler models are sufficient.

Using these components, source imaging solves the so-called forward and inverse problems. The forward problem reflects the situation, when all characteristics of a source are known, e.g., when simulating EEG/MEG. The forward problem is then to calculate the scalp EEG/MEG data which corresponds to the given neural activity, taking the volume conduction into account. The forward problem has a mathematically unique solution and can thus be calculated with the available data and models. The inverse problem corresponds to the clinical situation: an EEG/MEG pattern is recorded, and the task is to find the generating neural sources. However, in principle, many neural sources as well as their combinations could potentially generate the measured activity. Consequently, the inverse problem has no unique solution. To nevertheless provide an estimation, inverse methods iteratively calculate forward solutions with different assumptions for the different source characteristics, e.g., the location. Each estimate is then compared to the measured data. This optimization process continues until a best fit is achieved, which might however be impacted by data quality and correctness of assumptions. Methodological differences, such as the assumed source model and applied statistical procedures, give rise to a wide spectrum of inverse methods. Each approach has its own sensitivity and specificity for distinct aspects of activity, reminiscent of the range of MRI sequences, which also provide different perspectives on the same anatomy.

19.2.3 Clinical Applications of Source Imaging

Source imaging is widely applied in neurocognitive research. Clinical applications include epileptic focus localization and functional mapping. Accordingly, the technique is mainly applied in epilepsy, specifically in evaluation for epilepsy surgery, and brain tumor surgery. The following summary is focused on the application of EEG, respectively electric source imaging (ESI). Similar literature also exists for MEG which, however, is not discussed here.

19.2.3.1 Epilepsy

Source imaging resolves issues caused by source orientation and volume conduction. This results in several advantages for the evaluation of patients with epilepsy. First, the technique can achieve a higher degree of accuracy, especially in patients with structural alterations and with predominantly tangential sources. With only visual interpretation, distortion of the recorded signals and the large distances between negativity/phase reversal and neural generators can lead to inaccurate and erroneous interpretations. Second, by disentangling contributions of distinct areas to the EEG signal, source imaging can improve local signal-to-noise ratios (SNR). The improved SNR may then be utilized to detect subtle epileptic activity, e.g., low-amplitude IED from small cortical areas. Beyond spikes, sharp waves, and ictal rhythms, the improved SNR may also enable detection of high-frequency oscillations (HFO) with only scalp EEG. Third, the compensation of volume conduction further allows examination of functional connectivity. Application of these analyses on "sensor level" data, i.e., the recorded EEG traces, may be affected by spurious connectivity caused by volume conducted signals from identical sources. Even when these issues are addressed by selecting appropriate connectivity measures, it remains difficult to relate the results to anatomical locations as each trace is a mixture of signals from numerous intracranial sources.

Accuracy of ESI for Epileptic Focus Localization

The clinical value of using source imaging and exploiting the theoretical advantages has been investigated in many studies over the past decades. Many of these focus on the accuracy of ESI to localize the epileptic focus, as it enables its use for planning of epilepsy surgery and invasive recordings. In principle, the literature discerns two types of accuracy in this context: spatial and diagnostic accuracy. The former concerns the largely technical property of localizing a typically known source with preferably little error and spatial dispersion. The latter is more clinically oriented and describes the relationship between ESI results, treatment, and patient outcomes.

Studies on spatial accuracy of ESI most often utilize either simulations or comparison with invasive EEG recordings. Simulations have the advantage that all aspects of potentially influencing factors can be evaluated systematically. Such studies report a spatial accuracy of around 15 mm, depending on the applied recording method (HD-EEG, LD-EEG, MEG) and inverse solution [5]. Furthermore, they show that accuracy of head modeling, as well as choosing adequate conductivity values, has a significant impact on spatial accuracy [6]. Comparisons of ESI with simultaneous or separately recorded invasive EEG support these findings and report errors in the range of approx. 2–27 mm, depending on source depth, number of electrodes, as well as choice of ESI method and parameters [7, 8]. Both simulations and comparisons to invasive EEG stress the importance of the head model for the spatial accuracy of ESI. Especially large lesions, bone defects, resection volumes from previous surgery, and other alterations of the "standard" anatomy have a considerable impact. Accordingly, detailed, e.g., FEM models should be used in such cases, while simpler approximations, such as template head models, may be sufficient when there are no major structural alterations.

While spatial accuracy is an important prerequisite for any method used for localization, the diagnostic accuracy is the metric that ultimately determines the impact on patients' outcomes. Correspondingly, many studies have investigated the relationship between ESI localizations, surgical resection, and postoperative outcomes. The overall consensus is that focal, clustered ESI localizations completely contained in the resection are related to a high probability of postoperative seizure freedom or at least significant improvement [9–13]. Note the somewhat cumbersome wording of this important finding, which avoids commenting on the completeness of resection. The reason for this choice of words is that determination of the extent of sources using ESI (or almost any other diagnostic technique) is exceptionally difficult. The extent depicted in ESI reports, such as the size of a dipole cluster or the amount of cortex highlighted by a distributed source model, mostly depends on the signal-to-noise ratio (SNR). The true size of the underlying activated cortex certainly influences the ESI extent. However, especially on the level of the individual patient, it is practically impossible to discern its contribution from the impact of SNR. Dipole localizations or maxima of distributed solutions mark correspond to a center of gravity in some sense. If these are located in a resection volume, it is therefore highly likely that a substantial portion of the neural source indeed has been removed and that a good postoperative outcome is expected. Depending on the specific methodology and included etiologies, literature reports a sensitivity of 81–90% (percentage of cases with ESI in the resection and good postoperative outcome). Specificity (percentage of cases with ESI outside the resection and suboptimal or poor outcome) was found to be lower at 45–47% [14]. The latter could at least partially be explained by cases in whom limited SNR or excessive propagation leads to erroneous and scattered localizations. Even if these results are then not considered for resection, patients may still achieve postoperative seizure freedom if other diagnostic data from presurgical evaluation are robust.

Ictal Vs. Interictal ESI

ESI is often performed on interictal data. The technique can however be applied to ictal seizure activity in much the same way [15]. The preference for interictal activity is largely due to practical reasons. Artifacts occurring during a seizure, e.g., due to patient movement, may prevent application of ESI, while IEDs may be selected from artifact-free data. When longer recordings are available, a considerable amount of IEDs is available in contrast to usually only a few seizures. Furthermore, seizures may show rapid evolution during the onset. Correspondingly, choice of the time points to analyze is crucial. When good quality data is available with activity at the seizure onset for few seconds however, ictal ESI provides valuable results. Patients in whom these are concordant with the remaining spectrum of presurgical diagnostic methods have a significantly higher likelihood to achieve postoperative seizure freedom compared to patients with discordant findings times higher likelihood ratio [9]. In comparison to interictal ESI, results of ictal ESI often correlate better to the resection in postoperatively seizure-free patients (odds ratio 4.0 vs. 7.9) [14]. These findings reflect the well-known zones concept [16]: IEDs are generated in the irritative zone, which may be identical, overlapping, or completely distinct from the seizure onset zone. As the latter is more strongly associated with the epileptogenic zone, i.e., the region that must be removed to abolish the seizures, ictal ESI should indeed provide more accurate results with adequate data quality. However, it should be noted that interictal ESI already shows good performance to guide epilepsy surgery or invasive recordings [14, 17]. ESI overcomes some limitations of visual interpretation, i.e., the issues of source orientation and volume conduction. Furthermore, IED averaging is often applied to increase SNR and thus gain access to early components of the patterns, which may enable localizations even before considerable propagation may occur. Correspondingly, ESI may diminish the difference between ictal and interictal activity and their significance to localize the epileptogenic zone in many, albeit not all, cases.

Comparison to Other Methods and Additional Value

EEG and its analysis with source imaging techniques is only one of many diagnostic methods applied in presurgical evaluation of epilepsy surgery. Especially neuroimaging and specifically structural MRI plays a crucial role in identifying potential epileptogenic lesions. PET and SPECT detect areas with altered glucose metabolism and blood flow. Combined with video-EEG monitoring and neuropsychological evaluation, the diagnostic spectrum provides a comprehensive arsenal of tools [18]. However, especially facing limited time and personal and financial resources, it is not viable to apply all methods in every patient. Correspondingly, whether ESI provides nonredundant clinical value beyond the information provided by other techniques is a relevant question.

Most evidence is indirect, based on the spatial and diagnostic accuracy of the technique. The data, however, show a favorable picture for ESI. ESI based on HD-EEG shows diagnostic accuracy comparable to or exceeding MRI, PET, and even ictal SPECT [10, 13]. In a prospective study by Brodbeck and colleagues [10], for example, sensitivity and specificity of HD-ESI were 84% and 88%, respectively. MRI showed a high sensitivity of 76.3%, however a more limited specificity of 52.9%. While MRI enables detection of lesions, it cannot prove their epileptogenicity. Furthermore, structural imaging can only rarely if at all depict the extent not of the lesion itself but of the seizure onset or epileptogenic zone. In addition, structural imaging rarely, if ever, shows the extent of the seizure onset or epileptogenic zone itself but rather the extent of the lesion. Correspondingly, resection of only the lesion does not guarantee postoperative seizure freedom. PET and ictal SPECT also showed good performance, but again, with limitation: PET 68.7% sensitivity, 43.8% specificity; ictal SPECT 57.7% and 46.7%. Results were better in patients with temporal lobe epilepsy (TLE) compared to extratemporal lobe epilepsy (ETLE): 91.7 vs. 75.0% sensitivity.

Only a few studies have focused on the additional, nonredundant value of ESI. A typical design of such studies is a two-stage result presentation and decision-making process. First, all findings of presurgical evaluation are presented and discussed by a multidisciplinary team, resulting in a first decision on viability and extent of surgery or the need for invasive recordings. Then, source imaging results are presented. Whether these then change the initial decision is taken as evidence for nonredundant value. Finally, the additional information is compared to postoperative outcomes for evaluation of their validity. Such studies report additional, nonredundant value in 20–34% of cases [12, 19] with confirmed validation by postoperative outcomes or invasive EEG findings in the majority of these. Correspondingly, ESI not only provides diagnostic accuracy comparable to other diagnostic methods, but it also provides additional and valid information for a relevant portion of patients (Fig. 19.1).

LD- vs. HD-EEG

For visual EEG interpretation, additional electrodes are used to provide an overall better coverage and thus sensitivity for epileptic activity, e.g., by adding temporal basal electrodes [20, 21]. Increasing the overall number of EEG electrodes beyond the recommended IFCN electrode array [20] may further improve IED detection rates, although the limited available data on this aspect suggests only a modest effect

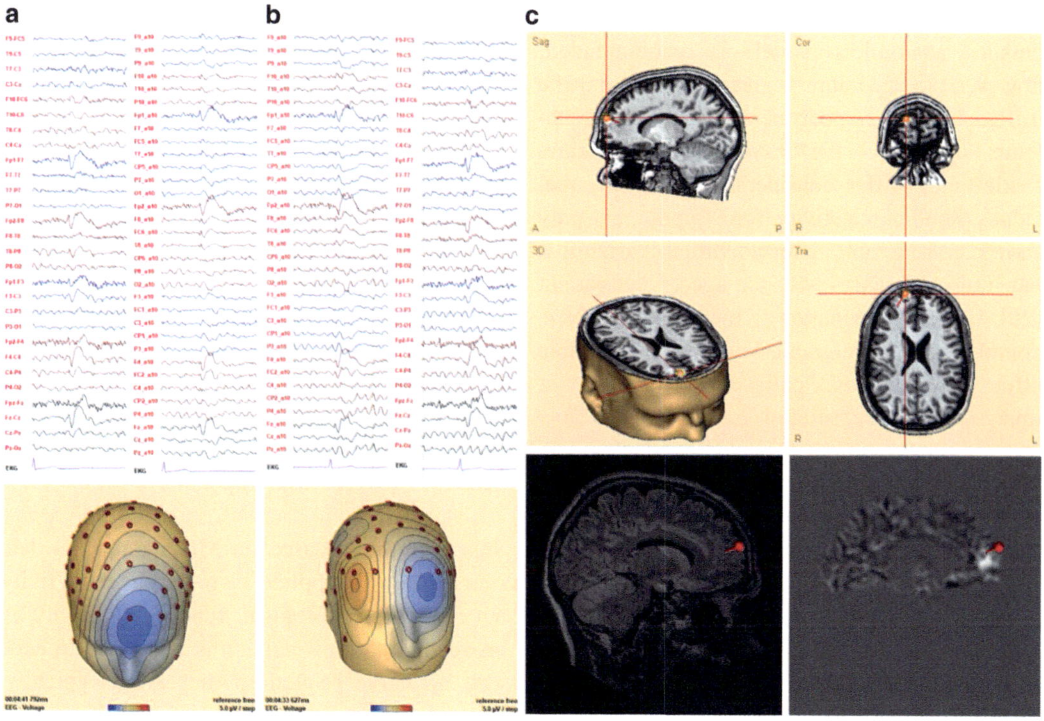

Fig. 19.1 Example of source imaging results of a 25-year-old male patient with focal epilepsy due to a right fronto-polar focal cortical dysplasia (FCD) IIb. (**a**, **b**) Examples of interictal epileptic discharges (IED) in bipolar and average-reference montages. Voltage maps at the IED peak are shown underneath the EEG traces. Maximum negativity in average-reference montages and voltage maps are located over the midline or the left fronto-polar region, i.e., not over the location of the lesion. (**c**) Results of source imaging using an equivalent current dipole (ECD) model applied to an IED average. The dipole localization is located directly at the FCD, as illustrated by 3D FLAIR and MRI post-processing (junction map, MAP07 software, showing statistically significant alterations in the right fronto-polar region). Resection of the FCD resulted in seizure freedom (Engel 1) at a follow-up of 2 years

[22]. Source imaging techniques however can leverage the additional data of HD-EEG with at least 64 electrodes to achieve a higher degree of accuracy [10, 23–25]. More electrodes further increase spatial accuracy but with diminishing returns [26]. Nevertheless, low-density EEG (LD-EEG) should not be underestimated in its ability to provide helpful diagnostic information. In the large study by Brodbeck and colleagues [10], LD-EEG achieved a diagnostic performance similar to MRI, PET, and ictal SPECT. Some recent studies have contrasted the additional information provided by HD-EEG with the influence of high SNR by evaluating larger numbers of IEDs [22, 27, 28]. Especially long-term video-EEG may provide hundreds and thousands of patterns over the recording duration of several days. Averaging these significantly reduces background noise and unrelated brain activity. The averaged patterns then show smooth topographies which can be sampled by a comparably small number of electrodes. The high SNR enables high spatial accuracy of source imaging, an effect that exceeds the impact of higher numbers of EEG electrodes [22].

Limitations

Although ESI provides useful information for planning of epilepsy surgery, the technique has limitations due to technical, methodological, and neurophysiological reasons. In addition to the quality and SNR of the recorded data, ESI is limited by the accuracy of the volume conductor model. The use of individual MRIs to set up detailed anatomical models is recommended; however, inaccurate segmentations of the different tissue compartments may cause problems. Furthermore, the "standard" tissue conductivities may differ considerably from the real values. Skull conductivity, for example, not only shows considerable interindividual variability, but it also decreases with the age of the patient [29]. Exact determination of individual values is complex and not possible without considerable effort and resources. Spatial accuracy is further impacted by co-registration procedures to transfer electrode positions, volume conductor, and MRI for visualization in a common coordinate system. Any error that occurs in this step will impact the accuracy of source imaging and limit the achievable spatial resolution.

Source imaging provides localization results for almost any pattern. Whether these are clinically relevant strongly depends on their specificity, i.e., whether they in fact represent epileptic activity and not just artifacts or normal variants. This decision remains in the hands of the physician, although source imaging provides supporting tools to resolve this issue [30]. Finally, while ESI can extract quite useful value from interictal activity, it cannot completely remove the functional difference between the irritative and the epileptogenic zone. Here also, the final interpretation remains with the physician.

Novel Developments

Development on improving source imaging for epileptic focus localization has been quite active in recent years. Methodological studies have introduced novel variants of inverse solutions, more accurate volume conductor models, and approaches combining EEG with other modalities, e.g., MEG or fMRI [31, 32]. Introduction of automated methods facilitate the practical application of ESI by minimizing time-consuming manual processing steps [25, 33]. These also allow analysis of large numbers of IEDs with limited effort. The high SNR of such data reduces the importance of high numbers of electrodes and allows analysis of early time points within the IED to avoid propagation.

Technical advances have also made it possible to analyze components of the EEG that are not obvious to the human eye. Oscillatory markers, ranging from infraslow activity to beta, gamma, and high-frequency oscillations, show associations with epileptic foci [34, 35]. Network measures investigate cortical communication and identify network hubs that appear to be related to irritative and seizure onset zones. Finally, machine learning and artificial intelligence may provide not only further automation but also access to subtle features of the scalp EEG.

19.2.4 Functional Mapping

Source imaging based on MEG and EEG data cannot only be applied to pathological activity but also to physiological, functional activity. A broad spectrum of paradigms is applied to activate networks related to the tasks. Typically, these consist of many repetitions to allow for a robust SNR. The overall approach is similar and often identical to ERP (evoked response) studies. However, the analysis then does not exclusively rely on amplitude and latency differences or topographic changes in sensor space but extracts characteristics of the involved neuronal populations in source space. This technique is applied widely in neurocognitive research [36–38] and complements other neuroimaging methods, e.g., fMRI and PET, with its high temporal resolution.

In clinical contexts, source imaging is applied for functional mapping, i.e., the determination of location and—to some degree—extent of regions and networks related to brain function. Especially sensorimotor cortex, verbal language, and memory functions are evaluated. The results are then used to plan epilepsy or tumor surgery. While many studies have examined methods and clinical value of magnetic source imaging (MSI), literature on ESI for functional brain mapping is sparse [39, 40]. The reasons for these disparities are not entirely clear. However, comparisons between MEG and EEG show that the sensitivity of EEG to neocortical sources, such as sensorimotor and language-related cortex, is significantly lower than that of MEG [41, 42]. Accordingly, functional mapping using ESI may provide limited accuracy for these applications, although this may be addressed by longer recordings and potentially higher numbers of electrodes. In fact, studies utilizing HD-EEG for localization of the somatosensory cortex indeed show a quite good concordance with fMRI and direct cortical stimulation [39, 42]. EEG studies on lateralization of the language dominant hemisphere show promising results, albeit do not employ source imaging so far [39, 43, 44]. Studies on using ESI to evaluate memory functions for presurgical mapping are currently not available [39].

19.2.5 Other Applications

In addition to epileptic focus localization and functional mapping, further clinical applications of source imaging are being explored. For example, ESI techniques are applied to differentiate patients with mild cognitive impairment and Alzheimer's disease from healthy controls by estimating the sources of the P300 event-related potentials [45]. In addition to this determination of sources and their activation, ESI enables investigation of functional connectivity while controlling for spurious connectivity caused by volume conduction. In patients with Alzheimer's disease (AD), disruption of functional connectivity has been identified, which could potentially serve as a neurophysiological biomarker for the disease [46]. With such markers of altered neuronal processing, early detection of AD may be possible even before the onset of clear dementia symptoms. Furthermore, such objective markers may facilitate development of therapies by providing gradual correlates of disease severity and improvement [47].

In patients after stroke, ESI-based approaches allow prediction of motor and language improvement by imaging network mechanisms and their plasticity [48, 49]. These approaches also utilize ESI to circumvent volume conduction to extract a robust measure of connectivity, as well as spatially resolved estimates of oscillatory activity. Ultimately, such markers could be used for targeted training for rehabilitation, e.g., by a personalized neurofeedback training [50]. Performing such analyses on the sensor level would face the challenge to separate components of interest from the physiological background. For example, alpha band synchrony seems to represent a surrogate marker of cognitive and motor performance in patients after ischemic stroke [49, 50]. However, for utilization for diagnosis and treatment, it must be differentiated from the posterior alpha of the visual system, which is a nontrivial task without taking volume conduction into account.

19.2.6 Conclusions

Source imaging bridges the gap between neurophysiology and neuroimaging. It determines the likely neural sources of measured EEG/MEG activity, considering the influence of source orientation and volume conduction. This results in improved spatial accuracy and allows evaluation of functional connectivity. Source imaging is used to localize epileptic foci for planning invasive recordings and epilepsy surgery. The technique can also be used to map functional cortex, although this is currently rarely used in clinical practice. Applications in other areas, such as dementia and poststroke rehabilitation, are being developed and may expand the range of practical applications of source imaging in the future.

19.3 EEG Combined with Functional MRI

19.3.1 Rationale for Concurrent EEG and fMRI Acquisition

Arguably, our knowledge on EEG and its link with human behavior, brain states, and health arguably surpass that of functional MRI (fMRI), a tomographic technique primarily sensitive to changes in regional blood flow [51]. For instance, the corpus of original research articles on EEG is roughly twice as large as that on fMRI. Although this partly reflects EEG's longer history, more articles were published on EEG than on fMRI since the latter's advent, around 1990. Equally, it would also seem difficult to argue strongly that we understand how EEG arises, and its role or function, better than fMRI. Is EEG analogous to the noise produced by a combustion engine (no air: no oxygen, no noise, and no combustion either) or more fundamentally functional? However, most would agree that EEG is more closely linked to brain function, which results from electrochemical processes, while fMRI reflects the activity of a kind of connected supply and evacuation system: the former's activity reflects the latter's more than vice versa, making EEG more fundamental, especially in an idealized, healthy brain.

EEG is commonly considered the most direct, noninvasive record of the brain's activity. Localization of EEG patterns is crucial, especially in clinical contexts like epilepsy, where source localization has progressed significantly over the last three decades, revealing extensive generative networks with confidence for certain activities, such as interictal epileptiform discharges [52]. However, these techniques depend on modelling assumptions, unlike fMRI, where signal localization is integral to data collection via spatial encoding. Despite fMRI signals being less directly related to the sources of interest, their localization is based on simpler physics and engineering principles, with accuracy largely independent of the source's location and complexity. Conversely, electrophysiological recordings have been used to improve our understanding of the physiology of the main fMRI contrast mechanism: the blood oxygen level-dependent (BOLD) effect [53–55]. Questions arise about which features of an electrophysiological signal best reflect BOLD fluctuations and what can be inferred about neuronal activity from BOLD patterns [56, 57].

This section will discuss recent technical advancements in recording of EEG during fMRI, followed by a review of its main application area: BOLD mapping of electrophysiological phenomena, with emphasis on epileptic activity recorded on the scalp (intracranial EEG combined with fMRI will be discussed in a dedicated subsection).

19.3.2 The Technology of Concurrent Scalp Electrophysiological Recordings during fMRI: Subject Safety and Data Quality

The combination of the strong and/or time-varying electromagnetic fields used in MRI and unsurpassed sensitivity of electrophysiological recording equipment (and overlapping bandwidths) is a very challenging data acquisition process: subject safety and data quality [58]. This section focuses on MRI scanner static field strengths of 1.5 and 3 T; for EEG-fMRI at 7 T, see Jorge et al. [59].

The primary physical phenomena at play are electromagnetic induction and magnetization, and secondarily, mechanical forces (projectile effects) and heating through the Joule effect. Although the presence of EEG equipment in the MRI scanner can interfere with the latter's performance, causing artifacts in the images [60–62], satisfactory off-the-shelf solutions are available commercially.

However, the interactions between the MRI scanner fields and the circuit formed by the EEG recording equipment and subject have an important impact on EEG data quality and give rise to specific safety issues that require careful, continued consideration. The main sources of artifacts on intra-MRI EEG are the heartbeat and the time-varying electromagnetic (EM) fields used for image acquisition [60, 61]. These will be discussed in greater detail in a dedicated section below.

Concerning safety, firstly, the introduction of any object or device that does not form an integral part of the MRI scanner into the controlled area represents a theoretical risk of injury due to the possibility that it becomes a magnetized projectile and therefore requires careful prior consideration and evaluation [62]. We will not consider this issue further in this chapter, as it falls under general MRI safety guidance. Secondly, a number of potential safety risks arise when connecting a powered device to a person placed in the MRI environment [62]. It is worth mentioning that to our knowledge, no adverse incidents have been formally reported linked to the recording of EEG inside MRI scanner, with thousands of recordings performed to date [58]. However, there have been reports of injury linked to ECG [58], which is typically performed during EEG-fMRI to help reduce the impact of heartbeat-related artifacts on EEG, as discussed below. Safety of EEG-fMRI has been investigated by empirical experiments on test objects and computational simulation experiments designed to identify the sources of, and quantify, the most important health hazards. This is a complex issue due to the variety of types and designs of sources of the EM fields; especially, radio frequency (RF, in the tens of MHZ range) transmit coils and time-varying magnetic field gradient (in the kHz range) coils and possible multichannel EEG lead–subject configurations in relation to those same sources: a priori, every transmitting RF coil design produces specific electrical and magnetic field patterns to which the EEG leads and subject, with their own variability, are exposed and therefore requires specific investigation.

In brief, investigations on standard clinical MR instruments have resulted in a consensus that RF-induced heating in the leads and electrodes in contact with the body constitute the greatest health hazard and require specific solutions [62–65]. Furthermore, it is advisable to restrict any claims made by investigators about the risk levels revealed by their experiments to the specific setup, equipment, and conditions of the same experiments and in particular to avoid generalizing to other MRI scanners, special gradient field, RF coils, or electrophysiological recording equipment [62, 65]. A pragmatic solution to the problem posed by the variety of RF transmit coils is to limit the investigations to the most standard head-only or whole-body designs. The availability of commercial EEG recording systems with specific MRI operational conditions statements in compliance with regulatory requirements has helped to make the technique available across a range of standard clinical 1.5 and 3 T scanners.

19.3.2.1 Intra-MRI EEG Data Quality

Intra-MRI EEG data quality is degraded by exposure of the EEG recording circuit (head-EEG leads-amplifier) to the MRI scanner electromagnetic fields due to either the time-varying nature of the field or fields, or movement of the circuit (head and/or EEG leads), or both, resulting in nonneuronal, induced currents in the circuit. Four types of artifacts can result from this phenomenon, related to body/head motion, heartbeat, MR image acquisition, and, in some MRI scanners, cryogen compressor pump activity [66]. We will not discuss the latter in more detail, except to say that it can reduced satisfactorily using post-processing methods or eliminated completely by turning the pump off during scanning [67, 68].

Motion-Related Artifact

Here, we are referring to the effects of motion of the whole head and therefore also on the EEG electrodes and leads connected to it. Any movement of the EEG recording circuit (head-EEG leads-amplifier) exposed to the MRI scanner main static magnetic field ("B_0") will result in induced voltages in the circuit (Faraday's law), altering the EEG. Careful consideration of this mechanism is important for the following reasons (in addition to the well-documented detrimental effects of head motion on fMRI sensitivity): firstly, the induced voltages can be large, in proportion to the rate of motion, and potentially irregular; secondly, some of the most effective and widely used post-processing methods for the heartbeat- and MR image acquisition-related artifacts rely relatively strongly on the assumption that these artifacts are repetitive (see the following two sections). In addition, there may be an additional implicit assumption that the circuit's position is relatively fixed in relation to the MRI scanner in those correction methods.

As with all artifacts discussed here, the methods to address this issue fall into two main categories: reduction at the source and correction using signal processing techniques [60, 61]. The former includes head immobilization using an inflatable cushion; limiting the length of the EEG leads to a minimum, lead twisting, and placement along the scanner bore central axis. In addition, modified EEG caps with either a so-called reference layer [69] or wire loops [68] have been proposed to effectively measure the induced voltages close to the electrodes (note that the wire loops method also helps to reduce the cryogen compressor pump-related artifact). High-precision head motion video tracking has also been used to provide an independent input for artifact correction [70, 71].

Heartbeat-Related Artifact

This artifact is typically visible in most channels (as soon as the subject is placed in the MRI scanner bore) in the form of a distortion that accompanies each heartbeat (Fig. 19.2). It arises through exposure of the EEG recording circuit to the MRI scanner B_0 field via the following

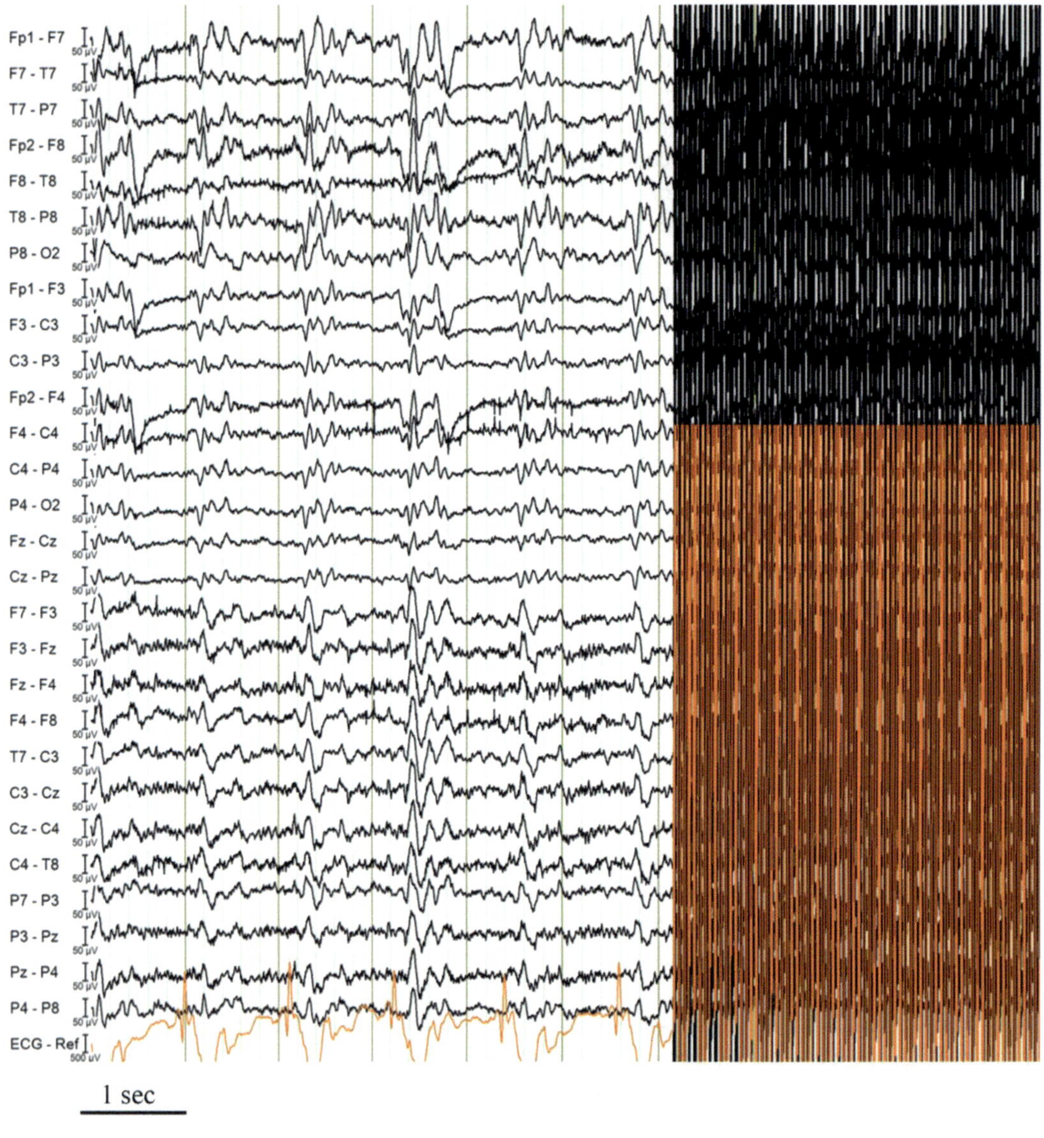

Fig. 19.2 Illustration of the heartbeat- and scanning-related artifacts on intra-MRI scalp EEG. Scanning starts approximately 6 s into the EEG segment, which is unprocessed. The (distorted) ECG trace is shown at the bottom, characterized by a frequency of around 1 beat per second. One can clearly see the heartbeat artifact in all EEG channels, which varies between channels and over time in shape and amplitude (relatively low in F_7–T_7; high in F_{p2}–F_7). Scanning causes the EEG (and ECG) to be unreadable due to the large distortion caused by switching gradient fields. Note that these artifacts do not reflect amplifier saturation, which is avoided by their special design. A scale at the bottom marks one-second intervals

two possible heartbeat-related mechanisms [62]: induction of an electromotive force due to pulsation of the body (torso and head) and scalp in the vicinity of blood vessels and voltage fluctuations in electrodes resulting from the Hall effect in pulsating underlying blood flow. The first of these two mechanisms is essentially the same as the effect of bulk head motion described in the previous section, except for the scalp pulsation component, which is localized. Therefore, any method designed to reduce or correct the effects of bulkhead motion on EEG quality should help address the heartbeat-related artifacts [60, 61].

Pulsatile head motion seems to be the dominant effect [72], and the artifact's magnitude can vary considerably between electrode locations and subjects [73]. The variability of this artifact's primary source (the heartbeat) over time (from one beat to the next) means that this intra-MRI EEG artifact remains a difficult challenge.

The two most widely used port-processing artifact reduction methodologies are average artifact subtraction (AAS) [73] and optimal basis set (OBS) [74], which are sometimes combined to maximize signal quality. These rely on ECG recording of sufficient quality to allow the reliable detection of the heartbeat. The basic principle of AAS is that the periodic heartbeat-related artifact is relatively stable and uncorrelated with the neuronal signal and therefore can be estimated by (channel-wise) averaging time-locked with the heartbeat; this running average can then be subtracted from the recording to reveal the neuronal signal: the "true" EEG (Fig. 19.3).

OBS attempts to account for inter-heartbeat variability. AAS and OBS are implemented in publicly available freeware and commercial platforms which all require user interventions and can be very time-consuming, with cost implications for the technique's clinical appli-

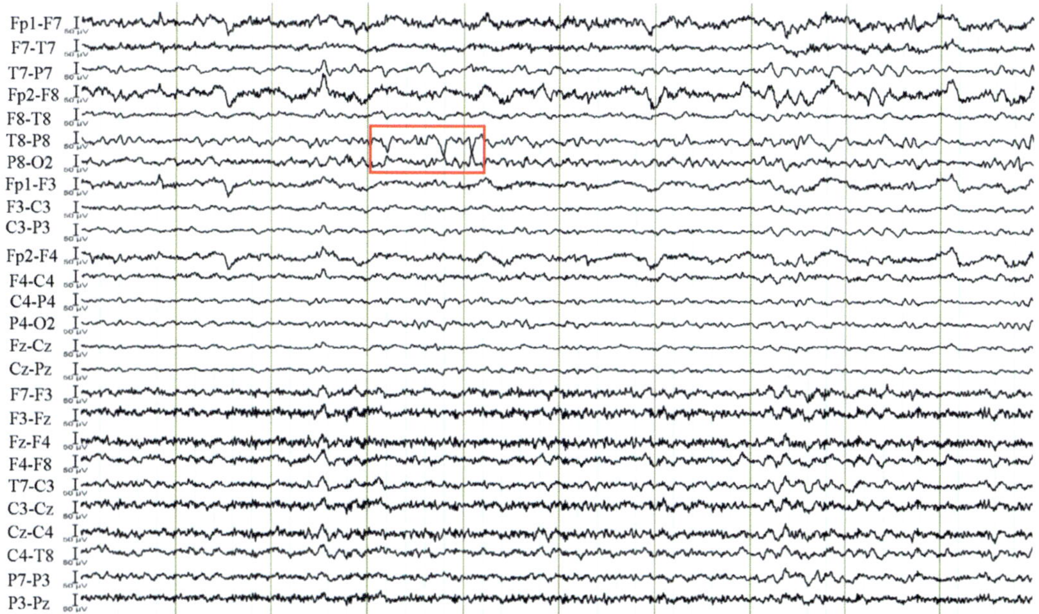

Fig. 19.3 Illustration of intra-MRI scalp EEG after correction of the heartbeat- and scanning-related artifacts. Data recorded at 3 T during fMRI scanning using MR-compatible EEG amplifier system (BrainAmp MRþ; Brain Products, Gilching, Germany). Same recording as in Fig. 19.2. The AAS method was used for heartbeat- [73] and scanning-related [76] artifact correction. A spike can be seen at P8 (red box). The horizontal axis represents time, marked with a scale of 1 second

cability. Any sufficiently large deviations from the idealized scenario of a perfectly immobile subject with a perfectly regular heartbeat will result in some residual artifact whatever method is used. Nonetheless, for investigators willing to overcome those hurdles, EEG data of sufficient quality can be obtained for many applications, such as the identification of IED [75] and the investigation of brain rhythms [75].

MRI Scanning (Gradient Switching)-Related EEG Artifact

In principle, two sources of EM fields can give rise to induced voltages in the EEG recording circuit during the MR image acquisition process (assuming that the subject and recording system are static, in which case the scanner's main field is an additional source): switching magnetic gradients and radio frequency (RF) pulses [76], with the former identified as the dominant effect [77]. The important features of this artifact are (see Fig. 19.2) the magnitude of the induced voltages, which can exceed the physiological signal manifold, and its spectral profile, which covers a wide bandwidth: from 1/TR, where TR is the MRI sequence repetition time (~1 s) to the kHz range for the gradients and tens of MHz for the RF. Typically, any standard MRI scanning obliterates the EEG making it unreadable irrespective of the type of EEG amplifier used. For some applications and EEG recording systems (e.g., that do not saturate during MR image acquisition), "interleaved" acquisition schemes, whereby gaps in the fMRI scanning allow artifact-free EEG to be recorded, can be an adequate solution [78–80]. However, for most applications, this is considered too limiting in terms of EEG and/or fMRI coverage or quality, and therefore uninterrupted, good quality EEG and fMRI acquisitions are sought. Based on the assumptions of a periodic image acquisition-related artifact (reflecting the MRI acquisition sequence) and purely additive physiological signal, hardware specifically designed for intra-MRI recording is available that is capable of faithfully capturing the distorted signal, including the physiological (neuronal) component, enabling its subsequent correction [81]. The main features of such systems are front-end low-pass filtering, an exceptionally wide dynamic range, and in some systems, digitization synchronized with the MRI scanning sequence to ensure consistent sampling of the artifact [61].

Using appropriate EEG hardware allows the repetitive MRI acquisition-related artifact to be captured accurately and be removed using the same averaged artifact subtraction methodology as for the heartbeat-related artifact [76]. Just as for the heartbeat artifact, motion, and particularly transient events, can cause the shape of the artifact to change abruptly and therefore deviate from the running average, resulting in incorrect subtraction and therefore larger residual artifacts [60]. Although the AAS method can account for some variation in artifact shape, this capability is limited. OBS, and other post-processing methods, can also be employed to account for instantaneous variations in the artifact [61]. This problem has been addressed at the source using motion tracking and prospective correction, by adjusting the MR image acquisition parameters dynamically to attempt to compensate for the effects of large motion events on the shape of the artifact [82].

Artifact Minimization and Reduction Methods: Conclusions

Intra-MRI EEG of quality sufficient for the identification of activity such as IED can be obtained by following recommended data acquisition setup, recording equipment, and post-processing methods [60, 61] (see Fig. 19.3). The dissemination of recommendations for good practice [60, 61] may lead to a further general improvement in intra-MRI EEG data quality. The nature of the problem makes it unlikely to be fully resolved soon. Therefore, it would be helpful to better identify residual artifacts. Although anecdotal evidence suggests experienced EEG readers can easily distinguish intra-MRI EEG from normal EEG, a more rigorous and systematic comparison, such as an intra-MRI EEG atlas illustrating artifacts, would be useful. As mentioned above, intra-MRI EEG artifact correction can be time-consuming, with a real impact on the technique's appeal. Although there has been a partially successful attempt to automate intra-MRI EEG artifact correction [83], improved automation combined with best practices for data acquisition would greatly advance the technique, likely leading to wider clinical adoption.

19.3.3 The Scientific Utility of Concurrent EEG-fMRI Acquisitions

EEG-fMRI is a form of fMRI: a tool to obtain images or maps of brain activity and therefore primarily designed to answer "Where?" questions. Concurrent EEG-fMRI acquisitions are most appropriate for the study of phenomena or effects that cannot be reliably reproduced across multiple mono-modal data acquisitions (i.e., EEG-only session followed by fMRI only). Examples include intrinsically electrographic (i.e., that can only be detected on EEG) phenomena such as interictal epileptiform discharges (IED; "epileptic spikes"), spontaneous (unpredictable) events, or events considered for their individual features (e.g., variations in the features of individual evoked potentials). This includes most epileptic seizures (possible exceptions being reflex seizures), spontaneous brain rhythms, and the individual responses to repeated stimuli (e.g., for the study of the sources of inter-response variability). In such circumstances, EEG-fMRI is the best possible tool to address the question: what is the BOLD pattern associated with event "X"?

Conversely, concurrent EEG-fMRI acquisitions can be used to study spontaneous BOLD signal fluctuations (e.g., at rest), to address questions such as the following: what does the EEG look like in relation to BOLD pattern "Y"? We will not deep well further on this interesting problem here; we refer the reader to [84, 85] for examples.

19.3.3.1 fMRI Mapping of Electrophysiological Phenomena Recorded on the Scalp

The above question can be rephrased as an imaging one: are there brain regions of increased and/or decreased BOLD signal in relation to an EEG phenomenon of interest? This is the question addressed by the vast majority of studies involving concurrent EEG-fMRI data in humans and, in particular, the localization of the generators of epileptic discharges. In this sense, EEG-fMRI is simply a "functional" imaging technique whereby EEG is used to identify brain events or states, analogous to the stimuli or tasks employed in functional (or cognitive) mapping fMRI studies.

EEG-fMRI Mapping of Epileptiform Activity: Methodology

The first EEG recordings inside a MRI scanner to be reported in the literature were performed to map epileptic activity recorded on EEG [86, 87]. In brief, the EEG-fMRI method consists of [88] recording EEG (using special equipment described previously) in a patient lying (generally at rest: without any stimuli presented, or task assigned, to them) with fMRI data acquisition proceeding either continuously or with brief interruptions, the post-acquisition identification on the EEG of any pathological activity of interest that occurs during the EEG-fMRI session, the calculation of the predicted time course of the EEG activity-related BOLD changes (typically based on a hemodynamic convolution model) for inclusion in a general linear model (GLM) of the fMRI time-series data for voxel-wise estimation (over the whole brain), and finally, the application of an appropriate statistical significance threshold, as typically performed in cognitive fMRI studies. The hypothesis underlying such experiments is that the resulting BOLD maps reveal regions where brain activity, as expressed in the BOLD effect, is correlated with the occurrence of the recorded EEG activity, shifted by the assumed hemodynamic delay.

The GLM is most widely used approach to fMRI time-series data analysis for the purpose of mapping BOLD changes associated with any given event or phenomenon of interest captured during the scanning session, such as epileptic activity. Typically, in addition to the effects of interest, such a model attempts to account for nuisance effects (often referred to as confounds) such as motion, physiological noise, etc.

GLM Building: Effects of Interest

Assuming that the quality of the intra-MRI EEG data is adequate (see above), the identification of epileptic events is arguably the most challenging and important step of the process because of its subjectivity and the choices that are required to be made by the investigator. The necessarily limited scan duration (typically up to 40 min) and wide range of epileptiform activity frequency (rate of occurrence) in patients with epilepsy mean that patient selection has been a primordial consideration: although a minimum "spiking rate" recruitment criterion (based on previously recorded clinical EEG data) has been employed, sessions devoid of epileptic activity are a common occurrence. Before discussing how such data can be analyzed usefully, we describe the main issues faced by the investigator for EEG-fMRI datasets in which epileptiform activity has occurred. Any misidentification of epileptic activity, be it inclusion of non-epileptic activity or noninclusion or misclassification of genuine epileptic activity, will result in reduced BOLD mapping efficacy due to a mismatch between the predicted and observed BOLD changes in the involved brain region(s) [89]. For example, for cases with multiple types of epileptiform discharges (different fields and/or waveform characteristics), we expect a different BOLD pattern to be associated with each type because one must assume that the discharges of a given type arise from a specific generator, different from the generators of the other discharges. This issue raises another important modelling and interpretation issue for cases showing a variety of epileptiform discharges, namely, the nature of the EEG-fMRI test's result: given that a statistical BOLD map is associated with each event type, what is the test's result? Most investigators have chosen to select and consider a single BOLD map as the test's result: the map obtained for the type of discharge that they consider most clinically relevant. While this approach can be justified (e.g., for some clinical studies), it does not remove the benefits of properly accounting for all potential sources of fMRI signal variation that can be identified, be they epileptic or not (physiological noise, motion, etc.), for the purpose of creating a GLM. In some studies, all BOLD maps derived from the different EEG discharge types are reported and considered for interpretation [90–93].

The predicted time course of the BOLD changes (for each discharge type capture on the intra-MRI EEG) is a mathematical construct. Its creation is a crucial element of BOLD mapping of epileptiform activity because it embodies the investigators' hypotheses about the nature of the recorded events and their relationship with BOLD signal fluctuations [94]. It requires the classification of each event into categories conceived as representing individual instances of the activity of the same generator. The mathematical representation of each event is another important choice, often neglected in discussions; typically, IEDs are represented as zero-duration "stick" (or delta) functions of equal amplitude. The benefits of epileptiform discharge parameterization have been investigated but can be very time-consuming [95]. Electrical source imaging, also applied throughout the intra-MRI recording, has also been used to increase the technique's sensitivity and specificity [96]. The next fundamental element of this process is the choice of HRF (hemodynamic response function, or perhaps more appropriately in the context of spontaneous brain activity: hemodynamic coupling convolution kernel) for the EEG events of interest. There is no consensus on the optimal approach to this problem: some authors simply assume that epileptic events can be treated like brief external stimuli and therefore use a standard HRF (e.g., the Friston so-called canonical HRF with a hemodynamic delay of around 6 s), while others favor more flexible models such as using a series of standard HRFs with a range of hemodynamic delays (an expanded hemodynamic kernel or basis set) [94, 97, 98]. A trade-off solution is to use a hemodynamic basis set consisting of the canonical and two of its derivatives that allows for some deviation from the canonical [99]. These can be taken as representing less stringent causal assumptions about the relationship between hemodynamic changes and IED occurrence as recorded on scalp EEG, and it has been suggested that the canonical HRF may be suboptimal in pathological tissue [100, 101]. It is worth noting that the standard (canonical) HRF of cognitive fMRI is a mathematical model derived from a relatively small number of stimulus-based experiments in healthy subjects and that substantial intersubject and interregional variations in shape and delay have also been reported in healthy tissue [101]. Importantly, the use of expanded hemodynamic basis sets complicates the statistics and choice of contrasts at the GLM estimation stage (possible need for correction for multiple tests or the use of multi-effects F-contrasts, depending on the details of the modelling), and therefore, the "canonical HRF" assumption can be seen as a convenient simplification. It is important to note that such a simplification corresponds to a simpler hypothesis; the question is whether it is justified.

In cases devoid of epileptiform activity on the EEG recorded during fMRI, two approaches have been explored: firstly, the analysis of the fMRI data without any reference to the EEG using so-called data-driven techniques such as ICA and regional measures of co-fluctuation (signal homogeneity, coherence, etc.) that aim to reveal pathological BOLD patterns [102]. While this has the clear potential benefit of reducing the need for concurrent EEG recordings and has provided interesting observations that may lead us one day to be able "read" fMRI like an EEG, it effectively amounts to transposing the problems of EEG interpretation onto the fMRI time-series data. Secondly, it has been shown that information derived from EEG recorded outside the MRI scanner, such as IEDs captured in long-term EEG for clinical purposes, can be used to reveal BOLD signal variations apparently linked to specific, potentially epileptic source activity [103, 104]. For example, IED-specific topographic maps obtained from clinical EEG have been used as a "template" to estimate intra-MRI IED-likeness over time, hypothesizing that BOLD changes may reflect sub-detection threshold source activity locally. In this approach, the effect of interest consists of the strength of correlation between the topography of the ongoing EEG and that of the IED-derived template, with hemodynamic convolution as for cases with IED on the intra-MRI EEG. We note that this approach can also be used for recordings with IED [104].

GLM Building: Nuisance Effects

The GLM framework is conceived to account for the possible effects of all observed factors on the dependent variable (in this case the fMRI signal). While there is no agreed consensus on this aspect of EEG-fMRI data modelling in epilepsy, most investigators include subject motion as a nuisance effect due to its importance. Motion is usually estimated based on fMRI volume time-series realignment and represented as three position and three rotation variables, although some investigators add so-called scan nulling (or erasing/blanking) to account for particularly large motion events [98, 105]. This approach, which is quite specific to this type of fMRI application, can be justified on two grounds: firstly, motion tends to be greater in patients than healthy volunteers; secondly, in general, each patient (and dataset), particularly in a presurgical context, is unique and the results of the analysis most relevant at the individual level (rather than as part of cognitive group studies, for example, in which subjects can be considered equivalent and therefore can be substituted in case of data quality issues). Therefore, greater effort on post-acquisition data processing and analysis may be employed to extract as much epileptic activity-specific information as possible from the dataset. For the same reasons, some investigators also include physiological noise effects in the GLM and other non-epileptic activity, including derived from the concurrently recorded ECG and video of the patient's face and limbs (e.g., eye blinks, head jerking, etc.) [106, 107].

Mapping Effects of Interest: Contrasts and Statistical Significance Thresholding

The GLM consisting of the effects of interest and nuisance is typically estimated using standard statistical methods. The contribution of every modelled effect to the fMRI signal in a GLM is estimated. In addition, parameter estimates for different effects can be combined ("contrasts") to assess differential or compound effects, for example, F-contrasts across the hemodynamic basis set consisting of the canonical HRF, and its two derivatives have been employed to account for possible deviations from standard (e.g., "canonical") hemodynamic response function.

Considering the residual variance (over time), a voxel-wise t-test can be applied for mapping each effect of interest, subject to appropriate statistical significance thresholding given the large multiplicity of tests (i.e., voxels). A survey of the interictal EEG-fMRI shows a wide range of statistical significance thresholding practices, with different combinations of minimum cluster sizes (zero, 5, 10, etc. voxels) and voxel Z scores (corresponding to probability of false positive $p = 0.05$ or 0.001, corrected or uncorrected for multiple comparisons). Although this makes comparisons between studies much more difficult, there may be some justification for this practice through the following reasoning: in clinical contexts, the results of all localization tests for an individual patient are usually interpreted together and in full knowledge of the patient's clinical record. In addition to the implied consideration of prior hypotheses provided by the results of the other tests, the fact that agreement or disagreement of the tests might itself be a key factor in the clinical management process, above the actual localizing information, may reduce the relevance of the scientific convention for the assessment of statistical significance ($p = 0.05$) for each test taken individually.

In general, as is the case for most functional MRI, the interpretation of the resulting statistical maps contains a subjective element. For example, numerous studies on the potential clinical utility of scalp EEG-fMRI of IED evaluate its localizing power by comparing it to independent information, such as the results of icEEG investigations at a lobar level based on visual anatomical labelling of the IED-related BOLD clusters [108, 109]. We note the rare use of digital atlases for automated labelling; for example, the assessment of the putative involvement of the default mode network (DMN) in relation to IED in TLE [110] might benefit from the use of more objective methodologies, such as ICN_Atlas [111].

EEG-fMRI Epileptiform Activity Mapping Methodology: Summary

We have shown how different approaches have been proposed for the analysis of EEG-fMRI data for mapping epileptic activity: choices at the EEG feature extraction, IED-related BOLD time course predictor construction, and nuisance effect selection stages. This can be summarized as follows: these choices result in different GLMs, and different GLMs ask different questions from the data. In other words, different GLMs that embody different assumptions and hypotheses can be applied to the same dataset. In addition, the possibility of multiple clinically relevant effects of interest (i.e., type of epileptiform discharge), each resulting in a different BOLD map, amounts to layers of complexity that make the technique challenging. As shown in the following, a consensus is gradually emerging on key aspects of the technique's clinical potential.

19.3.3.2 Applications of Scalp EEG-fMRI in Epilepsy

Generalized Epilepsy

The hallmark feature of generalized epilepsy and to be more specific, of idiopathic generalized epilepsies (IGE), is the generalized spike and wave discharge (GSW) and seizures that are generalized (bilateral, symmetrical, and synchronous) from the onset [112]. These striking characteristics have led to a number of theories on their origin, all involving the thalamus in the onset and/or maintenance of GSW [113], that are difficult to test conclusively in humans. Therefore, in view of some electro-clinical features of IGE that may be considered favorable to fMRI investigations (including the clear EEG signature and ictal semiology), this was the subject of investigations aimed at shedding new light on the localization of the generators of GSW.

Early small case series revealed patterns of cortical deactivation, including the posterior cingulate [114], predominance of thalamic activation and cortical deactivation [115, 116], and in one particularly striking case, a pattern of strong and widespread cortical deactivation and thalamic activation during three absence seizures [117]. More formal assessments of group effects linked to GSW demonstrated a common pattern dominated by cortical deactivation (particularly involving the parietal cortex, suggestive of DMN involvement) and thalamic activation [115, 117–122] (Fig. 19.4). See Carney et al. [118] for a more detailed discussion.

Focal Epilepsy

In patients with recurring focal seizures that continue to disrupt their lives despite the most advanced drug treatments, surgical intervention may be considered. The identification of a surgical target, typically referred to as the epileptogenic zone (EZ), remains a challenging problem in many patients with severe, drug-resistant epilepsy [123]. Therefore, there is a continuing need to develop new presurgical localization methods. The idea of using fMRI for this purpose arose in the 1990s, soon after the development of fMRI, and was based on imaging epileptic seizures observed during scanning without EEG [124, 125]. Although this methodology has limitations, in particular the difficulty of detecting the seizure onset, it showed that some clinical focal seizures are associated with detectable BOLD changes [126], a finding that was later confirmed using video-EEG-fMRI [127, 128]. However, in most patients, seizures are relatively rare and often characterized by substantial body movement which, combined with the challenges associated with the MRI scanning environment, makes ictal fMRI impractical as a clinical tool in the current state of technology. Following early work allowing the interleaved acquisition of EEG and fMRI [86, 87, 129, 130], the advent of technology for intra-MRI EEG recordings of sufficient quality throughout the fMRI acquisition (and without significant health hazards) made "continuous EEG-fMRI" possible around the turn of the millennium [31, 81]. In the best possible circumstances, this technique allows the study of the relationship between the two types of signals with no other constraints or limitations than those of each individually.

The first question that arises when considering performing fMRI of interictal epileptic activity was: are IED associated with observable BOLD signal changes? [87, 131, 132]. As numerous studies using standard resting-state fMRI acquisition protocols with concurrent EEG have demonstrated since the early 2000s, the answer to this question is complex. Despite the lack of uniformity across investigators with regard to most methodological aspects, such as patient recruitment (usually based on relatively high rates of IED), EEG and fMRI data acquisition and processing (e.g., EEG artifact reduction), and fMRI data analysis (modelling) and interpretation, a few general conclusions can be reached: there is no definitive rule linking the amount or type of epileptiform activity recorded on the intra-MRI EEG and the expected magnitude, or perhaps more appropriately, the likelihood of revealing statistically significant BOLD signal changes. While there is a tendency for more active EEGs to increase the likelihood of BOLD changes being revealed [133], many studies have reported cases of either absence of (statistically significant) BOLD changes related to substantial amounts of epileptiform discharges or the opposite: significant BOLD changes linked to relatively small amounts of epileptiform activity (Fig. 19.5). Taking sensitivity to be the proportion of cases who undergo EEG-fMRI in whom IED-related BOLD changes are revealed irrespective of localization, reported values fall in the range 40–70% [134, 135]. However, this includes cases of total absence of epileptic activity on the intra-MRI EEG, meaning that no conventional GLM analysis can be performed, and therefore, the result in such a case is null. As mentioned previously, the application of the IED topography-based [104] method has been shown to greatly increase the technique's sensitivity, in part by allowing IED-negative EEG-fMRI datasets to be analyzed [103, 136].

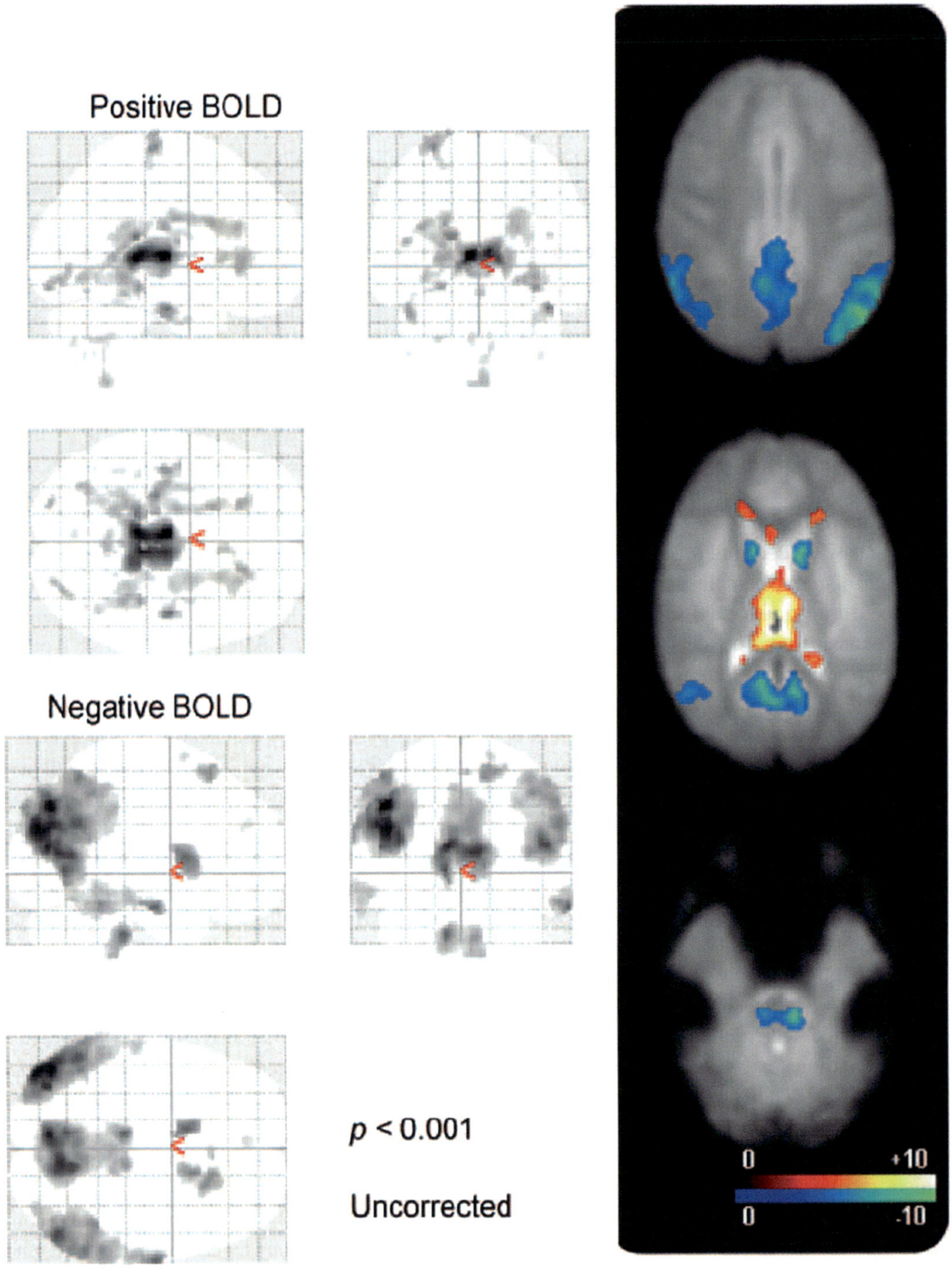

Fig. 19.4 Second-level group analysis of 10 subjects with absence seizures. Glass brain blood oxygen level dependent (BOLD) signal change for positive and negative BOLD response are shown ($p < 0.001$ uncorrected). Color map images of positive BOLD (red to white: 0 to +10) and negative BOLD (blue to green: 0 to −10) displayed on the group mean EPI image. (Adapted with permission from Carney et al. [118])

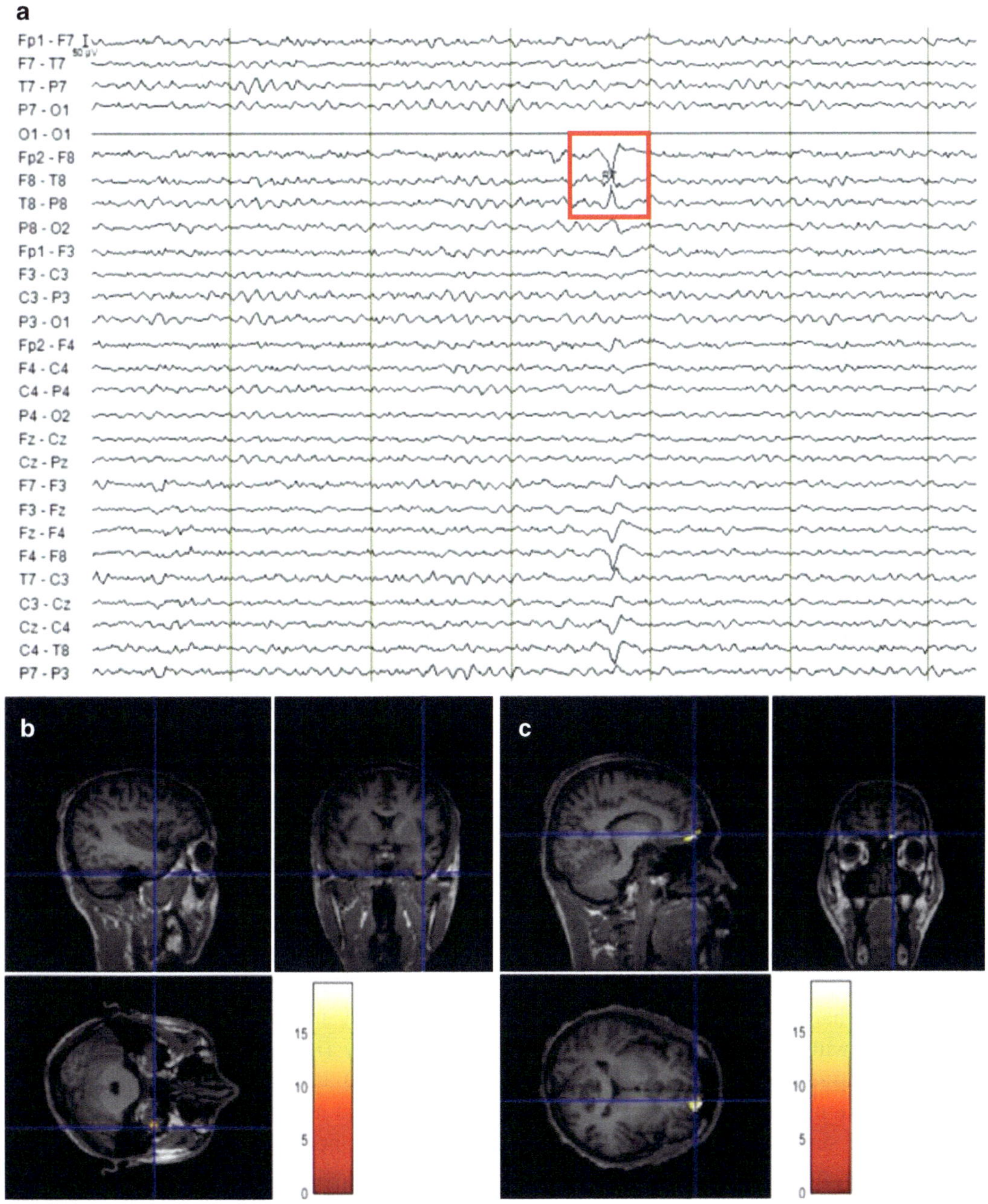

Fig. 19.5 Example of interictal scalp EEG-fMRI results for a patient with right hippocampal sclerosis. A 10-min resting-state EEG-fMRI data acquisition was performed. Three IEDs were identified and used as an effect of interest in the GLM. (A) One of the IEDs on intra-MRI EEG (red box). The resulting activation map consisted of four significant BOLD activation clusters, two of which are illustrated here (statistical significance threshold: $p = 0.05$ corrected for family-wise error; cluster size threshold ≥ 5 voxels). (B) Cluster located in the right temporal lobe, location of the presumed EZ (presurgical assessment). (C) Cluster in the right frontal lobe containing the global statistical maximum. The other two clusters were located in the left frontal lobe. In relation to the results of other, independent presurgical localization tests, namely, invasive video-EEG telemetry in this case, this EEG-fMRI result can be classified as having "Some Concordance," corresponding to "one or more activation clusters, but not the one containing the global statistical maximum, located in the same lobe as the EZ, with additional clusters located in other lobes" [90] (in contrast, an "Entirely concordant" BOLD map would be one in which all clusters are located within the same lobe as the EZ). This patient subsequently underwent a right anterior temporal lobectomy which resulted in total seizure freedom at a follow-up of 9 years (which can be taken to mean that the "presumed EZ" is confirmed)

Quantifying sensitivity in terms of providing clinically relevant information is an even more challenging proposition [137, 138]. Whether IED-related BOLD maps reveal the location of the IED generators is the most directly pertinent question given the very nature of the technique [96]; the fMRI mapping of IED-related BOLD changes. In this relatively limited sense, the literature demonstrates a degree of agreement, which can be interpreted as EEG and fMRI providing complementary information [96, 139]. However, the most clinically relevant question in drug-resistant cases is how the regions of significant IED-related BOLD changes relate to the EZ. To date, there is anecdotal evidence and no substantial prospective studies on the predictive power of interictal EEG-fMRI for surgical outcome. However, two prospective studies assessing the impact of the EEG-fMRI findings on the clinical management process have demonstrated its potential in that regard, e.g., by guiding icEEG electrode implantation plans [90, 138].

Most studies evaluating the technique's EZ localizing capability have done so retrospectively as case series, by comparing the IED-related BOLD maps to various forms of independently derived localization data. As alluded to previously, conceptually, this is the same problem as evaluating the role of IED mapping using EEG and/or MEG in presurgical assessment of patients with drug-resistant epilepsy, which shares the issue of lack of a proper ground truth with all presurgical imaging tests [140]. Two options are available: intracranial EEG localization and, in cases who have undergone curative surgery with good outcome, the resected brain area as identified on postoperative MRI. Numerous "concordance schemes" for assessing the degree of agreement between EEG-fMRI and the chosen independent localization have been proposed, each typically based on multiple and often subjective considerations such as whether to give greater importance to the BOLD cluster containing the most significantly activated voxel (Fig. 19.5), whether to consider the sign of the BOLD changes, and whether a BOLD cluster is co-localized with the IED field [133, 138, 141–144]. An important consideration for cases with multiple IED types is whether the interpretation should be limited to the map related to the "most clinically relevant" IEDs. Some general conclusions that emerge from this work are IED-related BOLD maps are often complex, consisting of multiple areas of activation (and/or deactivation), often located in multiple lobes, while there is no consensus on concordance schemes, relatively high degrees of concordance are reported, suggesting that in many cases the technique does indeed indicate the localization of the EZ; there is no clear indication of the etiologies or localizations for which the technique is more or less promising; it remains unclear whether the technique's greatest clinical potential is indicating whether surgery or presurgical assessment using more invasive means should proceed and, if so, whether and in what cases it can best provide localizing information either to guide subsequent invasive EEG recordings or curative surgery. Current evidence suggests that these questions deserve proper evaluation. Given the often complex nature of the BOLD maps mentioned above, and the variety of proposed concordance schemes, it may be useful to reframe this problem more formally in terms of identifying the relevant features of those maps for the clinical problem at hand.

The application of EEG-fMRI to the study of ictal activity in focal epilepsy has been much more limited for obvious reasons, including patient safety (in the constrained space of the scanner bore) and data quality (due to possible seizure-related movement). In a series of 50 patients specifically recruited, 20 patients had typical seizures identified either on intra-MRI EEG and video, EEG alone, or video alone [128]. Using a modelling strategy based on ictal phases, this study revealed BOLD changes specifically related to the preictal phase, seizure onset and evolution, and intrinsic connectivity networks (including the default mode network) recruited during seizures in a significant proportion of patients. Ictal onset phase-related BOLD changes are often co-localized with the independently defined seizure onset zone.

19.3.4 Concurrent Intracranial EEG-fMRI (icEEG-fMRI)

Concurrent scalp–intracranial EEG recordings in patients with epilepsy reveal the much greater richness of epileptiform activity captured by brain-indwelling electrodes [145]. In clinical epileptology, depth and stereo-EEG are considered by many to be the best method for the identification and characterization of the sources (that can often appropriately be described as forming a network) and propagation of epileptic activity: seizures and epi-ictal phenomena and IED. On the other hand, the technique's volume of sensitivity is limited to the immediate vicinity of each contact, combined with the limited number of probes that can be inserted due to a range of surgical constraints. In view of fMRI's almost complete brain coverage, there has been interest in combining the two techniques to investigate the whole-brain hemodynamic correlates of epileptic activity captured with unsurpassed sensitivity. Investigations of the safety aspects of icEEG-fMRI were performed to determine under which conditions icEEG probes in current clinical use could be used to record signals during fMRI scanning performed on standard MRI scanners [65, 146–148]. Up to now, there are only a few dozen icEEG-fMRI datasets reported in the literature, at 1.5 T [149–151] and 3 T [152–154]. Given the aforementioned relative abundance and variety of epileptic activity recorded intracranially, the investigator is faced with an even greater BOLD modelling challenge. As expected, the sensitivity of icEEG-fMRI is far superior to that of scalp EEG-fMRI, in part due to the much greater abundance of IED; on the other hand, it is important to keep in mind that the sensitivity of fMRI might not match the exquisite sensitivity of icEEG, and in particular depth recordings, and therefore, the hemodynamic changes associated with very localized IED may be difficult to detect. Nonetheless, similarly to scalp EEG-fMRI investigations of IED, the BOLD patterns show varying degrees of complexity, in terms of extent and relationship with the presumed or confirmed epileptogenic zone [153–155].

The icEEG-fMRI technique and availability of resting and task-related data with brain-indwelling electrodes in humans also provides opportunities for the study of the relationship, locally and network-wise, between electrophysiological data recorded with unsurpassed sensitivity and specificity and BOLD signals in humans [156–158].

19.3.5 Conclusion

Concurrent electrophysiological and fMRI data acquisitions have the unique capacity to study changes in activity (i.e., regional blood flow) over the whole brain linked to spontaneous and unpredictable events such as IED and epileptic seizures (especially when combined with video). Many of the BOLD patterns thus revealed are complex and widespread, and we must now investigate their meaning further to determine their true clinical utility, be it in terms of contribution to the surgical planning process or more generally as a predictor of treatment outcome.

19.4 EEG Combined with Metabolic Images

19.4.1 Rationale

Over the last 40 years, positron emission tomography (PET) brain imaging has significantly advanced our understanding of brain functions, both under normal and diseased conditions [159]. PET's capability to pinpoint metabolic shifts in the brain and accurately measure their distribution has made it an invaluable tool in neuroimaging. At the heart of a PET scan is the radiotracer: this compound's biological pathway is tracked to observe specific brain phenomena. Neurodegenerative diseases universally impair neuronal activity, leading to reduced metabolic energy. The radiotracer fluorodeoxyglucose (18F-FDG) is frequently used to detect these deficits, and it's well-known that various diseases manifest distinctive patterns of diminished 18F-FDG uptake. Since differentiating between many neurological disorders is often difficult, the development of molecular imaging techniques for identifying these pathologies is a vital research focus, supporting the creation of new therapeutic treatments. Though 18F-FDG remains the most commonly used radiopharmaceutical, it is not ideal for brain scans due to its high natural cortical uptake and limited specificity. This has driven the emergence of new radiotracers, each with its own benefits and drawbacks, which have expanded the possibilities of brain imaging [160, 161]. As new radiotracers continue to emerge, there is renewed interest in PET systems designed specifically for brain imaging, especially when combined with magnetic resonance. The ultimate goal of integrating PET and MRI into a single machine is to seamlessly incorporate PET detectors into the MRI scanner, enabling the simultaneous collection of data. This fusion would produce combined functional and structural images with superior soft tissue contrast, excellent anatomical spatial resolution, and precise fusion of temporal and spatial data. Additionally, since MRI provides functional insights [162], such as through blood oxygenation level-dependent imaging (the BOLD signal, see previous paragraph on EEG-fMRI), a hybrid PET/MRI system could offer multifunctional, in vivo insights into physiological processes. However, both PET and MRI have temporal limitations. EEG stands out in terms of temporal resolution, offering superior time course analysis compared to other imaging methods. PET, MRI, and EEG each contribute unique insights into the brain's structure, physiology, metabolism, and function. Traditionally, data from each of these modalities have been acquired sequentially and analyzed in parallel, often using software to co-register the images. However, a trimodal approach—simultaneous PET–MR–EEG imaging [163]—would allow for real-time observation of multiple facets of the same process, under consistent physiological and psychological conditions. This combined approach provides a more comprehensive view, capturing the relationship between different brain functions and offering complementary anatomical, physiological, metabolic, and functional insights.

19.4.2 How Does Simultaneous PET-EEG Work?

The simultaneous PET-EEG technique combines positron emission tomography (PET) and electroencephalography (EEG) to provide a detailed view of brain function. While PET/CT systems are currently the most widespread, the development of hybrid PET/MR systems for clinical use, coupled with MR-compatible EEG equipment, offers a promising opportunity to collect structural MRI and EEG-monitored PET data simultaneously. PET/MR systems function by placing removable PET detector inserts within the bore of the MR scanner. These components are connected via a single patient table, allowing sequential imaging with both modalities in the same frame of reference through a 180° rotation of the scanner table between the MRI and PET scanner. The EU-funded TRIMAGE project (www.trimage.eu), initiated in late 2013 [159], aims to develop a trimodal, cost-effective PET/MR/EEG imaging tool that utilizes advanced technology surpassing current standards. The TRIMAGE scanner adopts a fully integrated solution with the PET scanner positioned between the gradient coils and the radiofrequency coil to optimize space and performance. EEG signals are recorded using commercially available, MR-compatible scalp EEG systems with 64–128 channels (Brain Product, Germany). A study completed in 2017 explored and quantified the impact of MR-compatible EEG cap components on PET images during simultaneous trimodal studies [164]. This study investigated how the EEG cap affects PET image quality and quantification during simultaneous measurements. The brain-PET-MR emission images of the Iida phantom with [18F]-fluorodeoxyglucose, as well as human subjects with an EEG cap, did not show significant artifacts from the EEG cap, even though the attenuation correction did not account for the cap itself. The quality of EEG data remains relatively unaffected by the PET system, with the main artifacts in EEG data during simultaneous PET-MR acquisitions being related to the MRI environment, such as magnetic field interference and gradient switching.

These artifacts can be corrected using advanced signal processing techniques similar to those applied in EEG-fMRI studies (see previous EEG-fMRI section). Although this integrated PET/MR/EEG system is in its early stages, preliminary studies have demonstrated the feasibility of acquiring data from each modality simultaneously during extended sessions (up to 70 min) with high-quality results from each type of image.

19.4.3 Clinical Applications

18FDG-PET is commonly used in epilepsy field because of its ability to reveal areas of interictal cerebral hypometabolism associated with epileptic activity and epileptogenic lesions [165, 166]. It is particularly useful in patients with apparently non-lesional and temporal lobe epilepsy [166, 167]. In clinical setting, EEG monitoring before and during uptake of the PET tracer is a common practice for detecting subclinical epileptic seizures that could cause false-negative results by confounding interictal hypometabolism with ictal hypermetabolism [168]. The occurrence of EEG abnormalities can influence FDG uptake especially in case of frequent spike activities or focal polymorphic slow waves [169, 170]. In addition, in case of seizures, the area of hypermetabolism varies according to the intensity, duration, migration, and timing of ictal events. For all these reasons, PET without corroborating EEG findings might misidentify the underlying epileptogenic area. Nevertheless, this information can be achieved only if EEG is recorded at the same time of PET study. In a cohort of 46 epilepsy children studied with routine EEG before, during, and after FDG tracer injection, areas of hypometabolism correlates with the frequency of interictal discharges and severity of slow-wave activity, and the review of the EEG at the time of PET allows to interpret findings and minimize false lateralization. In a recent study [171], FDG-PET with EEG monitoring was used to investigate 12 patients with focal drug-resistant epilepsies and interictal spikes on EEG. FDG-PET/EEG was included together with high-density EEG, EEG-fMRI, and structural MRI acquisition in a long quadrimodal single session of approximately 2 h length. EEG was recorded at 1 kHz using a 256-channel MR-compatible system (Electrical Geodesics Inc., Eugene, OR), and PET/MRI was acquired using an Ingenuity TF PET-MR scanner (Philips Healthcare, Cleveland, OH). EEG was acquired simultaneously during uptake of the PET tracer, and patients were requested to remain at rest with eyes closed to avoid spurious hypermetabolism in the auditory or visual cortices. EEG traces were reviewed for interictal spikes and/or to document potential subclinical epileptic seizures that could modify metabolism and alter the interpretation of PET maps. To note, the EEG cap was removed before PET scanning after tracing uptake to avoid the potential effect of the electrodes on the PET images and to maximize patient comfort. A clear advantage of a multimodal approach like the one described is to guarantee the same medication level and medical condition for all modalities. A more recent development of 18FDG-PET allows to acquire PET scans dynamically over consecutive series of time windows for about 60 min. The advantage compared to traditional "static" 18FDG-PET which measures cerebral metabolism within a single time window of about 15 min is the possibility to capture the kinetics of glucose wash-in, metabolism, and wash-out from the point of injection. This new technology utilizes CT/PET scanners and implies a noninvasive estimation of blood input function that calibrates the calculation of FDG uptake. A proof-of-concept study explores the sensitivity of dynamic 18FDG-PET in nine patients with focal epilepsies with better performances compared to static 18FDG-PET [172]. The recent and more advanced simultaneous PET/MR/EEG technology within the TRIMAGE framework has not been implemented in epilepsy field so far. It has been preliminary used in schizophrenic patients to explore the relationships between neurotransmission with PET (particularly serotoninergic and glutamatergic) and evoked potentials (mismatch negativity, P300, and NoGo-Anteriorization) [173].

19.4.4 Other Applications

In addition to epilepsy research, the simultaneous combination of EEG and PET has garnered interest for exploring brain function in healthy individuals. Specifically, integrating and comparing neurometabolic and functional data from PET and resting-state fMRI with neuroelectric measurements has enabled researchers to investigate neurovascular–metabolic coupling. This new methodology holds promise for establishing individual physiological and pathological profiles across various neuroscience research fields. Using the simultaneous MR-PET-EEG system within the TRIMAGE framework, researchers correlated electrical sources of different EEG frequency bands (particularly δ, θ, and α) with glucose metabolism via PET, functional connectivity measures from resting-state fMRI, and MRS spectra. This approach was used to better characterize the brain's default mode network (DMN) in healthy males [174]. A year later, the same team applied this methodology to correlate EEG microstates (stable EEG scalp topographies) with metabolic and functional measures from the DMN in healthy subjects [163]. More recently, the MR-PET-EEG system was utilized to investigate physiological associations between EEG microstates, activity of different resting-state networks (beyond the DMN), glucose metabolism, and neuroreceptor binding availability (non-displaceable binding potential, BPND). This was achieved using various radiotracers to explore excitatory/inhibitory neurotransmission [163]. These findings, although limited to studies on healthy subjects, demonstrate the feasibility of trimodal acquisition and data analysis. They lay the groundwork for applying such multimodal measures as biomarkers for diagnosis, disease staging, treatment response, and monitoring of neuropsychiatric disorders.

19.4.5 Conclusion

Simultaneous PET-EEG imaging provides a unique opportunity to observe both metabolic processes and electrical activity in the brain concurrently. This integrated approach offers valuable insights into the mechanisms underlying various neurological conditions and cognitive processes. The PET-EEG technique typically utilizes a hybrid MR/PET scanner combined with MR-compatible EEG equipment. Despite its promise, this method faces challenges, such as managing artifacts and optimizing the co-registration of data from different modalities. However, ongoing advancements in technology and signal processing are steadily improving the quality and reliability of the data obtained from these combined modalities. The TRIMAGE international project has been a significant milestone in this field, developing an advanced tri-modal imaging tool that integrates PET, MRI, and EEG. This project has highlighted the feasibility and advantages of combining these modalities, particularly in enhancing the understanding of brain function in healthy individuals and in the diagnosis and treatment of schizophrenia. While the clinical application of PET-EEG is still in its early stages, epilepsy stands out as a promising field for its use.

19.5 EEG Combined with Neurostimulation and Neuromodulation Techniques

19.5.1 Rationale

EEG provides real-time, high-resolution insights into the brain's electrical activity, allowing for precise monitoring of neural responses to various interventions. When integrated with neurostimulation and neuromodulation techniques, such as transcranial magnetic stimulation (TMS) or transcranial electrical stimulation (TES), EEG enables researchers and clinicians to tailor stimulation protocols based on the brain's current state and activity patterns. This synergy enhances the efficacy of neuromodulation by ensuring that stimulation is delivered at optimal times and locations. Furthermore, the combination of EEG with these techniques facilitates the assessment of how neurostimulation alters brain function dynamically allowing to make inferences on local excitability and area-to-area functional connectivity in the nervous system [175–177]. This integrated approach not only deepens our understanding of neural mechanisms but also holds promise for personalized and more effective interventions for a range of neurological and psychiatric disorders.

19.5.2 TMS-EEG

Transcranial magnetic stimulation (TMS) has proven to be an effective, noninvasive tool for probing the human brain [178]. This technique uses a magnetic field for indirect electrical stimulation of the brain. A strong, very brief, magnetic field is delivered to the brain via a transducing coil. The changing magnetic field induces a time-varying electric field (E-field) in the cortex. Depending on the orientation of the E-field with respect to the geometry of the cortex and cortical neurons, the E-field leads to a depolarization of axons in the stimulated brain area. Because of the short pulse duration, the temporal resolution of TMS is sub-milliseconds, which allows for real-time modulation of the brain. The spatial extent of the cortical area stimulated by TMS depends on the coil geometry, stimulus intensity, target area, and, therefore, coil to cortex distance [179, 180]. As magnetic fields attenuate rapidly with distance and as the induced E-field approaches zero at the center of the head, TMS stimulates superficial cortical layers more strongly than deeper layers. However, the induced neuronal activity depends also on other aspects (like the position and orientation of neuronal structures and membrane characteristics). In summary, besides stimulating the target area and surrounding tissues, TMS indirectly activates synaptically interconnected sites, a feature exploited in brain connectivity studies [181].

The brain activity evoked by TMS can be recorded with different neuroimaging techniques such as EEG, fMRI, near-infrared spectroscopy, and PET [182]. However, the most successful and thus commonly used combination has been with EEG because it is a widespread method, is less expensive than other neuroimaging techniques, and is technically the least complicated to be combined online with TMS. The combination of TMS with EEG has been relevant for addressing fundamental neuroscientific questions in new ways. In particular, the two techniques complement each other, in that causal information provided by TMS overcomes the correlational nature of EEG data, whereas the ability to record from the whole scalp provides a global picture of the brain activity generated by the E-field. One of the main advantages of using TMS-EEG is that outcome measures, derived from EEG responses to TMS (i.e., evoked potentials or brain oscillations), can be used as a neurophysiological marker of excitability or connectivity for any brain area, including the regions where TMS does not generate a proxy of cortical/corticospinal excitability, such as motor evoked potentials (MEPs) or phosphenes [183]. When a TMS pulse is applied to the cortex, time-locked depolarization of underlying neurons is obtained, as well as trans-synaptic activation of local and distal cortical networks. This activity can be recorded through the skull by EEG electrodes placed on the scalp: where the summation of synaptic potentials produces a series of positive and negative deflections visible in the EEG signal, termed the TMS-evoked potential (TEP). TEP is a measure of cortical reactivity: changes in its amplitude and latency reflect changes in cortical activity of the stimulated region [184]. Although the precise origin of TEPs remains to be determined, they are thought to involve the spatial and temporal summation of excitatory and inhibitory postsynaptic potentials originating from the activity of a large population of cortical pyramidal neurons and interneurons [185, 186]. TEP potential is composed of a series of peaks and troughs at specific latencies and can last 300 ms or more (Hill et al. 2016). It has been deeply described from primary motor cortex (M1) and dorsolateral prefrontal cortex (DMPFC), while for other non-motor cortical regions, it is less characterized [184]. Different methodological approaches exist to analyze TEP, the choice of which is highly dependent on the goal of the study and the driven hypothesis.

These methods are not described here, but a recent review article provided a comprehensive and critical assessment of these approaches [184]. TMS effects on brain activity can be further investigated in the frequency domain. When a cortical area is perturbed by TMS, the neuronal response as measured by EEG tends to oscillate at a specific natural frequency [187–189]. Therefore, TMS-EEG can be used to manipulate and investigate brain rhythms by measuring the impact of a TMS pulse on EEG oscillatory activity and associated behavioral effects [190]. The same methods used to study EEG oscillations can be used in TMS-triggered oscillations [191, 192]. The study of cortical oscillations has provided important insights in the effects of plasticity-inducing protocols on brain activity in healthy subjects, in the effect of specific treatments in clinical populations, and the role of a specific oscillatory network in the pathophysiology of clinical disorders [193].

Recording EEG during TMS presents technical challenges since the strong electrical field generated by TMS can saturate recording amplifiers. However, recent advancements in amplifier technology have enabled the development of TMS-compatible EEG equipment that can operate in high, time-varying magnetic fields without being saturated. The typical setup for acquiring TMS-EEG data includes (a) TMS devices and coils, (b) TMS-compatible EEG amplifiers, and (c) TMS-compatible electrodes. It's also highly recommended to integrate a neuronavigation system to maintain the coil's position on the target with consistent orientation and angle during each session, especially for longitudinal studies. In terms of EEG electrodes, only a few may be necessary for monitoring background brain activity or response amplitude variations, but high-resolution recordings using 50–100 electrodes are essential for determining brain source distributions [178]. The choice of electrode type (active or passive) can also influence outcomes, along with active masking procedures to counter the TMS coil click, which might interfere with the recorded responses. Additionally, the use of mono- or biphasic pulse waveforms can critically affect study results. Beyond technical improvements that limit artifacts during recording, offline signal processing for artifact removal is also necessary to ensure high-quality data. Although there are various open-source toolboxes available for this purpose, the diversity of approaches and absence of a standardized method currently limit the widespread clinical adoption of TMS-EEG [194, 195].

19.5.2.1 TMS-EEG in Epilepsy

In recent years, TMS-EEG has emerged as a promising new biomarker in epilepsy, with the potential to improve the diagnostic and prognostic capabilities of traditional EEG. Unlike conventional EEG, TMS-EEG offers an objective, operator-independent analysis of data and serves as an active paradigm for exploring brain function. Evidence suggests that active paradigms can uncover aspects of brain states, such as transitions to epileptic states, that are not easily detected with passive recordings [195]. In focal epilepsy, TMS-EEG has reliably identified the epileptogenic zone [196] and has revealed cortical hyperexcitability in gray matter heterotopias [197]. In genetic generalized epilepsy, it has provided insights into epilepsy's pathophysiological mechanisms, serving as a sensitive and specific diagnostic marker, and predicting responses to antiseizure medications [198]. Another application of TMS-EEG involves monitoring the acute effects of magnetic stimulation on epileptiform discharges and seizures, especially in drug-resistant cases. Low-frequency repetitive TMS (rTMS, <1 Hz) has been shown to reduce cortical excitability, while high-frequency stimulation enhances it, resembling long-term depression (LTD) and long-term potentiation (LTP), respectively. Despite promising results regarding rTMS's ability to reduce seizures [199, 200], recent narrative and Cochrane reviews [201, 202] highlight the need for more uniform methodologies and consistent reporting before the technique can be fully validated for long-term outcomes. Nevertheless, the method is considered safe and painless, showing great promise as a clinical tool for epilepsy.

19.5.2.2 Other Applications

TMS-EEG holds potential in conditions such as disorders of consciousness, psychiatric diseases, and neurodegenerative disorders like Alzheimer's and Parkinson's. In disorders of consciousness, TMS-EEG has been useful for assessing disrupted cortical information processing without requiring active patient participation, making it ideal for identifying conditions like minimally conscious state (MCS) and unresponsive wakefulness syndrome (UWS) [203, 204]. The perturbational complexity index (PCI) derived from TMS-EEG has shown high sensitivity in detecting MCS and prognostic value in UWS patients. Its diagnostic accuracy aligns closely with findings from FDG-PET, indicating its potential as a diagnostic tool [205, 206].

In psychiatric conditions, TMS-EEG has revealed changes in cortical inhibition and gamma oscillations in schizophrenia, offering new insights into the disorder's pathophysiology and potential for objective diagnostic and prognostic markers [207, 208]. In mood disorders, TMS-EEG has helped to assess neurophysiological responses to antidepressants and may contribute to personalized treatment approaches [209]. It has also shown promise in monitoring abnormal cortical reactivity and connectivity in substance use disorders and ADHD, aiding in therapeutic intervention and predicting treatment outcomes [184].

In Alzheimer's disease (AD), early TMS-EEG studies suggest it could help predict the progression from mild cognitive impairment (MCI) to AD by monitoring TMS-induced activity alterations [210]. TMS-EEG may also aid in the early detection of Parkinson's disease, although clinical application is still in its infancy [211].

Recent evidence supports using real-time EEG state-dependent TMS to reduce interindividual variability in therapeutic outcomes, offering potential applications in stroke, depression, Parkinson's, and epilepsy [212, 213]. For example, real-time EEG stimulation targeting specific brain oscillations has shown promising results in stroke and depression, though further research is needed before these techniques can be widely implemented in clinical practice [214].

19.5.3 TES-EEG

Transcranial electrical stimulation (TES) is designed to modulate neuronal circuits in the brain using various approaches, including constant and alternating currents. TES includes techniques such as transcranial direct current stimulation (tDCS), transcranial alternating current stimulation (tACS), and transcranial random noise stimulation (tRNS). These methods apply weak electrical currents to the scalp to either enhance or inhibit neuronal activity. Among these, tDCS is the most commonly used, delivering low-intensity direct currents (~1–2 mA) through electrodes placed on the scalp. tACS, on the other hand, applies sinusoidal currents to examine cortical oscillation patterns, while tRNS delivers fluctuating currents across a range of frequencies (0.1–640 Hz) [182, 215]. These techniques are characterized by simplicity, low costs, and portability, as they require only electrodes and a stimulator. The neuromodulatory effects of tDCS are well-established. Research suggests that neurons respond to membrane polarization changes induced by tDCS, resulting in reduced spontaneous neuronal firing rates following cathodal stimulation, while anodal stimulation has the opposite effect [216]. Similarly, tRNS can alter cortical excitability through membrane polarization mechanisms, with the advantage of not being constrained by current flow direction [215]. tACS enables the delivery of oscillatory currents at specific frequencies, facilitating cortical oscillation entrainment [217] and providing insights into cortical binding during different brain functions [218].

TES and EEG are highly compatible, sharing similar spatial and temporal characteristics, such as the ability to record or stimulate brain activity at millisecond-scale temporal resolution and millimeter-to-centimeter spatial resolution. This compatibility facilitates real-time monitoring and adjusting TES parameters based on EEG signals, known as closed-loop brain state-dependent stimulation [182]. Combining TES and EEG allows for a deeper understanding of human brain dynamics that surpasses the insights provided by either technique alone. Co-registering TES and

EEG provides high temporal resolution data regarding TES-induced cortical changes and assesses how TES affects distant neural regions. The spread of TES effects to connected areas can be tracked via simultaneous EEG recordings, enabling a broader understanding of brain-wide activations. In experimental contexts, this approach allows for direct observation of how TES modulates brain regions activated during specific tasks or how different brain states influence cortical connectivity [182, 216]. However, applying TES during EEG recordings presents challenges such as electrode placement and TES-related artifacts. A detailed review of these issues can be found elsewhere [219], and guidelines are available on how to mitigate these artifacts [220, 221]. Newer devices also enable EEG recordings before, during, and after TES application [222]. Recent studies have even explored the feasibility of combining TES (tDCS and tACS) with intracranial EEG recordings in patients with drug-resistant focal epilepsy [223]. Most studies so far have employed EEG in an "offline" mode, recording cortical electrophysiological data (such as evoked potentials and frequency bands) before and after TES to evaluate its neuromodulatory effects across different cortical regions. In epilepsy research, EEG also serves as a neuronavigation tool for identifying scalp areas with functional abnormalities (e.g., higher IED incidence) to guide electrode placement.

19.5.4 Conclusion

Both TMS and TES can be combined with EEG, either in a sequential or simultaneous manner. Electrophysiological measurements play a key role in determining the location, timing, and method of stimulation, thereby enhancing the precision and effectiveness of noninvasive transcranial brain stimulation and modulation protocols. Overall, recent technological advancements are facilitating the execution of experiments that would have been extremely challenging in the past. Although technical challenges may arise, the combination of TMS/TES and EEG is a highly promising approach for various research objectives, as it enables a direct examination of how these techniques impact cortical connectivity and reactivity.

Acknowledgements We would like to thank Saurabh Joshi and Boyuan Song for their help with some of the figures.

References

1. Liu S, Cai W, Liu S, Zhang F, Fulham M, Feng D, et al. Multimodal neuroimaging computing: a review of the applications in neuropsychiatric disorders. Brain Inform. 2015;2(3):167–80.
2. Correa N, Adali T, Li YO, Calhoun V. Canonical correlation analysis for data fusion and group inferences. IEEE Signal Process Mag. 2010;27(4):39–50.
3. Buzsaki G, Anastassiou CA, Koch C. The origin of extracellular fields and currents--EEG, ECoG, LFP and spikes. Nat Rev Neurosci [Internet] 20120518th ed. 2012;13(6):407–20. Available from: https://www.ncbi.nlm.nih.gov/pubmed/22595786
4. McCann H, Pisano G, Beltrachini L. Variation in reported human head tissue electrical conductivity values. Brain Topogr [Internet]. 20190503rd ed. 2019;32(5):825–58. Available from: https://www.ncbi.nlm.nih.gov/pubmed/31054104.
5. Hedrich T, Pellegrino G, Kobayashi E, Lina JM, Grova C. Comparison of the spatial resolution of source imaging techniques in high-density EEG and MEG. Neuroimage [Internet]. 20170613th ed. 2017;157:531–44. Available from: https://www.ncbi.nlm.nih.gov/pubmed/28619655
6. Vorwerk J, Cho JH, Rampp S, Hamer H, Knösche TR, Wolters CH. A guideline for head volume conductor modeling in EEG and MEG. Neuroimage. 20140625th ed. 2014;100:590–607.
7. Unnwongse K, Rampp S, Wehner T, Kowoll A, Parpaley Y, von Lehe M, et al. Validating EEG source imaging using intracranial electrical stimulation. Brain Commun [Internet]. 20230207th ed. 2023;5(1):fcad023. Available from: https://www.ncbi.nlm.nih.gov/pubmed/36824389.
8. Mikulan E, Russo S, Parmigiani S, Sarasso S, Zauli FM, Rubino A, et al. Simultaneous human intracerebral stimulation and HD-EEG, ground-truth for source localization methods. Sci Data [Internet]. 20200428th ed. 2020;7(1):127. Available from: https://www.ncbi.nlm.nih.gov/pubmed/32345974.
9. Beniczky S, Lantz G, Rosenzweig I, Akeson P, Pedersen B, Pinborg LH, et al. Source localization of rhythmic ictal EEG activity: a study of diagnostic accuracy following STARD criteria. Epilepsia [Internet]. 20130814th ed. 2013;54(10):1743–52.

Available from: https://www.ncbi.nlm.nih.gov/pubmed/23944234
10. Brodbeck V, Spinelli L, Lascano AM, Wissmeier M, Vargas MI, Vulliemoz S, et al. Electroencephalographic source imaging: a prospective study of 152 operated epileptic patients. Brain [Internet]. 2011;134(Pt 10):2887–97. Available from: https://www.ncbi.nlm.nih.gov/pubmed/21975586.
11. Coito A, Biethahn S, Tepperberg J, Carboni M, Roelcke U, Seeck M, et al. Interictal epileptogenic zone localization in patients with focal epilepsy using electric source imaging and directed functional connectivity from low-density EEG. Epilepsia Open [Internet]. 20190414th ed. 2019;4(2):281–92. Available from: https://www.ncbi.nlm.nih.gov/pubmed/31168495
12. Duez L, Tankisi H, Hansen PO, Sidenius P, Sabers A, Pinborg LH, et al. Electromagnetic source imaging in presurgical workup of patients with epilepsy: A prospective study. Neurology [Internet]. 20190104th ed. 2019;92(6):e576–86. Available from: https://www.ncbi.nlm.nih.gov/pubmed/30610090
13. Lascano AM, Perneger T, Vulliemoz S, Spinelli L, Garibotto V, Korff CM, et al. Yield of MRI, high-density electric source imaging (HD-ESI), SPECT and PET in epilepsy surgery candidates. Clin Neurophysiol [Internet]. 20150509th ed. 2016;127(1):150–5. Available from: https://www.ncbi.nlm.nih.gov/pubmed/26021550.
14. Sharma P, Seeck M, Beniczky S. Accuracy of Interictal and ictal electric and magnetic source imaging: A systematic review and meta-analysis. Front Neurol [Internet]. 20191203rd ed. 2019;10:1250. Available from: https://www.ncbi.nlm.nih.gov/pubmed/31849817.
15. van Mierlo P, Vorderwulbecke BJ, Staljanssens W, Seeck M, Vulliemoz S. Ictal EEG source localization in focal epilepsy: review and future perspectives. Clin Neurophysiol [Internet]. 20200815th ed. 2020;131(11):2600–16. Available from: https://www.ncbi.nlm.nih.gov/pubmed/32927216.
16. Rosenow F, Luders H. Presurgical evaluation of epilepsy. Brain [Internet]. 2001;124(Pt 9):1683–700. Available from: https://www.ncbi.nlm.nih.gov/pubmed/11522572.
17. Mouthaan BE, Rados M, Boon P, Carrette E, Diehl B, Jung J, et al. Diagnostic accuracy of interictal source imaging in presurgical epilepsy evaluation: A systematic review from the E-PILEPSY consortium. Clin Neurophysiol [Internet]. 20190208th ed. 2019;130(5):845–55. Available from: https://www.ncbi.nlm.nih.gov/pubmed/30824202.
18. Nascimento FA, Friedman D, Peters JM, Bensalem-Owen MK, Cendes F, Rampp S, et al. Focal epilepsies: update on diagnosis and classification. Epileptic Disord [Internet]. 20230417th ed. 2023;25(1):1–17. Available from: https://www.ncbi.nlm.nih.gov/pubmed/36938903.
19. Foged MT, Martens T, Pinborg LH, Hamrouni N, Litman M, Rubboli G, et al. Diagnostic added value of electrical source imaging in presurgical evaluation of patients with epilepsy: A prospective study. Clin Neurophysiol [Internet]. 20190816th ed. 2020;131(1):324–9. Available from: https://www.ncbi.nlm.nih.gov/pubmed/31466846.
20. Seeck M, Koessler L, Bast T, Leijten F, Michel C, Baumgartner C, et al. The standardized EEG electrode array of the IFCN. Clin Neurophysiol [Internet]. 20170717th ed. 2017;128(10):2070–7. Available from: https://www.ncbi.nlm.nih.gov/pubmed/28778476.
21. Rosenzweig I, Fogarasi A, Johnsen B, Alving J, Fabricius ME, Scherg M, et al. Beyond the double banana: improved recognition of temporal lobe seizures in long-term EEG. J Clin Neurophysiol [Internet]. 2014;31(1):1–9. Available from: https://www.ncbi.nlm.nih.gov/pubmed/24492440.
22. Bach Justesen A, Foged MT, Fabricius M, Skaarup C, Hamrouni N, Martens T, et al. Diagnostic yield of high-density versus low-density EEG: the effect of spatial sampling, timing and duration of recording. Clin Neurophysiol [Internet]. 20190822nd ed. 2019;130(11):2060–4. Available from: https://www.ncbi.nlm.nih.gov/pubmed/31541983.
23. Lantz G, Grave de Peralta R, Spinelli L, Seeck M, Michel CM. Epileptic source localization with high density EEG: how many electrodes are needed? Clin Neurophysiol [Internet]. 2003;114(1):63–9. Available from: https://www.ncbi.nlm.nih.gov/pubmed/12495765.
24. Ntolkeras G, Tamilia E, AlHilani M, Bolton J, Ellen Grant P, Prabhu SP, et al. Presurgical accuracy of dipole clustering in MRI-negative pediatric patients with epilepsy: validation against intracranial EEG and resection. Clin Neurophysiol [Internet]. 20210326th ed. 2022;141:126–38. Available from: https://www.ncbi.nlm.nih.gov/pubmed/33875376.
25. Vorderwulbecke BJ, Carboni M, Tourbier S, Brunet D, Seeber M, Spinelli L, et al. High-density Electric Source Imaging of interictal epileptic discharges: How many electrodes and which time point? Clin Neurophysiol [Internet]. 20201015th ed. 2020;131(12):2795–803. Available from: https://www.ncbi.nlm.nih.gov/pubmed/33137569.
26. Sohrabpour A, Lu Y, Kankirawatana P, Blount J, Kim H, He B. Effect of EEG electrode number on epileptic source localization in pediatric patients. Clin Neurophysiol [Internet]. 20140711th ed. 2015;126(3):472–80. Available from: https://www.ncbi.nlm.nih.gov/pubmed/25088733.
27. Heide EC, van de Velden D, Garnica Agudelo D, Hewitt M, Riedel C, Focke NK. Feasibility of high-density electric source imaging in the presurgical workflow: effect of number of spikes and automated spike detection. Epilepsia Open [Internet]. 20230601st ed. 2023;8(3):785–96. Available from: https://www.ncbi.nlm.nih.gov/pubmed/36938790.
28. Spinelli L, Baroumand AG, Vulliemoz S, Momjian S, Strobbe G, van Mierlo P, et al. Semiautomatic interictal electric source localization based on long-

term electroencephalographic monitoring: a prospective study. Epilepsia [Internet]. 20230215th ed. 2023;64(4):951–61. Available from: https://www.ncbi.nlm.nih.gov/pubmed/36346269.
29. Antonakakis M, Schrader S, Aydin Ü, Khan A, Gross J, Zervakis M, et al. Inter-subject variability of skull conductivity and thickness in calibrated realistic head models. NeuroImage. 2020;223:117353.
30. Kural MA, Tankisi H, Duez L, Sejer Hansen V, Udupi A, Wennberg R, et al. Optimized set of criteria for defining interictal epileptiform EEG discharges. Clinical Neurophysiology [Internet]. 2020;131(9):2250–4. Available from: https://linkinghub.elsevier.com/retrieve/pii/S1388245720304089.
31. Rosenkranz K, Lemieux L. Present and future of simultaneous EEG-fMRI. MAGMA. 2010;23(5–6):309–16.
32. Aydin Ü, Vorwerk J, Dümpelmann M, Küpper P, Kugel H, Heers M, et al. Combined EEG/MEG can outperform single modality EEG or MEG source reconstruction in Presurgical epilepsy diagnosis. PLoS One. 2015;10(3):e0118753.
33. van Mierlo P, Strobbe G, Keereman V, Birot G, Gadeyne S, Gschwind M, et al. Automated long-term <scp>EEG</scp> analysis to localize the epileptogenic zone. Epilepsia Open. 2017;2(3):322–33.
34. Thomschewski A, Hincapié AS, Frauscher B. Localization of the epileptogenic zone using high frequency oscillations. Front Neurol. 2019;10
35. Ikeda A, Takeyama H, Bernard C, Nakatani M, Shimotake A, Daifu M, et al. Active direct current (DC) shifts and "red slow": two new concepts for seizure mechanisms and identification of the epileptogenic zone. Neurosci Res. 2020;156:95–101.
36. Lopes da Silva F. EEG and MEG: relevance to neuroscience. Neuron [Internet]. 2013;80(5):1112–28. Available from: https://www.ncbi.nlm.nih.gov/pubmed/24314724.
37. Zhang X, Lei X, Wu T, Jiang T. A review of EEG and MEG for brainnetome research. Cogn Neurodyn [Internet]. 20131122nd ed. 2014;8(2):87–98. Available from: https://www.ncbi.nlm.nih.gov/pubmed/24624229.
38. Zhang H, Zhou QQ, Chen H, Hu XQ, Li WG, Bai Y, et al. The applied principles of EEG analysis methods in neuroscience and clinical neurology. Mil Med Res [Internet] 20231219th ed. 2023;10(1):67. Available from: https://www.ncbi.nlm.nih.gov/pubmed/38115158.
39. Coolen T, Dumitrescu AM, Bourguignon M, Wens V, Urbain C, De Tiège X. Presurgical electromagnetic functional brain mapping in refractory focal epilepsy. Z Epileptol. 2018;31(3):203–12.
40. Kreidenhuber R, De Tiege X, Rampp S. Presurgical functional cortical mapping using electromagnetic source imaging. Front Neurol [Internet]. 20190613th ed. 2019;10:628. Available from: https://www.ncbi.nlm.nih.gov/pubmed/31249552.
41. Piastra MC, Nussing A, Vorwerk J, Clerc M, Engwer C, Wolters CH. A comprehensive study on electroencephalography and magnetoencephalography sensitivity to cortical and subcortical sources. Hum Brain Mapp [Internet]. 20201106th ed. 2021;42(4):978–92. Available from: https://www.ncbi.nlm.nih.gov/pubmed/33156569.
42. Klamer S, Elshahabi A, Lerche H, Braun C, Erb M, Scheffler K, et al. Differences between MEG and high-density EEG source localizations using a distributed source model in comparison to fMRI. Brain Topogr [Internet]. 20141009th ed. 2015;28(1):87–94. Available from: https://www.ncbi.nlm.nih.gov/pubmed/25296614.
43. Nix KC, Oh A, Goad BS, Wu W, Lucas MV, Baumer FM. Detection of language lateralization using spectral analysis of EEG. J Clin Neurophysiol [Internet]. 2024;41(4):334–43. Available from: https://www.ncbi.nlm.nih.gov/pubmed/38710040.
44. Trimmel K, Sachsenweger J, Lindinger G, Auff E, Zimprich F, Pataraia E. Lateralization of language function in epilepsy patients: A high-density scalp-derived event-related potentials (ERP) study. Clin Neurophysiol [internet]. 20170112th ed. 2017;128(3):472–9. Available from: https://www.ncbi.nlm.nih.gov/pubmed/28160753.
45. Tsolaki AC, Kosmidou V, Kompatsiaris IY, Papadaniil C, Hadjileontiadis L, Adam A, et al. Brain source localization of MMN and P300 ERPs in mild cognitive impairment and Alzheimer's disease: a high-density EEG approach. Neurobiol Aging [Internet] 20170329th ed. 2017;55:190–201. Available from: https://www.ncbi.nlm.nih.gov/pubmed/28461101.
46. Hata M, Kazui H, Tanaka T, Ishii R, Canuet L, Pascual-Marqui RD, et al. Functional connectivity assessed by resting state EEG correlates with cognitive decline of Alzheimer's disease – an eLORETA study. Clin Neurophysiol [Internet] 20151019th ed. 2016;127(2):1269–78. Available from: https://www.ncbi.nlm.nih.gov/pubmed/26541308.
47. Kim J, Jeong M, Stiles WR, Choi HS. Neuroimaging modalities in Alzheimer's disease: diagnosis and clinical features. Int J Mol Sci [Internet]. 20220528th ed. 2022;23(11). Available from: https://www.ncbi.nlm.nih.gov/pubmed/35682758.
48. Nicolo P, Rizk S, Magnin C, Pietro MD, Schnider A, Guggisberg AG. Coherent neural oscillations predict future motor and language improvement after stroke. Brain [Internet]. 20150709th ed. 2015;138(Pt 10):3048–60. Available from: https://www.ncbi.nlm.nih.gov/pubmed/26163304.
49. Dubovik S, Ptak R, Aboulafia T, Magnin C, Gillabert N, Allet L, et al. EEG alpha band synchrony predicts cognitive and motor performance in patients with ischemic stroke. Behav Neurol [Internet]. 2013;26(3):187–9. Available from: https://www.ncbi.nlm.nih.gov/pubmed/22713421
50. Mottaz A, Solca M, Magnin C, Corbet T, Schnider A, Guggisberg AG. Neurofeedback training of alpha-band coherence enhances motor performance. Clin Neurophysiol [Internet]. 20141206th ed.

2015;126(9):1754–60. Available from: https://www.ncbi.nlm.nih.gov/pubmed/25540133.
51. Deichmann R, Nöth U, Merola A, Weiskopf N. The basics of functional magnetic resonance imaging. In: EEG – fMRI. Cham: Springer International Publishing; 2022. p. 49–77.
52. Sohrabpour A, Cai Z, Ye S, Brinkmann B, Worrell G, He B. Noninvasive electromagnetic source imaging of spatiotemporally distributed epileptogenic brain sources. Nat Commun. 2020;11(1):1946.
53. Ogawa S, Lee TM, Kay AR, Tank DW. Brain magnetic resonance imaging with contrast dependent on blood oxygenation. Proc Natl Acad Sci. 1990;87(24):9868–72.
54. Buxton RB, Uludağ K, Dubowitz DJ, Liu TT. Modeling the hemodynamic response to brain activation. NeuroImage. 2004;23:S220–33.
55. Friston KJ, Jezzard P, Turner R. Analysis of functional MRI time-series. Hum Brain Mapp. 1994;1(2):153–71.
56. Logothetis NK, Pauls J, Augath M, Trinath T, Oeltermann A. Neurophysiological investigation of the basis of the fMRI signal. Nature [Internet]. 2001;412(6843):150–7. Available from: http://www.ncbi.nlm.nih.gov/pubmed/11449264.
57. Ekstrom AD. Regional variation in neurovascular coupling and why we still lack a Rosetta stone. Philos Trans R Soc B Biol Sc. 2021;376(1815):20190634.
58. Hawsawi HB, Allen PJ, Warbrick T, Störmer R, Iannotti G, Grouiller F, et al. EEG instrumentation and safety in the MRI environment. In: EEG – fMRI. Cham: Springer International Publishing; 2022. p. 141–66.
59. Jorge J, Grouiller F, Ipek Ö, Stoermer R, Michel CM, Figueiredo P, et al. Simultaneous EEG–fMRI at ultra-high field: artifact prevention and safety assessment. NeuroImage. 2015;105:132–44.
60. Bullock M, Jackson GD, Abbott DF. Artifact reduction in simultaneous EEG-fMRI: A systematic review of methods and contemporary usage. Front Neurol. 2021;12
61. Warbrick T. Simultaneous EEG-fMRI: what have we learned and what does the future hold? Sensors. 2022;22(6):2262.
62. Lemieux L, Allen PJ, Franconi F, Symms MR, Fish DR. Recording of EEG during fMRI experiments: patient safety. Magn Reson Med. 1997;38(6):943–52.
63. Carmichael DW, Pinto S, Limousin-Dowsey P, Thobois S, Allen PJ, Lemieux L, et al. Functional MRI with active, fully implanted, deep brain stimulation systems: safety and experimental confounds. NeuroImage. 2007;37(2):508–17.
64. Negishi M, Abildgaard M, Laufer I, Nixon T, Constable RT. An EEG (electroencephalogram) recording system with carbon wire electrodes for simultaneous EEG-fMRI (functional magnetic resonance imaging) recording. J Neurosci Methods. 2008;173(1):99–107.
65. Carmichael DW, Thornton JS, Rodionov R, Thornton R, McEvoy AW, Ordidge RJ, et al. Feasibility of simultaneous intracranial EEG-fMRI in humans: A safety study. NeuroImage. 2010;49(1):379–90.
66. Mullinger K, Debener S, Coxon R, Bowtell R. Effects of simultaneous EEG recording on MRI data quality at 1.5, 3 and 7 tesla. Int J Psychophysiol. 2008;67(3):178–88.
67. Rothlübbers S, Relvas V, Leal A, Murta T, Lemieux L, Figueiredo P. Characterisation and reduction of the EEG artefact caused by the helium cooling pump in the MR environment: validation in epilepsy patient data. Brain Topogr. 2015;28(2):208–20.
68. van der Meer JN, Pampel A, Van Someren EJW, Ramautar JR, van der Werf YD, Gomez-Herrero G, et al. Carbon-wire loop based artifact correction outperforms post-processing EEG/fMRI corrections—a validation of a real-time simultaneous EEG/fMRI correction method. NeuroImage. 2016;125:880–94.
69. Daniel AJ, Smith JA, Spencer GS, Jorge J, Bowtell R, Mullinger KJ. Exploring the relative efficacy of motion artefact correction techniques for EEG data acquired during simultaneous fMRI. Hum Brain Mapp. 2019;40(2):578–96.
70. Steinbrenner M, McDowell A, Centeno M, Moeller F, Perani S, Lorio S, et al. Camera-based prospective motion correction in paediatric epilepsy patients enables EEG-fMRI localization even in high-motion states. Brain Topogr. 2023;36(3):319–37.
71. Maziero D, Stenger VA, Carmichael DW. Unified retrospective EEG motion educated artefact suppression for EEG-fMRI to suppress magnetic field gradient artefacts during motion. Brain Topogr. 2021;34(6):745–61.
72. Mullinger KJ, Havenhand J, Bowtell R. Identifying the sources of the pulse artefact in EEG recordings made inside an MR scanner. NeuroImage. 2013;71:75–83.
73. Allen PJ, Polizzi G, Krakow K, Fish DR, Lemieux L. Identification of EEG events in the MR scanner: the problem of pulse artifact and a method for its subtraction. Neuroimage [Internet]. 1998;8(3):229–39. Available from: http://www.ncbi.nlm.nih.gov/pubmed/9758737.
74. Niazy RK, Beckmann CF, Iannetti GD, Brady JM, Smith SM. Removal of FMRI environment artifacts from EEG data using optimal basis sets. NeuroImage. 2005;28(3):720–37.
75. Scheeringa R, Laufs H. Brain rhythms. In: EEG – fMRI. Cham: Springer International Publishing; 2022. p. 377–404.
76. Allen PJ, Josephs O, Turner R. A method for removing imaging artifact from continuous EEG recorded during functional MRI. Neuroimage [Internet]. 2000;12(2):230–9. Available from: http://www.ncbi.nlm.nih.gov/pubmed/10913328
77. Garreffa G, Carnı̀ M, Gualniera G, Ricci GB, Bozzao L, De Carli D, et al. Real-time MR artifacts filtering during continuous EEG/fMRI acquisition. Magn Reson Imaging. 2003;21(10):1175–89.
78. Stern. Simultaneous EEG and fMRI of the alpha rhythm. Neuroreport. 2002;13:2487–92.

79. Bonmassar G, Schwartz DP, Liu AK, Kwong KK, Dale AM, Belliveau JW. Spatiotemporal brain imaging of visual-evoked activity using interleaved EEG and fMRI recordings. NeuroImage. 2001;13(6):1035–43.
80. Symms MR, Allen PJ, Woermann FG, Polizzi G, Krakow K, Barker GJ, et al. Reproducible localization of interictal epileptiform discharges using EEG-triggered fMRI. Phys Med Biol. 1999;44(7):N161–8.
81. Lemieux L, Salek-Haddadi A, Josephs O, Allen P, Toms N, Scott C, et al. Event-related fMRI with simultaneous and continuous EEG: description of the method and initial case report. NeuroImage. 2001;14(3):780–7.
82. Maziero D, Velasco TR, Hunt N, Payne E, Lemieux L, Salmon CEG, et al. Towards motion insensitive EEG-fMRI: correcting motion-induced voltages and gradient artefact instability in EEG using an fMRI prospective motion correction (PMC) system. NeuroImage. 2016;138:13–27.
83. Mayeli A, Al Zoubi O, Henry K, Wong CK, White EJ, Luo Q, et al. Automated pipeline for EEG artifact reduction (APPEAR) recorded during fMRI. J Neural Eng. 2021;18(4):0460b4.
84. Wirsich J, Ridley B, Besson P, Jirsa V, Bénar C, Ranjeva JP, et al. Complementary contributions of concurrent EEG and fMRI connectivity for predicting structural connectivity. NeuroImage. 2017;161:251–60.
85. Laufs H. Multimodal analysis of resting state cortical activity: what does EEG add to our knowledge of resting state BOLD networks? NeuroImage. 2010;52(4):1171–2.
86. Ives JR, Warach S, Schmitt F, Edelman RR, Schomer DL. Monitoring the patient's EEG during echo planar MRI. Electroencephalogr Clin Neurophysiol. 1993;87(6):417–20.
87. Warach S, Ives JR, Schlaug G, Patel MR, Darby DG, Thangaraj V, et al. EEG-triggered echo-planar functional MRI in epilepsy. Neurology. 1996;47(1):89–93.
88. Wirsich J, Bagshaw AP, Guye M, Lemieux L, Bénar CG. Experimental design and data analysis strategies. In: EEG – fMRI. Cham: Springer International Publishing; 2022. p. 267–322.
89. Flanagan D, Abbott DF, Jackson GD. How wrong can we be? The effect of inaccurate mark-up of EEG/fMRI studies in epilepsy. Clin Neurophysiol. 2009;120(9):1637–47.
90. Markoula S, Chaudhary UJ, Perani S, De Ciantis A, Yadee T, Duncan JS, et al. The impact of mapping interictal discharges using EEG-fMRI on the epilepsy presurgical clinical decision making process: a prospective study. Seizure. 2018;61:30–7.
91. Bonaventura CD, Vaudano AE, Carnì M, Pantano P, Nucciarelli V, Garreffa G, et al. EEG/fMRI study of ictal and interictal epileptic activity: methodological issues and future perspectives in clinical practice. Epilepsia. 2006;47(SUPPL. 5):52.
92. Pedreira C, Vaudano AE, Thornton RC, Chaudhary UJ, Vulliemoz S, Laufs H, et al. Classification of EEG abnormalities in partial epilepsy with simultaneous EEG-fMRI recordings. NeuroImage. 2014;99:461.
93. Vaudano AE, Ruggieri A, Vignoli A, Avanzini P, Benuzzi F, Gessaroli G, et al. Epilepsy-related brain networks in ring chromosome 20 syndrome: An EEG-fMRI study. Epilepsia. 2014;55(3):403.
94. Salek-Haddadi A, Diehl B, Hamandi K, Merschhemke M, Liston A, Friston K, et al. Hemodynamic correlates of epileptiform discharges: an EEG-fMRI study of 63 patients with focal epilepsy. Brain Res. 2006;1088(1):148–66.
95. Walz JM, Pedersen M, Omidvarnia A, Semmelroch M, Jackson GD. Spatiotemporal mapping of epileptic spikes using simultaneous EEG-functional MRI. Brain. 2017;140(4):998–1010.
96. Vulliemoz S, Thornton R, Rodionov R, Carmichael DW, Guye M, Lhatoo S, et al. The spatio-temporal mapping of epileptic networks: combination of EEG–fMRI and EEG source imaging. NeuroImage. 2009;46(3):834–43.
97. Bagshaw AP, Aghakhani Y, Bénar C, Kobayashi E, Hawco C, Dubeau F, et al. EEG-fMRI of focal epileptic spikes: analysis with multiple haemodynamic functions and comparison with gadolinium-enhanced MR angiograms. Hum Brain Mapp. 2004;22(3):179–92.
98. Lemieux L, Salek-Haddadi A, Lund TE, Laufs H, Carmichael D. Modelling large motion events in fMRI studies of patients with epilepsy. Magn Reson Imaging. 2007;25(6):894–901.
99. Salek-Haddadi A, Lemieux L, Merschhemke M, Diehl B, Allen PJ, Fish DR. EEG quality during simultaneous functional MRI of interictal epileptiform discharges. Magn Reson Imaging. 2003;21(10):1159–66.
100. Masterton RAJ, Harvey AS, Archer JS, Lillywhite LM, Abbott DF, Scheffer IE, et al. Focal epileptiform spikes do not show a canonical BOLD response in patients with benign rolandic epilepsy (BECTS). NeuroImage. 2010;51(1):252–60.
101. Aguirre GK, Zarahn E, D'Esposito M. The variability of human, BOLD hemodynamic responses. Neuroimage. 1998;8(4):360–9.
102. Rodionov R, De Martino F, Laufs H, Carmichael DW, Formisano E, Walker M, et al. Independent component analysis of interictal fMRI in focal epilepsy: comparison with general linear model-based EEG-correlated fMRI. NeuroImage. 2007;38(3):488–500.
103. Chatzistefanidis D, Huang D, Dümpelmann M, Jacobs J, Schulze-Bonhage A, LeVan P. Topography-related EEG-fMRI in surgically confirmed epileptic foci: a comparison to spike-related EEG-fMRI in clinical practice. Brain Topogr. 2021;34(3):373–83.
104. Grouiller F, Thornton RC, Groening K, Spinelli L, Duncan JS, Schaller K, et al. With or without spikes: localization of focal epileptic activity by simultaneous electroencephalography and func-

tional magnetic resonance imaging. Brain [Internet]. 2011;134(10):2867–86. Available from: https://academic.oup.com/brain/article-lookup/doi/10.1093/brain/awr156.
105. Power JD, Schlaggar BL, Petersen SE. Recent progress and outstanding issues in motion correction in resting state fMRI. NeuroImage. 2015;105:536–51.
106. Chaudhary UJ, Rodionov R, Carmichael DW, Thornton RC, Duncan JS, Lemieux L. Improving the sensitivity of EEG-fMRI studies of epileptic activity by modelling eye blinks, swallowing and other video-EEG detected physiological confounds. NeuroImage. 2012;61(4):1383–93.
107. Ruggieri A, Vaudano AE, Benuzzi F, Serafini M, Gessaroli G, Farinelli V, et al. Mapping (and modeling) physiological movements during EEG-fMRI recordings: the added value of the video acquired simultaneously. J Neurosci Methods. 2015;239:223.
108. Thornton R, Vulliemoz S, Rodionov R, Carmichael DW, Chaudhary UJ, Diehl B, et al. Epileptic networks in focal cortical dysplasia revealed using electroencephalography-functional magnetic resonance imaging. Ann Neurol. 2011;70(5):822–37.
109. Coan AC, Campos BM, Beltramini GC, Yasuda CL, Covolan RJM, Cendes F. Distinct functional and structural MRI abnormalities in mesial temporal lobe epilepsy with and without hippocampal sclerosis. Epilepsia. 2014;55(8):1187–96.
110. Laufs H, Hamandi K, Salek-Haddadi A, Kleinschmidt AK, Duncan JS, Lemieux L. Temporal lobe interictal epileptic discharges affect cerebral activity in "default mode" brain regions. Hum Brain Mapp. 2007;28(10):1023–32.
111. Kozák LR, van Graan LA, Chaudhary UJ, Szabó ÁG, Lemieux L. ICN_Atlas: automated description and quantification of functional MRI activation patterns in the framework of intrinsic connectivity networks. NeuroImage. 2017;163:319–41.
112. Polack PO, Guillemain I, Hu E, Deransart C, Depaulis A, Charpier S. Deep layer somatosensory cortical neurons initiate spike-and-wave discharges in a genetic model of absence seizures. J Neurosci. 2007;27(24):6590–9.
113. Meeren H, Van Luijtelaar G, Lopes Da Silva F, Coenen A. Evolving concepts on the pathophysiology of absence seizures: the cortical focus theory. Arch Neurol. 2005;62(3):371–6.
114. Archer JS, Abbott DF, Waites AB, Jackson GD. fMRI "deactivation" of the posterior cingulate during generalized spike and wave. NeuroImage. 2003;20(4):1915–22.
115. Labate A, Briellmann RS, Abbott DF, Waites AB, Jackson GD. Typical childhood absence seizures are associated with thalamic activation [Internet]. Available from: www.brain.org.au/iBrain
116. Guo JN, Kim R, Chen Y, Negishi M, Jhun S, Weiss S, et al. Impaired consciousness in patients with absence seizures investigated by functional MRI, EEG, and behavioural measures: a cross-sectional study. Lancet Neurol. 2016;15(13):1336–45.
117. Hamandi K, Salek-Haddadi A, Laufs H, Liston A, Friston K, Fish DR, et al. EEG-fMRI of idiopathic and secondarily generalized epilepsies. NeuroImage. 2006;31(4):1700–10.
118. Carney PW, Masterton RAJ, Harvey AS, Scheffer IE, Berkovic SF, Jackson GD. The core network in absence epilepsy: differences in cortical and thalamic BOLD response. Neurology. 2010;75(10):904–11.
119. Vaudano AE, Laufs H, Kiebel SJ, Carmichael DW, Hamandi K, Guye M, et al. Causal hierarchy within the thalamo-cortical network in spike and wave discharges. PLoS One. 2009;4(8):e6475.
120. Pugnaghi M, Carmichael DW, Vaudano AE, Chaudhary UJ, Benuzzi F, Di Bonaventura C, et al. Generalized spike and waves: effect of discharge duration on brain networks as revealed by BOLD fMRI. Brain Topogr. 2014;27(1)
121. Benuzzi F, Mirandola L, Pugnaghi M, Farinelli V, Tassinari CA, Capovilla G, et al. Increased cortical BOLD signal anticipates generalized spike and wave discharges in adolescents and adults with idiopathic generalized epilepsies. Epilepsia. 2012;53(4):622–30.
122. Benuzzi F, Ballotta D, Mirandola L, Ruggieri A, Vaudano AE, Zucchelli M, et al. An EEG-fMRI study on the termination of generalized spike-and-wave discharges in absence epilepsy. PLoS One. 2015;10(7):e0130943.
123. Ryvlin P, Cross JH, Rheims S. Epilepsy surgery in children and adults. Vol. 13, The Lancet Neurology. Lancet Publishing Group; 2014. p. 1114–26.
124. Detre JA, Alsop DC, Aguirre GK, Sperling MR. Coupling of cortical and thalamic ictal activity in human partial epilepsy: demonstration by functional magnetic resonance imaging. Epilepsia. 1996;37(7):657–61.
125. Detre JA, Sirven JI, Alsop DC, O'Connor MJ, French JA. Localization of subclinical ictal activity by functional magnetic resonance imaging: correlation with invasive monitoring. Ann Neurol. 1995;38(4):618–24.
126. Krings T, Töpper R, Reinges MHT, Foltys H, Spetzger U, Chiappa KH, et al. Hemodynamic changes in simple partial epilepsy: a functional MRI study. Neurology. 2000;54(2):524.
127. Chaudhary UJ, Duncan JS, Lemieux L. Mapping hemodynamic correlates of seizures using fMRI: a review. Hum Brain Mapp. 2013;34(2):447–66.
128. Chaudhary UJ, Carmichael DW, Rodionov R, Thornton RC, Bartlett P, Vulliemoz S, et al. Mapping preictal and ictal haemodynamic networks using video-electroencephalography and functional imaging. Brain. 2012;135(Pt 12):3645–63.
129. Krakow K, Allen P, Symms M, Fish DL. Imaging of interictal epileptiform discharges using spike-triggered fMRI. Int J Bioelectromagn [Internet]. 1999;1(1):96–101. Available from: http://ijbem.net/volume1/number1/pdf/ijbem_a96-101.pdf
130. Symms MR, Allen PJ, Woermann FG, Polizzi G, Krakow K, Barker GJ, et al. Reproducible localiza-

tion of interictal epileptiform discharges using eeg-triggered fMRI. Phys Med Biol. 1999;44(7):161–8.
131. Krakow K, Messina D, Lemieux L, Duncan JS, Fish DR. Functional MRI activation of individual interictal epileptiform spikes. NeuroImage. 2001;13(3):502–5.
132. Krakow K, Lemieux L, Messina D, Scott CA, Symms MR, Duncan JS, et al. Spatio-temporal imaging of focal interictal epileptiform activity using EEG-triggered functional MRI. Epileptic Disord. 2001;3(2):67–73.
133. Ikemoto S, Pana R, von Ellenrieder N, Gotman J. Electroencephalography–functional magnetic resonance imaging for clinical evaluation in focal epilepsy. Epilepsia Open. 2023;9:84.
134. Gotman J, Pittau F. Combining EEG and fMRI in the study of epileptic discharges. Epilepsia. 2011;52 Suppl 4(Suppl 4):38–42.
135. Coan AC, Chaudhary UJ, Grouiller F, Campos BM, Perani S, De Ciantis A, et al. EEG-fMRI in the presurgical evaluation of temporal lobe epilepsy. J Neurol Neurosurg Psychiatry [Internet]. 2016;87(6):642–9. Available from: http://jnnp.bmj.com/lookup/doi/10.1136/jnnp-2015-310401.
136. Elshoff L, Groening K, Grouiller F, Wiegand G, Wolff S, Michel C, et al. The value of EEG-fMRI and EEG source analysis in the presurgical setup of children with refractory focal epilepsy. Epilepsia. 2012;53(9):1597–606.
137. Khoo HM, von Ellenrieder N, Zazubovits N, He D, Dubeau F, Gotman J. The spike onset zone: the region where epileptic spikes start and from where they propagate. Neurol Int 2018;91(7):e666–e674. Available from: http://www.neurology.org/lookup/doi/10.1212/WNL.0000000000005998.
138. Abdallah C, Hedrich T, Koupparis A, Afnan J, Hall JA, Gotman J, et al. Clinical yield of electromagnetic source imaging and hemodynamic responses in epilepsy: validation with intracerebral data. Neurology. 2022;98(24):E2499–511.
139. Vulliemoz S, Lemieux L, Daunizeau J, Michel CM, Duncan JS. The combination of EEG source imaging and EEG-correlated functional MRI to map epileptic networks. Epilepsia. 2010;51(4):491–505.
140. Duncan JS, Trimmel K. Advanced neuroimaging techniques in epilepsy. Vol. 35, Current Opinion in Neurology. Lippincott Williams and Wilkins; 2022. p. 189–95.
141. Pittau F, Dubeau F, Gotman J. Contribution of EEG/fMRI to the definition of the epileptic focus. Neurology. 2012;78(19):1479–87.
142. Vaudano AE, Avanzini P, Tassi L, Ruggieri A, Cantalupo G, Benuzzi F, et al. Causality within the epileptic network: an EEG-fMRI study validated by intracranial EEG. Front Neurol. 2013;4
143. An D, Fahoum F, Hall J, Olivier A, Gotman J, Dubeau F. Electroencephalography/functional magnetic resonance imaging responses help predict surgical outcome in focal epilepsy. Epilepsia. 2013;54(12):2184–94.
144. Vaudano AE, Mirandola L, Talami F, Giovannini G, Monti G, Riguzzi P, et al. fMRI-based effective connectivity in surgical remediable epilepsies: a pilot study. Brain Topogr. 2021;34(5):632–50.
145. Tao JX, Ray A, Hawes-Ebersole S, Ebersole JS. Intracranial EEG substrates of scalp EEG Interictal spikes. Epilepsia. 2005;46(5):669–76.
146. Carmichael DW, Vulliemoz S, Rodionov R, Thornton JS, McEvoy AW, Lemieux L. Simultaneous intracranial EEG–fMRI in humans: protocol considerations and data quality. NeuroImage. 2012;63(1):301–9.
147. Boucousis SM, Beers CA, Cunningham CJB, Gaxiola-Valdez I, Pittman DJ, Goodyear BG, et al. Feasibility of an intracranial EEG–fMRI protocol at 3T: risk assessment and image quality. NeuroImage. 2012;63(3):1237–48.
148. Fujita Y, Khoo HM, Hirayama M, Kawahara M, Koyama Y, Tarewaki H, et al. Evaluating the safety of simultaneous intracranial electroencephalography and functional magnetic resonance imaging acquisition using a 3 Tesla magnetic resonance imaging scanner. Front Neurosci. 2022;16
149. Vulliemoz S, Carmichael DW, Rosenkranz K, Diehl B, Rodionov R, Walker MC, et al. Simultaneous intracranial EEG and fMRI of interictal epileptic discharges in humans. NeuroImage. 2011;54(1):182–90.
150. Chaudhary UJ, Centeno M, Thornton RC, Rodionov R, Vulliemoz S, McEvoy AW, et al. Mapping human preictal and ictal haemodynamic networks using simultaneous intracranial EEG-fMRI. Neuroimage Clin. 2016;11:486–93.
151. Ridley B, Wirsich J, Bettus G, Rodionov R, Murta T, Chaudhary U, et al. Simultaneous intracranial EEG-fMRI shows inter-modality correlation in time-resolved connectivity within Normal areas but not within epileptic regions. Brain Topogr. 2017;30(5):639–55.
152. Cunningham CBJ, Goodyear BG, Badawy R, Zaamout F, Pittman DJ, Beers CA, et al. Intracranial EEG-fMRI analysis of focal epileptiform discharges in humans. Epilepsia. 2012;53(9):1636–48.
153. Tehrani N, Wilson W, Pittman DJ, Mosher V, Peedicail JS, Aghakhani Y, et al. Localization of interictal discharge origin: A simultaneous intracranial electroencephalographic–functional magnetic resonance imaging study. Epilepsia. 2021;62(5):1105–18.
154. George AG, Beers CA, Wilson W, Mosher V, Pittman DJ, Dykens P, et al. Mesial temporal lobe spiking reveals distinct patterns of blood oxygen level-dependent functional magnetic resonance imaging activation using simultaneous intracranial electroencephalography–functional magnetic resonance imaging. Epilepsia. 2024;65(8):2295–307.
155. Chaudhary UJ, Centeno M, Carmichael DW, Diehl B, Walker MC, Duncan JS, et al. Mapping epileptic networks using simultaneous intracranial EEG-fMRI. Front Neurol. 2021;12

156. Murta T, Chaudhary UJ, Tierney TM, Dias A, Leite M, Carmichael DW, et al. Phase–amplitude coupling and the BOLD signal: a simultaneous intracranial EEG (icEEG) – fMRI study in humans performing a finger-tapping task. NeuroImage. 2017;146:438–51.
157. Murta T, Hu L, Tierney TM, Chaudhary UJ, Walker MC, Carmichael DW, et al. A study of the electro-haemodynamic coupling using simultaneously acquired intracranial EEG and fMRI data in humans. NeuroImage. 2016;142:371–80.
158. Carmichael DW, Vulliemoz S, Murta T, Chaudhary U, Perani S, Rodionov R, et al. Measurement of the mapping between intracranial EEG and fMRI recordings in the human brain. Bioengineering. 2024;11(3):224.
159. Masturzo L, Carra P, Erba PA, Morrocchi M, Pilleri A, Sportelli G, et al. Monte Carlo characterization of the trimage brain PET system. J Imaging. 2022;8(2)
160. Zimmer L, Luxen A. PET radiotracers for molecular imaging in the brain: past, present and future. NeuroImage. 2012;61(2):363–70.
161. D'Souza MM, Sharma R, Tripathi M, Panwar P, Jaimini A, Mondal A. Novel positron emission tomography radiotracers in brain tumor imaging. Indian Journal of Radiology and Imaging. 2011;21(03):202–8.
162. Bandettini PA. Twenty years of functional MRI: the science and the stories. NeuroImage. 2012;62(2):575–88.
163. Rajkumar R, Farrher E, Mauler J, Sripad P, Régio Brambilla C, Rota Kops E, et al. Comparison of EEG microstates with resting state fMRI and FDG-PET measures in the default mode network via simultaneously recorded trimodal (PET/MR/EEG) data. Hum Brain Mapp. 2021;42(13):4122–33.
164. Rajkumar R, Rota Kops E, Mauler J, Tellmann L, Lerche C, Herzog H, et al. Simultaneous trimodal PET-MR-EEG imaging: do EEG caps generate artefacts in PET images? PLoS One. 2017;12(9):e0184743.
165. Van Paesschen W, Dupont P, Sunaert S, Goffin K, Van Laere K. The use of SPECT and PET in routine clinical practice in epilepsy. Curr Opin Neurol. 2007;20(2):194–202.
166. Chassoux F, Rodrigo S, Semah F, Beuvon F, Landre E, Devaux B, et al. FDG-PET improves surgical outcome in negative MRI Taylor-type focal cortical dysplasias. Neurology. 2010;75(24):2168–75.
167. Carne RP, O'Brien TJ, Kilpatrick CJ, MacGregor LR, Hicks RJ, Murphy MA, et al. MRI-negative PET-positive temporal lobe epilepsy: a distinct surgically remediable syndrome. Brain. 2004;127(10):2276–85.
168. Barrington SF, Koutroumanidis M, Agathonikou A, Marsden PK, Binnie CD, Polkey CE, et al. Clinical value of "ictal" FDG-positron emission tomography and the routine use of simultaneous scalp EEG studies in patients with intractable partial epilepsies. Epilepsia. 1998;39(7):753–66.
169. Bittar RG, Olivier A, Sadikot AF, Andermann F, Comeau RM, Cyr M, et al. Localization of somatosensory function by using positron emission tomography scanning: a comparison with intraoperative cortical stimulation. J Neurosurg. 1999;90(3):478–83.
170. Altay EE, Fessler AJ, Gallagher M, Attarian HP, Dehdashti F, Vahle VJ, et al. Correlation of severity of FDG-PET Hypometabolism and Interictal Regional Delta slowing in temporal lobe epilepsy. Epilepsia. 2005;46(4):573–6.
171. Grouiller F, Delattre BMA, Pittau F, Heinzer S, Lazeyras F, Spinelli L, et al. All-in-one interictal presurgical imaging in patients with epilepsy: single-session EEG/PET/(f)MRI. Eur J Nucl Med Mol Imaging. 2015;42(7):1133–43.
172. Seshadri V, Zarroli KA, Schetlick RS, Massey JC, Reyes JM, Eluvathingal Muttikal TJ, et al. Dynamic FDG-PET in localization of focal epilepsy: a pilot study. Epilepsy Behav. 2021;122
173. Del Guerra A, Ahmad S, Avram M, Belcari N, Berneking A, Biagi L, et al. TRIMAGE: a dedicated trimodality (PET/MR/EEG) imaging tool for schizophrenia. Eur Psychiatry. 2018;(50):7–20.
174. Shah NJ, Arrubla J, Rajkumar R, Farrher E, Mauler J, Kops ER, et al. Multimodal fingerprints of resting state networks as assessed by simultaneous Trimodal MR-PET-EEG imaging. Sci Rep. 2017;7(1):6452.
175. Massimini M, Ferrarelli F, Huber R, Esser SK, Singh H, Tononi G. Breakdown of cortical effective connectivity during sleep. Science. 1979;309(5744):2228–32.
176. Komssi S, Kähkönen S, Ilmoniemi RJ. The effect of stimulus intensity on brain responses evoked by transcranial magnetic stimulation. Hum Brain Mapp. 2004;21(3):154–64.
177. Komssi S, Savolainen P, Heiskala J, Kähkönen S. Excitation threshold of the motor cortex estimated with transcranial magnetic stimulation electroencephalography. Neuroreport. 2007;18(1):13–6.
178. Hernandez-Pavon JC, Veniero D, Bergmann TO, Belardinelli P, Bortoletto M, Casarotto S, et al. TMS combined with EEG: recommendations and open issues for data collection and analysis. Vol. 16, Brain Stimulation. Elsevier Inc; 2023. p. 567–93.
179. Deng ZD, Lisanby SH, Peterchev AV. Electric field depth–focality tradeoff in transcranial magnetic stimulation: simulation comparison of 50 coil designs. Brain Stimul. 2013;6(1):1–13.
180. Romero MC, Davare M, Armendariz M, Janssen P. Neural effects of transcranial magnetic stimulation at the single-cell level. Nat Commun. 2019;10(1):2642.
181. Ilmoniemi RJ, Virtanen J, Ruohonen J, Karhu J, Aronen HJ, Näätänen R, et al. Neuronal responses to magnetic stimulation reveal cortical reactivity and connectivity. Neuroreport. 1997;8(16):3537–40.
182. Bergmann TO, Karabanov A, Hartwigsen G, Thielscher A, Siebner HR. Combining non-invasive transcranial brain stimulation with neuroimag-

ing and electrophysiology: current approaches and future perspectives. NeuroImage. 2016;(140):4–19.
183. Miniussi C, Thut G. Combining TMS and EEG offers new prospects in cognitive neuroscience. Brain Topogr. 2010;22:249–56.
184. Tremblay S, Rogasch NC, Premoli I, Blumberger DM, Casarotto S, Chen R, et al. Clinical utility and prospective of TMS–EEG. Vol. 130, Clinical Neurophysiology. Elsevier Ireland Ltd; 2019. p. 802–44.
185. Hill AT, Rogasch NC, Fitzgerald PB, Hoy KE. TMS-EEG: A window into the neurophysiological effects of transcranial electrical stimulation in non-motor brain regions. Neurosci Biobehav Rev. 2016;64:175–84.
186. Kirschstein T, Köhling R. What is the source of the EEG? Clin EEG Neurosci. 2009;40(3):146–9.
187. Rosanova M, Casali A, Bellina V, Resta F, Mariotti M, Massimini M. Natural frequencies of human corticothalamic circuits. J Neurosci. 2009;29(24):7679–85.
188. Thut G, Miniussi C, Gross J. The functional importance of rhythmic activity in the brain. Curr Biol. 2012;22(16):R658–63.
189. Vallesi A, Del Felice A, Capizzi M, Tafuro A, Formaggio E, Bisiacchi P, et al. Natural oscillation frequencies in the two lateral prefrontal cortices induced by transcranial magnetic stimulation. NeuroImage. 2021;227:117655.
190. Veniero D, Gross J, Morand S, Duecker F, Sack AT, Thut G. Top-down control of visual cortex by the frontal eye fields through oscillatory realignment. Nat Commun. 2021;12(1):1757.
191. Premoli I, Bergmann TO, Fecchio M, Rosanova M, Biondi A, Belardinelli P, et al. The impact of GABAergic drugs on TMS-induced brain oscillations in human motor cortex. NeuroImage. 2017;163:1–12.
192. Thut G, Bergmann TO, Fröhlich F, Soekadar SR, Brittain JS, Valero-Cabré A, et al. Guiding transcranial brain stimulation by EEG/MEG to interact with ongoing brain activity and associated functions: a position paper. Clin Neurophysiol. 2017;128(5):843–57.
193. Assenza G, Capone F, di Biase L, Ferreri F, Florio L, Guerra A, et al. Oscillatory activities in neurological disorders of elderly: biomarkers to target for Neuromodulation. Front Aging Neurosci. 2017;9
194. Rogasch NC, Sullivan C, Thomson RH, Rose NS, Bailey NW, Fitzgerald PB, et al. Analysing concurrent transcranial magnetic stimulation and electroencephalographic data: a review and introduction to the open-source TESA software. NeuroImage. 2017;147:934–51.
195. Julkunen P, Kimiskidis VK, Belardinelli P. Bridging the gap: TMS-EEG from lab to clinic. J Neurosci Methods. 2022;369
196. Valentin A, Arunachalam R, Mesquita-Rodrigues A, Garcia Seoane JJ, Richardson MP, Mills KR, et al. Late EEG responses triggered by transcranial magnetic stimulation (TMS) in the evaluation of focal epilepsy. Epilepsia. 2008;49(3):470–80.
197. Shafi MM, Vernet M, Klooster D, Chu CJ, Boric K, Barnard ME, et al. Physiological consequences of abnormal connectivity in a developmental epilepsy. Ann Neurol. 2015;77(3):487–503.
198. Kimiskidis VK, Tsimpiris A, Ryvlin P, Kalviainen R, Koutroumanidis M, Valentin A, et al. TMS combined with EEG in genetic generalized epilepsy: A phase II diagnostic accuracy study. Clin Neurophysiol. 2017;128(2):367–81.
199. Kimiskidis VK, Kugiumtzis D, Papagiannopoulos S, Vlaikidis N. Transcranial magnetic stimulation (TMS) modulates epileptiform discharges in patients with frontal lobe epilepsy: a preliminary EEG-TMS study. Int J Neural Syst. 2013;23(1):1250035.
200. Hsu WY, Cheng CH, Lin MW, Shih YH, Liao KK, Lin YY. Antiepileptic effects of low frequency repetitive transcranial magnetic stimulation: a meta-analysis. Epilepsy Res. 2011;96(3):231–40.
201. Walton D, Spencer DC, Nevitt SJ, Michael BD. Transcranial magnetic stimulation for the treatment of epilepsy. Vol. 2021, Cochrane Database of Systematic Reviews. Wiley; 2021.
202. Gefferie SR, Jiménez-Jiménez D, Visser GH, Helling RM, Sander JW, Balestrini S, et al. Transcranial magnetic stimulation-evoked electroencephalography responses as biomarkers for epilepsy: a review of study design and outcomes. Vol. 44, Human Brain Mapping. Wiley; 2023. p. 3446–60.
203. Sarasso S, D'Ambrosio S, Fecchio M, Casarotto S, Viganò A, Landi C, et al. Local sleep-like cortical reactivity in the awake brain after focal injury. Brain. 2020;143(12):3672–84.
204. Luppi AI, Cain J, Spindler LRB, Górska UJ, Toker D, Hudson AE, et al. Mechanisms underlying disorders of consciousness: bridging gaps to move toward an integrated translational science. Neurocrit Care. 2021;35(S1):37–54.
205. Bodart O, Gosseries O, Wannez S, Thibaut A, Annen J, Boly M, et al. Measures of metabolism and complexity in the brain of patients with disorders of consciousness. Neuroimage Clin. 2017;14:354–62.
206. Arai N, Nakanishi T, Nakajima S, Li X, Wada M, Daskalakis ZJ, et al. Insights of neurophysiology on unconscious state using combined transcranial magnetic stimulation and electroencephalography: a systematic review. Neurosci Biobehav Rev. 2021;131:293–312.
207. Li X, Honda S, Nakajima S, Wada M, Yoshida K, Daskalakis ZJ, et al. TMS-EEG research to elucidate the pathophysiological neural bases in patients with schizophrenia: a systematic review. J Pers Med. 2021;11(5):388.
208. Kim HK, Blumberger DM, Daskalakis ZJ. Neurophysiological biomarkers in schizophrenia—P50, mismatch negativity, and TMS-EMG and TMS-EEG. Front Psych. 2020;11
209. Dhami P, Atluri S, Lee J, Knyahnytska Y, Croarkin PE, Blumberger DM, et al. Neurophysiological mark-

ers of response to theta burst stimulation in youth depression. Depress Anxiety. 2021;38(2):172–84.

210. Ferreri F, Guerra A, Vollero L, Ponzo D, Määtta S, Könönen M, et al. TMS-EEG biomarkers of amnestic mild cognitive impairment due to Alzheimer's disease: a proof-of-concept six years prospective study. Front Aging Neurosci. 2021;13

211. Maidan I, Zifman N, Hausdorff JM, Giladi N, Levy-Lamdan O, Mirelman A. A multimodal approach using TMS and EEG reveals neurophysiological changes in Parkinson's disease. Parkinsonism Relat Disord. 2021;89:28–33.

212. Zrenner C, Desideri D, Belardinelli P, Ziemann U. Real-time EEG-defined excitability states determine efficacy of TMS-induced plasticity in human motor cortex. Brain Stimul. 2018;11(2):374–89.

213. Baur D, Galevska D, Hussain S, Cohen LG, Ziemann U, Zrenner C. Induction of LTD-like corticospinal plasticity by low-frequency rTMS depends on prestimulus phase of sensorimotor μ-rhythm. Brain Stimul. 2020;13(6):1580–7.

214. Hussain SJ, Hayward W, Fourcand F, Zrenner C, Ziemann U, Buch ER, et al. Phase-dependent transcranial magnetic stimulation of the lesioned hemisphere is accurate after stroke. Brain Stimul. 2020;13(5):1354–7.

215. Yang C, Jung B, Lee SH. Transcranial electrical stimulation: clinical implication and practice for treatment of psychiatric illness. Clin Psychopharmacol Neurosci. 2023;22:391.

216. Miniussi C, Brignani D, Pellicciari MC. Combining transcranial electrical stimulation with electroencephalography: a multimodal approach. Clin EEG Neurosci. 2012;43(3):184–91.

217. Kanai R, Paulus W, Walsh V. Transcranial alternating current stimulation (tACS) modulates cortical excitability as assessed by TMS-induced phosphene thresholds. Clin Neurophysiol. 2010;121(9):1551–4.

218. Pogosyan A, Gaynor LD, Eusebio A, Brown P. Boosting cortical activity at beta-band frequencies slows movement in humans. Curr Biol. 2009;19(19):1637–41.

219. Siebner HR, Bergmann TO, Bestmann S, Massimini M, Johansen-Berg H, Mochizuki H, et al. Consensus paper: combining transcranial stimulation with neuroimaging. Brain Stimul. 2009;2(2):58–80.

220. Gebodh N, Esmaeilpour Z, Adair D, Chelette K, Dmochowski J, Woods AJ, et al. Inherent physiological artifacts in EEG during tDCS. NeuroImage. 2019;185:408–24.

221. Wunder S, Hunold A, Fiedler P, Schlegelmilch F, Schellhorn K, Haueisen J. Novel bifunctional cap for simultaneous electroencephalography and transcranial electrical stimulation. Sci Rep. 2018;8(1):7259.

222. Tashiro S, Siebner HR, Charalampaki A, Göksu C, Saturnino GB, Thielscher A, et al. Probing EEG activity in the targeted cortex after focal transcranial electrical stimulation. Brain Stimul. 2020;13(3):815–8.

223. Simula S, Makhalova J, Pizzo F, Garnier E, Damiani G, Mercadal B, et al. Impact of transcranial electrical stimulation on simultaneous stereoelectroencephalography recordings: a randomized sham-controlled study. Clin Neurophysiol. 2024;166:211–22.

20 New Perspectives in Electroencephalography (EEG) Analysis: From Functional Connectivity to Artificial Intelligence

Gian Marco Duma, Alberto Danieli, and Paolo Bonanni

20.1 Introduction

This chapter provides and overview of emerging perspectives in the advanced analyses of the electroencephalography (EEG) signal. These analyses offer new tools and insights for understanding the evolving applications of EEG technology. However, a full detailed description of mathematical, physical, and engineering background is beyond the scope of this chapter. Here, we will present sufficient information to construct a conceotual framework for understanding how novel EEG analytical methods are advancing knowledge in neurophysiology and neurocognition. While a detailed exploration of each topic is not feasible here, the intention is to stimulate further inquiry through the scientific literature. Methodology and applications are deeply interconnected; in fact, novel methods open the window to the possibility of investigating additional elements never explored before. At the same time, there is a deep commingling between basic research and clinical applications. Thanks to the basic research, we could infer how a specific system in certain conditions normally works. This can be translated into the identification of altered mechanisms in clinical conditions. However, the inverse could be also feasible. In this scenario, it is very hard, and probably not correct, to segregate novel methods, basic research, and clinical and technological application in different sections of the chapters. However, we will exploit the categorization bias in which mental categories help memorization and understanding [1]. To this aim, the chapter is organized as follows: (i) basic knowledge of advanced processing methods making the EEG a neuroimaging tool, (ii) EEG as a tool investigating large-scale brain dynamics, (iii) potential impact of novel EEG analysis methods on clinical practice, and (iv) EEG as mediating brain–computer–machine interface.

20.2 Electrical Source Imaging (ESI): The EEG as a Neuroimaging Technique

When thinking about EEG, the most direct mental image is a net of electrodes positioned on the scalp, where the difference of the electric potential between these electrodes, represented as a

G. M. Duma · A. Danieli · P. Bonanni (✉)
Scientific Institute IRCCS E. Medea, Unit of Epilepsy and Clinical Neurophysiology, Conegliano, TV, Italy
e-mail: gianmarco.duma@lanostrafamiglia.it; alberto.danieli@lanostrafamiglia.it; paolo.bonanni@lanostrafamiglia.it

F. Brigo, O. Mecarelli (eds.), *EEG: The First 100 Years*,
https://doi.org/10.1007/978-3-031-86878-8_20

function of time, is the so-called electroencephalogram [2]. The neurophysiology of the EEG generators has been extensively addressed in the previous chapters. However, as a quick reminder, we summarize that EEG electrodes are recording activity coming from postsynaptic (excitatory and/or inhibitory) potentials along the apical dendrites of the pyramidal cells of the gray matter, normally oriented to the scalp [3]. Indeed, the activity of a single neuron is not strong enough to pass through all the layers between the brain and the scalp. The electrodes record the summed activity of a large number of neurons which are synchronously electrically active [4]. Therefore, a single electrode is recording the summed activity coming from different sources of the brain. For this reason, the inference that a sensor is capturing the activity directly coming from the underlying cortex is essentially incorrect. The analysis of EEG waveforms has been performed for decades, helping clinicians and physiologists in the understanding of neuronal mechanisms. However, novel technologies are opening the possibility to exploit scalp-derived EEG to localize cortical generators and to investigate whole-brain dynamics. The process of the identification of the cortical generators from scalp EEG data is called source reconstruction or electrical source imaging (ESI) [5]. Given the resemblance of the neuronal assemblies to the electric circuits, important improvements of the understanding of the brain behavior were possible by translationally applying physics and engineering knowledge to describe and predict neuronal activity. Let us assume a classic problem of physics: given an electrical generator inside a surface, we need to calculate the electrical field in every point of the surrounding surface. This can be done by applying the Poisson's equation [6] which provides a solution linking the potentials at any position in a volume conductor and the applied current sources. The conductor volume would be the head, including all the layers, while the current source would obviously be the brain. In this scenario, we could compute a priori the electric potential, derived by the brain activity, recorded by each electrode positioned in any point of the scalp. This is the so-called forward problem. At first, between the brain and the electrodes, there are multiple layers mitigating and deflecting the electric field of the neural activity. Each layer has a different level of conductivity. Second, the shape of the brain is not homogeneous but is characterized by sulci and gyri. Third and probably the most important, the brain has an incredible number of potential generators (neurons); how many of them should be modeled?

The first and easiest volume conductor models of the human head and representation of the brain as a generator consisted of a homogeneous sphere [7]. However, the skull tissue has a lower conductivity as compared to that of scalp and brain tissue. As a possible solution, the three-shell concentric spherical head model was introduced. In this model, the inner sphere represents the brain, the intermediate layer represents the skull, and the outer layer represents the scalp. The approximation of the brain and head to a sphere is far from capturing the real shape of the brain and therefore the real processes generating the electric field. For this reason, novel forward models were introduced, going by the name of *realistic head models*. These models take into account the proper shape of the brain and the head by creating a surface representation composed of small triangles. This triangular surface is technically called a mesh (see Fig. 20.1).

One of the most renowned realistic method is the boundary element method (BEM), [8] which calculates the field of the three surfaces: brain–skull interface, skull–scalp interface, and the outer surface, with the limitation that each tissue should be considered isotropic, namely, the conductivity is equal in each of its points. A more precise but computational expensive method is the finite element method (FEM) [9] which uses a tetrahedral mesh with five layers, namely, scalp, skull, white matter, gray matter, and cerebrospinal fluid. This method does not assume isotropic conductivity in the layers, and it represents one of the most precise methods. A fundamental step for the model precision is to generate the mesh of the head and brain surfaces using the individual anatomy derived from T1 MRI [10]. The process of layer extraction (skull, white matter, gray matter) is defined *segmentation* [11]. Additionally, it is

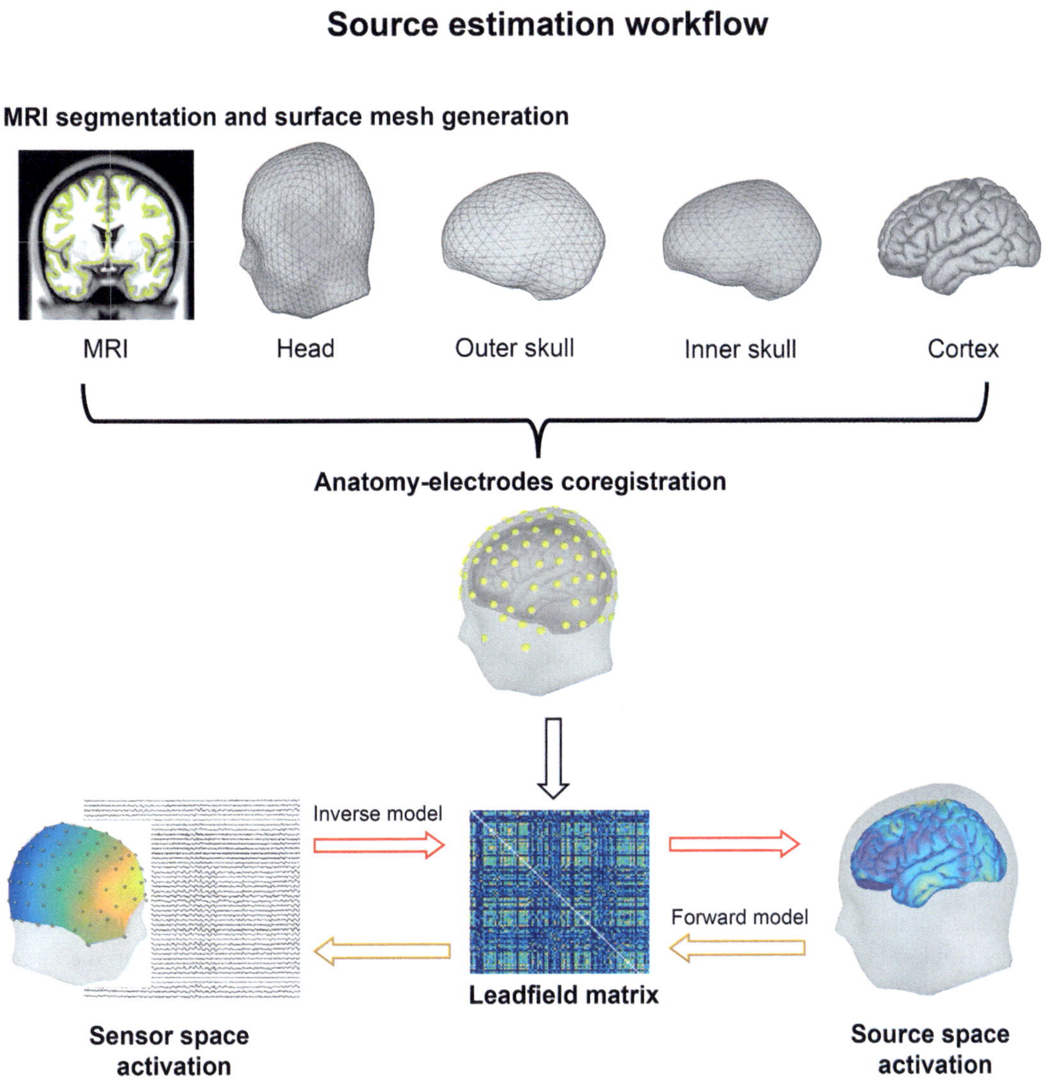

Fig. 20.1 Electrical source imaging analysis workflow. The present figure illustrates the analysis workflow necessary to perform electrical source imaging. The upper panel displays the result of the segmentation process of the T1 MRI and the derived triangular mesh generation of the necessary layers to solve the forward model. The lower panel illustrates how it is possible, thanks to the leadfield/gain matrix, to pass from the sensor to source space in order to estimate the intensity of the activation of the cortical dipoles

fundamental to digitize the EEG cap electrode position on the individual head and overlap it to the individual anatomy, in line with the referential points, namely, inion, nasion, left and right ear, anterior and posterior commissure, and inter-hemispheric point. This process is called coregistration [12]. We will see later in this section, the importance of using individual anatomy for the precision of the source estimation. For a full explanation of the mathematical modelling involved in the forward model, please see Hallez et al. [13].

Once the forward model is computed, this information can be used to calculate the *inverse model*, which will provide the estimation of the activation intensity and the position of the dipoles generating the EEG signal.

The inverse problem is often defined as an ill-posed problem due to the nonuniqueness of its solution. In fact, a large number of brain activity configurations could generate the same pattern of the electric field recorded at the sensor level. Therefore, some constraints or assumptions are necessary to converge to a unique solution explaining the data.

In this light, two source estimation approaches can be identified: dipole fitting methods and distributed sources.

In the dipole fitting, the position and amplitude of one to a few equivalent current dipoles (ECD) are estimated over relatively short time windows. In contrast, in the distributed models, the location (and typically, the orientation) of a large number of dipoles is fixed, and it is modeled as the brain mesh described before. The brain is divided into a fixed number of dipoles. The larger the number, the more the model precision. These models require the estimation of the amplitude of a vast number of dipoles (in the order of tens of thousands) in a fixed grid (dipole position and orientation) at each time point [14]. A trade-off between a good representation of the brain and the computational and data storing costs is necessary. Modern distributed methods use between 8 and 30 k dipoles, namely, the vertices in the mesh surface (see Fig. 20.1).

The dipole fitting approaches can be adequate when the number of brain regions expected to be active is relatively small. However, they are not adequate to capture complex dynamics over extended periods of time (resting-state recordings or event related epochs). In contrast, distributed source models, which are less user dependent, can be flexibly applied to all experimental conditions. They yield a time-resolved solution, with the corresponding source activation each time point. This allows using the source activity information similarly to a functional magnetic resonance imaging (fMRI). Some of the most famous inverse models are the weighted minimum norm estimate (wMNE) or low-resolution electromagnetic tomography, better known as LORETA. However, there are a multitude of methods present in literature, and this is not the correct forum to address the details of each of them. We recommend for a more in-depth study of the solutions of the inverse problem external and dedicated reviews [15–17]. Despite the numerous methods available, from a theoretical perspective, the distributed approaches are able to uplift the potential use of the EEG to a neuroimaging tool, providing information about the cortical dynamics and their spatial location in the gray matter. Moreover, differently from the fMRI, the EEG signal is a direct measure of the electrical activity, strictly linked to a precise underlying physiology. For this reason, it could provide more precise characterization of the neural processes underlying the recording. However, there are some limitations in the usage of the EEG as a neuroimaging tool. At first, the spatial resolution of ESI is strictly related to the number of electrodes of the EEG. High-density montage (64, 128, 256 electrodes) is required for a reliable estimation of cortical generators. Indeed, the ESI spatial resolution increases as a function of the number of electrodes, even though the difference between 128 and 256 electrodes montage remains limited [18]. Moreover, the spatial localization is influenced by the usage of the individual anatomy [19]. Finally, the most important limitation is the lower capability of ESI, with respect to fMRI, of identifying activity in the deep brain areas (e.g., hippocampus). These structures have a distinct neuronal architecture which may cause them to produce weaker signals in comparison to cortical structures [20, 21]. Moreover, the larger distance of deep areas from the recording sites attenuates the activity detection by the electrodes [22]. Nonetheless, recent evidence suggests that novel methods are capable of detecting subcortical activity with high-density EEG improving therefore the spatial resolution of EEG-derived source activity [23, 24]. The new methods and also the development of computational optimized open tools based on MATLAB (e.g., EEGLAB, BrainStorm, and FieldTrip) [25–27] or Python (e.g., MNE) [28] for ESI are opening new venues of application of EEG in the research and clinical world, increasing exponentially the understanding of the brain as a dynamical system. In fact, having an estimated time-resolved cortical activation allows the study of the functional interaction between brain regions. In the next chapter, we will provide a general overview of the vast land of functional connectivity, which will provide the basis for the understanding of the novel venture of EEG-derived technologies.

20.3 Foundation of Functional Connectivity with EEG

It is not easy to condensate in one explanatory sentence the vast concept of functional connectivity. Broadly speaking, the functional connectivity (FC) in neuroscience investigates the relationship/synchronization in the activity between neural elements from micro (cellular level) to macro (brain regions) scales [29]. However, ESI spatial resolution is ~1.1/1.5 cm [30, 31]; therefore, it is only possible to make reliable inference on large-scale FC. The FC can be also conceptualized as a statistical interdependence between the dynamic activity of neural units that in the framework of the EEG can be ascribed to electrodes or cortical areas of the source reconstructed activity. Based on the definition proposed, FC is not able to track the direction of the information flow across elements. The allowed inference is to establish the presence and the strength of a relationship between regions, without knowledge on the type of the interaction. Conversely, effective-FC, which is based on autoregressive multivariate (e.g., granger causality) or mechanistic models (e.g., dynamic causal modeling), includes the directionality of the interaction elucidating about the underlying mechanisms of information flow in the brain [32]. In the framework of noninvasive investigation like the EEG, the strength of FC characterizes the brain functioning at large scales. From this point, the brain can be conceptualized as a network. A network is defined by interacting elements, namely, the nodes and their connections or edges. The FC can be computed at the scalp level, using the electrodes as nodes. However, limited knowledge on the brain functional organization can be derived from scalp-based FC. In fact, the statistical interdependence between two electrodes can be driven by the fact that they are recording the same cortical region whose activity, due to volume conduction, is synchronously represented in two or more electrodes, especially if spatially contiguous [33]. The computation of FC at the electrode level can be performed under certain conditions and interpretative limitations that will be later discussed. In fact, if the research purpose is to generate inference on the functional organization of the brain, it is recommended to perform FC on source-derived activity [34]. The ESI represents a spatial deconvolution separating the contribution of each generator to the scalp-recorded EEG. However, the larger dimensionality of source activity compared to scalp-EEG posits a computational caveat. The FC is usually stored into a matrix of the N × N dimensions named as adjacency matrix, where N represents the number of the nodes. If we assume a cap of 128 electrodes, a scalp-EEG adjacency matrix will be 128 × 128, which does not represent a computation nor a storing problem. On the other hand, let us assume a cortical mesh of 15k vertices (dipoles). If we want to compute whole-brain connectivity, we should compute a 15k × 15k connectivity matrix, resulting in the storing, for a single recording, of 225 million elements (see Fig. 20.2a). The generalization of this approach to the study of a large population becomes extremely complicated. To solve this issue, usually, the source activity is spatially downsampled to a brain atlas prior FC analysis. The brain atlases parcel the brain into separated regions, based on cytoarchitectonic or functional organization. Each vertex/voxel of the brain surface/volume after the segmentation is attributed to a specific region of the atlas. Let us exploit one of the most famous atlas, namely, the Desikan–Killiany. This parcellation separates the brain into 68 cortical regions. The activity of a specific region can be obtained, for example, by averaging the time-series of the dipoles contained into a specific parcel. This would result in a reduction from 15k time-series, one for each dipole, to 68 time-series, one for each region. The computation of the FC would now result in a 68 × 68 adjacency matrix, solving the computation and storing problem (see Fig. 20.2b). Nonetheless, recent finding are developing methods to handle the large dimensionality of the full data matrix to compute connectivity prior the downsampling to an atlas, supporting that this procedure result in

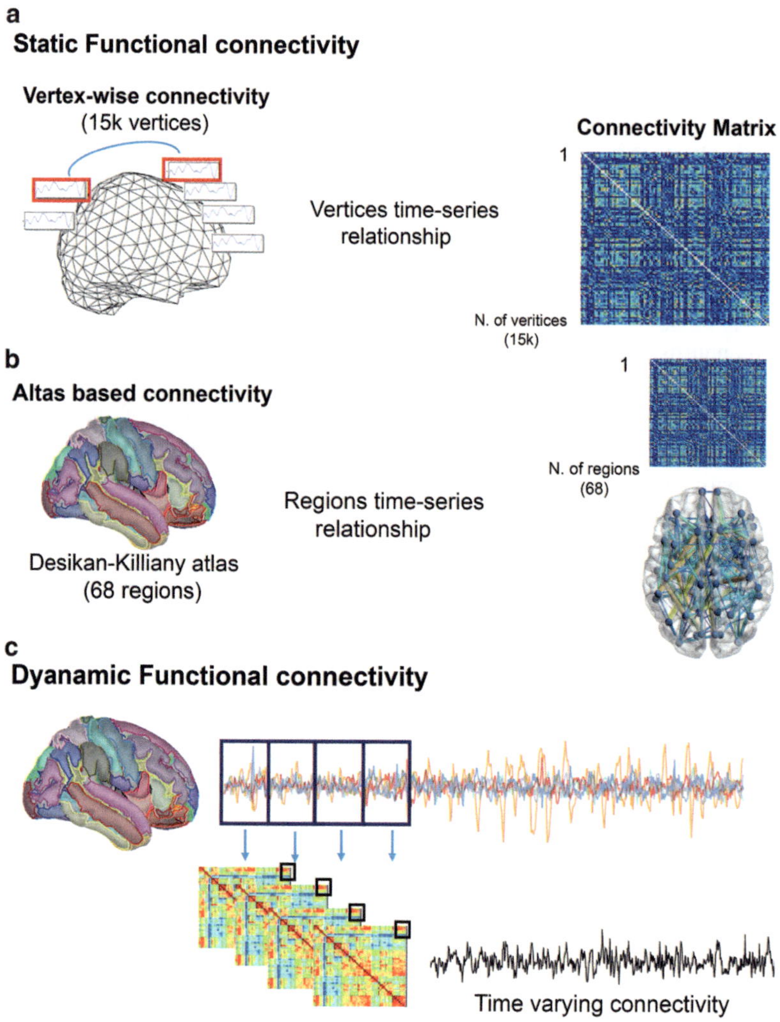

Fig. 20.2 Functional connectivity computation. The present figure provides a graphical representation of the basic principle of functional connectivity, from time-series to the connectivity/adjacency matrix. Panel A indicates how to compute connectivity across all the dipoles, resulting in number of dipoles × number of dipoles connectivity matrix. In the example, a mesh of 15k vertices was used, resulting in connectivity matrix of 225 million of entries. Panel B shows the region-based connectivity derived from a previous downsampling of the data to a specific brain parcellation. The figure illustrates the example of the Desikan–Killiany atlas with 68 regions, thus resulting in region × region connectivity matrix. Panel C shows the basic principle of dynamic functional connectivity, derived by a sliding window approach, resulting in multiple connectivity matrices. By considering a specific edge across all the matrices, it possible to trace the dynamic of the time-related interaction across two brain regions

more precise computation of FC [35]. Different types of atlases may be applied; however, in the light of the ESI spatial resolution, the usage of fine-grained atlas with more than 200 regions is not recommended.

Until now, the FC is a portrait of the relationship between brain regions, compressing into a single adjacency matrix the complex dynamics of interactions. However, neural oscillatory activity is the byproduct of the brain as a dynamical system, namely, changing over time. From this perspective, the interactions among regions will adapt to this time-varying nature of the cortical functional organization. Encapsulating this rich and complex dynamic into a single matrix is an oversimplification of the neuronal oscillatory phenomena. For this reason, methods of dynamic functional connectivity (dFC) were developed. Usually, these analyses use the concept of the sliding window. The signal is segmented into short pieces by applying a temporal window to the signal, moving from the beginning to the end of the recording of interest. For each movement of the window, an FC is computed, thus obtaining a time-varying strength of the connection between two nodes [36]. This approach results into a reduced time resolution based on the length of the window applied. Neuronal activity is a rhythmic variation over time of the voltage, which can be decomposed into sinusoidal element signals. A generic oscillatory activity is characterized by a frequency (number of oscillations in a specific time unit) and phase (angle-like quantity representing the fraction of the cycle covered up to time quantity). Additional methods leveraging the extraction of instantaneous phase allow the computation of the dFC with the same time resolution of the original signal [37]. Overall, a time-varying functional connectivity provides information about how the brain regions interact with each other across time and the patterns of their interaction (see Fig. 20.2c).

The purpose of this chapter was to present the conceptual foundation of functional connectivity. We purposely did not address all the possible metrics to compute FC and dFC. For a general overview of connectivity metrics, we refer to a further seminal article [29]. Instead, here, we focused on giving enough information to understand what it lurks behind the novel application of the EEG as a potential functional neuroimaging tool.

20.4 New Perspectives in Clinical Application of EEG

Previous chapters of this book extensively discuss the electrical nature of the signal that the EEG is recording. In this perspective, the most natural and direct application in the clinical world is to study populations characterized by the alteration of the electrical activity of the brain. In this scenario, one of the most renowned and diffused clinical conditions characterized by a dysregulation of the electrical brain activity is epilepsy. The EEG is fundamental for the epilepsy diagnosis and to identify signatures of clinically relevant electrical activation [38]. A difficult clinical question during the diagnostic process in epilepsy is the identification of the brain regions driving the electrical alterations. The research in the last decades has started the difficult quest of the localization of the epileptogenic zone (EZ) [39]. The progress of the ESI, discussed in the previous sections, opened a new application of the EEG as a noninvasive neuroimaging tool potentially able to identify the EZ.

20.4.1 EZ Identification via EEG-Derived ESI

The noninvasive identification of the EZ is a multidisciplinary question involving clinicians, mathematicians, physicists, and engineers. The first problem is what part of the EEG signal can be source reconstructed. The EEG signature of seizures and interictal epileptiform discharges (IEDs) allow for a crude estimation of the seizure onset zone (SOZ) and the onset zone of IEDs, which is termed the irritative zone (IZ). Both the SOZ and, to a lesser extent, the IZ are taken as a surrogate for the conceptual EZ. Overall, despite the widely available ictal EEG recordings, in most patients, noninvasive presurgical evaluation using ESI relies on interictal epileptic activity. Several works focused on the source activation of IEDs or high-frequency oscillations (HFO) as a proxy of the identification of the EZ [40–44]. However, the IED activity can be divided at least in three parts: onset, half-rise, and peak. A relevant question is which part of the IED provides larger information about the generating source. Converging evidence showed that the peak may represent the most representative point for an accurate localization of the source generators [43, 45]. Despite the larger volume of literature using IED-reconstructed activity, certain kinds of seizure, those in which patients do not produce wide and evident movement, can be successfully reconstructed via scalp EEG. This may provide the identification of the SOZ as a more direct proxy of the EZ in comparison to IZ [46]. While from a technical point of view, source estimation of the SOZ and the IZ can be efficiently performed under certain constraints, there is still one important question. How can we determine the precision of these methods? The only possible solution in humans are retrospective studies. Usually, the location of the ESI and the surgical resection is compared and related to the Engel class (surgery success) after the surgery. If the ESI localizes the source in the surgical hollow, in relation to seizure-free patients, it is possible to assume a positive outcome of the ESI. The last decades of research highlighted the fundamental roles of high-density EEG montage caps in an accurate estimation of the source as well as the usage of the individual anatomy to increase spatial resolution of the source reconstruction. In general, assuming the best possible configuration with a high-density montage, the individual MRI, and the sensor digitized location, the spatial precision in the reconstruction of is around 1.1/1.5 cm [30]. Among the multiple source inversion algorithms of the distributed methods, the novel nonlinear technique named maximum entropy on the mean provided the best spatial precision in the estimation of the source localization [47, 48]. As discussed, distributed methods use prior assumptions on the number and position of the dipoles. A recent method, conceptually similar to equivalent current dipoles, exploited a model named successive multivariate variational mode decomposition (SMVMD) in order to estimate the source without knowing position or their expression on the scalp [49]. This method outperformed linear distributed methods like LORETA or previous equivalent dipoles source techniques. In general, in this section, we observed that novel EEG analysis methods are increasing the successful application of ESI in the localization of cortical epileptiform activity generators with the purpose to inform surgical intervention. However, epilepsy is probably one of the best models for the theoretical conceptualization of a network disease. In the next chapter, we will show how the brain dynamics embedded in EEG-derived networks are able to inform more precise models reaching the level of the individualized medicine.

20.4.2 Functional Connectivity in the Understanding of Altered Neuronal Behavior in Clinical Conditions

20.4.2.1 Epilepsy

The rate of success of the epilepsy surgery ranges from 47% to around 65%, in relation to the clinical characteristics of the patients [50, 51]. It means that a large percentage of patients are not seizure free, neither after the resection of the presumed EZ. Nowadays, epilepsy is considered a network disorder [52]. In this perspective, the removal of the EZ may not be sufficient since the dynamic of the rest of the network may be still dysregulated, in relation to the presence of remaining perturbing nodes. In this scenario, the EEG-derived functional connectivity can provide multilevel information: (i) identification of etiopathological relevant mechanisms, (ii) relation between central nervous system organization and neurocognitive functioning, and (iii) identification of the perturbing nodes to be removed or about how the seizure phenomena spread in the brain network with the possibility of implementing less invasive surgical intervention like disconnections. To this purpose, both network configuration during epileptiform phenomena (i.e., seizures or IEDs) and intrinsic functional organization of the brain system (resting-state activity) can be considered informative. Ding and colleagues [53] investigated functional connectivity during an ictal period, showing that the EEG source reconstructed areas with highest outgoing information flow has the largest concordance with MRI lesions or hyperperfusion seen in ictal SPECT, within a precision of 15 mm. These results suggest that the inclusion of EEG-derived network information together with EEG source reconstruction is able to provide a more informative perspective as compared to ESI alone. This point of view is supported by evidence derived by one of the largest cohorts of ESI during the ictal period. More recent investigation, by including ESI and ESI+FC of 111 seizures from 27 patients, demonstrated that the inclusion of the strongest outgoing connection from the estimated SOZ during an ictal phenomena, estimated using EEG source-derived functional connectivity, increases SOZ identification accuracy of around 40% as compared to the solely performed ESI [54]. Functional connectivity may be dynamically applied to track network configuration changes across time. This was done from a recent work performing ESI on the averaged IEDs recorded with scalp high-density EEG. The regions expressing the activation amplitude peak and additional surrounding regions were extracted from the ESI map [55]. By applying time-resolved FC, it was observed that the IZ is a perturbing hub of the network, displaying an increase and decrease perturbation behavior time-locked to the ascending, peaking, and descending phase of the IEDs. Conversely, the surrounding regions did not display the same IED-related behavior. This method has a large potential to inform the surgical decision tree, using noninvasive information. Taken together, these findings suggest that the inclusion of FC during both ictal and interictal activity can be more informative of the exclusive source identification. However, seizure and spikes can be considered relatively rare events. For this reason, inferring information on the epileptogenic network from the basal brain functional configuration represents an important technical advantage since it does not require intensive data management related to long monitoring necessary for seizure and spike recordings. Recent works generalized the concept of neuronal avalanches, to the investigation of resting EEG activity in epilepsy [56]. The usage of neuronal avalanches (NA) was applied to investigate the nonlinear brain dynamic at large scale. This measure in fact is able to provide a detailed portrait of the diffusion of the activation bursts across regions, which can be therefore conceptualized as a novel measure of functional connectivity. The strength of this measure is that it can be applied to the raw signal without requiring intensive preprocessing or stationarity assumptions like power-or-phase-based connectivity measures. This novel approach with NA was able to identify in patients with temporal lobe epilepsy regions related to the seizure ignition and propagation. Relevantly, this was performed in a resting condition with no epileptiform activity (i.e., seizures or spikes). The clinical relevance of neuronal avalanches was also corroborated showing a coupling with brain morphological network in epilepsy and its link to the pharmacological treatment (number of anti-seizure medications) [57]. Despite the potential relevance in clinical application, inferring the localization of EZ from brain intrinsic organization is a challenging quest because it requires a high level of tailoring of the connectivity approach. To this purpose, a section of the Human Brain Project, the largest European brain project, ended after 10 years on September 1, 2023, dedicated an intensive effort to generate the highest level of personalization in brain research in epilepsy. This resulted in the virtual epileptic patient, which can be considered a first attempt to virtualize the functional and structural organization of an individual patient for EZ estimation [58]. This approach uses a generative model of brain activity, namely, the Epileptor, [59] which is able to simulate the transition from a pre-ictal to an ictal state and the return to the baseline. This is performed using the patient's specific brain shape and white matter fibers, to simulate how the activity spread in an individual anatomy. The present method allows to virtually test the clinical hypotheses of localization of the EZ. In fact, from the generative model of neural activity and the individual anatomy (shape and connections), it is possible to simulate both intracranial and scalp EEG. After the EEG/sEEG simulation, throughout statistical Bayesian approaches, the regions with the largest probability of being the epileptogenic or the propagation zones are estimated [60, 61]. This groundbreaking approach allows the virtualization of the patient opening the possibility of testing of different therapeutic approaches like region ablation or disconnections prior to the real application on the patients [62, 63]. In this perspective, this approach represents one of the highest points of personalized medicine and one of the most direct application of EEG-derived connectivity in clinical scenarios [64].

20.4.2.2 Machine Learning and Seizure Classification and Prediction

Epilepsy represents probably one of the most self-evident scenarios where EEG-derived information is fundamental to manage both the diagnosis and treatment. Until now, we showed how new analyses methods are assisting noninvasive identification of epileptogenic networks, reaching a new level of personalization. However, a fundamental problem for patients' daily life is represented by the stochastic-like nature of seizure occurrence. Indeed, chronobiology has identified potential circadian and annual rhythms of seizure occurrence [65]. Nonetheless, this kind of understanding is not accurate yet to forecast when an ictal event will manifest. The translational application of artificial intelligence may represent a potential solution to this problem. From the machine learning (ML) point of view, this is a twofold problem. First, a model necessitates to automatically recognize a seizure from the baseline activity. Secondly, in order to be useful in a real-life condition, it has to forecast when a seizure is going to occurrence. While both of the problems are complex, the solution to the seizure automatic recognition is actually more advanced as compared to the seizure forecasting. Importantly, the research efforts are directed to the generalizability of these methods in a user-friendly configuration. For this reason, most of the researchers solely used the scalp EEG without the necessity of reconstructing the source estimation. This scenario represents the perfect condition where to use FC at the scalp level. In fact, scientists in this research field simply use the FC as an informational feature helping the model learning.

Seizure automatic identification of epileptiform activity is a helpful tool in the diagnosis process. The misdiagnosis rate of epilepsy is around 30–40%, depending on the type of the specialization of the center and type of epilepsy [66, 67]. In this context, medical doctors may need support in the diagnostic process, and the automatic identification of epileptiform activity in the EEG signal represents an important advancement [68]. The scientific production on this topic has increased tremendously in the last decade. However, despite the large effort, very few studies were characterized by a solid methodological background with a large sample and the real potentiality of application in clinical scenarios. At first, a recent review evidenced that seizure detection related to wearable devices like EEG are reliable in the limited scenario of tonic–clonic seizures (GTCS) and focal-to-bilateral tonic–clonic seizures (FBTCS), limiting the theoretical generalizability of these methods [69]. The latest AI model for epileptiform activity detection named Standardized Computer-based Organized Reporting of EEG–Artificial Intelligence (SCORE-AI) [70] have proved to achieved human expert-level performance in fully automated interpretation of routine EEG, with potential and direct application in the patient's care. At the moment, the latter study represents the largest sample size trained for epileptiform activity recognition, and no seizure forecasting studies have reached such a large sample size. A potential explanatory factor is that seizure and epileptiform activity automatic recognition is less complex as compared to seizure forecasting. Several efforts have been made in temporal prediction of seizure occurrence. Studies exploiting functional connectivity have reported limited predictive performance from indices related to global network configuration like functional segregation, integration, or robustness [71, 72]. However, the dynamic reconfiguration of functional organization resulted to be a more sensitive measure increasing the success rate of seizure forecasting [73, 74]. These results suggest that complex spatial–temporal interactions in the large-scale epileptic brain network represent an important measure for seizure forecasting capturing features that would be missed by indicators designed for simple one-dimensional systems (see Fig. 20.3). Overall, EEG-derived source localization and functional connectivity may be beneficial in the identification of functionally altered areas (e.g., the epileptogenic zone) and cortical network alterations. While epilepsy is the most representative

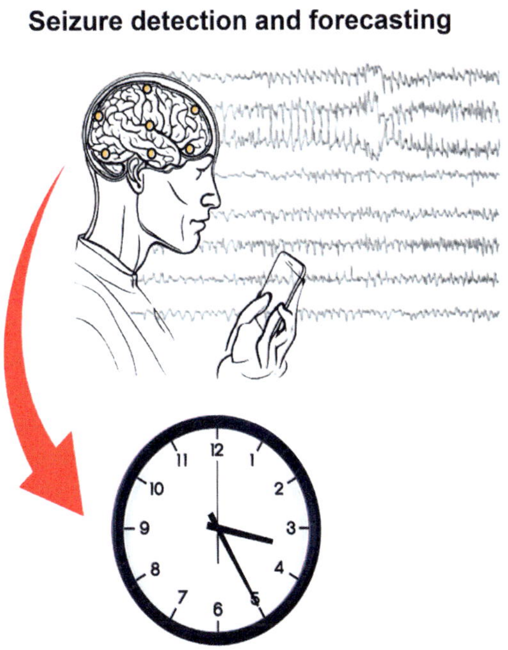

Fig. 20.3 Seizure detection and forecasting. The present figure is a graphical representation of the general concept of seizure detection and/or forecasting based on the EEG signal

condition in which EEG is applied, source activity and connectivity methods may be translated to several clinical conditions, for example, stroke or Alzheimer and Parkinson's diseases [75, 76].

20.4.3 EEG for Cognitive Phenomena Investigation

The EEG is characterized by high temporal resolution at the millisecond scale, and together with the improvement of the ESI methods, it allows a temporal fine-grained investigation of the neural organization scaffolding cognitive functions. The link between neural activity and cognition is a fascinating topic which has driven the last decades of cognitive neuroscience. By the investigation of EEG time-locked activity to specific stimuli, it is possible to track the time-dependent evolution of the EEG reflected in the event-related potentials (ERP). The ERP are obtained as an average of multiple EEG activity segments, time-locked to a specific stimulus and condition [77]. The neuroscientific literature linked brain waves morphological patterns with specific information processing steps. This allowed, for example, to trace the brain areas engagement in the memory creation, maintenance, and information recall, [78–80] as well as attention and cognitive control processes [81–83]. The evolution of the network neuroscience has improved behavior prediction by including spatiotemporal organization of brain connectivity. Recent evidence suggests that the intrinsic organization of the brain, namely, when it is not directly involved in stimulus processing, is able to predict the performance of multiple cognitive domains [84–87]. While this research topic was mainly driven by functional magnetic resonance (fMRI), the latest decade of research has generalized this analytical approach to the EEG derived FC [36, 88–90]. The investigation of the link between human behavior and neural organization is often reduced to computerized tasks in a laboratory setting, lacking ecological validity. From a theoretical and scientific perspective, human behavior is not uniquely the results of intrinsic neural activity but also of the environmental contest to which neural activity adapts. Therefore, it is very hard to efficiently predict human cognitive functioning by including exclusively brain activity. In fact, recent evidence suggested that brain–behavior correlation reaches around 35% when also including thousands of participants in the sample, resulting in more than 60% of unexplained variance of human behavior from central nervous system activity [91]. The brain is part of a more complex system, namely, the human body, which is surrounded by an environment. Therefore, the future of cognitive neuroscience will be linked with the inclusion of brain activity together with peripheral nervous system information as well as the investigation in an ecological setting [92–96]. In this light, EEG, thanks to its cost-effectiveness and portability, represents the election tool for the future investigations of brain–behavior link.

20.4.4 EEG in Brain–Computer Interfaces

One of the most recent applications of EEG-derived information are the brain–computer interfaces (BCIs). BCIs translate brain signals into commands in real time for control or communication with external devices [97]. In this scenario, BCI are characterized by a large applicability from rehabilitation to brain-based communication with assistive technology like wheelchairs. Broadly speaking, BCIs enable the brain-based communication with external interacting tools, which is indeed a futuristic venture of new technological advancements of neuroscience application. The efficient functioning of BCIs depends on the capability to comprehensively characterize brain functioning and correctly identify the user's mental state. To perform this sort of "mental reading" task, an important effort from the scientific community has been focused on improving the classification algorithms. Automatic learning processes (i.e., machine/deep learning) are indeed supporting BCI usability [98]. However, it is also fundamental to identify and to optimize the EEG signal features to be used. In fact, the growing evidence of the network organization of the brain activity suggest the improvement in the mental state decoding can be performed including the whole-brain functional configuration. There is an impressive body of research regarding EEG feature selection and decoding algorithms for BCI optimization. The present chapter will be limited to providing application examples of BCI, with special attention to clinical ones. For a most comprehensive review of BCI-related methods, we suggest external reviews [99, 100].

20.4.4.1 BCI in the Context of Rehabilitation

One of the largest applications of BCI is probably related to stroke rehabilitation [101]. In fact, after a stroke most patients require daily assistance to recover daily life functioning. Motor imagery (MI)-based BCI represents an alternative form to stimulate motor-related brain pathways [102]. MI-BCI refers to the act of imagining specific motor patterns without actually executing it. The theoretical background is that MI and motor execution rely on similar neural structures and processes [103, 104]. Importantly, MI can affect neuroplasticity changes after a period of training both in healthy and clinical populations, [105, 106] supporting the idea of using MI-BCI in rehabilitation. A recent meta-analysis showed that MI-BCI rehabilitation induced a larger increase of upper limb motor functioning, as compared to other therapies. In the theoretical framework of the brain as a network, MI-BCI rehabilitation produced changes in ipsilesional intra-hemispheric connectivity [107] as well as prefrontal area connectivity. Relevantly, the connectivity changes were related to motor outcome [108]. The detection of MI-dependent brain activity patterns is the mechanism supporting the mind-driven wheelchairs. The waves of activity recorded by the EEG are decoded by artificial intelligence algorithms and translated into movement of the wheelchair [108]. These represent the same mechanisms supporting the exoskeleton-mediated rehabilitation processes, in which movement intentions are decoded from the EEG and translated into an action of the exoskeleton supporting movement reinforcements after stroke, cerebral palsy, or other disorders requiring rehab processes [109] (Fig. 20.4).

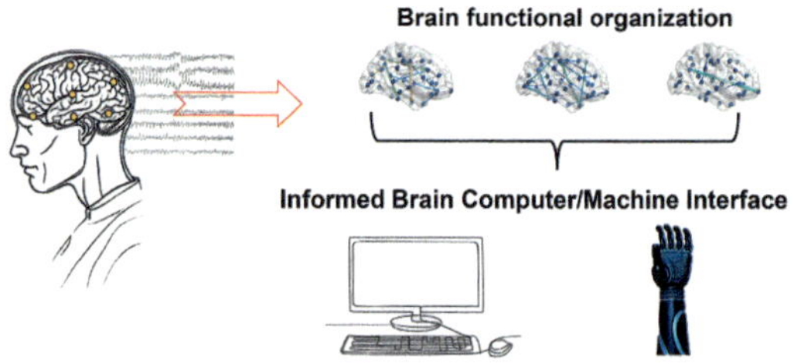

Fig. 20.4 Cortical dynamics-informed brain–computer interface. The figure represents an example of how from the EEG signal it is possible to perform electrical source imaging (ESI) and the ESI-derived connectivity, tracing whole-brain dynamics. These feature can be used to inform brain–computer–machine interface, for example, a robotic arm that can be used in rehabilitation

20.5 Conclusions

In this chapter, we showed how the novel advancement in the electromagnetic signal data processing evolved the usage of the EEG as a neuroimaging tool helping the investigation of the cortical brain dynamics and functional organization. Despite limitations in the spatial resolution of source identification, the usability, portability, and cost-effectiveness makes this tool highly flexible, for multipurpose application. In fact, as previously shown, EEG can be used in elucidating neural mechanisms in the clinical conditions, helping the identification of potential biomarkers. Additionally, it can be used to track the evolution of brain activity configuration scaffolding cognitive functions in healthy individuals. The dawn of AI enlarged the spectrum of the application of the EEG to the human–machine interactions via noninvasive decoding of brain states. Futuristic applications of EEG-derived signals have already been proposed, for example, in brain-controlled vehicles [110]. The introduction of Internet of Things is shaping the technology, increasing the possibility of remote control of a large set of tools. In this perspective, the EEG represents an election tool with ecological validity, helping to remotely connect humans and machines. We are aware we did not address every topic in depth. In fact, we aimed at providing an overview of the new application and evolution of EEG. In a dynamic world, it is difficult to predict all the potential evolutions. The information we provided in this chapter is intended to support navigation through a complex and broad topic such as EEG signal application.

References

1. Vlach HA. How we categorize objects is related to how we remember them: the shape bias as a memory bias. J Exp Child Psychol. 2016;152:12–30. https://doi.org/10.1016/j.jecp.2016.06.013.
2. Buzsáki G, Anastassiou CA, Koch C. The origin of extracellular fields and currents — EEG, ECoG, LFP and spikes. Nat Rev Neurosci. 2012;13(6):407–20. https://doi.org/10.1038/nrn3241.
3. Misulis KE. Generators of EEG activity. In: Abou-Khalil B, Misulis KE, Sonmezturk H, et al., editors. Atlas of EEG, seizure semiology, and management. Oxford University Press; 2022. https://doi.org/10.1093/med/9780197543023.003.0005.
4. Niedermeyer E, da Silva FHL. Electroencephalography: basic principles, clinical applications, and related fields. Lippincott Williams & Wilkins; 2005.
5. Michel CM, Brunet D. EEG source imaging: a practical review of the analysis steps. Front Neurol. 2019;10 https://doi.org/10.3389/fneur.2019.00325.
6. Helmholtz H. Ueber einige Gesetze der Vertheilung elektrischer Ströme in körperlichen Leitern, mit Anwendung auf die thierisch-elektrischen Versuche (Schluss.). Ann Phys. 1853;165(7):353–77. https://doi.org/10.1002/andp.18531650702.
7. Frank E. Electric potential produced by two point current sources in a homogeneous conducting sphere. J Appl Phys. 1952;23(11):1225–8. https://doi.org/10.1063/1.1702037.
8. He B, Musha T, Okamoto Y, Homma S, Nakajima Y, Sato T. Electric dipole tracing in the brain by means of the boundary element method and its accuracy. IEEE Trans Biomed Eng. 1987;BME-34(6):406–14. https://doi.org/10.1109/TBME.1987.326056.
9. Johnson CR. Numerical methods for bioelectric field problems. In: The biomedical engineering handbook; 1995. p. 162–80.
10. Brodbeck V, Spinelli L, Lascano AM, et al. Electroencephalographic source imaging: a prospective study of 152 operated epileptic patients. Brain. 2011;134(10):2887–97. https://doi.org/10.1093/brain/awr243.
11. Joshi M, Cui J, Doolittle K, et al. Brain segmentation and the generation of cortical surfaces. NeuroImage. 1999;9(5):461–76. https://doi.org/10.1006/nimg.1999.0428.
12. Koessler L, Cecchin T, Caspary O, Benhadid A, Vespignani H, Maillard L. EEG–MRI Co-registration and sensor labeling using a 3D laser scanner. Ann Biomed Eng. 2011;39(3):983–95. https://doi.org/10.1007/s10439-010-0230-0.
13. Hallez H, Vanrumste B, Grech R, et al. Review on solving the forward problem in EEG source analysis. J Neuro Eng Rehabil. 2007;4(1):46. https://doi.org/10.1186/1743-0003-4-46.
14. Baillet S, Mosher JC, Leahy RM. Electromagnetic brain mapping. IEEE Signal Process Mag. 2001;18(6):14–30. https://doi.org/10.1109/79.962275.
15. Asadzadeh S, Yousefi Rezaii T, Beheshti S, Delpak A, Meshgini S. A systematic review of EEG source localization techniques and their applications on diagnosis of brain abnormalities. J Neurosci Methods. 2020;339:108740. https://doi.org/10.1016/j.jneumeth.2020.108740.
16. Awan FG, Saleem O, Kiran A. Recent trends and advances in solving the inverse problem for EEG source localization. In: Inverse problems in science and engineering; 2019. Published online November 2, 2019. Accessed May 10, 2024. https://www.tand-

fonline.com/doi/abs/10.1080/17415977.2018.1490279.
17. Grech R, Cassar T, Muscat J, et al. Review on solving the inverse problem in EEG source analysis. J NeuroEng Rehabil. 2008;5(1):25. https://doi.org/10.1186/1743-0003-5-25.
18. Song J, Davey C, Poulsen C, et al. EEG source localization: Sensor density and head surface coverage. J Neurosci Methods. 2015;256:9–21. https://doi.org/10.1016/j.jneumeth.2015.08.015.
19. Henson RN, Mattout J, Phillips C, Friston KJ. Selecting forward models for MEG source-reconstruction using model-evidence. NeuroImage. 2009;46(1):168–76. https://doi.org/10.1016/j.neuroimage.2009.01.062.
20. Andersen LM, Jerbi K, Dalal SS. Can EEG and MEG detect signals from the human cerebellum? NeuroImage. 2020;215:116817. https://doi.org/10.1016/j.neuroimage.2020.116817.
21. Attal Y, Schwartz D. Assessment of subcortical source localization using deep brain activity imaging model with minimum norm operators: a MEG study. PLoS One. 2013;8(3):e59856. https://doi.org/10.1371/journal.pone.0059856.
22. Krishnaswamy P, Obregon-Henao G, Ahveninen J, et al. Sparsity enables estimation of both subcortical and cortical activity from MEG and EEG. Proc Natl Acad Sci. 2017;114(48):E10465–74. https://doi.org/10.1073/pnas.1705414114.
23. Hamid L, Habboush N, Stern P, et al. Source imaging of deep-brain activity using the regional spatiotemporal Kalman filter. Comput Methods Prog Biomed. 2021;200:105830. https://doi.org/10.1016/j.cmpb.2020.105830.
24. Seeber M, Cantonas LM, Hoevels M, Sesia T, Visser-Vandewalle V, Michel CM. Subcortical electrophysiological activity is detectable with high-density EEG source imaging. Nat Commun. 2019;10(1):753. https://doi.org/10.1038/s41467-019-08725-w.
25. Delorme A, Makeig S. EEGLAB: an open source toolbox for analysis of single-trial EEG dynamics including independent component analysis. J Neurosci Methods. 2004;134(1):9–21. https://doi.org/10.1016/j.jneumeth.2003.10.009.
26. Oostenveld R, Fries P, Maris E, Schoffelen JM. FieldTrip: open source software for advanced analysis of MEG, EEG, and invasive electrophysiological data. Intell Neurosci. 2011;2011:156869. https://doi.org/10.1155/2011/156869.
27. Tadel F, Baillet S, Mosher JC, Pantazis D, Leahy RM. Brainstorm: a user-friendly application for MEG/EEG analysis. Comput Intell Neurosci. 2011;2011:1–13. https://doi.org/10.1155/2011/879716.
28. Gramfort A, Luessi M, Larson E, et al. MNE software for processing MEG and EEG data. NeuroImage. 2014;86:446–60. https://doi.org/10.1016/j.neuroimage.2013.10.027.
29. Bastos AM, Schoffelen JM. A tutorial review of functional connectivity analysis methods and their interpretational pitfalls. Front Syst Neurosci. 2016;9 https://doi.org/10.3389/fnsys.2015.00175.
30. Kalogianni K, de Munck JC, Nolte G, Vardy AN, van der Helm FCT, Daffertshofer A. Spatial resolution for EEG source reconstruction—a simulation study on SEPs. J Neurosci Methods. 2018;301:9–17. https://doi.org/10.1016/j.jneumeth.2018.02.016.
31. Mégevand P, Spinelli L, Genetti M, et al. Electric source imaging of interictal activity accurately localises the seizure onset zone. J Neurol Neurosurg Psychiatry. 2014;85(1):38–43. https://doi.org/10.1136/jnnp-2013-305515.
32. Chiarion G, Sparacino L, Antonacci Y, Faes L, Mesin L. Connectivity analysis in EEG data: a tutorial review of the state of the art and emerging trends. Bioengineering. 2023;10(3):372. https://doi.org/10.3390/bioengineering10030372.
33. Nolte G, Bai O, Wheaton L, Mari Z, Vorbach S, Hallett M. Identifying true brain interaction from EEG data using the imaginary part of coherency. Clin Neurophysiol. 2004;115(10):2292–307. https://doi.org/10.1016/j.clinph.2004.04.029.
34. Baillet S. Magnetoencephalography for brain electrophysiology and imaging. Nat Neurosci. 2017;20(3):327–39. https://doi.org/10.1038/nn.4504.
35. Brkić D, Sommariva S, Schuler AL, et al. The impact of ROI extraction method for MEG connectivity estimation: practical recommendations for the study of resting state data. NeuroImage. 2023;284:120424. https://doi.org/10.1016/j.neuroimage.2023.120424.
36. Duma GM, Danieli A, Mattar MG, et al. Resting state network dynamic reconfiguration and neuropsychological functioning in temporal lobe epilepsy: an HD-EEG investigation. Cortex. 2022;157:1–13. https://doi.org/10.1016/j.cortex.2022.08.010.
37. Cabral J, Vidaurre D, Marques P, et al. Cognitive performance in healthy older adults relates to spontaneous switching between states of functional connectivity during rest. Sci Rep. 2017;7(1):5135. https://doi.org/10.1038/s41598-017-05425-7.
38. Tatum WO, Rubboli G, Kaplan PW, et al. Clinical utility of EEG in diagnosing and monitoring epilepsy in adults. Clin Neurophysiol. 2018;129(5):1056–82. https://doi.org/10.1016/j.clinph.2018.01.019.
39. Bartolomei F, Lagarde S, Wendling F, et al. Defining epileptogenic networks: contribution of SEEG and signal analysis. Epilepsia. 2017;58(7):1131–47. https://doi.org/10.1111/epi.13791.
40. Avigdor T, Abdallah C, von Ellenrieder N, et al. Fast oscillations >40 Hz localize the epileptogenic zone: an electrical source imaging study using high-density electroencephalography. Clin Neurophysiol. 2021;132(2):568–80. https://doi.org/10.1016/j.clinph.2020.11.031.
41. Fedele T, Ramantani G, Sarnthein J. High frequency oscillations as markers of epileptogenic tissue—End of the party? Clin Neurophysiol. 2019;130(5):624–6. https://doi.org/10.1016/j.clinph.2019.01.016.

42. Noorlag L, van Klink NEC, Kobayashi K, Gotman J, Braun KPJ, Zijlmans M. High-frequency oscillations in scalp EEG: a systematic review of methodological choices and clinical findings. Clin Neurophysiol. 2022;137:46–58. https://doi.org/10.1016/j.clinph.2021.12.017.
43. Strobbe G, Carrette E, López JD, et al. Electrical source imaging of interictal spikes using multiple sparse volumetric priors for presurgical epileptogenic focus localization. NeuroImage: Clin. 2016;11:252–63. https://doi.org/10.1016/j.nicl.2016.01.017.
44. Vorderwülbecke BJ, Baroumand AG, Spinelli L, Seeck M, van Mierlo P, Vulliémoz S. Automated interictal source localisation based on high-density EEG. Seizure. 2021;92:244–51. https://doi.org/10.1016/j.seizure.2021.09.020.
45. van de Velden D, Heide EC, Bouter C, Bucerius J, Riedel CH, Focke NK. Effects of inverse methods and spike phases on interictal high-density EEG source reconstruction. Clin Neurophysiol. 2023;156:4–13. https://doi.org/10.1016/j.clinph.2023.08.020.
46. Plummer C, Vogrin SJ, Woods WP, Murphy MA, Cook MJ, Liley DTJ. Interictal and ictal source localization for epilepsy surgery using high-density EEG with MEG: a prospective long-term study. Brain. 2019;142(4):932–51. https://doi.org/10.1093/brain/awz015.
47. Hedrich T, Pellegrino G, Kobayashi E, Lina JM, Grova C. Comparison of the spatial resolution of source imaging techniques in high-density EEG and MEG. NeuroImage. 2017;157:531–44. https://doi.org/10.1016/j.neuroimage.2017.06.022.
48. Heers M, Chowdhury RA, Hedrich T, et al. Localization accuracy of distributed inverse solutions for electric and magnetic source imaging of interictal epileptic discharges in patients with focal epilepsy. Brain Topogr. 2016;29(1):162–81. https://doi.org/10.1007/s10548-014-0423-1.
49. Sotudeh H, Sakhaei SM, Kazemitabar J. A novel brain source reconstruction using a multivariate mode decomposition. J Neural Eng. 2023;20(4):046004. https://doi.org/10.1088/1741-2552/acdffe.
50. de Tisi J, Bell GS, Peacock JL, et al. The long-term outcome of adult epilepsy surgery, patterns of seizure remission, and relapse: a cohort study. Lancet. 2011;378(9800):1388–95. https://doi.org/10.1016/S0140-6736(11)60890-8.
51. Vakharia VN, Duncan JS, Witt JA, Elger CE, Staba R, Engel J Jr. Getting the best outcomes from epilepsy surgery. Ann Neurol. 2018;83(4):676–90. https://doi.org/10.1002/ana.25205.
52. Scharfman HE, Kanner AM, Friedman A, et al. Epilepsy as a Network Disorder (2): what can we learn from other network disorders such as dementia and schizophrenia, and what are the implications for translational research? Epilepsy Behav. 2018;78:302–12. https://doi.org/10.1016/j.yebeh.2017.09.016.
53. Ding L, Worrell GA, Lagerlund TD, He B. Ictal source analysis: Localization and imaging of causal interactions in humans. NeuroImage. 2007;34(2):575–86. https://doi.org/10.1016/j.neuroimage.2006.09.042.
54. Staljanssens W, Strobbe G, Van Holen R, et al. EEG source connectivity to localize the seizure onset zone in patients with drug resistant epilepsy. NeuroImage Clin. 2017;16:689–98. https://doi.org/10.1016/j.nicl.2017.09.011.
55. Duma GM, Danieli A, Vettorel A, Antoniazzi L, Mento G, Bonanni P. Investigation of dynamic functional connectivity of the source reconstructed epileptiform discharges in focal epilepsy: a graph theory approach. Epilepsy Res. 2021;176:106745. https://doi.org/10.1016/j.eplepsyres.2021.106745.
56. Duma GM, Danieli A, Mento G, et al. Altered spreading of neuronal avalanches in temporal lobe epilepsy relates to cognitive performance: a resting-state HDEEG study. Epilepsia. 2023;64(5):1278–88. https://doi.org/10.1111/epi.17551.
57. Duma GM, Pellegrino G, Rabuffo G, et al. Altered spread of waves of activities at large scale is influenced by cortical thickness organization in temporal lobe epilepsy: a magnetic resonance imaging–high-density electroencephalography study. Brain Commun. 2024;6(1):fcad348. https://doi.org/10.1093/braincomms/fcad348.
58. Jirsa VK, Proix T, Perdikis D, et al. The virtual epileptic patient: individualized whole-brain models of epilepsy spread. NeuroImage. 2017;145:377–88. https://doi.org/10.1016/j.neuroimage.2016.04.049.
59. Jirsa VK, Stacey WC, Quilichini PP, Ivanov AI, Bernard C. On the nature of seizure dynamics. Brain. 2014;137(8):2210–30. https://doi.org/10.1093/brain/awu133.
60. Courtiol J, Guye M, Bartolomei F, Petkoski S, Jirsa VK. Dynamical mechanisms of interictal resting-state functional connectivity in epilepsy. J Neurosci. 2020;40(29):5572–88. https://doi.org/10.1523/JNEUROSCI.0905-19.2020.
61. Hashemi M, Vattikonda AN, Sip V, et al. The Bayesian Virtual Epileptic Patient: A probabilistic framework designed to infer the spatial map of epileptogenicity in a personalized large-scale brain model of epilepsy spread. NeuroImage. 2020;217:116839. https://doi.org/10.1016/j.neuroimage.2020.116839.
62. Kini LG, Bernabei JM, Mikhail F, et al. Virtual resection predicts surgical outcome for drug-resistant epilepsy. Brain. 2019;142(12):3892–905. https://doi.org/10.1093/brain/awz303.
63. Nissen IA, Millán AP, Stam CJ, et al. Optimization of epilepsy surgery through virtual resections on individual structural brain networks. Sci Rep. 2021;11(1):19025. https://doi.org/10.1038/s41598-021-98046-0.
64. Jirsa V, Wang H, Triebkorn P, et al. Personalised virtual brain models in epilepsy. Lancet Neurol. 2023;22(5):443–54. https://doi.org/10.1016/S1474-4422(23)00008-X.

65. Leguia MG, Andrzejak RG, Rummel C, et al. Seizure cycles in focal epilepsy. JAMA Neurol. 2021;78(4):454–63. https://doi.org/10.1001/jamaneurol.2020.5370.
66. Uldall P, Alving J, Hansen LK, Kibæk M, Buchholt J. The misdiagnosis of epilepsy in children admitted to a tertiary epilepsy centre with paroxysmal events. Arch Dis Child. 2006;91(3):219–21. https://doi.org/10.1136/adc.2004.064477.
67. Zaidi A, Clough P, Cooper P, Scheepers B, Fitzpatrick AP. Misdiagnosis of epilepsy: many seizure-like attacks have a cardiovascular cause. J Am Coll Cardiol. 2000;36(1):181–4. https://doi.org/10.1016/S0735-1097(00)00700-2.
68. Oto M (Meritxell). The misdiagnosis of epilepsy: Appraising risks and managing uncertainty. Seizure. 2017;44:143–6. https://doi.org/10.1016/j.seizure.2016.11.029.
69. Beniczky S, Wiebe S, Jeppesen J, et al. Automated seizure detection using wearable devices: A clinical practice guideline of the International League Against Epilepsy and the International Federation of Clinical Neurophysiology. Clin Neurophysiol. 2021;132(5):1173–84. https://doi.org/10.1016/j.clinph.2020.12.009.
70. Tveit J, Aurlien H, Plis S, et al. Automated interpretation of clinical electroencephalograms using artificial intelligence. JAMA Neurol. 2023;80(8):805–12. https://doi.org/10.1001/jamaneurol.2023.1645.
71. Geier C, Lehnertz K, Bialonski S. Time-dependent degree-degree correlations in epileptic brain networks: from assortative to dissortative mixing. Front Hum Neurosci. 2015;9 https://doi.org/10.3389/fnhum.2015.00462.
72. Mitsis GD, Anastasiadou MN, Christodoulakis M, Papathanasiou ES, Papacostas SS, Hadjipapas A. Functional brain networks of patients with epilepsy exhibit pronounced multiscale periodicities, which correlate with seizure onset. Hum Brain Mapp. 2020;41(8):2059–76. https://doi.org/10.1002/hbm.24930.
73. Fruengel R, Bröhl T, Rings T, Lehnertz K. Reconfiguration of human evolving large-scale epileptic brain networks prior to seizures: an evaluation with node centralities. Sci Rep. 2020;10(1):21921. https://doi.org/10.1038/s41598-020-78899-7.
74. Rings T, von Wrede R, Lehnertz K. Precursors of seizures due to specific spatial-temporal modifications of evolving large-scale epileptic brain networks. Sci Rep. 2019;9(1):10623. https://doi.org/10.1038/s41598-019-47092-w.
75. Briels CT, Schoonhoven DN, Stam CJ, de Waal H, Scheltens P, Gouw AA. Reproducibility of EEG functional connectivity in Alzheimer's disease. Alzheimers Res Ther. 2020;12(1):68. https://doi.org/10.1186/s13195-020-00632-3.
76. Das S, Puthankattil SD. Functional connectivity and complexity in the phenomenological model of mild cognitive-impaired Alzheimer's disease. Front Comput Neurosci. 2022;16 https://doi.org/10.3389/fncom.2022.877912.
77. Luck SJ. An introduction to the event-related potential technique. 2nd ed. MIT Press; 2014.
78. Beauchene C, Hinault T, Sarma SV, Courtney SM. Dynamic cognitive states explain individual variability in behavior and modulate with EEG functional connectivity during working memory. Comput Brain Behav. 2023;6(2):246–61. https://doi.org/10.1007/s42113-022-00153-2.
79. Duma GM, Mento G, Cutini S, et al. Functional dissociation of anterior cingulate cortex and intraparietal sulcus in visual working memory. Cortex. 2019;121:277–91. https://doi.org/10.1016/j.cortex.2019.09.009.
80. Jin W, Nobre AC, van Ede F. Temporal expectations prepare visual working memory for behavior. J Cogn Neurosci. 2020;32(12):2320–32. https://doi.org/10.1162/jocn_a_01626.
81. Capizzi M, Ambrosini E, Arbula S, Vallesi A. Brain oscillatory activity associated with switch and mixing costs during reactive control. Psychophysiology. 2020;57(11):e13642. https://doi.org/10.1111/psyp.13642.
82. Duma GM, Granziol U, Mento G. Should I stay or should I go? How local-global implicit temporal expectancy shapes proactive motor control: an hdEEG study. NeuroImage. 2020;220:117071. https://doi.org/10.1016/j.neuroimage.2020.117071.
83. Tafuro A, Ambrosini E, Puccioni O, Vallesi A. Brain oscillations in cognitive control: A cross-sectional study with a spatial stroop task. Neuropsychologia. 2019;133:107190. https://doi.org/10.1016/j.neuropsychologia.2019.107190.
84. Bola M, Borchardt V. Cognitive processing involves dynamic reorganization of the whole-brain network's functional community structure. J Neurosci. 2016;36(13):3633–5. https://doi.org/10.1523/JNEUROSCI.0106-16.2016.
85. Braun U, Schäfer A, Walter H, et al. Dynamic reconfiguration of frontal brain networks during executive cognition in humans. Proc Natl Acad Sci USA. 2015;112(37):11678–83. https://doi.org/10.1073/pnas.1422487112.
86. Sala-Llonch R, Arenaza-Urquijo EM, Valls-Pedret C, et al. Dynamic functional reorganizations and relationship with working memory performance in healthy aging. Front Hum Neurosci. 2012;6:152. https://doi.org/10.3389/fnhum.2012.00152.
87. Shine JM, Bissett PG, Bell PT, et al. The dynamics of functional brain networks: integrated network states during cognitive task performance. Neuron. 2016;92(2):544–54. https://doi.org/10.1016/j.neuron.2016.09.018.
88. Duma GM, Di Bono MG, Mento G. Grounding adaptive cognitive control in the intrinsic, functional brain organization: an HD-EEG resting state investigation. Brain Sci. 2021;11(11):1513. https://doi.org/10.3390/brainsci11111513.

89. Hakim N, Awh E, Vogel EK, Rosenberg MD. Inter-electrode correlations measured with EEG predict individual differences in cognitive ability. Curr Biol. 2021;31(22):4998–5008.e6. https://doi.org/10.1016/j.cub.2021.09.036.
90. Ujma PP, Bódizs R, Dresler M, et al. Multivariate prediction of cognitive performance from the sleep electroencephalogram. NeuroImage. 2023;279:120319. https://doi.org/10.1016/j.neuroimage.2023.120319.
91. Marek S, Tervo-Clemmens B, Calabro FJ, et al. Reproducible brain-wide association studies require thousands of individuals. Nature. 2022;603(7902):654–60. https://doi.org/10.1038/s41586-022-04492-9.
92. Chang M, Büchel D, Reinecke K, Lehmann T, Baumeister J. Ecological validity in exercise neuroscience research: a systematic investigation. Eur J Neurosci. 2022;55(2):487–509. https://doi.org/10.1111/ejn.15595.
93. Cristaldi FDP, Oosterwijk S, Barrett LF. Predictive processing and embodiment in emotion. In: The Routledge handbook of embodied cognition. 2nd ed. Routledge; 2024.
94. Mavros P, Austwick MZ, Smith AH. Geo-EEG: towards the use of eeg in the study of urban behaviour. Appl Spat Anal. 2016;9(2):191–212. https://doi.org/10.1007/s12061-015-9181-z.
95. Olsen S, Alder G, Williams M, et al. Electroencephalographic recording of the movement-related cortical potential in ecologically valid movements: a scoping review. Front Neurosci. 2021;15 https://doi.org/10.3389/fnins.2021.721387.
96. Westlin C, Theriault JE, Katsumi Y, et al. Improving the study of brain-behavior relationships by revisiting basic assumptions. Trends Cogn Sci. 2023;27(3):246–57. https://doi.org/10.1016/j.tics.2022.12.015.
97. Fouad MM, Amin KM, El-Bendary N, Hassanien AE. Brain computer interface: a review. In: Hassanien AE, Azar AT, editors. Brain-computer interfaces: current trends and applications. Springer International Publishing; 2015. p. 3–30. https://doi.org/10.1007/978-3-319-10978-7_1.
98. Zhang X, Ma Z, Zheng H, et al. The combination of brain-computer interfaces and artificial intelligence: applications and challenges. Ann Transl Med. 2020;8(11):712. https://doi.org/10.21037/atm.2019.11.109.
99. Gonzalez-Astudillo J, Cattai T, Bassignana G, Corsi MC, Fallani FDV. Network-based brain–computer interfaces: principles and applications. J Neural Eng. 2021;18(1):011001. https://doi.org/10.1088/1741-2552/abc760.
100. Hamedi M, Salleh SH, Noor AM. Electroencephalographic motor imagery brain connectivity analysis for BCI: a review. Neural Comput. 2016;28(6):999–1041. https://doi.org/10.1162/NECO_a_00838.
101. Cervera MA, Soekadar SR, Ushiba J, et al. Brain-computer interfaces for post-stroke motor rehabilitation: a meta-analysis. Ann Clin Transl Neurol. 2018;5(5):651–63. https://doi.org/10.1002/acn3.544.
102. Prasad G, Herman P, Coyle D, McDonough S, Crosbie J. Applying a brain-computer interface to support motor imagery practice in people with stroke for upper limb recovery: a feasibility study. J Neuroeng Rehabil. 2010;7(1):60. https://doi.org/10.1186/1743-0003-7-60.
103. Jeannerod M, Decety J. Mental motor imagery: a window into the representational stages of action. Curr Opin Neurobiol. 1995;5(6):727–32. https://doi.org/10.1016/0959-4388(95)80099-9.
104. Munzert J, Lorey B, Zentgraf K. Cognitive motor processes: the role of motor imagery in the study of motor representations. Brain Res Rev. 2009;60(2):306–26. https://doi.org/10.1016/j.brainresrev.2008.12.024.
105. Bakker M, De Lange FP, Helmich RC, Scheeringa R, Bloem BR, Toni I. Cerebral correlates of motor imagery of normal and precision gait. NeuroImage. 2008;41(3):998–1010. https://doi.org/10.1016/j.neuroimage.2008.03.020.
106. Cicinelli P, Marconi B, Zaccagnini M, Pasqualetti P, Filippi MM, Rossini PM. Imagery-induced cortical excitability changes in stroke: a transcranial magnetic stimulation study. Cereb Cortex. 2006;16(2):247–53. https://doi.org/10.1093/cercor/bhi103.
107. Pichiorri F, Morone G, Petti M, et al. Brain–computer interface boosts motor imagery practice during stroke recovery. Ann Neurol. 2015;77(5):851–65. https://doi.org/10.1002/ana.24390.
108. Sharma N, Baron JC, Rowe JB. Motor imagery after stroke: relating outcome to motor network connectivity. Ann Neurol. 2009;66(5):604–16. https://doi.org/10.1002/ana.21810.
109. Colucci A, Vermehren M, Cavallo A, et al. Brain–computer interface-controlled exoskeletons in clinical neurorehabilitation: ready or not? Neurorehabil Neural Repair. 2022;36(12):747–56. https://doi.org/10.1177/15459683221138751.
110. Zhang Z, Han S, Yi H, et al. A brain-controlled vehicle system based on steady state visual evoked potentials. Cogn Comput. 2023;15(1):159–75. https://doi.org/10.1007/s12559-022-10051-1.

21 From Analog to Digital and Beyond: The Future of Electroencephalogram (EEG)

Marco Cursi and Massimo Filippi

21.1 Introduction: The Evolving World of EEG Technology

The transition from primitive observations of neural phenomena in frogs to the complexities of contemporary digital electroencephalogram (EEG) instrumentation reflects the persistent quest to unravel the enigmatic workings of the human brain. This chapter embarks on an engaging exploration of the past and prospective evolution of EEG technology, navigating through historical epochs and observing the progression from analog to digital domains and beyond.

Here is a general overview of the topics covered throughout this chapter.

From Frogs to Transistors: The Evolution of Analog EEG Instrumentation: In the first section, we delve into the early days of EEG, where pioneering researchers observed electrical activities in frogs, laying the foundation for the analog instrumentation that would follow. We unravel the particulars of the initial instruments that provided glimpses into the mysterious world of brain waves.

Brain Waves in Bytes: From Analog to Digital EEG: The transition from analog to digital EEG marks a significant turning point in the history of neuroscientific exploration. In this section, we navigate through the transformation of brain waves into bytes, exploring the technological leaps that have revolutionized EEG instrumentation. The marriage of precision and computational power has opened unprecedented possibilities for capturing, analyzing, and interpreting the electrical activity of the brain.

Mindful Tech: Unveiling the Potential of Wearable EEG Systems: The present era witnesses the emergence of wearable EEG systems, pushing the boundaries of neurotechnology beyond the confines of laboratories. We explore how advancements in electrode and amplifier technology, coupled with the quest for portability and user-friendliness, are shaping the landscape of EEG applications in everyday life.

Connecting Thoughts to Actions: The Frontier of Brain–Computer Interface: The convergence of neuroscience and technology brings us to the frontier of brain–computer interface (BCI). Here, we unravel the threads connecting thoughts to actions, exploring how users can control external devices seamlessly through the power of their minds. The potential applications in healthcare,

M. Cursi (✉)
Clinical Neurophysiology Unit, Department of Neurology, IRCCS San Raffaele Scientific Institute, Milan, Italy
e-mail: cursi.marco@hsr.it

M. Filippi
Vita-Salute San Raffaele University, Neurology Unit, Neurorehabilitation Unit, MS Center, Neuroimaging Research Unit, Division of Neuroscience, IRCCS San Raffaele Scientific Institute, Milan, Italy
e-mail: filippi.massimo@hsr.it

F. Brigo, O. Mecarelli (eds.), *EEG: The First 100 Years*,
https://doi.org/10.1007/978-3-031-86878-8_21

accessibility, and beyond are nothing short of revolutionary.

Mind Melding: Bridging Minds Through the Wonders of Brain-to-Brain Interfaces: Beyond individual interfaces, we enter the realm of collective consciousness with brain-to-brain interfaces. This section explores the visionary concepts of mind melding, where the boundaries between individual minds blur, opening possibilities for shared cognition and communication.

This chapter provides a glimpse into the vast possibilities shaping the future of digital EEG, with the understanding that the information presented is a curated exploration rather than an exhaustive account.

21.2 From Frogs to Transistors: The Evolution of Analog EEG Instrumentation

21.2.1 The Origins

As early as the sixth century BC, the Greek philosopher Thales of Miletus noted the peculiar ability of amber (electron, ἤλεκτρον, in Greek), to attract light objects when rubbed against sheep's fleece. This phenomenon, documented in various ancient manuscripts across cultures [1], set the stage for the exploration of electricity. However, it would take almost two millennia before the discovery of electricity within living organisms.

The origins of electrophysiology can be traced back to the Italian physician Luigi Galvani (1737–1798). He developed a frog nerve–muscle preparation capable of detecting static electricity. Galvani proposed a theory suggesting that living beings inherently possessed electricity generated by the brain and transmitted to muscles through nerves. Galvani published the results of his research on "animal electricity" in 1791 in the famous *Commentarius* [2]. The instruments used by Galvani in his studies essentially consisted of the friction machine and the Leyden jar. The German physicist Otto von Guericke (1602–1686) developed the friction machine, the first electrostatic device capable of producing electricity, in 1672. The Leyden jar, a device capable of storing static electricity (an ancestor of the capacitor), was introduced in 1745 by the Dutch scientist Pieter van Musschenbroek of Leiden (1692–1761). The friction machine could be successfully used in combination with the Leyden jar: the electricity generated by the former could be stored in the latter and used as needed.

The Italian chemist and physicist Alessandro Volta (1755–1832) initially accepted Galvani's results but soon turned against him, incorrectly attributing all of Galvani's findings to "metallic electricity" resulting from the inadvertent use of two different metals that had directly stimulated the muscles of frogs [3]. For this reason, for the next half century the idea of "animal electricity" was not generally accepted, although it had been independently confirmed [4, 5].

21.2.2 Parallel Advances in Electromagnetism and Physiology

However, Alessandro Volta's divergence from Galvani's ideas contributed significantly to the field of electricity. Volta introduced inventions such as the voltaic pile, perpetual electrophorus, capacitor, electrometric balance, and condenser electrometer.

In 1820, the Danish physicist Hans Christian Ørsted (1777–1851) discovered the interrelation between electricity and magnetism when he observed a compass needle deflecting near a wire carrying an electric current. Ørsted announced his great discovery to the world in a circular letter in Latin, *Experimenta circa effectione conflictus electric in acum magneticam* [6]. This pivotal moment laid the groundwork for the theory of electromagnetism.

Concurrently, the German physicist Georg Ohm (1787–1854) and the English chemist and physicist Michael Faraday (1791–1867) developed crucial theories on the relationship between current and voltage (Ohm's law) and on the interaction between magnetic field and current (Faraday's law of induction) [7, 8]. The first studies on electricity were essentially carried out using the electroscope, an instrument capable of measuring weak electrical potentials by exploiting the repulsive movement of two thin gold leaves.

Leopoldo Nobili (1784–1835), an Italian physicist, made further strides in electrophysiology in 1828 by using an improved astatic galvanometer to measure the "frog current," validating Galvani's earlier studies [9]. Carlo Matteucci (1818–1868), another Italian physicist and physiologist, continued investigating "animal electricity," presenting fundamental results on nervous mechanisms responsible for electric discharges in torpedoes in 1837 [10].

The widespread acceptance of intrinsic electrical activity in nerve and muscle was solidified by the German physiologist Emil Du Bois-Reymond (1818–1896) in his two-volume book *Untersuchungen uber Thierische Elektricitat* [11]. Du Bois-Reymond assembled a new galvanometer with increased sensitivity, able to measure the action potential preceding muscle contraction. He also developed clay nonpolarizable electrodes, used in later years to record EEG in animals and humans. After Matteucci, Du Bois-Reymond has been often regarded as the founder of electrophysiology.

Advancements in galvanometer technology continued in 1858 with William Thomson (1824–1907), known as Lord Kelvin, and later in 1882 with Jacques-Arsène d'Arsonval (1851–1940). The operating principle of d'Arsonval's galvanometer was based on Thomson's syphon recorder, which was used for telegraphy on transatlantic cables. This represented a fundamental innovation compared to galvanometers in use between 1850 and 1880: the magnet was fixed, while the coil, suspended by a wire, was movable. It's noteworthy that the motion of d'Arsonval's galvanometer laid the foundation for pen-based oscillographic systems used in electroencephalographs until the twentieth century.

The electroscope too was improved during this period. The capillary mercury electroscope was perfected by Gabriel Lippmann (1846–1922) in 1873 and used in 1880 by Waller to record the electrocardiogram from both dogs and humans.

21.2.3 The First EEGs

In 1875, the British physician and physiologist Richard Caton (1842–1926) succeeded in recording potentials from the brain of rabbits and monkeys, using a mirror galvanometer with DuBois-Reymond nonpolarizable electrodes [12]. For his studies, Caton has been recognized as the pioneer of EEG.

The Polish scientist Adolf Beck (1863–1939) made significant contributions to neurophysiological research by recording spinal potentials from frogs and detecting a modification in the spontaneous activity of the canine cerebral cortex in response to visual and sensory stimuli. The results of his work, published in 1890 [13], contain detailed information about the instrumentation used in his experiments, primarily a modified version of Hermann's galvanometer. In 1908, moving his interests on cerebrocerebellar relationship, he began to use the string galvanometer developed in Holland by Willem Einthoven (1860–1927). This instrument, introduced in 1901, was capable of recording potentials without distortions, earning Einthoven the Nobel Prize in 1924. The string galvanometer gradually replaced the use of the capillary electroscope and the d'Arsonval galvanometer.

A transformative moment arrived in the early 1900s, when Hans Berger (1873–1941) began exploring the electrical activity of the brain. Despite initial efforts using the Lippmann capillary electrometer, in 1924, he made the first human EEG recordings, using Edelmann's string galvanometer. His recordings, requiring photographic development, marked the beginning of human EEG studies. Berger faced challenges, experimenting with different electrodes (needle, cup, and plate) and instruments. By 1929, he used a Siemens double-coil galvanometer, more sensitive than previous devices, for EEG recordings. His findings, including the discovery of 10-Hz alpha waves, were met with skepticism, and Berger himself doubted their authenticity. Despite initial disbelief, Berger's work laid the foundation for EEG research, revealing fundamental brain wave patterns and paving the way for future advancements in neuroscience [14, 15].

21.2.4 The Advent of Differential Amplifier

The German engineer Jan Friedrich Toennies (1902–1970), credited with building the first ink-writing oscillograph or "neurograph" for recording brain potentials, made significant contributions to neuroscience. During his fellowship at the Rockefeller Foundation in 1932, he developed a differential amplifier, a milestone shared with Brian Matthews (see Box 21.1). Toennies' report in 1933 not only detailed the design and application of the amplifier but also laid out fundamental concepts of differential recording, considering impedance and the effects of volume conduction [16].

Box 21.1 The Differential Amplifier

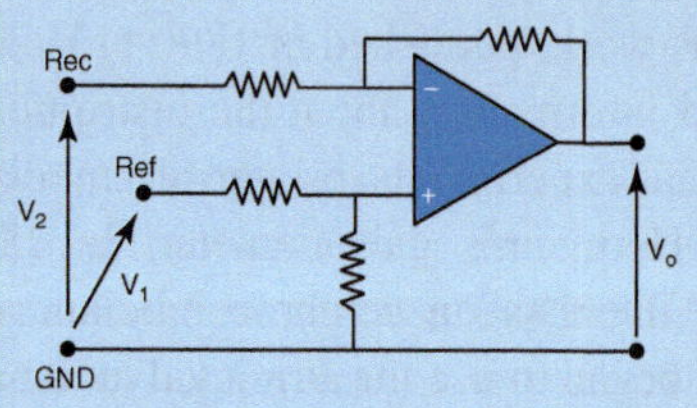

In an ideal differential amplifier, shown on the top figure, the output voltage V_0 is obtained by multiplying the difference of the two input voltages $(V_1 - V_2)$ by the differential gain A_D:

$$V_0 = A_D (V_1 - V_2)$$

This is the significant advantage of differential amplifiers because, in the presence of common noise on both inputs, it is eliminated through subtraction.
In practical scenarios, the output voltage also includes a factor due to the common mode $(V_1 + V_2)/2$, resulting in the equation:

$$V_0 = A_D (V_1 - V_2) + A_C (V_1 + V_2)/2$$

where the term A_C represents the Common mode gain. In this case, a common noise would also be amplified.

The effectiveness of a differential amplifier is determined by the relationship between the differential gain A_D and the common-mode gain A_C. A higher A_D compared to A_C indicates a better-performing amplifier. The Common Mode Rejection Ratio (CMRR), expressed in dB as

$$CMRR = (A_D / A_C)_{dB} = 20 \log_{10} (A_D / A_C)$$

measures the amplifier's ability to reject common-mode signals. A higher CMRR, typically greater than 100 dB (meaning that A_D is 100,000 times higher than A_C) signifies a superior differential amplifier.

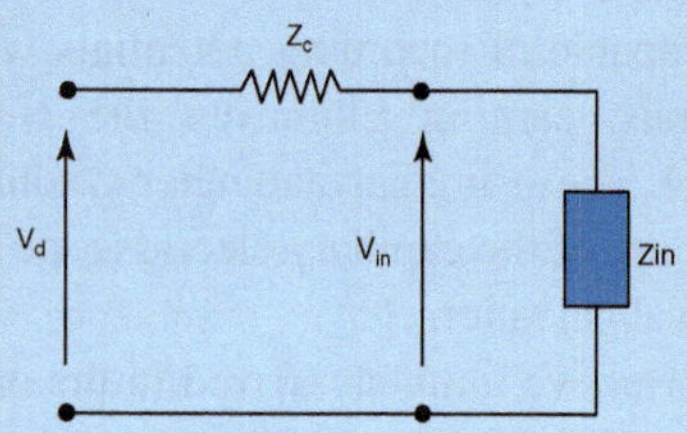

A schematic representation of the acquisition system is shown on the bottom figure. Here Z_C denotes the electrode-skin impedance, while Z_{in} refers to the input impedance of the amplifier. Our focus is directed towards the desired recorded voltage (V_d) and the effectively measured voltage (V_{in}). Applying Kirchhoff's laws to this straightforward circuit, we derive the relationship between V_d and V_{in}:

$$V_{in} / V_d = (1 + Z_C / Z_{in})^{-1}$$

Optimal performance (meaning $V_{in} \approx V_d$) is achieved with higher input impedances (Z_{in}>20 MΩ) and lower electrode-skin impedances (Z_C<SKΩ).

In Cambridge, England, Edgar Douglas Adrian (1889–1977), also known as "Lord Adrian," played a pivotal role in advancing EEG research after discovering Hans Berger's work in 1933. Adrian, a prominent neuroscientist, had previously collaborated with Dr. Detlev Bronk to develop a method for recording single neuron action potentials using a capillary electrometer and a vacuum tube amplifier (invented by Fleming in 1904 and then used as an amplifier by Lee De Forest in 1907). This technology allowed them to study afferent impulses in peripheral nerves responding to various stimuli. Collaborating with electrical engineer Brian Matthews (1906–1986), known as "Sir Brian," they aimed to adapt their instrumentation for EEG studies.

They predominantly used an inkwriting oscillograph, developed around 1926 for electrocardiography. This system incorporated a chart drive mechanism powered by a clock motor and a modified loudspeaker to audibly represent raw EEG waves. Matthews introduced the use of differential input amplifiers to electrophysiology, creating a three-channel system with balanced differential amplifiers for each channel. Adrian and Matthews' use of multiple differential channels significantly advanced EEG and neuronal physiology concepts. They demonstrated the spatial and temporal relationships between signals by recording simultaneously and independently from different brain areas. Their detailed reports conclusively affirmed the reality of EEG and the "Berger rhythm," confirming its cortical origin and distinguishing it from muscle, eye, or other sources of artifact [17]. Adrian and Matthews' contributions marked a crucial phase in EEG development, providing a solid foundation for subsequent research in neuroscience [18].

21.2.5 EEG in North America

In the realm of electrophysiological instrumentation development and application, Dr. Alexander Forbes (1882–1965), a physician with a physics background (he worked on submarine detection during World War I), significantly contributed to the field while at Harvard Medical School. In the early 1920s, Forbes pioneered the use of a vacuum tube amplifier in recording nerve action potentials [19]. Forbes' laboratory published detailed technical reports that explored tube amplifier properties and systematically designed high-quality instrumentation, addressing considerations such as frequency compensation, shielding, balancing, and gain stability.

Forbes continued his electrophysiological work into the 1930s, collaborating with Hallowell Davis (1896–1992) and Frederick Gibbs (1903–1992) in the exploration of EEG. Meanwhile, at Washington University in St. Louis, physiologists Gasser and Erlanger made significant contributions, including the development of the first cathode-ray tube (CRT) oscilloscope. They received their first tube from Dr. Arthur Holly Compton (1892–1962), head of the Physics Department, and developed amplifiers similar to Forbes' design. They produced photographic records by placing sensitized paper against the CRT display.

Herbert Henri Jasper (1906–1999), collaborating with Leonard Carmichael (1898–1973), played a pivotal role in advancing electroencephalography (EEG) research in North America [20]. In 1933, Jasper secured a grant from the Rockefeller Foundation to establish a clinical and experimental EEG laboratory at the Bradley Hospital in Rhode Island. In 1934, they used Westinghouse four-channel mirror oscillographs to make the first human EEG recordings west of the Atlantic. Their amplifiers featured differential, battery-powered inputs using the push–pull design attributed to Matthews, allowing independent bipolar derivations. Jasper's group devoted significant effort to electrode development. Initially using liquid-filled electrodes, they later adopted 5-mm chlorided silver "hats" with felt-covered brims, attached to the head with collodion and filled with electrode jelly. These electrodes, with properties comparable to modern scalp electrodes, allowed bipolar recordings from various brain regions.

The principles and operational conventions established by Jasper's group remain remarkably unchanged today. Their work laid the foundation for the first epilepsy monitoring and EEG localization of epileptogenic foci, leading to surgical treatment of seizures. The comprehensive approach undertaken by Jasper's team in the 1930s continues to influence contemporary EEG techniques.

Jasper moved to Montreal to collaborate with Wilder Penfield (1891–1976). In 1939, a pioneering Laboratory of Electroencephalography was established, equipped with sound-shielded rooms featuring observation windows, operating facilities, and modern instrumentation. Over the course of that year, approximately 1000 EEG recordings were carried out on epileptic patients in the laboratory. The findings from this study were subsequently published in a book in 1941 [21]. Chap. XIV of this publication, authored by Jasper, delineates the 10–20 International System for electrode placement. A captivating narrative detailing the evolution of EEG at the Montreal Neurological Institute is provided in a historical paper by Jasper [22].

In 1949, Herbert Jasper and John Hunter presented a method for concurrently capturing EEG and clinical events through the use of a movie camera, establishing the foundation for video EEG systems [23]. Their publication makes mention of a photographic method detailed by Dr. Robert Schwab at Harvard in 1938, which utilized two cameras with synchronizing marks to streamline the creation of a unified composite film.

Alfred Lee Loomis (1887–1975), the second published American electroencephalographer, emerged shortly after Jasper's confirmation of Berger's EEG findings. Working in a private laboratory in Tuxedo Park, New York, Loomis's group used an innovative switch box linking 12 silver disk electrodes to 6 channels of amplification, forming one of the earliest headboxes. The four-stage, push–pull tube amplifiers exhibited a frequency response of 0.3–46 Hz, boasting high impedance and common-mode rejection. The system featured six EEG pens and seven additional pens, including an event marker. Additional amplifiers made possible the recordings of heart, respiration, and movement, providing detailed recordings during extended sleep sessions. Loomis' contributions laid the groundwork for modern sleep studies, offering summaries of various brain rhythms throughout the night [24, 25].

Hallowell Davies (1896–1992) at the Harvard Medical School led a third American laboratory, reporting human EEG recordings. The American neurologist Frederic Andrews Gibbs (1903–1992) and his wife and technician Erna Leonhardt Gibbs (1904–1987) joined the Davies' group in 1934. Under the guidance of William Gordon Lennox (1884–1960), they documented seizures using a single-channel inkwriting EEG recorder and firstly described the "3 Hz spike-and-wave" of petit mal [26].

21.2.6 EEG Devices Become Commercial

After a trip to the European EEG laboratories, in collaboration with engineer Albert Grass (1919–1992), Dr. and Mrs. Gibbs perfected the EEG recording system, taking inspiration from Toennies' neurograph and from the system developed by Matthews and Adrian. In 1935, Grass produced the first commercial electroencephalographer, the Grass Model I, with three differential channels and an inkwriter recording on rolls of paper. A perfected system, the Grass Model II, was produced in 1938, in two different configurations: four or six channels. The "console" Model III appeared in 1946, allowing 8-channel as well as 16-channel design, presented a special electrode selector, useful to create and modify bipolar montages.

Ellen H. Robinson Grass played a pivotal role in advancing neuroscience during the mid-twentieth century. She forged vital connections between Harvard physiologists and her husband Albert. Actively participating in various clinical and basic neuroscience studies, she laid the groundwork for our modern understanding of epilepsy and EEG. In 1945, Mrs. Grass founded the Grass Instrument Company, contributing to the advancement of physiological technology. Additionally, she championed the development of professional societies such as ASET, the American EEG Society, the American Epilepsy Society, and the Epilepsy Foundation of America, further propelling progress in these areas [27].

In 1947, at Bell Laboratories, Walter Brattain (1902–1987) and John Bardeen (1908–1991), under the guidance of William Shockley (1910–1989), invented the transistor, representing one of the fundamental elements of modern electronics. The American EEG manufacturer Franklin Offner, starting his activity in 1935, produced in 1950s the first system based on transistorized EEG amplifiers: the "Type T," marking a significant leap in EEG technology.

Between the years 1940 and 1947, many manual and mechanical types of frequency analyzers became available. Dr. John Knott (1911–1993), from the group established by Lee Edward Travis (1896–1987) at the University of Iowa, was the first scientist publishing a report on EEG frequency analysis [28].

In 1953, Norman Jefferis "Jeff" Holter (1914–1983) revolutionized ambulatory patient monitoring by developing a telemetry system for recording single-channel EKGs. This innovation laid the groundwork for what would become the "Holter EKG," a long-term, single-channel cassette recorder introduced in 1961 [29]. As the need for similar ambulatory monitoring systems for EEG became apparent, researchers faced various technical challenges. These included accommodating the multiple channels required for EEG recordings, adjusting amplifiers for the differing signal amplitudes of EEG compared to EKG, ensuring prolonged recording durations (at least 24 hours), and implementing systems for reviewing recorded data. In the early 1970s, Marson and McKinnon tackled these challenges by developing a portable four-channel cassette recorder primarily for industrial use [30]. Building upon their work, Ives and Woods demonstrated its feasibility for ambulatory EEG recording a few years later [31]. This pivotal step paved the way for the commercialization of ambulatory EEG recording technology. In 1979, the Oxford Medilog 4–24, equipped with the HDX-82 on-head preamplifier chip, was introduced as the first commercial ambulatory four-channel EEG cassette recorder. Complemented by a display unit (PMD12) for visual data analysis, this system marked a significant advancement in EEG monitoring technology.

21.2.7 Conclusions

The intricate journey from Galvani's observations to the commercial analog EEG devices underscores the collaborative efforts, innovative thinking, and technological milestones that have shaped our understanding of the brain's electrical activity. As we reflect on this evolutionary path, visually summarized in Fig. 21.1, it becomes evident that the rich history of analog EEG technology has paved the way for the sophisticated methodologies and tools in use today.

For a deeper exploration of the topics covered in this section, we encourage to consult the comprehensive papers by Thomas Collura [32] and Albert Grass [33]. Additionally, John S. Ebersole's first chapter in the book *Ambulatory EEG Monitoring* provides a thorough examination of the history of Holter EEG [34]. These resources offer valuable insights and perspectives that complement the overview presented here.

Year		Publication/Device/EEG	Scientists
1775		Perpetual electrophorus	A. Volta
1780		Electroscope	B. Franklin
1791		Commentarius [2]	L. Galvani
1800		Voltaic pile	A. Volta
1820		Ørsted experiment [6]	H.C. Ørsted
1827		Ohm's law [7]	G. Ohm
1828		About galvanometers [9]	L. Nobili
1832		Faraday's law [8]	M. Faraday
1837		Electric discharges in torpedoes [10]	C. Matteucci
1848		Investigations into animal electricity [11]	E. Du Bois-Reymond
1850		Inductorium	E. Du Bois-Reymond
1873		Capillary electrometer	G. Lippmann
1875		First EEGs from rabbit and monkey brains	R. Caton
1875		Electric currents of the brain [12]	R. Caton
1882		D'Arsonval galvanometer	J.A. D'Arsonval
1883		Diode	T. Edison
1890		Currents of the nerve centers [13]	A. Beck
1897		Duddell oscillograph	W. Duddell
1901		String galvanometer	W. Einthoven
1907		Electron tube	L. De Forest
1920		String galvanometer amplifier	A. Forbes
1922		Cathode ray oscilloscope	J. Erlanger and H. S. Gasser
1924		First human EEG	H. Berger
1926		Inkwriting EKG	B. Mattews
1929		Human EEG description [14]	H. Berger
1934		Differential amplifier	B. Mattews
1934		First human EEG in the East USA	A. Forbes, H. Davis, and F. Gibbs
1934		First human EEG in the West USA	H. Jasper and L. Carmichael
1935		Brain rhythms during sleep [24]	A.L. Loomis, E.N. Harvey, and G. Hobart
1935		Grass Model I EEG device	A. Grass
1938		Grass Model II EEG device	A. Grass
1941		The 10-20 system [21]	H. Jasper
1946		Grass Model III EEG device	A. Grass
1949		First Video-EEG	H. Jasper and J. Hunter
1949		Cinematographic technique in EEG [23]	J. Hunter and H. Jasper
1949		Transistor	W. Brattain, J. Bardeen, and W. Shockley
1958		Offner Type T EEG device	F. Offner
1979		Oxford Medilog 4-24 ambulatory EEG device	Oxford Instruments, UK

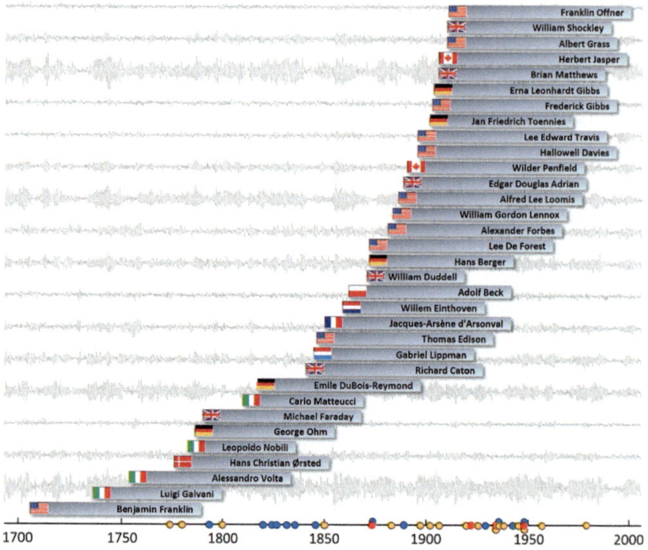

Fig. 21.1 Charting the technological odyssey of EEG. The table on the left highlights key milestones in the advancement of electromagnetism and electrophysiology, crucial for the evolution of EEG technology. Each significant achievement is attributed to pioneering scientists, with foundational publications marked by blue blocks, electric and electronic devices by yellow, and groundbreaking EEG recordings by red. The chart on the right visually traces these pivotal milestones. Each gray block denotes the lifespan of notable scientists, with flags indicating their countries of origin. The colored dots on the timeline are referred to the publications (blue), device inventions (yellow), and pioneering EEG recordings (red)

21.3 Brain Waves in Bytes: From Analog to Digital EEG

21.3.1 Pilot Experiments in Data Transmission

The demand for improved communication has been a driving force in advancing technology, especially in the realm of data converters. To extend communication capabilities beyond spoken language, various methods emerged for interaction over long distances, such as drum beating and smoke signaling. The foundational principles of encoding and decoding are evident in these early communication techniques. The late 1800s saw significant progress in electrical science, marking the beginning of the electronic communication era. This period witnessed the true origins of data conversion, notably with the advent of the telegraph and telephone.

In communication engineering, the preliminary experiments on time-division multiplexing (TDM) in telephony prompted questions about how frequently it was necessary to sample a continuous-time signal. Attempts to multiplex multiple analog signals onto a single telephone line began shortly after the early commercial successes with telegraphy in the 1840s.

Willard Miner carried out experiments using an electromechanical rotating commutator to multiplex analog telephone conversations onto a single pair of wires, patenting his method in 1903 after years of pilot experiments [35]. Miner determined the required sampling rate experimentally, assuming it would be at least somewhat larger than the upper frequency components of speech. However, due to the lack of practical electrical components at the time, Miner's work on sampling and TDM had no immediate follow-up. Nonetheless, it underscored the necessity of sampling a signal at a minimum rate to preserve information.

21.3.2 The Birth of the Sampling Theorem

Theoretical communication engineers did not address the sampling problem until surprisingly late (for a comprehensive exploration of the history of the sampling theorem, refer to Stankovic's work [36]). In the mid-1920s, Harry Nyquist (1889–1976) studied telegraph signaling to determine the maximum rate over a channel with a given bandwidth. His findings, published in two classic papers in 1924 and 1928 [37, 38], summarized the principles of the sampling theorem, defining the so-called Nyquist frequency: the sample rate must be at least twice the signal's bandwidth, to avoid aliasing. The Russian Vladimir Aleksandrovich Kotelnikov (1908–2005) independently formulated a similar theorem, proving it for low-pass and band-pass signals in a 1933 study [39].

In 1937, the English scientist Alec Harley Reeves (1902–1971) expanded on Miner's system by incorporating a quantizer, laying the groundwork for pulse code modulation (PCM). PCM consists in three consecutive steps: sampling, quantizing, and coding (see Box 21.2). Sampling is done at higher than the Nyquist frequency, quantizing rounds off the sample amplitude to the closest allowed level; encoding represents any allowed level as a unique binary code word. With the aim of minimizing quantization noise, Reeves used a quantizer with a high number of levels [40]. The landmark paper of 1949 by Claude Shannon (1916–2001) solidified the PCM theory [41]. The sampling theorem was also discussed in the 1949 Japanese book *Hakei Densō* (波形伝送: *Waveform Transmission*) [42] by Isao Someya, leading to the term "Someya's theorem" in some Japanese publication.

Interestingly, the English mathematician Edmund Taylor Whittaker (1873–1956) can be credited as the first one addressing the sampling theorem for all bandlimited functions, with his results published in 1915 [43], well before the theoretical studies by Nyquist and Shannon. The various names attributed to the sampling theorem in the literature (Shannon, Nyquist, Kotelnikov, Whittaker, Someya) have sparked discussions about its origins, even though it is almost universally accepted to refer to the theorem using the names Shannon and Nyquist.

Box 21.2 The Analog to Digital Conversion

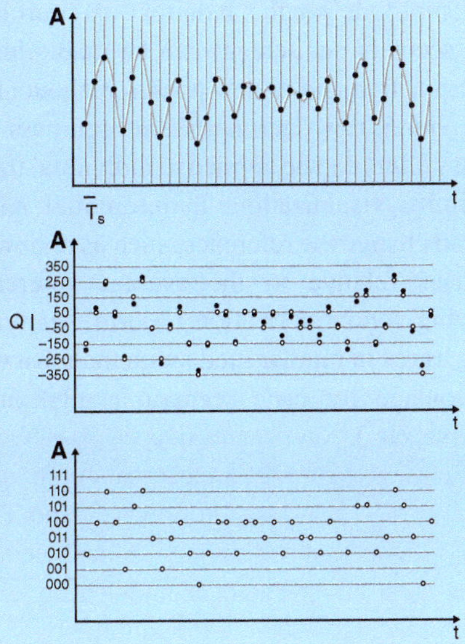

Sampling

An analog signal (solid line) is defined continuously over time and amplitude. The signal is sampled at distinct time points (black dots) with a sampling period T_s. The sampling frequency (f_s = $1/T_s$), expressed in Hertz, determines the number of samples per second and must adhere to the Nyquist-Shannon sampling theorem, ensuring it is greater than twice the highest frequency in the analog signal.

Quantization

The quantization stage divides the range of amplitudes (e.g. −350 ÷ +350) into discrete levels, usually a power of 2 (e.g., $8=2^3$ levels). The quantization interval (Q) represents the distance between two adjacent levels. The amplitude values of samples (black dots) are approximated to the nearest level (white dots), introducing a quantization error (maximum error = Q/2), which increases with larger intervals between levels.

Coding

The coding step associates a binary code with each quantization level. The number of bits in the code corresponds to the exponent of 2 used to determine the number of levels (in this example, n_{bits}=3 coding for 2^3 = 8 levels). The sequence of binary codes is termed Pulse Code Modulation (PCM).

Nevertheless, this historical account exposes a common pattern observed in theoretical challenges within the realms of technology. Initially, practitioners propose practical guidelines or rules of thumb. Subsequently, theoreticians work toward formulating a comprehensive solution applicable to general cases. Eventually, it comes to light that mathematicians had already resolved the underlying mathematical problem long before the practitioners and theoreticians, showcasing a recurrent sequence in problem-solving methodologies [44].

21.3.3 EEG Becomes Digital

The first analog-to-digital conversion of the EEG was done at the Data Processing Laboratory (DPL) of the Brain Research Institute, University of California, Los Angeles (UCLA), in 1961, under the direction of Dr. Ross Adey (1922–2004). DPL stood as the first electronic and computer laboratory designed specifically to advance automated technology for the study of the nervous system, making pioneering contributions in analog to digital conversion, automated data acquisition and preprocessing, frequency analysis, interactive graphics, modeling, and simulation. In the early 1970s, a connection among three computers (a DEC PDP-12, an IMLAC PDS-lD, and an IBM 360-91) was set at the DPL, forming a network of distributed processors for interactive graphics programs. DPL collaborated with US Air Force and National Aeronautics and Space Administration (NASA) to study the effects of environmental stresses, similar to that experimented in space flight, on the brain mechanisms. Dr. Adey studied the relation between EEG and behavioral states of wakefulness, alertness, and decision-making, using implanted electrodes in normal behaving animals. In this epoch, when CAT scan just emerged and MRI was still far away, engineer Thelma Estrin (1924–2014), Director of DPL, coordinated a simulation program, called BRAINMAPS, to aid stereotaxic neurosurgeons to determine the best trajectory for approaching a targeted brain structure [45]. A fascinating overview of the activities carried out at DPL is reported in Thelma Estrin's historical paper, published in 1987 [46].

Digital EEG entered the commercial landscape in the 1990s, marking the onset of remarkable advancements that have since exceeded expectations.

21.3.4 Advantages of Digital EEG

After undergoing digital conversion, EEG opens the door to a multitude of applications, spanning from straightforward to intricate processes. It is essential to highlight that all manipulations applied to EEG traces are nondestructive, ensuring the possibility to revert to the original recording. Typically, EEG signals are acquired in referential mode with open filters, so they are ready to be analyzed in the most convenient way. Some key advantages of digital EEG are exposed below:

Customization of Visualization Parameters Real-time adjustments to sensitivity (trace amplitude on the monitor) and scrolling speed (seconds per screen) can be made during both acquisition and review phases without altering the stored raw data. Different high-pass and low-pass filters can be applied to individual traces to optimize visualization. In referential mode, users can change the reference, such as reviewing the traces relative to the average reference, employing source derivation (Hjorth), or representing traces in bipolar mode with the most suitable montage for each scenario (longitudinal, transverse, etc.).

Precision Measurements Accurate measurements of graph element amplitude and time intervals between events can usually be made using measurement cursors. Trace epochs can be selected for display in a dedicated analysis window, offering information in the time or frequency domain, including the power spectrum of the selected channel. Topographic maps can be displayed, showing potential distribution on the scalp at specific instants (amplitude maps) or power distribution within certain frequency bands (frequency maps).

Time–Frequency Analysis Some systems allow the display of information derived from time-varying spectral analysis of EEG. Various graphs show the temporal trend of spectral parameters such as changes in absolute or relative power, spectral edge frequency, mean dominant frequency, coherence between channels, and interhemispheric symmetry indices. To evaluate the time course of the entire power spectrum of one or more channels, compressed spectral arrays (CSA) or density spectral arrays (DSA) can also be displayed.

Storage Digital EEG data is typically stored locally on devices such as hard disks, SSDs, or flash drives and/or remotely on file servers. Reviewing software often includes integrated databases, streamlining the retrieval of patient traces and facilitating comparisons between exams conducted at different times. The availability of ample storage space enables the recording of large quantities of data, particularly in long-term EEG monitoring (LTM). In LTM, additional polygraphic channels and video recordings can be incorporated, with dedicated software ensuring seamless synchronization between different signals. This comprehensive approach enhances diagnostic capabilities and enables in-depth analysis of EEG recordings over extended periods [47].

Standard File Formats While manufacturers often use proprietary file formats to store EEG traces, most of the EEG systems allow exporting data in standard formats, to facilitate data interchange between different systems. Examples include European standards EDF and EDF+ [48, 49], international standards BIDS [50], and DICOM [51].

Data Transmission over the Internet Taking advantage from the current data transmission rates, sending EEG traces over the Internet is seamless. This capability allows telemedicine applications, remote consultations, or data sharing among research groups around the world. Security measures, including authentication, encryption, masking, and fractional storage, must be ensured for sensitive data [52, 53].

Software for EEG Analysis Data exported from an EEG recorder can be thoroughly examined using external toolboxes. Freeware applications, such as EEGLAB [54], FieldTrip [55], BrainStorm [56], LORETA [57], and its advancements sLORETA [58] and eLORETA [59], enable a diverse range of EEG analyses. These include simple visualization and preprocessing, automatic artifact detection, independent component analysis, source localization using standard head models or based on real MRI imaging, functional and effective connectivity assessments, statistical analysis, and much more (Fig. 21.2).

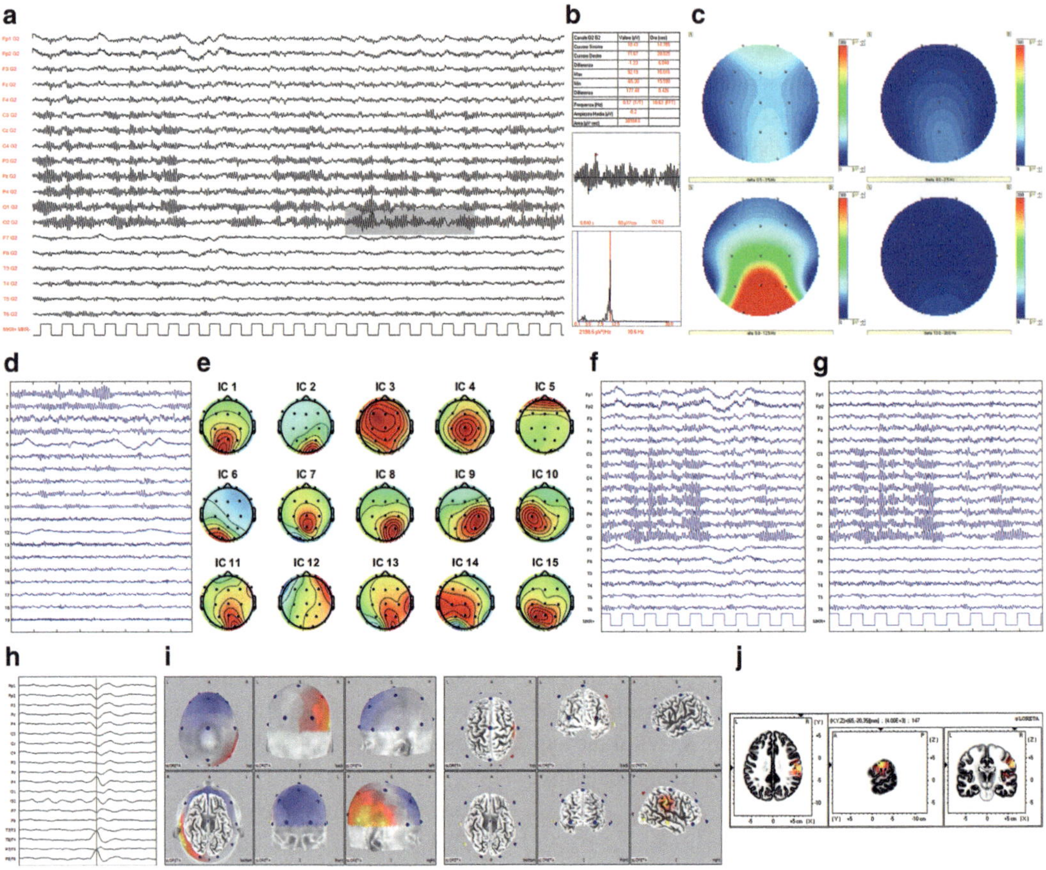

Fig. 21.2 Examples of EEG processing. (**a**) A 25-s EEG page is displayed, with a selected epoch on channel O2 (gray background). (**b**) Information about the selected epoch is reported; in particular, the power spectrum is shown in the lower box. (**c**) Topographic maps indicating the power distribution of the four classical spectral bands (delta, theta, alpha, and beta) referred to the EEG displayed in (**a**). (**d**) Time course of component activations obtained by independent component analysis (ICA) carried out with EEGLAB on the EEG reported in (**a**). Components 5 and 12 seem to follow the trend of ocular artifacts. (**e**) Maps of the first 15 independent components: the topographic distribution of components 5 and 12 seems to confirm the ocular origin (vertical and horizontal, respectively) of the activity. (**f, g**) Effect of ICA correction. Removing components 5 and 12 from the EEG reported in (**f**) results in the cleaned traces displayed in (**g**). (**h**) A 1-second EEG epoch containing an epileptic spike uploaded in sLORETA. (**i, j**) Topographic amplitude map over the scalp of the cursor marked epileptic peak displayed in (**h**), and the corresponing source localizazion on the cortical surface. (**j**) Source localization displayed on the three plane slices, relative to maximum sLORETA values

21.4 Mindful Tech: The Promise of Wearable EEG Systems

21.4.1 EEG Comes Out of the Lab

Advancements in electrode and amplifier technology, along with improved health standards, have allowed for noninvasive and high-quality EEG recordings. Nowadays, applying conductive gel between electrodes and the scalp is typically all that has needed, with active electrodes and shielding techniques further enhancing signal quality. Recent developments have even introduced alternative electrode types, such as saline-based or dry electrodes, eliminating the need for gel. Additionally, there is a noticeable trend toward miniaturization in EEG amplifiers, enabling wearable and portable devices. These advancements aim to move EEG research beyond the confines of the laboratory, empowering real-world applications unhindered by the limitations of older, bulkier machines.

To characterize a wearable/wireless EEG system, several crucial aspects need consideration [60, 61]. Key features include the number of channels, electrode types, data acquisition, and transmission capabilities.

21.4.2 Number of Channels

The number of EEG channels in a wearable system primarily hinges on the specific application requirements. It is crucial to determine whether the electrodes need to cover the entire surface of the scalp or focus on particular regions, such as the visual, motor, or somatosensory areas. In research settings, a minimum of 19 EEG channels is typically considered adequate for comprehensive scalp coverage. However, for more targeted applications that only necessitate exploration of specific regions, a smaller number of electrodes may suffice. The market offers a range of wearable devices, featuring diverse solutions, from straightforward 4-channel systems to more advanced 128-channel devices. Most of them come with a preestablished position of the electrodes, while others (nonintegrated systems) offer the possibility of customizing the electrode locations. The reference electrode is typically positioned on the earlobes or mastoids; alternatively, it may be placed at a midline point, such as Fz or Cz. Additionally, some systems incorporate bipolar polygraphy channels, including electro-oculogram, electromyography, electrocardiogram, and respiration. Moreover, sensors embedded in the headset, such as gyroscopes and/or accelerometers, can offer insights into the patient's head movements.

An innovative EEG acquisition technique, designed to enhance the convenience of EEG measurements, is Ear-EEG. This method involves recording EEG activity using electrodes positioned in or around the ear. One significant advantage is the use of hair-free areas, promoting optimal electrode–skin contact, leading to improved signal quality. Additionally, the absence of flashy helmets contributes to greater user comfort. The acquisition system can be attached to the ear, ensuring easy wearability and enabling user interaction through simple touch. While the practicality of Ear-EEG for everyday applications is promising, it is important to consider the limited dimensions of the recording area. This technique is particularly adept at capturing activity from the temporal regions, albeit at the expense of coverage in more distant areas. Despite this limitation, the Ear-EEG system holds great potential for user-friendly, real-life applications. To learn more about this topic, see the review of Kaongoen and colleagues [62].

21.4.3 Electrode Type

Typical electrodes used in EEG recordings are Ag/Ag-Cl disks, either clipped into a cap or fixed to the scalp with adhesive paste. These electrodes necessitate the use of conductive gel to ensure optimal conduction and minimize electrode–skin impedance. However, the use of gel has drawbacks, including time-consuming preparation, limited recording periods due to gel drying, and the need for assistance from trained personnel during application.

To address these issues, dry electrodes have been developed, eliminating the need for gel. Dry electrodes use pressure to establish skin contact, making them quicker to apply. However, they can cause discomfort, such as headaches, due to the pressure exerted on the scalp. Various shapes of dry electrodes exist, with fingered electrodes being among the most prevalent. Unlike disc electrodes, which may be hindered by hair, fingered electrodes enable direct skin contact by displacing the hair. Some advanced dry electrode devices incorporate ultra-low-noise amplifiers directly on the electrode (known as active electrodes), with some models featuring a separate amplifier for each finger for balanced recordings, even if some fingers do not make contact. For passive electrodes, shielding can be used on electrode cables to reduce external interferences during the transmission of the signal from the electrodes toward the amplifier.

Despite the advantages, dry electrodes face challenges. Signal quality may be compromised compared to traditional disc electrodes, as the impedance of electrodes is inversely proportional to the contact surface, resulting in noisier recordings in case of fingered electrodes. Discomfort and noise issues are, therefore, notable downsides of dry electrodes.

An intermediate category comprises water-based or wet electrodes, which are also Ag/Ag-Cl disks surrounded by a spongy material. Before use, these electrodes are soaked in a saline solution, eliminating the need for skin degreasing and conductive gel. However, like dry electrodes, wet electrodes have a drawback, as they dry out over time, limiting recording durations.

Fortunately, nonintegrated commercial devices offer flexibility in electrode selection, allowing users to adapt recordings to specific application needs.

The latest advancement in dry electrodes involves ultrathin film devices with conductive materials like metal or graphene, deposited using inkjet printing on temporary tattoo paper. These tattoo electrodes offer excellent skin contact, a seamless interface, and an almost imperceptible connection between the human body and electronics. Notably, this technology allows for long-lasting recordings, as the tattoo is perforable and accommodates hair growth [63, 64].

21.4.4 Data Acquisition and Transmission

Wearable EEG systems, similar to traditional laboratory counterparts, feature amplifiers characterized by parameters such as maximum input range, bandwidth, input impedance, and common mode rejection ratio (CMRR). Regarding analog–digital conversion, key considerations encompass the sampling frequency and the number of bits used for quantization.

Market-available systems exhibit a wide range of input ranges, with an average bandwidth between 0 and 100 Hz. These systems boast high input impedances exceeding 1 GigaOhm, a CMRR surpassing 100 dB, and a standard sampling frequency ranging between 200 and 500 Hz, accompanied by 16 or 24 quantization bits.

As technology advances, researchers and engineers actively pursue the miniaturization of EEG amplifiers to meet the demand for more portable, user-friendly, and less intrusive EEG systems. This significant trend is driven by the desire to develop smaller and lighter amplifiers without compromising signal quality. Miniaturized EEG amplifiers play a pivotal role in the development of portable and wearable EEG devices, especially beneficial for applications requiring mobility, such as monitoring brain activity during daily activities, sports, or sleep.

In addition to their reduced size, these amplifiers are designed to seamlessly integrate into users' daily routines, offering discretion and comfort. This discreet design allows users to wear them without causing discomfort or drawing attention, making them suitable for various real-world scenarios. Furthermore, advancements in power efficiency are crucial for wearable devices, ensuring a longer battery life and reducing the need for frequent recharging. For instance, the smallest and lightest commercial eight-channel wearable amplifier, measuring a mere 40x40x20 mm and weighing only 27 grams, exemplifies the strides made in achieving both portability and efficiency.

Wearable systems often feature wireless connectivity, commonly using protocols such as Wi-Fi or Bluetooth, allowing them to transmit recorded signals seamlessly. Additionally, many of these devices come equipped with on-board data storage, often utilizing SD cards. This built-in storage proves to be valuable in scenarios where there are issues with the wireless connection. In the event of a communication interruption, the data is securely stored and can be retrieved for retransmission once the connection is restored. This redundancy ensures data integrity and continuity even under challenging communication conditions.

21.4.5 Applications of Wearable EEG Systems

The use of wearable devices extends across various realms of neurotechnological research. Clinical applications of such systems range from detecting epileptic seizures [65] to analyzing sleep patterns [66], allowing patients to carry the acquisition system with them for continuous EEG data recording during their normal lives.

At present, wearable EEG systems operate predominantly as one-way data recorders. However, the next steps of this EEG technology would tend to establish bidirectional communication, facilitating both data sensing and action. This advancement holds promise for data-driven feedback, optimizing treatment timing and efficacy, potentially reducing treatment frequency [67]. While closed-loop, data-responsive treatments are emerging, such as adaptive deep brain stimulation for Parkinson's disease [68] and closed-loop electrical stimulators for epilepsy [69], current implementations primarily rely on implanted sensors. These implants offer better signal properties and less interference sources compared to scalp EEG, limiting their application to select severe cases due to surgical costs and practical considerations. Noninvasive alternatives stand poised to bridge the gap between conventional pharmaceutical approaches and invasive electroceuticals, albeit with significant design challenges and regulatory barriers [70]. Nonetheless, leveraging EEG for targeted, personalized interventions represents a pivotal direction in electrode and instrumentation research for the future.

While clinical applications remain predominant, wearable EEG technology is increasingly finding applications in everyday contexts. Some examples of these applications are tracking of mental workload [71], drivers' vigilance monitoring [72], and emotion recognition [73].

Wearable systems also hold promise for brain–computer interfaces, enabling users to control external devices through their brain activity, such as spellers, robotic systems, or video games.

21.5 Connecting Thoughts to Actions: The Frontier of Brain–Computer Interface

21.5.1 Linking Brains and Machines

A brain–computer interface (BCI), also known as a brain–machine interface (BMI) or direct neural interface (DNI), is a communication pathway that allows the exchange of information between the brain and an external device, typically a computer or a robotic apparatus. The primary goal of BCIs is to enable individuals to control external devices or communicate with computers using their brain activity (usually revealed by EEG).

In recent decades, BCI has increasingly captured the attention of the scientific community. Scientists worldwide have actively contributed to advancing this field by creating a multitude of tools and methodologies to capture, interpret, and translate neural signals into commands that a computer or device can understand. This remarkable progress owes much to the parallel development of technology, providing researchers with essential devices crucial for any analytical endeavor. During the last 30 years, the once fanciful notion of communication between humans and the external world via BCI interfaces has evolved from science fiction to contemporary reality.

21.5.2 An Overview of BCI

The first studies on BCI began in the 1970s at UCLA, with the pioneering paper of Professor Jacques Vidal [74], universally recognized as the inventor of BCI. Since then, the striking progress in computer technology, machine learning, and neuroscience has facilitated the creation of different BCI systems [75–77].

The first applications of BCI were developed with the aim of helping people with disabilities, for example, for communication purposes in locked-in patients [78]. More recently, systems have been developed for neurorehabilitation of patients suffering from stroke [79] or other chronic pathologies [80], or to control smart wheelchairs [81] or exoskeletons [82]. Other "less clinical" applications are aimed at video games [83], virtual or augmented reality [84], neuromarketing [85], home automation [86], or emotional state recognition [87].

In BCI applications, surface EEG stands out as the most convenient and widely adopted method for recording brain activity. While alternative neural signals, such as fNIRS [88], offer insights into blood oxygen consumption, they suffer from delayed responses and limited temporal resolution. While fMRI plays a crucial role in informing BCI research, development, and understanding of brain function [89], it is not typically used in real-time BCI applications, due to its limited temporal resolution and impracticality for real-time feedback. Deep recordings from implanted electrodes, or electrocorticography (ECoG) derived from cortical grid electrodes, yield stable signals but involve an invasive approach [90].

Recently, the startup Neuralink has garnered attention for developing its first implantable multielectrode recording system in a human brain [91]. On January 24, 2024, Elon Musk, the founder of the startup, announced the first surgical implantation of the Neuralink device in a tetraplegic subject [92]. Neuralink's brain chip consists of an array of 1024 microelectrodes inserted into the brain to record the activity of individual neurons. Although this has sparked significant interest among the general public, the lack of detailed information suggests that, at present, effective results obtained by Neuralink's neural interface does not represent a significant advancement compared to traditional BCI systems proposed in the past [93, 94].

Given these factors, surface EEG remains the preferred method in BCI due to its affordability, high temporal resolution, and straightforward setup. Moreover, EEG-based BCI systems can be enhanced by incorporating supplementary information from additional channels such as EMG, EOG, EKG, eye tracking, accelerometers, and similar sources.

Typically, commercial BCI systems are intended for one specific purpose. To overcome this limitation, the scientists from the Wadsworth Center have developed a general-purpose BCI platform called BCI2000 [95]. BCI2000 can integrate different brain signals, signal processing methods, output devices, and operating protocols. BCI2000 is an open-source software system, and it is free for noncommercial use.

Another open-source software platform is OpenVIBE, created by the French researchers Yann Renard and Anatole Lécuyer [96]. The platform provides simplified visual programming tools to implement BCI protocols and embedded tools for visualization and feedback based on virtual reality and 3D displays.

21.5.3 Different Approaches to BCI

Different methods can be employed to steer a BCI system. Broadly, these approaches can be classified based on whether they rely on evoked responses, such as P300 or visual evoked potentials, or spontaneous EEG activity, like mu or beta sensorimotor rhythms (SMRs). The acquired brain signals need to be processed to extract meaningful information: subsequent to the recording stage, relevant features are extracted from the obtained signals through diverse procedures in time and/or frequency domain, including spatial filtering, voltage amplitude measurements, and spectral analyses. During the feature extraction phase, it is crucial to pay special attention to the presence of any artifacts that may be caused by events or environmental interferences or by factors more closely related to the patient's own activity, such as muscular/eye artifacts, movement-related artifacts, or poor electrode contact. The extracted features are subsequently translated into precise device commands, tailored for specific BCI applications, employing both linear methods (such as statistical analysis) and nonlinear algorithms (such as neural networks). The translation algorithms should exhibit adaptive behavior, dynamically adjusting to user characteristics; for instance, if the signal feature is mu rhythm, the algorithm should adapt to the individual user's specific range of mu rhythm amplitude and frequency. A second level of adaptation should consider the time-varying behavior of the feature of interest, attributed, for example, to physiological variations during the day, such as fatigue, sleepiness, or other influencing factors.

21.5.3.1 The P300 Approach

Farwell and Donchin [78] pioneered the use of a P300-based BCI. The P300 event-related potential (ERP) is typically observable when an individual is exposed to a sequence of standard visual, auditory, or tactile stimuli interspersed with deviant stimuli, known as "oddball" stimuli, capturing their attention. Their study involved participants observing a 6 × 6 matrix of items (letters and symbols), focusing on a target item while the matrix's rows and columns flashed randomly (Fig. 21.3). As the average of responses to the target item was different from that to the other items, the BCI identified the subject's desired selection by detecting the disparity in average responses. This system enabled subjects to spell words and its reliance on the P300 potential, indicative of attention rather than mere gaze direction, suggested applicability for individuals lacking eye movement control. Auditory stimuli present a feasible alternative to visual cues and can be beneficially used for individuals with visual impairments [97]. More recent P300-based BCIs rely not only on classical P300 paradigms but also on other ERPs elicited by face recognition or tactile stimulation [98, 99].

21.5.3.2 The Motor Imagery Approach

The initial application of SMRs for BCI control was proposed in 1991 [100]. SMRs, encompassing mu (8–12 Hz) and beta (18–30 Hz) oscillations, are recorded over sensorimotor cortices and exhibit changes in amplitude during movement, imagined movement, or preparation for movement [101]. The suitability of mu and/or beta rhythms as signal features for EEG-based communication is supported by several factors. Movement or preparation for movement typically leads to a decrease in mu and beta rhythms, especially contralateral to the movement, termed as "event-related desynchronization" (ERD). Conversely, an increase in rhythm, known as "event-related synchronization" (ERS), occurs after movement and during relaxation. Importantly for BCI applications, both ERD and ERS can occur without actual movement, also manifesting during motor imagery. Individuals

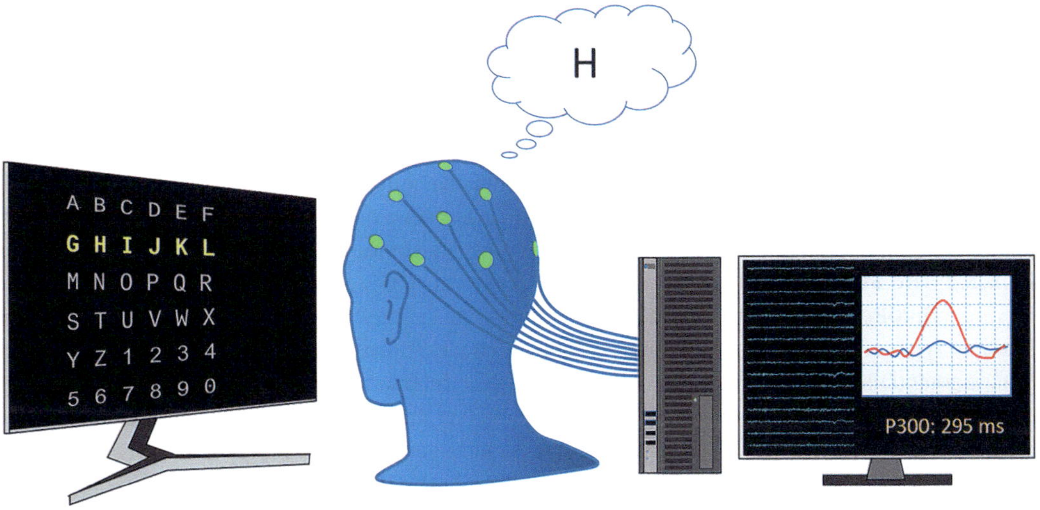

Fig. 21.3 In the P300-based BCI approach, the subject selects a letter within a matrix displayed on the screen to control a speller (e.g., choosing the letter "H"). The subject gazes at the target letter while the rows and columns of the matrix light up alternately. The flashing of the row or column containing the selected letter triggers a P300 ERP in the brain. The BCI software recognizes this P300 ERP after averaging across multiple repetitions. The intersection of the row and column that evoked the P300 determines the chosen letter, enabling the subject to pilot the speller accurately

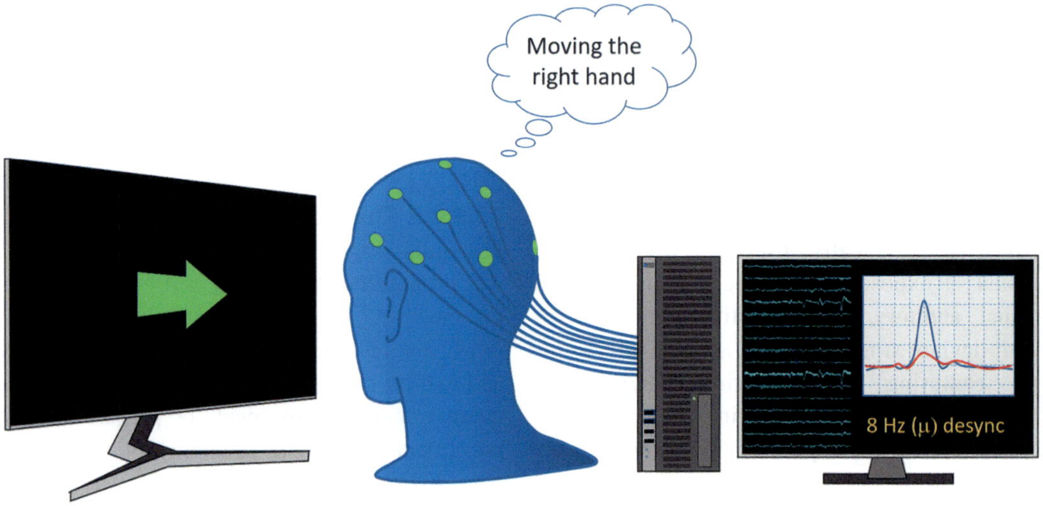

Fig. 21.4 In the motor imagery approach, the subject envisions moving their right or left hand, influencing the movement of an object on the screen. For instance, when imagining right-hand movement, an arrow pointing to the right is displayed. Throughout the task, the BCI software analyzes the left and right sensorimotor rhythms, detecting a reduction in the power of the mu rhythm contralateral to the imagined hand movement. This mu rhythm desynchronization is indicative of specific brain activity patterns. In real time, this information is harnessed to precisely control the movement of the object on the screen

can acquire the skill to control SMR amplitudes, manipulating them to move a cursor toward targets on a screen or perform various computer-based tasks. Spectral analysis measures SMR amplitude and users learn to modulate it, enabling cursor movement toward the intended target (Fig. 21.4). Successive studies by various research groups have further refined and expanded this BCI method, demonstrating the capability of multidimensional prosthesis control [102].

21.5.3.3 The Steady-State Visual Evoked Potential Approach

Another type of BCI uses steady-state visual evoked potentials (SSVEPs) recorded over occipital cortex in response to lights that flash repeatedly [103]. In this approach, a subject views several lights that each flash at a different frequency. When the subject focuses attention on one particular light, EEG spectral analysis shows increased power at the frequency of the photic driving response over visual cortex. The BCI detects this and performs the action represented by that light, i.e., moving a cursor on the screen (Fig. 21.5). SSVEP-based BCIs can support high rates of information transfer [104]. Moreover, SSVEP-BCI approach requires little or no training, making it a great candidate for real-life applications.

21.5.4 Feedback in BCI

Feedback to the users in a BCI system is a crucial component that provides information about the outcome of their actions or the current state of the system [105]. The goal of feedback is to enhance the users' understanding and awareness, allowing them to adjust their mental strategies and improve their interaction with the BCI. Feedback can be delivered through various sensory modalities, including visual, auditory, haptic, or multimodal feedback [106].

Providing feedback in real time is crucial for the users to adjust their mental state or strategy as they interact with the BCI. Feedback is essential for error correction. If the BCI misinterprets the user's intention or if there's uncertainty in the decoding process, clear feedback allows the user

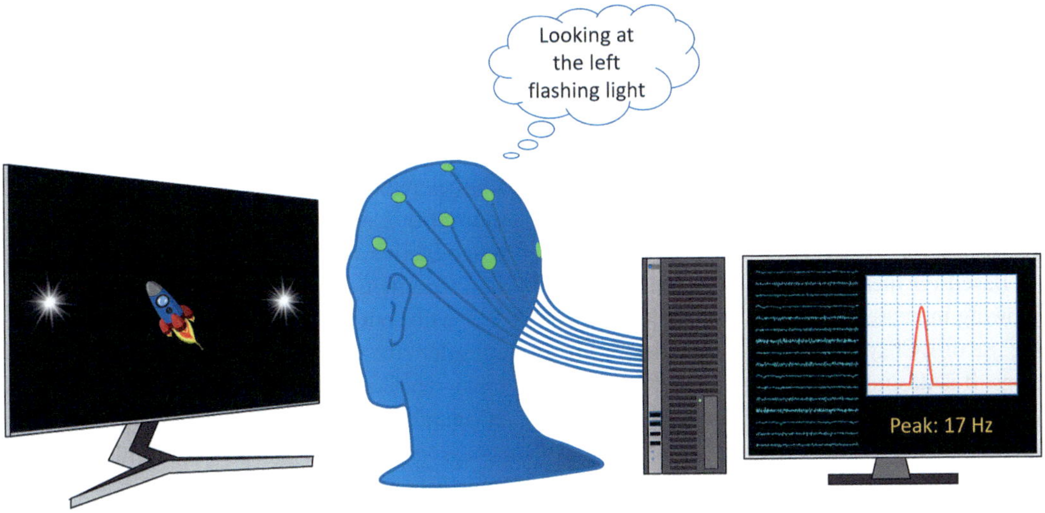

Fig. 21.5 In the SSVEP-based approach, the subject directs the movement of an object on the monitor by fixating on one of the flashing lights, typically positioned in the upper, lower, right, and left regions of the screen. Each light flashes at a distinct frequency. Through spectral analysis of EEG, the BCI system detects the photic driving effect, reflecting the synchronization of occipital rhythms with the frequency of the intermittent light. In this example, the subject gazes at the left 17 Hz flashing light (the right one is flashing at a different frequency), resulting in a prominent 17 Hz peak in the EEG spectrum. This frequency information is leveraged in real time to control the movement of the depicted spaceship, causing it to rotate to the left on the screen

to recognize and correct errors. Additionally, confirmation feedback reassures the user that their commands are being accurately detected.

Feedback is often used during the training phase of a BCI. Users can learn to modulate their brain signals more effectively when they receive feedback about the consequences of their mental actions. Adaptive BCIs may continue to provide feedback as the system learns and adapts to changes in the user's brain activity.

Engaging and informative feedback can enhance the user's motivation and overall experience. Positive feedback for successful actions can reinforce the user's mental strategies, encouraging them to continue refining their control over the BCI.

21.5.5 BCI in Rehabilitation

Over the past two decades, the scientific community has increasingly explored the tangible potential of utilizing BCI for rehabilitating patients affected by stroke and other central nervous system pathologies. Unlike BCI systems designed for communication and control, BCI for rehabilitation need only be effective in improving the recovery of specific functions, without requiring virtually flawless performance.

Various BCI applications have been proposed to contribute to functional recovery [107]. One promising use of BCI systems focuses on reinforcing neural plasticity, thereby consolidating the utilization of spared neural connections [108, 109].

Additionally, efforts are directed toward enhancing motor capabilities in patients with spinal cord lesions. A neurorehabilitation protocol developed at the Neurorehabilitation Laboratory in São Paulo, Brazil, demonstrated improved recovery rates in patients with spinal cord injuries when locomotor rehabilitation is integrated with BCI system training sessions. This system also incorporates a tactile feedback exoskeleton [110].

A pivotal feature extensively employed for BCI system control is motor imagery, analyzed through SMR desynchronization. This category of BCI systems has the capability to enhance the motor imagery process itself. The rationale for employing BCI-facilitated motor imagery for rehabilitation lies in its potential to activate neural systems crucial for actual movement, making it a potentially effective therapy for dysfunction resulting from stroke. Given that brain lesions may impede imagery, methods facilitating such mental exercises could prove instrumental in enhancing recovery [111, 112].

Based on literature data, there is a clear and growing increase in research studies focusing on the use of BCI systems for rehabilitation applications. It is desirable that progress in this technology can be successfully employed in the near future to help the functional recovery of individuals with strokes and other chronic neuromuscular pathologies.

21.5.6 Conclusions

The history of BCI has shown a remarkable evolution in relatively few years, rooted in the pioneering work of UCLA in the 1970s to its current status as a tangible reality with various applications. With feature extraction algorithms based on P300, SMR, and SSVEP, the adaptability of BCI extends well beyond traditional communication and control, now permeating various aspects of our daily lives, from healthcare and neurorehabilitation to entertainment and home automation. The ongoing convergence of neuroscience, computer technology, and machine learning propels BCI research toward a future where the seamless interaction between the human brain and external devices becomes increasingly refined and accessible.

However, despite the significant strides made, there are still some obstacles preventing BCIs from seamlessly transitioning from labs to everyday life, such as the convenience factor. Setting up an EEG acquisition system is often quite laborious: the lengthy process of placing EEG caps, extended calibration times, cable clutter, etc. slightly surpass the average user's acceptance threshold and are particularly relevant for people with disabilities. The evolution of wearable systems with a limited number of dry and wireless electrodes will help overcome these obstacles, making BCI systems more usable into daily life.

Thus, as we envision a future where these obstacles are resolved, BCIs hold the promise of becoming an integral part of our lives. Open-source platforms facilitate collaboration and innovation, emphasizing a growing focus on rehabilitative applications, poised to contribute significantly to improving life.

Finally, but not least, it would be appropriate to consider the ethical implications surrounding brain–machine interaction. Ensuring the security of neural data is crucial to prevent unauthorized access or potential misuse of highly sensitive information. BCIs have the potential to enhance cognitive abilities, raising questions about the ethical implications of using such technology for purposes beyond therapeutic or rehabilitative needs. Continuous interaction with BCIs might have psychological effects, and the long-term impact on mental health should be monitored. Addressing these ethical concerns requires collaboration among scientists, ethicists, policymakers, and the broader public to establish guidelines, regulations, and societal norms that ensure the responsible development and use of BCI technology.

21.6 Mind Melding: Bridging Minds Through the Wonders of Brain-to-Brain Interfaces

21.6.1 Science Fiction or Reality?

Mind-to-mind communication, once confined to the realm of science fiction, has become a focal point of scientific inquiry in recent years. In the past decade, scientists have redirected their focus toward an innovative extension of the neural interface concept, aiming to establish direct communication between two brains through noninvasive approaches, bypassing the peripheral nervous system. This technology, reminiscent of Arthur C. Clarke's renowned 1997 science fiction novel *3001: The Final Odyssey* [113], is known as brain-to-brain interface (B2BI). In its basic implementation, a B2BI system requires recording and decoding of the brain activity of one subject, termed the "Sender," as seen in BCI applications. Subsequently, this recorded brain activity should be encoded and transcribed into the brain of a second subject, referred to as the "Receiver," utilizing a neurostimulation system; this transmission phase can be named the computer–brain interface (CBI).

21.6.2 First Experiments

The initial exploration into decoding brain activity associated with motor imagery and subsequently stimulating the same brain in response to the performed action was carried out in 2011, at the Duke University [114]. In this brain-machine-brain experiment, a monkey was engaged in an exploration task, manipulating a cursor on a screen through a BMI system. Upon reaching the target, intracortical microstimulation of the primary somatosensory cortex was employed to provide tactile feedback.

Groundbreaking research in B2BI was carried out a few years later, by linking the brains of two rats [115], aiming for real-time transfer of sensorimotor information with behavioral significance. In this B2BI setup, a designated "encoder" rat engaged in different sensorimotor tasks, making choices between either tactile or visual stimuli. Concurrently, segments of cortical activity from the encoder rat were transmitted in real-time to corresponding cortical regions in a separate "decoder" rat through intracortical microstimulation. The overall performance reached by the rat pair ranged from 64% to 69%, depending on the task.

Although improperly qualified as "telepathic communication" in the press, two reports demonstrated that direct B2BI between two physically distant humans could efficiently work. The first human B2BI, which integrated EEG and transcranial magnetic stimulation (TMS), was reported by Rao and colleagues from the Department of Computer Science & Engineering at the University of Washington in 2014 [116]. The experiment was structured around a video game where participants engaged in firing a cannon to intercept rockets targeting a city. The game involved two roles: the Sender and the Receiver, three pairs of players participated to the study. The Sender's task was to mentally simulate a hand movement to initiate the "Fire" command. The B2BI system identified motor imagery through SMRs desynchronization and transmitted the "Fire" command via the Internet to a TMS device. This, in turn, induced a hand movement in the Receiver, ultimately triggering the firing of the cannon, a visual event observed by the Sender on their screen. The average accuracy reached 83.3% in a pair of players, while it was lower in the other two pairs. In a concurrent study during the same period [117], a B2BI system was reported, connecting two subjects situated in geographically very distant locations: the Sender (BCI side) in Thiruvananthapuram, Kerala, India, and the Receiver (CBI side) in Strasbourg, France. The experimental protocol encompassed the utilization of motor imagery for BCI and TMS over visual cortical areas for CBI, sending the control messages over the Internet.

21.6.3 More Recent Studies

Jiang and colleagues introduced a compelling B2BI system, marking the first instance of connecting more than two individuals collaboratively to address a problem [118]. Each session involved three participants: two Senders and one Receiver. The system involves EEG for recording signals from the Sender subjects and TMS for stimulating the Receiver subject. The study, modeled after a video game Tetris-like, tasked Senders with deciding whether to rotate a falling block before it occupies the line below. Sender decisions were coded through an SSVEP protocol, utilizing a flashing light at 17 Hz for the "Rotate" command and 15 Hz for the "Do not rotate" command. These decisions were transmitted to the Receiver through TMS stimulations on the occipital regions, inducing a phosphene for the "Rotate" command and no phosphene for the "Do not rotate" command. The Receiver was equipped with TMS, for receiving the decision from the Senders, and EEG to perform the action required (Fig. 21.6). The study, involved five

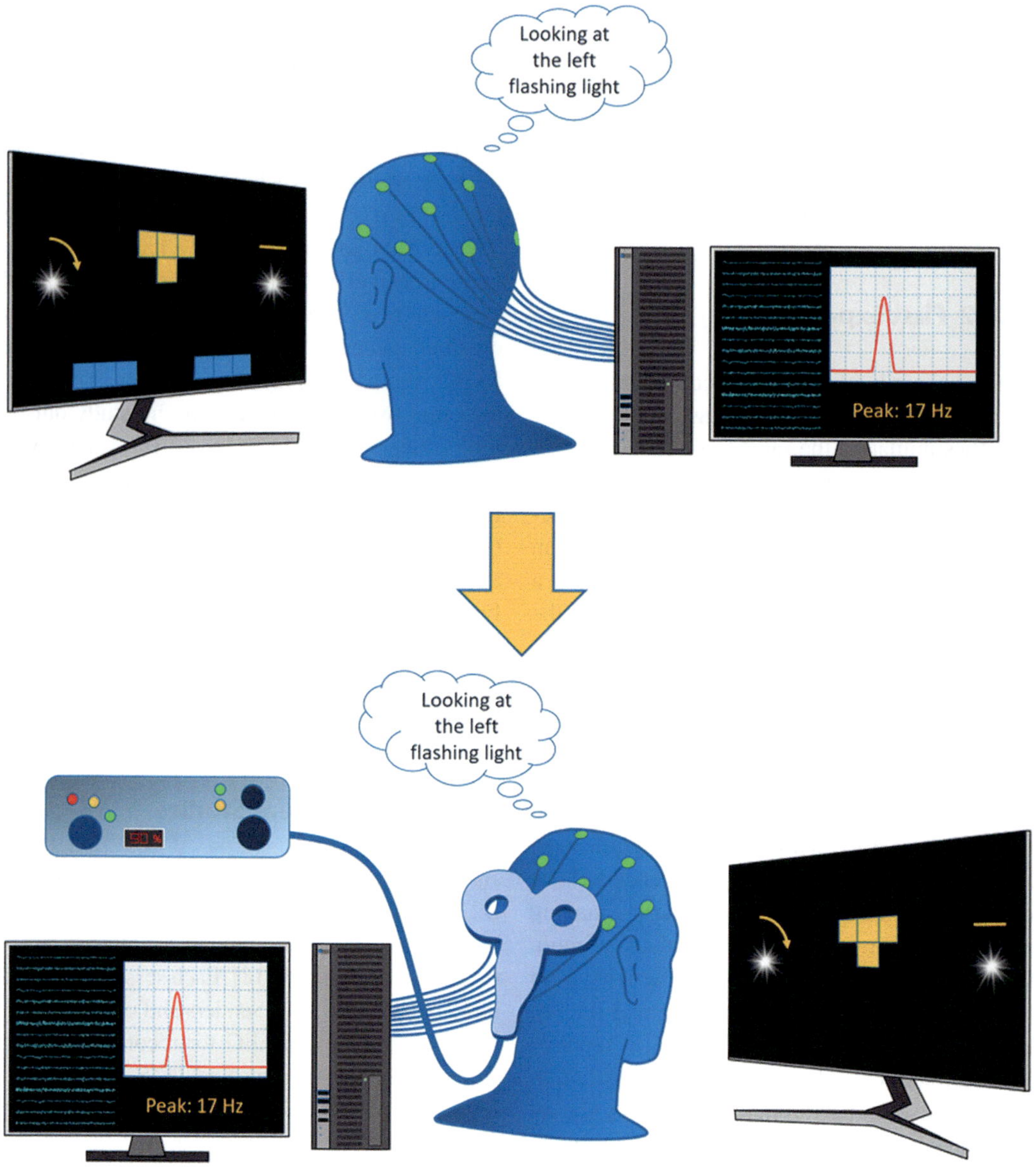

Fig. 21.6 A B2BI example (a simplified version of the experiment by Jiang et al., 2019). On the top, the Sender plays a Tetris-like game, deciding whether the falling block should be rotated or not to fill the gap on the bottom brick line. To make this decision, the Sender employs a steady-state visual evoked potential (SSVEP) BCI approach, encoding the command "Rotate" by looking at the left flashing light or "Do not rotate" by focusing on the right light, which flashes at a different frequency. The decoded command is then transmitted over the Internet to a system controlling a magnetic stimulator. On the bottom, a magnetic stimulus is delivered over the occipital areas of the Receiver, eliciting either a perception of rotation or no rotation in the form of a phosphene. In response to this stimulation, the Receiver decides the action to be performed, using the same SSVEP approach, by observing the left or right flashing lights on their screen, determining whether to rotate or not rotate the falling block. Importantly, the Receiver does not have visibility of the bottom line of bricks and must rely solely on the Sender's decision

groups, achieving an impressive average accuracy rate of 81.25%.

In 2017, Lee and collaborators introduced an alternative noninvasive approach for transmitting information to the Receiver's brain [119]. In this study, the Sender engaged in a motor imagery task, and the associated changes in SMRs, specifically the mu rhythm (8–10 Hz), were detected. These changes were then used to control the movement of a cursor toward a target appearing on the left or right side of a computer screen. Upon reaching the target, a signal was transmitted to another computer via the Internet, activating a focused ultrasound (FUS) device. This device selectively stimulated either the right or left hand somatosensory area of the Receiver, inducing a unilateral tactile sensation on the Receiver's hand and thereby establishing hemispheric B2BI.

In a recent review by Takeuchi [120], exploring the efficacy of noninvasive brain stimulation for enhancing teacher–student interaction, the potential application of a B2BI system for physiotherapy application is proposed. The initial hypothesis, substantiated by existing literature, suggests that brain–brain synchrony during teaching–learning interactions may lead to or result from behavioral entrainment, cognitive alignment, and affective contagion, all of which could enhance the learning process. As such, one might consider recording the physiotherapist's brain activity and, through concurrent transcranial alternating current stimulation (tACS), induce synchronized activity in the patient's brain.

21.6.4 Ethical Concerns

Current B2BI applications are limited to recognizing and transmitting binary yes/no-like information, lacking the capacity for more intricate communication or "mind reading." Despite this limitation, it is prudent to proactively address ethical aspects during technological development, especially in anticipation of future, more complex applications [121]. A B2BI system has been envisioned involving multiple individuals connected through B2BI and a cloud-based server, resembling current social networks [118]. This technology could prove beneficial in scenarios requiring collaboration among specific groups to achieve common goals. However, the emergence of complex ethical issues in multi-person B2BI, encompassing safety, shared control, responsibility, privacy, identity, self-concept, and the extended mind, demands careful consideration [122]. Ethical challenges in these scenarios are likely to be more intricate due to the nuanced distribution of roles and implications for users' self-concept and autonomy [123].

The risk of spreading information derived from brain recordings without the subject's explicit agreement is a genuine concern in multi-person B2BI networks. Securing informed consent from each participant becomes crucial in mitigating this risk. In this context, Senders should have the autonomy to decide the type of information permissible for transmission, while Receivers should possess the ability to specify the kind of information they wish to receive, from whom, and when. Privacy concerns become pronounced when neural data circulating within the network could potentially unveil a participant's health condition or mental state. The threat of "neurocrime," including scenarios of "brain-hacking" involving the illicit access and manipulation of neural information and computation, appears as a potential challenge within a brain network [124].

The discussion extends to potential future applications, encompassing games, personal enhancements, user state monitoring, cryptography, and silent commands, particularly in military contexts. The envisaged use of B2BI in military operations challenges conventional notions of responsibility [125]. Should B2BI-connected soldiers be considered fully responsible for their actions? Do they possess the freedom to act independently? Can they authenticate thoughts and memories generated while connected to the interface? Answering these questions is imperative before contemplating the deployment of B2BI in military settings.

In navigating the evolving landscape of B2BI, an ethically grounded approach is paramount, ensuring the responsible development and deployment of this technology as we continue to unlock its vast potential for collaboration, communication, and human enhancement.

21.7 Conclusions: Navigating the Landscape of EEG Technology

The journey through the evolution of EEG instrumentation, from its origins with ancient observations to the modern advancements in B2BI, reveals a remarkable trajectory marked by collaboration, innovation, and ethical considerations.

From the foundational work of pioneers like Luigi Galvani and Hans Berger to the commercialization of EEG devices and the birth of digital EEG, each milestone has propelled the field forward, expanding the possibilities to better understand the human brain behavior and its complexities.

The commercialization of EEG devices marked a significant milestone in the field, enabling broader access to EEG technology. However, the true revolution came with the invention of the transistor, leading to transistorized amplifiers and sophisticated recording systems. This ushered in an era of digital EEG, with the first analog-to-digital conversion conducted at UCLA's Data Processing Laboratory in 1961. Digital EEG offered unparalleled versatility and precision in data visualization, analysis, safe storage, and transmission, revolutionizing EEG research and clinical practice.

The emergence of wearable EEG systems represents another milestone, extending EEG research beyond traditional laboratory settings. Advancements in electrode and amplifier technology have facilitated noninvasive and high-quality recordings, while innovations like dry electrodes and miniaturized amplifiers have enabled wearable and portable devices. These systems find applications in various fields, from clinical settings to real-world scenarios like monitoring drivers' vigilance and controlling external devices through BCI.

The frontier of BCI technology further expands the possibilities of EEG applications, enabling communication between the brain and external devices. BCI has transcended its origins as a tool for assisting individuals with disabilities, evolving into a multifaceted discipline with applications in healthcare, entertainment, and beyond. However, ethical implications such as data security and psychological effects necessitate careful consideration and collaboration among stakeholders.

Similarly, B2BI technology, once confined to science fiction, is now a subject of scientific exploration. While current applications are limited, the potential for collaboration, communication, and human enhancement is vast. However, ethical concerns regarding safety, privacy, autonomy, and responsibility must be addressed to ensure responsible development and deployment of this technology.

The statement made by Zifkin and Avanzini in 2009 remains pertinent today: the EEG, a simple yet revolutionary method for studying brain activity, is likely to benefit from future technological advancements. It's conceivable that we have yet to reach its full theoretical potential [126]. Moreover, the evolution of EEG technology signifies more than just technological progress; it reflects our ongoing effort to understand the complexities of the human brain. By focusing on innovation while respecting ethical standards, we can use EEG technology to improve human health, communication, and collaboration in ways we never thought possible.

References

1. Wilson JV, Reynolds EH. Texts and documents. Translation and analysis of a cuneiform text forming part of a Babylonian treatise on epilepsy. Med Hist. 1990;34:185–98.
2. Galvani L. Aloysii Galvani De viribus electricitatis in motu musculari commentarius. Bononiae: Ex Typographia Instituti Scientiarium; 1791.
3. Volta A, Banks J, Basire J, Royal Society (Great Britain), United States. Patent Office. Scientific Library former owner D (1800) On the electricity excited by the mere contact of conducting substances of different kinds.

4. Aldini G. De animali electricitate dissertationes duæ. 1794.
5. von Humboldt A. Versuche über die gereizte Muskel- und Nervenfaser: nebst Vermuthungen über den chemischen Process des Lebens in der Thier- und Pflanzenwelt. Posen/Berlin: Bei Decker und Compagnie/Bei Heinrich August Rottmann; 1797.
6. Ørsted HC, Pedersen PO, Burndy Library donor D. Experimenta circa effectum conflictus electrici in acum magneticam. Hafniae. Typis Schultzianis; 1820.
7. Ohm GS, Kniestädt JGF, Kiesel former owner D, Burndy Library donor D (1827) Die galvanische Kette. Berlin : Bei T.H. Riemann.
8. Faraday M. Experimental Researches in Electricity. Philos Trans R Soc Lond. 1832;122:125–62.
9. Nobili L. Comparaison entre les deux galvanomètres les plus sensibles, la grenouille et le multiplicateur a deux aiguilles, suivie de quelques résultats nouveaux. Annales de Chimie et de Physique. 1828;38:225–45.
10. Moruzzi G. The electrophysiological work of Carlo Matteucci. 1964. Brain Res Bull. 1996;40:69–91.
11. Du Bois-Reymond EH. Untersuchungen über thierische elektricität. Berlin: G. Reimer; 1848.
12. Caton R. The electric currents of the brain. Br Med J. 1875;2
13. Beck A. Die Ströme der Nervencentren. Centralbl Physiol. 1890;4:572–3.
14. Berger H. Über das Elektrenkephalogramm des Menschen. Archiv f Psychiatrie. 1929;87:527–70.
15. Gloor P. The work of Hans Berger. Electroencephalogr Clin Neurophysiol. 1969;27: 649.
16. Toennies JF. Die Ableitung bioelektrischer Effekte vom uneroffneten Schädel. Physikalische Behandlung des Problems. J Psychol Neurol. 1933;45:154–71.
17. Adrian ED, Matthews BHC. The Berger rhythm: potential changes from the occipital lobes in man. Brain: A. J Neurol. 1934;57:355–85.
18. Compston A. The Berger rhythm: potential changes from the occipital lobes in man. Brain. 2010;133:3–6.
19. Forbes A, Thacher C. Amplification of action currents with the electron tube in recording with the string galvanometer. Am J Physiol. 1920;52:400–9.
20. Jasper HH, Carmichael L. Electrical potentials from the intact human brain. Science. 1935;81:51–3.
21. Jasper HH. Electroencephalography. In: Penfield W, Erickson TC, editors. Epilepsy and cerebral localization: A study of the mechanism, treatment and prevention of epileptic seizures. Springfield: CC Thomas; 1941.
22. Jasper HH. History of the early development of electroencephalography and clinical neurophysiology at the Montreal Neurological Institute: the first 25 years 1939-1964. Can J Neurol Sci. 1991;18:533–48.
23. Hunter J, Jasper HH. A method of analysis of seizure pattern and electroencephalogram; a cinematographic technique. Electroencephalogr Clin Neurophysiol. 1949;1:113.
24. Loomis AL, Harvey EN, Hobart GA. Potential rhythms of the cerebral cortex during sleep. Science. 1935;81:597–8.
25. Loomis AL, Harvey EN, Hobart GA. Distribution of disturbance-patterns in the human electroencephalogram with special reference to sleep. J Neurophysiol. 1938;1:413–30.
26. Gibbs FA, Lennox WG, Gibbs EL. The electroencephalogram in diagnosis and in localization of epileptic seizures. Arch Neurol Psychiatr. 1936;36:1225–35.
27. Pedley TA. Ellen R. Grass lecture: present at the beginning. Ellen Grass and the evolution of modern concepts regarding EEG and epilepsy. Am J Electroneurodiagnostic Technol. 2007;47:241–56.
28. Knott JR, Correll RE, Shepherd JN. Frequency analysis of electroencephalograms of stutterers and nonstutterers. J Speech Hear Res. 1959;2:74–80.
29. Holter NJ. New method for heart studies. Science. 1961;134:1214–20.
30. Marson GB, McKinnon JB. A miniature tape recorder for many applications. Control Instrum. 1972;4:46–7.
31. Ives JR. 4-channel 24 hour cassette recorder for long-term EEG monitoring of ambulatory patients. Electroencephalogr Clin Neurophysiol. 1975;39:88–92.
32. Collura TF. History and evolution of electroencephalographic instruments and techniques. J Clin Neurophysiol. 1993;10:476–504.
33. Grass AM. The electroencephalographic heritage until 1960. Am J EEG Technol. 1984;24:133–73.
34. Ebersole JS. The history of ambulatory EEG—A personal perspective. In: Tatum WO, editor. Ambulatory EEG monitoring. New York: Demos Medical; 2017.
35. Miner WM. Multiplex Telephony. US Patent Office 745734; 1903.
36. Stankovic RS, Astola JT, Karpovsky MG. Some historic remarks on sampling theorem. In: Proceedings of the 2006 International TICSP Workshop on Spectral Methods and Multirate Signal Processing, SMMSP2006, Florence, Italy, 2–3 September 2006. TICSP Series. J. Astola; 2006, pp. 163–170.
37. Nyquist H. Certain factors affecting telegraph speed. Bell Syst Tech J. 1924;3:324–46.
38. Nyquist H. Certain topics in telegraph transmission theory. Trans Am Inst Electr Eng. 1928;47:617–44.
39. Kotelnikov VA. On the transmission capacity of the 'ether' and of cables in electrical communications. In: Proceedings of the first All-Union Conference on the technological reconstruction of the communications sector and the development of low-current engineering. Moscow; 1933.
40. Reeves AH. Electric signaling system. US Patent Office 2272070; 1939.
41. Shannon C. Communication in the presence of noise. Proc I.R.E. 1949;37:10.

42. Someya I. Hakey Denso (Waveform Transmission). Tokyo: Shyukyo Ltd.; 1949.
43. Whittaker ET. XVIII.—On the functions which are represented by the Expansions of the Interpolation-Theory. Proc R Soc Edinburgh. 1915;35:181–94.
44. Luke HD. The origins of the sampling theorem. IEEE Commun Mag. 1999;37:106–8.
45. Estrin T, Sclabassi R, Buchness R. Computer graphic applications to neurosurgery. In: Proc First World Conference on Medical Information (MEDINFO); 1974. p. 831–6.
46. Estrin T. The UCLA Brain Research Institute data processing laboratory. In: Proceedings of ACM conference on History of medical informatics. Bethesda: ACM Press; 1987. p. 75–83.
47. Tatum WO. Long-term EEG monitoring: a clinical approach to electrophysiology. J Clin Neurophysiol. 2001;18:442–55.
48. van de Velde M, van den Berg-Lenssen MM, van Boxtel GJ, Cluitmans PJ, Kemp B, Gade J, Thomsen CE, Värri A. Digital archival and exchange of events in a simple format for polygraphic recordings with application in event related potential studies. Electroencephalogr Clin Neurophysiol. 1998;106:547–51.
49. Kemp B, Olivan J. European data format 'plus' (EDF+), an EDF alike standard format for the exchange of physiological data. Clin Neurophysiol. 2003;114:1755–61.
50. Pernet CR, Appelhoff S, Gorgolewski KJ, Flandin G, Phillips C, Delorme A, Oostenveld R. EEG-BIDS, an extension to the brain imaging data structure for electroencephalography. Sci Data. 2019;6:103.
51. Lang C, Winkler S, Koren J, Huber M, Kluge T, Baumgartner C. DICOM® integrated EEG data: A first clinical implementation of the new DICOM standard for neurophysiology data. Clin Neurophysiol. 2023;155:107–12.
52. Shahzadi R, Anwar SM, Qamar F, Ali M, Rodrigues JJPC, Alnowami M. Secure EEG signal transmission for remote health monitoring using optical chaos. IEEE Access. 2019;7:57769–78.
53. Wen D, Jiao W, Li X, Wan X, Zhou Y, Dong X, Lan X, Han W. The EEG signals encryption algorithm with K-sine-transform-based coupling chaotic system. Inf Sci. 2023;622:962–84.
54. Delorme A, Makeig S. EEGLAB: an open source toolbox for analysis of single-trial EEG dynamics including independent component analysis. J Neurosci Methods. 2004;134:9–21.
55. Oostenveld R, Fries P, Maris E, Schoffelen J-M. FieldTrip: Open source software for advanced analysis of MEG, EEG, and invasive electrophysiological data. Comput Intell Neurosci. 2011;2011:156869.
56. Tadel F, Baillet S, Mosher JC, Pantazis D, Leahy RM. Brainstorm: a user-friendly application for MEG/EEG analysis. Comput Intell Neurosci. 2011;2011:879716.
57. Pascual-Marqui RD, Michel CM, Lehmann D. Low resolution electromagnetic tomography: a new method for localizing electrical activity in the brain. Int J Psychophysiol. 1994;18:49–65.
58. Pascual-Marqui RD. Standardized low-resolution brain electromagnetic tomography (sLORETA): technical details. Methods Find Exp Clin Pharmacol. 2002;24 Suppl D:5–12.
59. Pascual-Marqui RD, Lehmann D, Koukkou M, et al. Assessing interactions in the brain with exact low-resolution electromagnetic tomography. Philos Trans A Math Phys Eng Sci. 2011;369:3768–84.
60. Casson AJ. Wearable EEG and beyond. Biomed Eng Lett. 2019;9:53–71.
61. Niso G, Romero E, Moreau JT, Araujo A, Krol LR. Wireless EEG: A survey of systems and studies. NeuroImage. 2023;269:119774.
62. Kaongoen N, Choi J, Woo Choi J, Kwon H, Hwang C, Hwang G, Kim BH, Jo S. The future of wearable EEG: a review of ear-EEG technology and its applications. J Neural Eng. 2023;20:051002. https://doi.org/10.1088/1741-2552/acfcda.
63. Kabiri Ameri S, Ho R, Jang H, Tao L, Wang Y, Wang L, Schnyer DM, Akinwande D, Lu N. Graphene electronic tattoo sensors. ACS Nano. 2017;11:7634–41.
64. Ferrari LM, Ismailov U, Badier J-M, Greco F, Ismailova E. Conducting polymer tattoo electrodes in clinical electro- and magneto-encephalography. npj Flex Electron. 2020;4:1–9.
65. Li W, Wang G, Lei X, Sheng D, Yu T, Wang G. Seizure detection based on wearable devices: A review of device, mechanism, and algorithm. Acta Neurol Scand. 2022;146:723–31.
66. Mohamed M, Mohamed N, Kim JG. Advancements in wearable eeg technology for improved home-based sleep monitoring and assessment: A review. Biosensors (Basel). 2023;13:1019.
67. Broccard FD, Mullen T, Chi YM, et al. Closed-loop brain-machine-body interfaces for noninvasive rehabilitation of movement disorders. Ann Biomed Eng. 2014;42:1573–93.
68. Little S, Pogosyan A, Neal S, et al. Adaptive deep brain stimulation in advanced Parkinson disease. Ann Neurol. 2013;74:449–57.
69. Stanslaski S, Afshar P, Cong P, Giftakis J, Stypulkowski P, Carlson D, Linde D, Ullestad D, Avestruz A-T, Denison T. Design and validation of a fully implantable, chronic, closed-loop neuromodulation device with concurrent sensing and stimulation. IEEE Trans Neural Syst Rehabil Eng. 2012;20:410–21.
70. Bressler S, Neely R, Yost RM, Wang D, Read HL. A wearable EEG system for closed-loop neuromodulation of sleep-related oscillations. J Neural Eng. 2023;20:056030.
71. Kutafina E, Heiligers A, Popovic R, Brenner A, Hankammer B, Jonas SM, Mathiak K, Zweerings J. Tracking of mental workload with a mobile EEG sensor. Sensors (Basel). 2021;21:5205.

72. Zhang G, Etemad A. Capsule attention for multimodal EEG-EOG representation learning with application to driver vigilance estimation. IEEE Trans Neural Syst Rehabil Eng. 2021;29:1138–49.
73. Topic A, Russo M, Stella M, Saric M. Emotion recognition using a reduced set of EEG channels based on holographic feature maps. Sensors (Basel). 2022;22:3248.
74. Vidal JJ. Toward direct brain-computer communication. Annu Rev Biophys Bioeng. 1973;2:157–80.
75. Wolpaw JR, Birbaumer N, McFarland DJ, Pfurtscheller G, Vaughan TM. Brain–computer interfaces for communication and control. Clin Neurophysiol. 2002;113:767–91.
76. McFarland DJ, Wolpaw JR. EEG-based brain-computer interfaces. Curr Opin Biomed Eng. 2017;4:194–200.
77. Kawala-Sterniuk A, Browarska N, Al-Bakri A, Pelc M, Zygarlicki J, Sidikova M, Martinek R, Gorzelanczyk EJ. Summary of over fifty years with brain-computer interfaces-A review. Brain Sci. 2021;11:43.
78. Farwell LA, Donchin E. Talking off the top of your head: toward a mental prosthesis utilizing event-related brain potentials. Electroencephalogr Clin Neurophysiol. 1988;70:510–23.
79. van Dokkum LEH, Ward T, Laffont I. Brain computer interfaces for neurorehabilitation—its current status as a rehabilitation strategy post-stroke. Ann Phys Rehabil Med. 2015;58:3–8.
80. Tayebi H, Azadnajafabad S, Maroufi SF, Pour-Rashidi A, Khorasanizadeh M, Faramarzi S, Slavin KV. Applications of brain-computer interfaces in neurodegenerative diseases. Neurosurg Rev. 2023;46:131.
81. Rebsamen B, Guan C, Zhang H, Wang C, Teo C, Ang MH, Burdet E. A brain controlled wheelchair to navigate in familiar environments. IEEE Trans Neural Syst Rehabil Eng. 2010;18:590–8.
82. Ma R, Chen Y-F, Jiang Y-C, Zhang M. A new compound-limbs paradigm: Integrating upper-limb swing improves lower-limb stepping intention decoding from EEG. IEEE Trans Neural Syst Rehabil Eng. 2023;31:3823–34.
83. Beveridge R, Wilson S, Callaghan M, Coyle D. Neurogaming with Motion-Onset Visual Evoked Potentials (mVEPs): Adults versus teenagers. IEEE Trans Neural Syst Rehabil Eng. 2019;27:572–81.
84. Putze F, Vourvopoulos A, Lécuyer A, Krusienski D, Bermúdez I, Badia S, Mullen T, Herff C. Editorial: brain-computer interfaces and augmented/virtual reality. Front Hum Neurosci. 2020;14:144.
85. Aldayel M, Ykhlef M, Al-Nafjan A. Deep learning for EEG-based preference classification in neuromarketing. Appl Sci. 2020;10:1525.
86. Kosmyna N, Tarpin-Bernard F, Bonnefond N, Rivet B. Feasibility of BCI control in a realistic smart home environment. Front Hum Neurosci. 2016;10:416.
87. Murugappan M, Ramachandran N, Sazali Y. Classification of human emotion from EEG using discrete wavelet transform. JBiSE. 2010;03:390–6.
88. Khan H, Khadka R, Sultan MS, Yazidi A, Ombao H, Mirtaheri P. Unleashing the potential of fNIRS with machine learning: classification of fine anatomical movements to empower future brain-computer interface. Front Hum Neurosci. 2024;18:1354143.
89. Leinders S, Vansteensel MJ, Piantoni G, et al. Using fMRI to localize target regions for implanted brain-computer interfaces in locked-in syndrome. Clin Neurophysiol. 2023;155:1–15.
90. Levett JJ, Elkaim LM, Niazi F, Weber MH, Iorio-Morin C, Bonizzato M, Weil AG. Invasive brain computer interface for motor restoration in spinal cord injury: A systematic review. Neuromodulation. 2023;S1094-7159(23):00754–7.
91. Abrams Z. Breakthroughs in brain implants. IEEE Pulse. 2023;14:7–10.
92. Mac R. Neuralink implanted a device in a patient's brain, Elon Musk says. The New York Times; 2024.
93. Drew L. Elon Musk's Neuralink brain chip: what scientists think of first human trial. Nature. 2024; https://doi.org/10.1038/d41586-024-00304-4.
94. Drew L. Neuralink brain chip: advance sparks safety and secrecy concerns. Nature. 2024;627:19.
95. Schalk G, McFarland DJ, Hinterberger T, Birbaumer N, Wolpaw JR. BCI2000: a general-purpose brain-computer interface (BCI) system. IEEE Trans Biomed Eng. 2004;51:1034–43.
96. Renard Y, Lotte F, Gibert G, Congedo M, Maby E, Delannoy V, Bertrand O, Lécuyer A. Openvibe: An open-source software platform to design, test, and use brain–computer interfaces in real and virtual environments. Presence Teleop Virt. 2010;19:35–53.
97. Klobassa DS, Vaughan TM, Brunner P, Schwartz NE, Wolpaw JR, Neuper C, Sellers EW. Toward a high-throughput auditory P300-based brain-computer interface. Clin Neurophysiol. 2009;120:1252–61.
98. Jin J, Allison BZ, Kaufmann T, Kübler A, Zhang Y, Wang X, Cichocki A. The changing face of P300 BCIs: a comparison of stimulus changes in a P300 BCI involving faces, emotion, and movement. PLoS One. 2012;7:e49688.
99. Eidel M, Kübler A. Identifying potential training factors in a vibrotactile P300-BCI. Sci Rep. 2022;12:14006.
100. Wolpaw JR, McFarland DJ, Neat GW, Forneris CA. An EEG-based brain-computer interface for cursor control. Electroencephalogr Clin Neurophysiol. 1991;78:252–9.
101. Pfurtscheller G, Aranibar A. Evaluation of event-related desynchronization (ERD) preceding and following voluntary self-paced movement. Electroencephalogr Clin Neurophysiol. 1979;46:138–46.

102. Yuan H, He B. Brain-computer interfaces using sensorimotor rhythms: current state and future perspectives. IEEE Trans Biomed Eng. 2014;61:1425–35.
103. Middendorf M, McMillan G, Calhoun G, Jones KS. Brain-computer interfaces based on the steady-state visual-evoked response. IEEE Trans Rehabil Eng. 2000;8:211–4.
104. Chen X, Wang Y, Nakanishi M, Gao X, Jung T-P, Gao S. High-speed spelling with a noninvasive brain-computer interface. Proc Natl Acad Sci USA. 2015;112:E6058–67.
105. Neuper C, Schlögl A, Pfurtscheller G. Enhancement of left-right sensorimotor EEG differences during feedback-regulated motor imagery. J Clin Neurophysiol. 1999;16:373–82.
106. Fleury M, Lioi G, Barillot C, Lécuyer A. A survey on the use of haptic feedback for brain-computer interfaces and neurofeedback. Front Neurosci. 2020;14:528.
107. McFarland DJ, Daly J, Boulay C, Parvaz M. Therapeutic applications of BCI technologies. Brain Comput Interfaces (Abingdon). 2017;47:37–52.
108. Dobkin BH. Brain-computer interface technology as a tool to augment plasticity and outcomes for neurological rehabilitation. J Physiol. 2007;579:637–42.
109. Daly JJ, Wolpaw JR. Brain-computer interfaces in neurological rehabilitation. Lancet Neurol. 2008;7:1032–43.
110. Nicolelis MAL, Alho EJL, Donati ARC, Yonamine S, Aratanha MA, Bao G, Campos DSF, Almeida S, Fischer D, Shokur S. Training with noninvasive brain-machine interface, tactile feedback, and locomotion to enhance neurological recovery in individuals with complete paraplegia: a randomized pilot study. Sci Rep. 2022;12:20545.
111. Pichiorri F, Morone G, Petti M, et al. Brain-computer interface boosts motor imagery practice during stroke recovery. Ann Neurol. 2015;77:851–65.
112. Jiang Y, Yin J, Zhao B, et al. Motor imagery brain-computer interface in rehabilitation of upper limb motor dysfunction after stroke. J Vis Exp. 2023; https://doi.org/10.3791/65405.
113. Clarke AC. 3001 The Final Odyssey. Del Rey Books USA; 1997.
114. O'Doherty JE, Lebedev MA, Ifft PJ, Zhuang KZ, Shokur S, Bleuler H, Nicolelis MAL. Active tactile exploration enabled by a brain-machine-brain interface. Nature. 2011;479:228–31.
115. Pais-Vieira M, Lebedev M, Kunicki C, Wang J, Nicolelis MAL. A brain-to-brain interface for real-time sharing of sensorimotor information. Sci Rep. 2013;3:1319.
116. Rao RPN, Stocco A, Bryan M, Sarma D, Youngquist TM, Wu J, Prat CS. A direct brain-to-brain interface in humans. PLoS One. 2014;9:e111332.
117. Grau C, Ginhoux R, Riera A, Nguyen TL, Chauvat H, Berg M, Amengual JL, Pascual-Leone A, Ruffini G. Conscious brain-to-brain communication in humans using non-invasive technologies. PLoS One. 2014;9:e105225.
118. Jiang L, Stocco A, Losey DM, Abernethy JA, Prat CS, Rao RPN. BrainNet: A multi-person brain-to-brain interface for direct collaboration between brains. Sci Rep. 2019;9:6115.
119. Lee W, Kim S, Kim B, Lee C, Chung YA, Kim L, Yoo S-S. Non-invasive transmission of sensorimotor information in humans using an EEG/focused ultrasound brain-to-brain interface. PLoS One. 2017;12:e0178476.
120. Takeuchi N. Perspectives on rehabilitation using non-invasive brain stimulation based on second-person neuroscience of teaching-learning interactions. Front Psychol. 2022;12:789637.
121. Perbal B. Knock once for yes, twice for no. J Cell Commun Signal. 2015;9:15–8.
122. Lyreskog DM, Zohny H, Savulescu J, Singh I. Merging minds: The conceptual and ethical impacts of emerging technologies for collective minds. Neuroethics. 2023;16:12.
123. Hildt E. Multi-person brain-to-brain interfaces: ethical issues. Front Neurosci. 2019;13:1177.
124. Ienca M, Haselager P. Hacking the brain: brain–computer interfacing technology and the ethics of neurosecurity. Ethics Inf Technol. 2016;18:117–29.
125. Latheef S. Brain to Brain Interfaces (BBIs) in future military operations; blurring the boundaries of individual responsibility. Monash Bioeth Rev. 2023;41:49–66.
126. Zifkin BG, Avanzini G. Clinical neurophysiology with special reference to the electroencephalogram. Epilepsia. 2009;50:30–8.

Index

F. Brigo, O. Mecarelli (eds.), *EEG: The First 100 Years*,
https://doi.org/10.1007/978-3-031-86878-8